Metallurgy: Understanding How, Learning whY

Studies in Honor of James D. Muhly

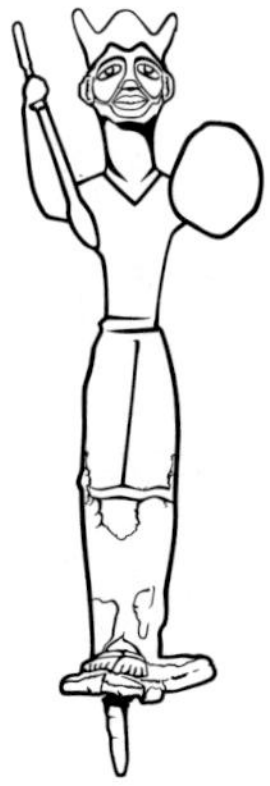

James D. Muhly, Pacheia Ammos, Crete, Greece, June 2007.

PREHISTORY MONOGRAPHS 29

Metallurgy:
Understanding
How,
Learning
whY

Studies in Honor of James D. Muhly

edited by

Philip P. Betancourt and Susan C. Ferrence

Published by
INSTAP Academic Press
Philadelphia, Pennsylvania
2011

Design and Production
INSTAP Academic Press

Printing
CRWGraphics, Pennsauken, New Jersey

Binding
Hoster Bindery, Inc., Ivyland, Pennsylvania

FSC www.fsc.org MIX Paper from responsible sources FSC® C006741

Library of Congress Cataloging-in-Publication Data

Metallurgy, understanding how, learning why : studies in honor of James D. Muhly / edited by Philip P. Betancourt and Susan C. Ferrence.
p. cm. -- (Prehistory monographs ; v. 29)
Includes bibliographical references.
ISBN 978-1-931534-57-4 (hardcover : alk. paper)
1. Metal-work, Prehistoric. 2. Bronze age--Cyprus. 3. Bronze age--Aegean Sea. 4. Bronze implements--Cyprus. 5. Bronze implements--Aegean Sea. 6. Cyrpus--Antiquities. 7. Aegean Sea--Antiquities. 8. Muhly, James David. I. Betancourt, Philip P., 1936- II. Ferrence, Susan C., 1974-
GN799.M4M48 2011
939'.37--dc23

2011017917

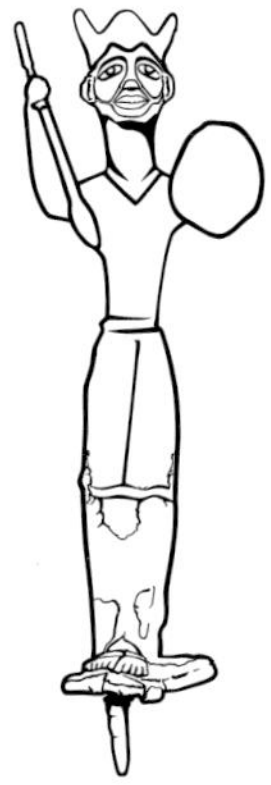

Table of Contents

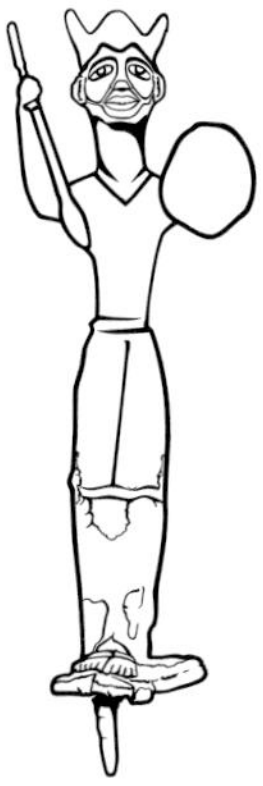

List of Tables in the Text

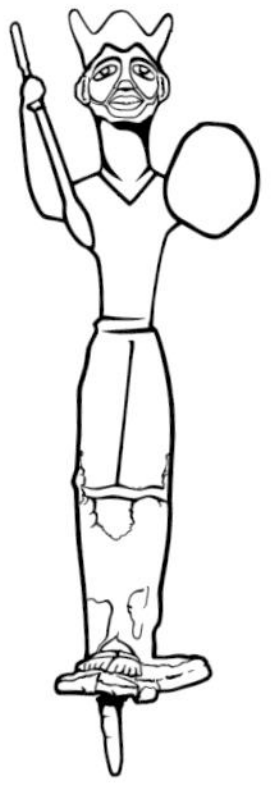

List of Figures in the Text

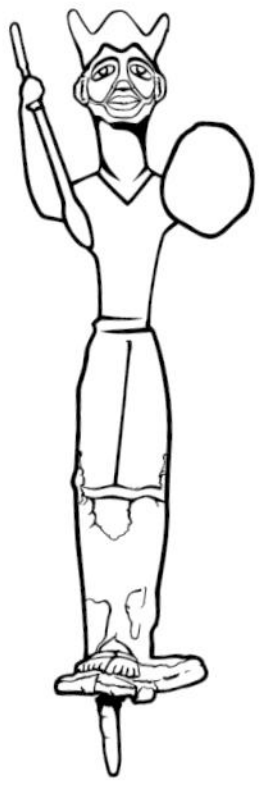

Life with Jim Muhly

Polymnia Muhly

James David Muhly was born in Minneapolis, Minn. on May 6, 1936, the middle child and only son of Gordon David and Violet Lucille Muhly. His mother was of Swedish descent, while his father had German roots and maintained a keen interest in the German connections of the Muhlys throughout his life.

Jim and his sisters, Carolyn and Mary, grew up in South Minneapolis, within walking distance of Minnehaha Falls, in a house that their maternal grandfather, a carpenter, had built. He attended the local public schools, graduating from Roosevelt High School in 1954. By all accounts, especially his own, he was a misfit, who had no interest in school activities beyond playing the trombone in the band. This does not mean that Jim is remembered as a quiet, lonesome child. Indeed the opposite seems to be the case. At the tender age of nine he had the distinction of being expelled from the Minnehaha Methodist church where his parents were lifelong members, and with his special group of like-minded friends had many hair-raising adventures, which, according to his mother, contributed heavily to the dazzingly white hue of her hair. Tamer activities, such as swimming, skating, ice hockey, canoeing, and hunting trips were also part of growing up in South Minneapolis, but for the Muhlys so was working in the Post Office where Gordon Muhly was employed all his working life. Jim worked there part-time during the school year and full time in the summers for many years, adding in the process some unusual idiomatic expressions to his vocabulary and acquiring the means to satisfy his lifelong passion for books and Classical music.

Jim enrolled at the University of Minnesota in 1954 as a Chemistry major, but he was soon enticed down the path to History, without ever losing his initial interest in science. He graduated in 1958. I met him in 1959 when we were both graduate students. Having graduated in Archaeology and History from the University of Athens, I had just become William MacDonald's third successive Greek assistant in a study of Messenian place names, an adventure that, as I thought at the time,was going to last for a year. Jim was in Ancient History and I was in Classical Studies, but this was a distinction of no great significance since both programs had few students, all of whom attended pretty much the same courses.

Being a part of this small group made the huge and, for a foreigner, rather scary campus seem a familiar environment. Attending seminars, sharing "flour" pies at the Coffman Union and drinking weak beer at campus hang-outs fostered close friendships that in many cases have endured all our lives. Everyone followed different career paths and ended up in different places, but even now, during our infrequent get-togethers recollections of Minnesota days sooner or later end up with stories about Tom B. Jones, who was the professor of Ancient History and the teacher who made the most vivid and lasting impression on us all. A legendary figure on the campus, he was endlessly discussed as much for his academic achievements in fields as diverse as Latin American History, Numismatics, or Assyriology as for his personality. Tom Jones, who was equally prone to terrorize his students as to support their aspirations, remained Jim's mentor to his death in 1999 and had a greater influence on Jim's interests and attitude toward research and academic achievement than any of his many other distinguished teachers.

Jim and I were married in Minneapolis in May 1961 and, after spending the summer in New York City on a fellowship from the American Numismatic Society, arrived in New Haven, CT, where he was to begin his studies in the Department of Near Eastern Languages and Literature at Yale. Courses were taught by scholars such as Albrecht Goetze, William Kelly Simpson, Franz Rosenthal, Maurice Pope, and Ferris Stevens, and the handful of students comprised an orthodox Israeli, a Dutch diplomat, a "mad" Hungarian, a Japanese Protestant minister, and Jim. Goetze was an inspiring teacher, who also taught his students how to do research. Jim still quotes the admonition: "never base an argument on unpublished research."

The years in New Haven were exhilarating, characterized at once by hard work and a feeling of freedom generated by a never again to be experienced minimalist life style. It was during this time that Jim made his first trip to Greece in the summer of 1962 and our daughter, Elizabeth, was born. The return to Minnesota where Jim got his first teaching job marked also the return to the family fold, giving our little girl the opportunity to become closer to her paternal grandparents and the rest of her American relatives than our other children were ever able to do.

It was between the two years at Minnesota that Jim spent his first full year in Greece, in 1965–1966, as a Fulbright student at the American School of Classical Studies at Athens, getting to know the country beyond Athens and Attica at a time when much that is now gone forever could still be seen and savored. The year in Greece also marked the watershed in Jim's career, since it was at the School that he met Mike Jameson, who was to offer him a position at the newly established Ancient History program at the University of Pennsylvania, working together with the Roman historian Robert Palmer.

We arrived in Philadelphia in August 1967 and settled in a large, old row house in West Philly that became our home for the next 30 years. To say that Penn in the late 1960's and 1970's was the perfect place for someone like Jim would be a serious understatement. The Oriental Studies Department where his formal appointment was located overflowed with great scholars in just about every field connected with the Near and Far East, as was also the case with History, Classical Archaeology, and Anthropology, while the University Museum served as a magnet that attracted a multitude of scholars from all over the world. It took constant work to keep up with students such as Jerry Rutter, Tamara Stech, Peter Kuniholm, or Harry Weiss, finish a dissertation (1969), and produce publications that would promote his career. For many years his family, to which two boys, Nicholas and Alexis, were added in 1968 and 1970, saw him only at dinner.

West Philly, however, was not a place where one felt isolated or lonely. Our neighborhood, on the westernmost edge of Penn territory, was a place where everybody, whether they were Penn people, old time residents, or new professionals, felt comfortable. Our next-door neighbors, Larry and Carme Dixon, were close friends long before Carme went to work for Penn, becoming for many years the soul of the Ancient History as well as the Classical Archaeology programs. It was a neighborhood well provided with children. The sound of their playing in the alley, summer and winter, is the part of life in West Philly that we have missed the most. Our own children, who ended up traveling all over the city to attend various public schools, grew up in an environment that taught them how to get along with all kinds of people and be at ease in diverse kinds of conditions (always excepting the legendary Greek "prison" camp that the boys attended one summer).

It is possible that we had settled into some sort of routine when two events took place in 1973 that changed our lives irreversibly: for Jim that event was meeting Bob Maddin, for me it was my first visit to the site of Syme in Crete.

Bob Maddin was at that time University Professor at Penn and a great authority on metallurgy. Jim's research interested him, because it opened up an aspect of his field that he had not explored before; for Jim association with Bob meant access to an unequaled scientific and technical expertise. Their close collaboration and friendship, which continue to this day, made their joint work not only rewarding in scholarly terms but also personally enjoyable.

Being a member of the Maddin and Muhly team, of which Tamara Stech was also part for many years, meant much travel to Greece, Turkey, Cyprus, Israel, Egypt, and even farther east to collect samples, examine new metallurgical discoveries, and confer with colleagues in the field, in the lab, or in the museum. In the process other joint projects were formulated, and enduring friendships were forged, especially with Vassos Karageorghis, Trude Dothan, Robert Merrillees, and Fulvia Lo Schiavo. Participation in numerous conferences, in some of which Jim served as co-editor of the proceedings, and publication of the results of the research conducted by the team, were inevitable consequences. In retrospect Jim's participation in excavation projects was a natural extension of these activities.

As Jim's teaching and administrative duties at Penn became heavier, especially after Mike Jameson's move to Stanford in 1976, and my own involvement in the Syme excavation and return to Graduate school consumed more and more time away from home, coordinating such activities with family obligations became increasingly difficult. Indeed, it was only with the assistance of our parents that such a life style could be sustained during the late 1970's and early 1980's.

Looking back to that period the years seem to have passed very swiftly in a blur of constant striving to meet deadlines of one sort or another. The events that stand out seem widely scattered—minor disasters (who could ever forget that Friday night at HUP Emergency after Alexis bit our dog and the dog bit him back), celebrations of new publications and awards, graduations and our daughter's wedding, and above all the Sabbaticals that gave us extra time to spend on research as well as on family time.

This phase of our lives came to an end in the early 1990's. Unlike the changes that took place in the early 1970's that seemed to have been effected overnight and were linked to specific events, the 90's evolved so gradually that transformation was hardly noticed. Jim became a full professor, our children grew into independent adults and more or less left home and we grew older. Penn also changed: many colleagues left or retired and new ones arrived; Oriental Studies became Asian and Middle Eastern Studies (AMES) and Classical Archaeology became part of the Graduate Group in Art and Archaeology of the Mediterranean World (AAMW). Despite the fact that in 1994 Jim was invited by Phil Betancourt to co-direct the excavation of the early metallurgical site of Chrysokamino in East Crete, an exciting project that has led to a long-term collaboration in the exploration and publication of other sites in this area, somehow life in Philadelphia came to seem rather flat and a radical change became a desirable option.

In 1997 Jim took early retirement and accepted the position of director at the American School of Classical Studies in Athens and we have stayed on in Greece ever since. The five years of his tenure at the School were enjoyable and as active as anyone could wish for, if somewhat lean in scholarly terms, as Jim's time was largely consumed by administration. There was still time for meeting friends passing through and making many new ones, for travel and for attending conferences and many, many lectures.

We are now officially retired and working harder than ever, since retirement has forced us to confront and try to fulfill all the long-standing obligations that had been put off because of lack of time. For Jim collaboration and contact with colleagues in the U.S., Cyprus, and Italy continues by means of e-mail, while the libraries of the American and other foreign schools and his own books meet most if not all research needs. Continuing work with Phil and Mary Betancourt in Crete has become a major part of Jim's life as a scholar and one of the few activities now capable of taking him away from his book-lined, paper-strewn office and his computer. Another factor that has become important in Jim's scholarly activity is his contact with young American and Greek students and colleagues, whose research is concerned with the many subjects that continue to be of vital interest to him. It is to them that Jim is fond of saying that the guiding principle he learned from the great teachers and scholars with whom he has worked is the belief that in order to be a good teacher it is necessary to be a good scholar, for the simple reason that one has to impart to students information and ideas that are not in textbooks.

Life in our neighborhood is pleasant most of the time. Jim loves the ready access to the seashore and enjoys being a familiar figure who is served automatically with his preferred newspapers and magazines or with his favorite bread. Nevertheless, even within the neighborhood and certainly when one ventures beyond it, one has to confront other, less pleasant aspects of life in Greece, which is never boring but all too often frustrating and not infrequently infuriating. We are currently debating whether we are, once again, due for a change, but have not yet made a firm decision.

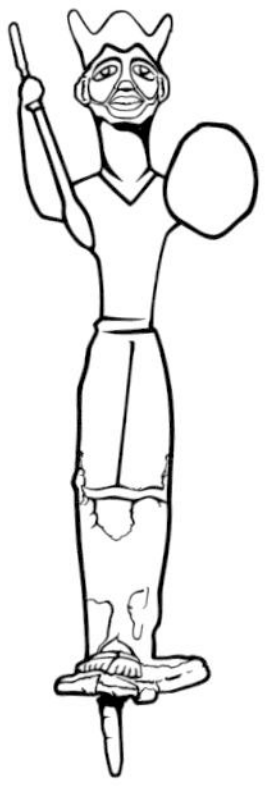

Bibliography of James D. Muhly

Degrees

1958 B.A., University of Minnesota.

1969 Ph.D., Yale University.

Publications

Muhly, J.D. 1965. Review of M.C. Astour, *Hellenosemitica: An Ethnic and Cultural Study in the West Semitic Impact on Mycenaean Greece*, Leiden, 1965, in *JAOS* 85, pp. 585–588.

———. 1970. "Homer and the Phoenicians," *Berytus* 19, pp. 19–64.

———. 1970. Review of W.S. Smith, *Interconnections in the Ancient Near East: A Study of the Relationships between the Arts of Egypt, the Aegean, and Western Asia*, New Haven, 1965, in *JAOS* 90, pp. 305–309.

———. 1971. Review of G. Daniel and J.D. Evans, "Chapter XXXVII: The Western Mediterranean," *CAH*, 2nd ed., vol. II, Cambridge, 1967, in *JAOS* 91, pp. 326–330.

———. 1971. Review of H.M. Hoenigswald and A. Senn, eds., *Indo-European and Indo-Europeans*, Philadelphia, 1970, in *AJA* 75, pp. 437–439.

———. 1971. Review of L.L. Orlin, *Assyrian Colonies in Cappadocia*, The Hague, 1970, in *AHR* 76, pp. 754–755.

———. 1972. Review of A.L. Oppenheim, R.H. Brill, and A. Von Saldern, *Glass and Glassmaking in Ancient Mesopotamia: An Edition of the Cuneiform Texts which Contain Instructions for Glassmakers with a Catalogue of Surviving Objects*, New York, 1970, in *JCS* 24 (1974), pp. 178–182.

———. 1972. "The Land of Alashiya: The History of Cyprus in the Late Bronze Age," in *Proceedings of the First International Congress of Cypriot Studies*, vol. I, V. Karageorghis, ed., Nicosia, pp. 201–219.

———. 1973. *Copper and Tin: The Distribution of Mineral Resources and the Nature of the Metals Trade in the Bronze Age* (*Transactions of the Connecticut Academy of Arts and Sciences* 43), Hamden, CT.

———. 1973. Review of I.E.S. Edwards, C.J. Gadd, and N.G.L. Hammond, eds., *Cambridge Ancient History*, 3rd ed., Vol. I, Part 1: *Prologmena and Prehistory*, Cambridge, 1971, in *CW* 66, pp. 305–307.

———. 1973. Review of I.E.S. Edwards, C.J. Gadd and N.G.L. Hammond, eds. *Cambridge Ancient History*, 3rd ed., Vol. I, Part 2: *Early History of the Middle East*, Cambridge, 1971, in *CW* 66, pp. 307–310.

———. 1973. Review of I.E.S. Edwards, C.J. Gadd and N.G.L. Hammond, eds., *Cambridge Ancient History*, 3rd ed., Vol. I, Part 2: *Early History of the Middle East*, Cambridge, 1971, in *JAOS* 93, pp. 576–578.

———. 1973. Review of W.B. Fisher, *The Middle East: A Physical, Social, and Regional Geography*, New York, 1971, in *IJMES* 4:2, pp. 242–243.

———. 1973. "Tin Trade Routes of the Bronze Age," *American Scientist* 61, pp. 404–413.

Muhly, J.D., and T.A. Wertime. 1973. "The Evidence for the Uses and Sources of Tin during the Bronze Age in the Near East," *WorldArch* 5, pp. 111–122.

Muhly, J.D. 1974. "Hittites and Achaeans: Ahhijawa Redomitus," *Historia* 23, pp. 129–145.

———. 1974. Review of M.J. Finley, *Early Greece: The Bronze and Archaic Ages*, New York, 1970, in *AHR* 79, pp. 78–79.

———. 1974–1975. Review of G.M.A. Hanfmann, *Letters from Sardis*, Cambridge, MA, 1972, and J.G. Pedley, *Ancient Literary Sources on Sardis*, Cambridge, MA, 1972, in *CW* 68, pp. 278–280.

———. 1974. "The Hittites and the Aegean World," *Expedition* 16, pp. 3–10.

Muhly, J.D., and R. Maddin. 1974. "Some Notes on the Copper Trade in the Ancient Mid-East," *Journal of Metals* 26:5, pp. 24–30.

Wheeler, T.S., R. Maddin, and J.D. Muhly. 1975. "Ingots and the Bronze Age Copper Trade in the Mediterranean: A Progress Report," *Expedition* 17:4, pp. 31–39.

Muhly, J.D. 1975. "Near Eastern Chronology and the Date of the Late Cypriot I Period," in *The Archaeology of Cyprus: Recent Developments*, N. Robertson, ed., New York, pp.76–89.

———. 1975. Review of *Acta of the 2nd International Colloquium on Aegean Prehistory. The First Arrival of Indo-European Elements in Greece*, Athens, 1972, in *AJA* 79, pp. 289–291.

———. 1975. Review of K. Branigan, *Aegean Metalwork of the Early and Middle Bronze Age*, Oxford, 1974, in *Materials Science and Engineering* 19, pp. 157–162.

———. 1975. Review of M. Gimbutas, *The Gods and Goddesses of Old Europe, 7000 to 3500 B.C.: Myths, Legends and Cult Images*, Berkeley, 1974 in *AHR* 80, pp. 616–617.

———. 1976. *Copper and Tin: The Distribution of Mineral Resources and the Nature of the Metals Trade in the Bronze Age*, 2nd ed. (*Transactions of the Connecticut Academy of Arts and Sciences* 46), Hamden, CT.

———. 1976. "The Ox-hide Ingots and the Development of Copper Metallurgy in the Late Bronze Age," *TUAS* 1, pp.10–14.

Muhly, J.D. 1976. Review of A.H. Jones, *Bronze Age Civilization: The Philistines and the Danites*, Washington, D.C., 1975, in *CW* 68, pp. 477–478.

———. 1976. Review of R. Drews, *The Greek Accounts of Eastern History*, Cambridge, MA, 1973, in *JNES* 35, pp. 41–43.

Muhly, J.D., T.S. Wheeler, and R. Maddin. 1976. "A Steel Tool of the Fourth Century B.C. from Al Mina in Syria," *Levant* 8, pp. 107–112.

Muhly, J.D. 1977. "Editorial," *Expedition* 19:4, pp. 2–3.

———. 1977. Review of I.E.S. Edwards, C.J. Gadd, N.G.L. Hammond, and E. Sollberger, eds., *Cambridge Ancient History*, 3rd ed., Vol. II, Part 1: *History of the Middle East and the Aegean Region c. 1800–1380 B.C.*, Cambridge, 1973, in *JAOS* 97, pp. 64–67.

———. 1977. Review of K. Branigan, *Aegean Metalwork of the Early and Middle Bronze Age*, Oxford, 1974, in *JNES* 36, pp. 153–157.

———. 1977. "The Copper Ox-hide Ingots and the Bronze Age Metals Trade," *Iraq* 39, pp. 73–82.

Muhly, J.D., T.S. Wheeler, and R. Maddin. 1977. "An Iron Adze of the Fifth–Fourth Centuries B.C. from Al Mina," *Levant* 9, pp. 156–161.

———. 1977. "How the Iron Age Began," *Scientific American* 237:4, pp. 122–131, 152.

———. 1977. "The Cape Gelidonya Shipwreck and the Bronze Age Metals Trade in the Eastern Mediterranean," *JFA* 4, pp. 353–362.

———. 1977. "Tin in the Ancient Near East: Old Questions and New Finds," Expedition 19:2, pp. 35–47.

Muhly, J.D. 1978. "Ancient Cartography," *Expedition* 20:2, pp. 26–31.

———. 1978. "New Evidence for Sources of and Trade in Bronze Age Tin," in *The Search for Ancient Tin*, A.D. Franklin, J.S. Olin, and T.A. Wertime, eds., Washington, D.C., pp. 43–48.

Muhly, J.D., P.P. Betancourt, R. Maddin, and T.S. Wheeler. 1978. "Metallurgy at Gournia," *MASCAJ* December, pp. 7–8.

Muhly, J.D., K.R. Maxwell-Hyslop, R. Maddin, and T.S. Wheeler. 1978. "An Iron Dagger from Tomb 240 at Tell Fara South," *Levant* 10, pp. 112–115.

Curtis, J.E., T.S. Wheeler, R. Maddin, and J.D. Muhly. 1979. "Neo-Assyrian Ironworking Technology," *PAPS* 123, pp. 369–390.

Muhly, J.D. 1979. "Cypriote Copper: Some Geological and Metallurgical Problems," in *Acts of the International Archaeological Symposium "The Relations between Cyprus and Crete, ca. 2000–500 B.C.,"* V. Karageorghis, ed., Nicosia, pp. 87–100.

———. 1979. "On the Shaft Graves at Mycenae," in *Studies in Honor of Tom B. Jones* (*Alter Orient und Altes Testament Bd.* 203), M.A. Powell Jr. and R.H. Sach, eds., Neukirchen-Vluyn, Germany, pp. 311–323.

———. 1979. "Possible Sources of Tin for the Bronze Age Aegean," *BICS* 26, pp. 122–123.

———. 1979. Review of N.K. Sandars, *The Sea Peoples: Warriors of the Ancient Mediterranean, 1250–1150 B.C.*, London, 1978, in *AJA* 83, pp. 355–356.

Muhly, J.D., R. Maddin, and T. Stech. 1979. "Mediterranean Trade in Copper and Tin in the Late Bronze Age," *AIIN* 26, pp. 139–152.

Muhly, J.D., R. Maddin, and T.S. Wheeler. 1979. "Ancient Metallurgy: Materials and Techniques," *Journal of Metals* 31/9, pp. 16–18.

Muhly, J.D. 1980. "Bronze Figurines and Near Eastern Metalwork," *IEJ* 30, pp. 148–161.

———. 1980. "Metals and Metallurgy in Crete and the Aegean at the Beginning of the Late Bronze Age," *TUAS* 5, pp. 25–36.

———. 1980. "New Research on Ancient Copper and Copper Alloys," in *Proceedings of the XVIth International Symposium on Archaeometry and Archaeological Prospection*, E.A. Slater and J.O. Tate, eds., Edinburgh, pp. 248–267.

———. 1980. Review of G. Rapp Jr. and S.E. Aschenbrenner, eds., *Excavations at Nichoria in Southwest Greece*. Vol. I: *Site, Environs, and Techniques*, Minneapolis, 1978, in *AJA* 84, pp. 101–102.

———. 1980. Review of J.C. Waldbaum, *From Bronze to Iron: The Transition from the Bronze Age to the Iron Age in the Eastern Mediterranean*, Göteborg, 1978, in *JHS* 100, pp. 262–264.

———. 1980. Review of J.D. Hawkins, ed., Trade in the Ancient Near East, London, 1977, in *JAOS* 100, pp. 173–175.

———. 1980. "The Bronze Age Setting," in *The Coming of the Age of Iron: Essays in Appreciation of Cyril Stanley Smith*, T. Wertime and J.D. Muhly, eds., New Haven, pp. 25–67.

———. 1980. "The Oxhide Ingots from Enkomi and Mathiati and Late Bronze Age Copper Smelting in Cyprus," *RDAC* 1980, pp. 84–99.

Muhly, J.D., R. Maddin, and T. Stech. 1980. "Copper Ingots and Cypriote Metallurgy," *RDAC* 1980, pp. 84–99.

———. 1980. "Distinguishing Artifacts of Native Copper," *JAS* 7, pp. 84–99.

———. 1980 [1981]. "Research at the Center for Ancient Metallurgy," *Paléorient* 6, pp. 111–119.

Stech-Wheeler, T., J.D. Muhly, K.R. Maxwell-Hyslop, and R. Maddin. 1981. "Iron at Taanach and Early Iron Metallurgy in the Eastern Mediterranean," *AJA* 85, pp. 245–268.

Muhly, J.D. 1981. Review of M. Gibson and R.D. Biggs, eds., *Seals and Sealing in the Ancient Near East*, Malibu, CA, 1977, in *JAOS* 101, pp. 399–401.

———. 1981. "The Origin of Agriculture and Technology – West or East Asia?" *Technology and Culture* 22, pp. 125–148.

———. 1982. "The Nature of Trade in the LBA Eastern Mediterranean: The Organization of the Metals Trade and the Role of Cyprus," in *Early Metallurgy in Cyprus, 4000–5000 B.C. Proceedings of the International Symposium, Larnaca, June 1–6, 1981*, J.D. Muhly, R. Maddin, and V. Karageorghis, eds., Nicosia, pp. 251–269.

———. 1982. "How Iron Technology Changed the Ancient World and Gave the Philistines a Military Edge," *Biblical Archaelogy Review* VII:6, pp. 40–54.

———. 1982. Review of K. Branigan, *Copper and Bronze Working in Early Bronze Age Crete*, Lund, 1968, in *BibO* 39, pp. 408–418.

———. 1983. "Gold Analysis and Sources of Gold in the Bronze Age," *TUAS* 8, pp. 1–14.

———. 1983. "Iron Objects from Masada: Metallurgical Studies. Armour Scales from Masada," *IEJ* 33, pp. 108–109.

———. 1983. "Kupfer," *Reallexikon der Assyriologie und vorderasiatischen Archäologie* VI:5–6, pp. 348–364.

———. 1983. "Lead Isotope Analysis and the Kingdom of Alashiya," *RDAC* 1983, pp. 210–218.

———. 1983. "On the Philistines," review of T. Dothan, *The Philistines and their Material Culture*, New Haven, CT, 1982, in *The Quarterly Review of Archaeology* 4:4, pp. 5–6.

Muhly, J.D., R. Maddin, and T.S. Stech. 1983. "Metal Working," in *Excavations at Athienou, Cyprus: 1971–1972* (*Qedem* 16), T. Dothan and A. Ben-Tor, eds., Jerusalem, pp. 132–138.

Muhly, J.D., R. Maddin, and J.C. Waldbaum. 1983. "Laminated Iron Tool," in *Metalwork from Sardis: The Finds through 1974* (*Archaeological Exploration of Sardis Monograph* 8), J.C. Waldbaum, ed., Cambridge, MA, pp 178–181.

Muhly, J.D., R. Knox, R. Maddin, and T. Stech. 1983. "Iron Objects from Masada: Metallurgical Studies," *IEJ* 33, pp. 97–107.

Muhly, J.D. 1984. "The Beginnings of Iron Metallurgy in Antiquity," *Qadmoniot* 17, pp. 2–11 (in Hebrew).

———. 1984. "The Role of the Sea Peoples in Cyprus during the LC III Period," in *Cyprus at the Close of the Late Bronze Age*, V. Karageorghis and J.D. Muhly, eds., Nicosia, pp. 39–56.

———. 1984. "Timna and King Solomon," review of H.G. Conrad and B. Rotherberg, eds., *Antikes Kupfer im Timna-Tal, Bochum*, 1980, in *BibO* 41, pp. 275–292.

Muhly, J.D., R. Maddin, T. Stech, and E. Brovarski. 1984. "Old Kingdom Models from the Tomb of Impy: Metallurgical Studies," *JEA* 70, pp. 33–41.

Muhly, J.D., R.H.R. Smith, R. Maddin, and T. Stech. 1984. "Bronze Age Steel from Pella," *CurrAnth* 25, pp. 234–236.

Muhly, J.D. 1985. "Beyond Typology: Aegean Metallurgy in Its Historical Context," in *Contributions to Aegean Archaeology: Studies in Honor of William A. McDonald*, N. Wilkie and W.D.E. Coulson, eds., Minneapolis, pp. 109–141.

———. 1985. "End of the Bronze Age," in *Ebla to Damascus: Art and Archaeology of Ancient Syria*, H. Weiss, ed., Washington, D.C., pp. 261–270.

———. 1985. "Lead Isotope Analysis and the Problem of Lead in Copper," *RDAC* 1985, pp. 78–82.

———. 1985. "Phoenicia and the Phoenicians," in *Biblical Archaeology Today. Proceedings of the International Congress on Biblical Archaeology*, J. Aviram, ed., Jerusalem, pp. 177–191.

———. 1985. Review of J.C. Waldbaum, *Metalwork from Sardis: The Finds Through 1974*, Cambridge, MA, 1983, in *CW* 79, pp. 339–340.

———. 1985. "Sources of Tin and the Beginnings of Bronze Metallurgy," *AJA* 89, pp. 275–291.

———. 1985. "The Late Bronze Age in Cyprus: A 25 Year Retrospect," in *Archaeology in Cyprus 1960–1986*, V. Karageorghis, ed., Nicosia, pp. 20–46.

Muhly, J.D., R. Maddin, and T. Stech. 1985. "Metallurgical Studies on Artifacts from the Tomb near 'Enan," *'Atiqot (English Series)* 17, pp. 75–82.

———. 1985. "The Analysis of Iron Artifacts from Palaepaphos-Skales," in *RDAC* 1985, pp. 192–202.

Muhly, J.D., D. Davis, R. Maddin, and T. Stech. 1985. "A Steel Pick from Mt. Adir in Palestine," *JNES* 44, pp. 41–51.

Muhly, J.D., R. Maddin, T. Stech, and E. Özgen. 1985. "Iron in Anatolia and the Nature of the Hittite Iron Industry," *AnatSt* 35, pp. 67–84.

Muhly, J.D. 1986. "Prehistoric Background Leading to the First Use of Metals in Asia," Bulletin of the Metals Museum (Sendai, Japan) 11, pp. 21–42.

———. 1986. "The Role of Cyprus in the Economy of the Eastern Mediterranean during the Second Millennium B.C.," in *Acts of the International Symposium "Cyprus between the Orient and the Occident,"* V. Karageorghis, ed., Nicosia, pp. 45–62.

Muhly, J.D., R. Maddin and T. Stech. 1986. "Copper Production at Kition in the Late Bronze Age," in *Kition* V: *The Pre-Phoenician Levels*, V. Karageorghis, ed., Nicosia, pp. 388–402.

Muhly, J.D, P. Åström, R. Maddin, and T. Stech. 1986. "Iron Artifacts from Swedish Excavations in Cyprus," *OpAth* 16, pp. 27–41.

Muhly, J.D. 1987. "Solomon, the Copper King. A Twentieth Century Myth," *Expedition* 29:2, pp. 38–47.

———. 1987. Review of R.D. Penhallurick, *Tin in Antiquity*, London, 1986, in *Archaeomaterials* 2, pp. 329–334.

———. 1987. Review of N. Marinatos, *Art and Religion in Thera*, Athens, 1984, in *Gnomon* 59, pp. 329–334.

Muhly, J.D., R. Maddin, and T. Stech. 1987. "An Iron Axe from Tell Qiri," in *Tell Qiri: A Village in the Jezreel Valley. Report of the Archaeological Excavations 1975–1977* (*Qedem* 24), A. Ben-Tor and Y. Portugali, eds., Jerusalem, pp. 244–245.

Muhly, J.D., F. Lo Schiavo, T. Stech, and R. Maddin. 1987. "Nuragic Metallurgy in Sardinia: Second Preliminary Report," in *Studies in Sardinian*

Archaeology III, M.S. Balmuth, ed., Oxford, pp. 179–187.

Muhly, J.D. 1988. "Concluding Remarks," in *Bronze-working Centres of Western Asia c. 1000–539 B.C.*, J. Curtis, ed., London, pp. 329–342.

———. 1988. "Report on Analysis of Fragment of Copper Oxhide Ingot (No. 189) from Maa-Palaeokastro," in *Excavations at* Maa-*Palaeokastro, 1979–1986*, V. Karageorghis and M. Demas, eds., Nicosia, pp. 471–473.

———. 1988. Review of S. Piggott, *The Earliest Wheeled Transport: From the Atlantic Coast to the Caspian Sea*, Ithaca, 1983, in *BASOR* 272, pp. 87–90.

———. 1988. "The Beginnings of Metallurgy in the Old World," in *The Beginnings of the Use of Metals and Alloys. Second International Conference*, Zhenghou, China, R. Maddin, ed., Cambridge, MA, pp. 2–20.

———. 1988. "The Wider World of Lead Ingots," *RDAC* 1988, pp. 263–265.

Muhly, J.D., A.B. Knapp, and P. Muhly. 1988. "To Hoard is Human: Late Bronze Age Metal Deposits in Cyprus and the Aegean," *RDAC* 1988, pp. 281–298.

Muhly, J.D., R. Maddin, and T. Stech. 1988. "Cyprus, Crete and Sardinia: Copper Ox-hide Ingots and the Bronze Age Metals Trade," *RDAC* 1988, pp. 281–298,

Muhly, J.D. 1989. "Çayönü Tepesi and the Beginnings of Metallurgy in the Old World," in *Archäometallurgie der Alten Welt. Beiträge zum Internationalen Symposium "Old World Archaeometallurgy,"* Heidelberg, 1987, A. Hauptmann, E. Pernicka, and G.A. Wagner, eds., Bochum, pp. 1–11.

———. 1989. Review of J. Wolters, *Die Granulation: Geschichte und Technik einer alten Goldschmiede-kunst*, Munich, 1983, and *C. Éluère, Les ors prehistorique*, Paris, 1982, in *AJA* 93, pp. 602–605.

———. 1989. "The Organization of the Copper Industry in Late Bronze Age Cyprus," in *Early Society in Cyprus*, E.J. Peltenburg, ed., Edinburgh, pp. 298–314.

Muhly, J.D., and P. Muhly. 1989. "Metal Artifacts," in *Excavations at Tel Michal*, Israel, Z. Herzog, G. Rapp Jr., and O. Negbi, eds., Minneapolis, pp. 267–295.

Muhly, J.D. 1990. "Black Athena versus Traditional Scholarship," *JMA* 3:1, pp. 83–110.

———. 1990. "Preface," *JMA* 3:1, pp. 53–55.

Muhly, J.D., and T. Stech. 1990. "Final Observations," in *Analisi metallurgiche e statistiche sui lingotti di rame della Sardegna* (*Quaderni* 17), F. Lo Schiavo and R. Maddin, eds., Ozieri, Sardinia, Italy, pp. 200–221.

Muhly, J.D., R. Maddin, and T. Stech. 1990. "The Metal Artifacts," in *Kinneret: Ergebnisse der Ausgrabungen auf dem Tell el'Orëme am See Gennesaret, 1982–1985*, V. Fritz, ed., Wiesbaden, pp. 159–175.

Muhly, J.D. 1991. "Copper in Cyprus: The Earliest Phase," in *Découverte du metal*, J.-P. Mohen and C. Éluère, eds., Paris, pp. 357–374.

———. 1991. "Egypt, the Aegean and Late Bronze Age Chronology in the Eastern Mediterranean: A Review Article," *JMA* 4:2, pp. 235–247.

———. 1991. "The Development of Copper Metallurgy in Late Bronze Age Cyprus," in *Bronze Age Trade in the Mediterranean* (*SIMA* 90), N.H. Gale, ed., Jonsered, pp. 180–196.

———. 1991. "The Iron from Kinneret," in *From Bloom to Knife. International Symposium of the Comité pour la sidérurgie ancienne de l'UISPP* (*Materiały Archeologiczne* XXVI), K. Radwanski, ed., Kraków, pp. 97–102.

Muhly, J.D., R. Maddin, and T. Stech. 1991. "Çayönü Tepesi: The Earliest Archaeological Metal Artifacts," in *Découverte du metal*, J.-P. Mohen and C. Éluère, eds., Paris, pp. 375–386.

Muhly, J.D., F. Begemann, Ö. Öztunali, E. Pernicka, S. Schmitt-Strecker, and G.A. Wagner. 1991. "The Bronze Metallurgy of Anatolia and the Question of Local Tin Sources," in *Archaeometry* '90, E. Pernicka and G. Wagner, eds., Basel, pp. 209–220.

Muhly, J.D. 1992. Review of P.M. McNutt, *The Forging of Israel: Iron Technology, Symbolism and Tradition in Ancient Society*, Sheffield, 1990, in *BiblArch* 55, pp. 153–154.

———. 1992. Review of P.M. McNutt, *The Forging of Israel: Iron Technology, Symbolism and Tradition in Ancient Society*, Sheffield, 1990, in *JAOS* 112, pp. 696–702.

———. 1992. "The Crisis Years in the Mediterranean World: Transition or Cultural Disintegration?" in *The Crisis Years: The 12th Century B.C.*, W.A. Ward and M.S. Joukowsky, eds., Dubuque, IA, pp. 10–26.

Muhly, J.D., and E. Pernicka. 1992. "Early Trojan Metallurgy and Metals Trade," in *Heinrich Schliemann: Grundlagen und Ergebnisse moderner Archäologie* 100 Jahre nach Schliemanns Tod, J. Herrmann, ed., Berlin, pp. 309–318.

Muhly, J.D. 1993. "Early Bronze Age Tin and the Taurus," *AJA* 97, pp. 239–253.

———. 1993. "Metalle. B: Archäologisch," in *Reallexikon der Assyriologie und vorderasiatische Archäologie* VIII:1–2, D.O. Edzard, ed., Berlin, pp. 119–136.

———. 1993. Review of E.N. Chernykh, *Ancient Metallurgy in the USSR: The Early Metal Age*, Cambridge, 1992, in *Geoarchaeology* 8:6, pp. 534–536.

———. 1995. “Lead Isotope and the Archaeologist,” *JMA* 8:1, pp. 54–58.

———. 1995. “Mining and Metalwork in Ancient Western Asia,” in *Civilizations of the Ancient Near East* III, J.M. Sasson, ed., New York, pp. 1501–1521.

———. 1995. Review of R.-B. Wartke, ed., *Handwerk und Technologie im alten Orient*, Mainz, 1994, in *AJA* 99, p. 753.

———. 1996. “The First Use of Metals in the Aegean,” in *The Copper Age in the Near East and Europe* (*13th International Congress of Prehistoric and Protohistoric Sciences, Series Colloquia* 10), B. Bagolini and F. Lo Schiavo, eds., Forli, Italy, pp. 75–84.

———. 1996. “The Significance of Metals in the Late Bronze Age Economy of Cyprus,” in *The Development of the Cypriot Economy: From the Prehistoric Period to the Present Day*, V. Karageorghis and D. Michaelides, eds., Nicosia, pp. 45–60.

———. 1997. “Metals: Typology and Technology; Artifacts of the Neolithic, Bronze and Iron Ages,” in *The Oxford Encyclopedia of Archaeology in the Near East*, vol. 4, E.M. Meyers, ed., Oxford, pp. 1–15.

———. 1997. “Recent Works in Archaeometallurgy,” *AJA* 101, pp. 771–773.

———. 1998. “An Appreciation,” in *Metallurgica Antiqua in Honour of Hans-Gert Bachmann and Robert Maddin* (*Der Anschnitt Beiheft* 8), T. Rehren, A. Hauptmann, and J.D. Muhly, eds., Bochum, pp. 12–14.

———. 1998. “Copper, Tin, Silver and Iron: The Search for Metallic Ores as an Incentive for Foreign Expansion,” in *Mediterranean Peoples in Transition: Thirteenth to Early Tenth Centuries B.C.E. in Honor of Trude Dothan*, S. Gitin, A. Mazar, and E. Stern, eds., Jerusalem, pp. 314–329.

———. 1998. “Metals and Metallurgy: Using Modern Technology to Study Ancient Technology,” in *Ancient Greek Technology. Proceedings of the 1st International Conference on Ancient Greek Technology, Thessaloniki, September 4–7, 1997*, T.P. Tassios, ed., Thessaloniki, pp. 27–33.

———. 1998. “On Re-Reading Helene Kantor,” in *The Aegean and the Orient in the Second Millennium* (*Aegaeum* 18), E.H. Cline and D. Harris-Cline, eds., Liège, pp. 211–214.

———. 1999. “Copper and Bronze in Cyprus and the Eastern Mediterranean,” in *The Archaeometallurgy of the Asian Old World*, V.C. Pigott, ed., Philadelphia, pp. 15–25.

———. 1999. “Metallurgy,” in *Encyclopedia of the Archaeology of Ancient Egypt*, K.A. Bard, ed., London, pp. 522–527.

———. 1999. “The Phoenicians in the Aegean,” in *MELETEMATA: Studies in Aegean Archaeology Presented to Malcolm H. Wiener as He Enters His 65th Year* II (*Aegaeum* 20), P.P. Betancourt, V. Karageorghis, R. Laffineur, and W.-D. Niemeier, eds., Liège, pp. 517–526.

Muhly, J.D., and R. Maddin. 1999. “Early Metalworking at Çayönü,” in *The Beginnings of Metallurgy. Proceedings of the International Conference, Bochum 1995* (*Der Anschnitt Beiheft* 9), A. Hauptmann, E. Pernicka, T. Rehren, and Ü. Yalçın, eds., Bochum, pp. 37–44.

Muhly, J.D., P.P. Betancourt, W.R. Farrand, C. Stearns, L. Onyshkevych, W.B. Hafford, and D. Evely. 1999. “Research and Excavation at Chrysokamino, Crete, 1995–1998,” *Hesperia* 68, pp. 343–370.

Muhly, J.D. 2000. *Crete 2000: A Centennial Celebration of American Archaeological Work on Crete (1900–2000),* J.D. Muhly and E. Sikla, eds., Athens.

———. 2000. “Excavation and Publication: The American School of Classical Studies at Athens, 1896–1999,” in *The Problem of Unpublished Excavations*, S. Hadjisavvas and V. Karageorghis, eds., Nicosia, pp. 79–88.

Muhly, J.D. 2001. Review of K. Aslihan Yener, *The Domestication of Metals: The Rise of Complex Metal Industries in Anatolia*, Leiden, 2000, in *AJA* 105, pp. 729–730.

Muhly, J.D., S.W. Manning, B. Weninger, A.K. South, B. Kling, P.I. Kuniholm, S. Hadjisavvas, D.A. Sewell, and G. Cadogan. 2001. “Absolute Age Range of the Late Cypriot IIC Period on Cyprus,” *Antiquity* 75, pp. 328–340.

Muhly, J.D. 2002. “Early Metallurgy in Greece and Cyprus,” in *Anatolian Metal* II (*Der Anschnitt Beiheft* 15), Ü. Yalçın, ed., Bochum, pp. 77–82.

———. 2002. Review of S. Swiny, R.H. Hohlfelder, and H. Wylde Swiny, eds., *Res Maritimae: Cyprus and the Eastern Mediterranean from Prehistory to Late Antiquity*, in *IEJ* 52, pp. 118–120.

———. 2003. “Archaeology and Archaeometry: Why We Need (and Should Want) to Work Together,” *METRON: Measuring the Aegean Bronze Age. Proceedings of the 9th International Aegean Conference, New Haven, April 18–21, 2002* (*Aegaeum* 24), K.P. Foster and R. Laffineur, eds., Liège, pp. 8–23.

———. 2003. “Archaeology in the Eastern Mediterranean: Where We Went Wrong in the Twentieth Century and How We Can Do Better in the Twenty-first,” in *One Hundred Years of American Archaeology in the Middle East. Proceedings of the American Schools of Oriental Research Centennial*

Celebration, Washington, D.C., April, 2000, D.R. Clark and V.H. Matthews, eds., Boston, pp. 19–32.

———. 2003. "Greece and Anatolia in the Early Iron Age: The Archaeological Evidence and the Literary Tradition," in *Symbiosis, Symbolism, and the Power of the Past: Canaan, Ancient Israel, and Their Neighbors from the Late Bronze Age through Roman Palestine. Proceedings of the Centennial Symposium, W.F. Albright Institute of Archaeological Research and American Schools of Oriental Research, Jerusalem, May 29–31, 2000*, W.G. Dever and S. Gitin, eds., Winona Lake, IN, pp. 23–35.

———. 2003. "Metalworking/Mining in the Levant," in *Near Eastern Archaeology: A Reader*, S. Richard, ed., Winona Lake, IN, pp. 174–183.

———. 2003. Review of C.F.E. Pare, ed., *Metals Make the World Go Round: The Supply and Circulation of Metals in Bronze Age Europe*, Oxford, 2000, in *AJA* 107, pp. 291–293.

———. 2003. "Trade in Metals in the Late Bronze Age and the Iron Age," in *Sea Routes: Interconnections in the Mediterranean 16th–6th c. B.C.*, N.C. Stampolidis and V. Karageroghis, eds., Athens, pp. 141–150.

Muhly, J.D., and T. Stech. 2003. "The Metallurgy of Ninevite 5," in *The Origins of North Mesopotamian Civilization Ninevite: 5 Chronology, Economy, Society* (*Subartu* IX), E. Rova and H. Weiss, eds., Turnhout, Belgium, pp. 417–428

Muhly, J.D. 2004. "Chrysokamino and the Beginnings of Metal Technology on Crete and in the Aegean," in *Crete Beyond the Palaces. Proceedings of the Crete 2000 Conference, Athens, July 10–12,* L.P. Day, M.S. Mook, and J.D. Muhly, eds., Philadelphia, pp. 283–289.

———. 2004. "Excavating Minoan Sites," review of J.L. Fitton, *Peoples of the Past: Minoans*, London, 2002, in *Archaeology Odyssey* 7:2, pp. 26–37.

———. 2004. Review of E.S. Elster and C. Renfrew, eds., *Prehistoric Sitagroi: Excavations in Northeast Greece, 1968–1970*. Vol. 2: *The Final Report* (*Monumenta Archaeologia* 20), Los Angeles, 2003, in *BMCR* 2004.6.21, pp. 1–6.

———. 2004. Review of G. Lanfranchi, M. Roaf, and R. Rollinger, eds., *Continuity of Empire (?): Assyria, Media, Persia. Proceedings of the International Meeting in Padua, 26–28 April 2001* (*History of the Ancient Near East Monographs* 5), Padua, 2003, in *BMCR* 2004.11.11, pp. 1–8.

———. 2005. "Cyprus and Copper for the World," in *Anatolian Metal* III (*Der Anschnitt Beiheft* 18), Ü. Yalçın, ed., Bochum, pp. 137–141.

———. 2005. "Kupfer und Bronze in der spätbronzezeitlichen Ägäis," in *Das Schiff von Uluburun – Welthandel vor 3000 Jahren*, Ü. Yalçın, C. Pulak, and R. Slotta, eds., Bochum, pp. 503–513.

———. 2005. "Mycenaeans Were There before the Israelites: Excavating the Dan Tomb," review of A. Biran and R. Ben-Dov, *Dan II: A Chronicle of the Excavations and the Late Bronze Age "Mycenaean" Tomb*, Jerusalem, 2002, in *Biblical Archaeology Review* 31:5, pp. 44–51.

———. 2005. Review of L. Steel, *Cyprus before History: From the Earliest Settlers to the End of the Bronze Age*, London, 2004, in *BMCR* 2005.09.83, pp. 1–7.

———. 2005. "Travelling Craftsmen: Love 'em or Leave 'em," in *EMPORIA: Aegeans in the Central and Eastern Mediterranean. Proceedings of the 10th International Aegean Conference, Athens, April 14–18, 2004*, vol. II (*Aegaeum* 25), R. Laffineur and E. Greco, eds., Liège, pp. 685–690.

———. 2006. "Anatolia," "Cilicia," "Levant," "Smyrna," and "Zoroaster," in *The Cambridge Dictionary of Classical Civilization*, G. Shipley, J. Vanderspoel, D. Mattingly, and L. Foxhall, eds., Cambridge, pp. 46–47, 200, 516–517, 828, 962.

———. 2006. "Chrysokamino in the History of Early Metallurgy," in *The Chrysokamino Metallurgy Workshop and Its Territory* (*Hesperia Suppl.* 36), P.P. Betancourt, Princeton, pp. 155–177.

———. 2006. Review of *Archaeological Field Survey in Cyprus: Past History, Future Potentials. Proceedings of a Conference Held by the Archaeological Research Unit of the University of Cyprus, December 1–2, 2000*, M. Iacovou, ed., Athens, 2004, in *BMCR* 2006.09.14, pp. 1–8.

———. 2006. Review of *Knossos: Palace, City, State. Proceedings of the Conference in Heraklion Organized by the British School at Athens and the 23rd Ephoreia of Prehistoric and Classical Antiquities of Heraklion, in November 2000, for the Centenary of Sir Arthur Evans's Excavations at Knossos* (*BSA Studies* 12), G. Cadogan, E. Hatzaki, and A. Vasilakis, eds., Athens, 2004, in *BMCR* 2006.09.17, pp. 1–6.

———. 2006. "Texts and Technology. The Beginnings of Iron Metallurgy in the Eastern Mediterranean," in *Ancient Greek Technology. Proceedings of 2nd International Conference on Ancient Greek Technology, Athens, October 17–21, 2005*, T.P. Tassios and C. Palyvou, eds., pp. 19–31.

Betancourt, P.P., and J.D. Muhly. 2006. "The Sistra from the Minoan Burial Cave at Hagios Charalambos," in

Timelines: Studies in Honour of Manfred Bietak II (*Orientalia Lovaniensia Analecta* 149), E. Czerny, I. Hein, H. Hunger, D. Melman, and A. Schwab, eds., Leuven, pp. 429–435.

———. 2007. Review of M. McConchie, *Archaeology at the North-East Anatolian Frontier* V: *Iron Technology and Iron-making Communities of the First Millennium* B.C., Louvain, 2004, in *BASOR* 347, pp. 112–114.

———. 2007. "The Crucibles from the Aghia Photia Cemetery," in *Metallurgy in the Early Bronze Age Aegean*, P.M. Day and R.C.P. Doonan, eds., Sheffield, pp. 146–153.

———. 2007. "The First Use of Metal on Minoan Crete," in *Metals and Mines: Studies in Archaeometallurgy*, S. La Niece, D. Hook, and P. Craddock, eds., London, pp. 97–102.

Kling, B., and J.D. Muhly, eds. 2007. *Joan du Plat Taylor's Excavations at the Late Bronze Age Mining Settlement at Apliki Karmallos, Cyprus*, Part I (*SIMA* 134:1), Sävedalen.

Betancourt, P. P., and J.D. Muhly. 2008. "The Sistra," in "Excavations in the Hagios Charalambos Cave: A Preliminary Report," P.P. Betancourt, C. Davaras, H.M.C. Dierckx, S.C. Ferrence, J. Hickman, P. Karkanas, P.J.P. McGeorge, J.D. Muhly, D.S. Reese, E. Stravopodi, and L. Langford-Verstegen, *Hesperia* 77, p. 577.

Muhly, J.D. 2008. "Agia Photia and the Cycladic Element in Early Minoan Metallurgy," in *Aegean Metallurgy in the Bronze Age*, I. Tzachili, ed., Athens, 2008, pp. 69–74.

———. 2008. "An Introduction to Minoan Archaeometallurgy," in *Aegean Metallurgy in the Bronze Age*, I. Tzachili, ed., Athens, pp. 35–41.

———. 2008. "Metal Deposits in the Aegean Region," in *Anatolian Metal* IV (*Der Anschnitt Beiheft* 21), Ü. Yalçın, ed., Bochum, pp. 67–75.

———. 2008. "The Metal Artifacts," in "Excavations in the Hagios Charalambos Cave: A Preliminary Report," P.P. Betancourt, C. Davaras, H.M.C. Dierckx, S.C. Ferrence, J. Hickman, P. Karkanas, P.J.P. McGeorge, J.D. Muhly, D.S. Reese, E. Stravopodi, and L. Langford-Verstegen, *Hesperia* 77, pp. 557–560.

———. 2009. "Oxhide Ingots in the Aegean and in Egypt," in *Oxhide Ingots in the Central Mediterranean*, F. Lo Schiavo, J.D. Muhly, R. Maddin, and A. Giumlia-Mair, eds., Rome, pp. 17–39.

———. 2009. "Perspective: Archaeology, History, and Chronology from Penn to the Present and Beyond," in *Tree-Rings, Kings, and Old World Archaeology and Environment: Papers Presented in Honor of Peter Ian Kuniholm*, S.W. Manning and M.J. Bruce, eds.m Oxford, pp. 3–11.

———. 2009. "The Origin of the Name 'Ionian'," in *Cyprus and the East Aegean: Intercultural Contacts from 3000 to 500 BC*, V. Karageorghis and O. Kouka, eds., Nicosia, pp. 23–30.

———. 2010. "History of Research," in *The Oxford Handbook of the Bronze Age Aegean (ca. 3000–1000 BC)*, E.H. Cline, ed., Oxford, pp. 3–10.

———. 2011. "Archaeometry and Shipwrecks: A Review Article," in *Expedition* 53:1. pp. 36–44.

———. 2011. "Bronze," "Gold," "Iron," "Metals," "Silver," and "Tin," in *The Homer Encyclopedia*, 3 vols., M. Finkelberg, ed., Chichester, UK, pp. 145–146, 321–323, 419–421, 515–516, 799–801, 879–881.

———. 2011 "Metals and Metallurgy," in *The Oxford Handbook of Ancient Anatolia, 10,000–323 BCE*, eds. S. R. Steadman and G. McMahon, Oxford, pp. 858–876

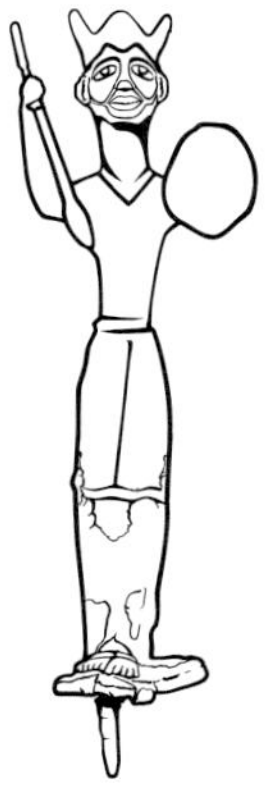

List of Abbreviations

Abbreviations for periodicals in the bibliographies of the individual articles follow the conventions of the *American Journal of Archaeology* 111.1 (2007), pp. 14–34.

AKR	excavation number, Akrotiri, Thera
cm	centimeter
dia.	diameter
EBA	Early Bronze Age
EC	Early Cycladic
EChal	Early Chalcolithic
ED-XRF	emission dispersive X-ray fluorescence
EH	Early Helladic
EM	Early Minoan
gr	gram
h.	height
HM	Herakleion Archaeological Museum
HNM	Hagios Nikolaos Archaeological Museum
L.	length
LBA	Late Bronze Age
LC	Late Cycladic or Late Cypriot
LChal	Late Chalcolithic
LH	Late Helladic
LM	Late Minoan
m	meter
MBA	Middle Bronze Age
MC	Midlle Cycladic
MChal	Middle Chalcolithic
MH	Middle Helladic
MM	Middle Minoan
NCSR	National Center for Scientific Research "Demokritos"
NM	National Archaeological Museum of Greece
NMD	Neolithic Museum, Diros, Mani
pers. comm.	personal communication
pers. obs.	personal observation
pres.	preserved

SEM/EDX	scanning electron microscopy and energy dispersive microanalyses
SM	Siteia Archaeological Museum
th.	thickness
w.	width
wt.	weight
XRD	X-ray diffractometry
th.	thickness
XRF	X-ray fluorescence spectrometry

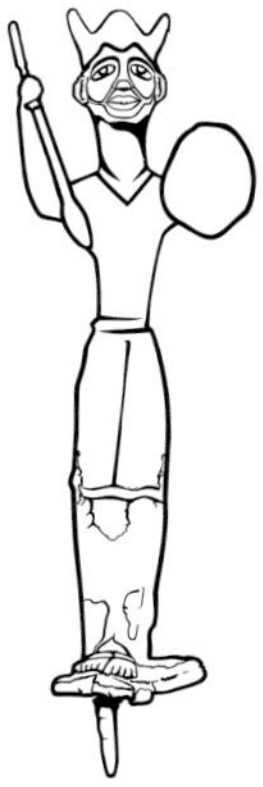

Introduction

Susan C. Ferrence

James D. Muhly is professor emeritus of Ancient Near Eastern History at the University of Pennsylvania in Philadelphia and director emeritus of the American School of Classical Studies at Athens in Greece. He has been a pioneering leader in the rarefied field of ancient metallurgy for over 45 years. His distinguished scholarship covers a wide geographic area from Italy to Mesopotamia and especially includes Cyprus and the Aegean. His knowledge of ancient literature and history, command of modern scholarship, and understanding of ever-evolving scientific analyses combine to form the basis of a long-standing and interdisciplinary academic career.

One of the highlights of his research was the publication of *Copper and Tin: The Distribution of Mineral Resources and the Nature of the Metals Trade in the Bronze Age* in 1973 (and subsequent second edition in 1976), which has been hailed as a benchmark by which other such works are measured. To quote Vince Pigott in this volume (p. 273): ". . . it proved to be . . . significant as an example *par excellence* for the successful integration of archaeological, metallurgical, ancient historical, and textual data."

In 1994 the Archaeological Institute of America gave the Pomerance Award for Scientific Contributions to Archaeology to Robert Maddin, James Muhly, and Tamara Stech. Their archaeological and scientific collaboration spanned almost 30 years and produced over 30 publications. Furthermore, Jim's body of research stands at 187 bibliographic entries and counting, an admirable number to say the least.

This volume of 28 chapters written by 38 authors of 8 different nationalities about ancient metallurgy spanning from Italy to western Asia exemplifies Jim's career of multi-national scholarly collaboration. Its title obviously reflects his—and all archaeologists'—innate curiosity about the ancient world, especially regarding the technological advances of prehistoric societies. *Metallurgy: Understanding How, Learning whY* is offered to a scholar, mentor, friend, and colleague who has influenced the lives, scientific research, and academic scholarship of the contributors represented in this festschrift.

Part I includes seven chapters on the metallurgy of Cyprus. Edgar Peltenburg presents 18 metal objects plus eight pieces of related evidence in his examination of the early history of metalwork on Cyprus. Alessandra Giumlia-Mair, Vasiliki Kassianidou, and George Papasavvas use X-ray fluorescence (XRF) to analyze seven miniature ingots from Cyprus, a program of scientific inquiry that yielded surprising results. Sophocles Hadjisavvas explores aspects of metallurgy at Alassa in Cyprus and how it relates to cult at the end of the Late Bronze Age. Vassos Karageorghis revisits the site of Athienou-*Pamboularin tis Koukkouninas* and suggests a new interpretation for a special deposit of interesting pottery and other unique finds. Vassiliki Kassianidou investigates the origins of pot bellows in Cyprus especially in relation to the site of Politiko-*Phorades*. Fulvia Lo Schiavo digs into the archives of the Florence Archaeological Museum to uncover the provenance of a miniature Cypriot tripod-stand. George Papasavvas discusses the manufacture and iconography of a metal figurine from Enkomi, which has been called the Ingot God, and explores a transformation in its meaning at the site.

Part II is comprised of seven chapters on the metallurgy of Crete. Mihalis Catapotis, Yannis Bassiakos, and Yiannis Papadatos present new scientific data resulting from a program of analysis that was undertaken to understand the role of copper production in eastern Crete at the juncture of the Final Neolithic and Early Minoan I periods. Calliope Galanaki, Yannis Bassiakos, and Vassilis Perdikatsis use three different types of elemental analyses to identify several metal objects from an Early Minoan I cemetery with Cycladic influence at Gournes. Jane Hickman discusses several aspects of the gold Dog Diadem from Mochlos—such as context, manufacture, date, and iconography—in her reconstruction of its original form, meaning, and life history. Keith Branigan adds four new specimens to his corpus of triangular daggers from prehistoric Crete, for a total of 98 pieces, and updates his theories on the topic. Philip Betancourt examines the iconography, meaning, and date of a gold ring from the burial cave of Hagios Charalambos, with implications for regional politics in eastern Crete during the Middle Bronze Age. Jean-Claude Poursat and Cécile Oberweiler apply fabric analyses and scanning electron microscopy (SEM) to clay crucibles, molds, bellows, and tuyères in their assessment of the proficient metalworking technology in use at Middle Minoan Malia. Jeffrey Soles describes a large metal sistrum that was excavated at Mochlos and details parallels from other sites in Crete in order to elucidate its context and meaning.

Part III contains six chapters on metallurgical technology. Zozi Papadopoulou characterizes the high level of specialized metalworking in southern Siphnos during the Early Bronze Age (EBA). Olga Philaniotou, Yannis Bassiakos, and Myrto Georgakopoulou investigate four known slag heaps on Seriphos in the Cyclades to shed light on copper smelting during the Early Bronze Age. Christos Doumas richly illustrates different types of tools—some of which are enigmatic—that come from several EBA Aegean sites and proposes that their function should be based in metallurgy. Anno Hein and Vassilis Kilikoglou discuss heat transfer in ceramics in relation to their examination of furnace fragments from Seriphos and Cyprus. Through the application of four types of scientific analyses, Andreas Hauptmann attempts to define "furnace conglomerate," which is a type of slag that was identified at Kition and Enkomi in Cyprus. Robert Maddin elucidates the reasoning behind the replacement of bronze with iron during the early Iron Age.

Part IV encompasses eight chapters about trade and interconnections in the history of metallurgy. Noël Gale uses lead isotope analysis to identify the Apliki mine region of Cyprus as the main source of ore for the production of most copper oxhide ingots. Zofia Stos-Gale analyzes lead isotope data in search of the origin of the earliest oxhide ingots, which are dated to Late Minoan IB and come from five sites on Crete. Reinhard Jung, Mathias Mehofer, and Ernst Pernicka attempt to provenance the raw material that was used to manufacture over 30 bronze objects that come from different parts of Italy and date to the Middle and Final Bronze Age. Bernard Knapp integrates cuneiform texts with archaeological and archaeometallurgical data to identify Alashia as Cyprus and place it within the larger eastern Mediterranean sphere of economic and political relations. Robert Merrillees critically reviews a recent volume and disagrees with its assertions regarding the identification of Cyprus with Alashiya. Aslihan Yener presents a battle ax from Alalakh in southern Turkey and suggests that its stylistic components are consistent with other objects that have been described as sacred weapons. Vincent Pigott reviews recent scholarship regarding the Bronze Age tin trade in southwestern Asia, particularly concerning the possible sources of tin. Cemal Pulak tells the story of the chance discovery of three oxhide ingots in southeastern Turkey and further postulates on their ancient riverine trade route.

PART I

Metallurgy of Cyprus

CHAPTER

1

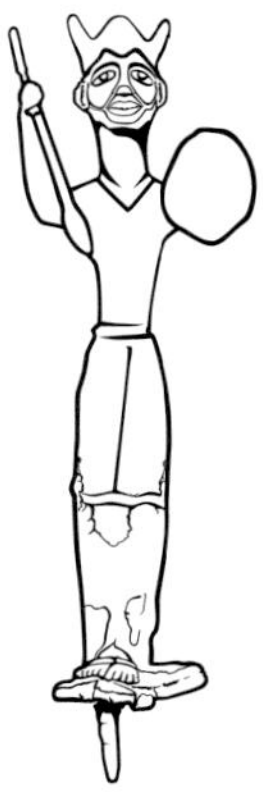

Cypriot Chalcolithic Metalwork

Edgar Peltenburg

In the most recent survey of the use of copper on the island of Cyprus before the Early Bronze Age, Noël Gale evaluated nine certain artifacts recovered from systematic excavations (Gale 1991). Since then, the tally has increased to 18 objects in addition to related evidence that bears on the early history of metalwork on an island renowned for its copper deposits (Table 1.1). James Muhly has been pre-eminent in elucidating the development of Cypriot metalworking in its wider technological, economic, social, and historical contexts (e.g., Muhly 1991). This note in honor of his many insights addresses a question that is germane to that discussion. Have recent additions altered reconstructions of the earliest stages of the production, consumption, and circulation of copper in Cyprus?

It is generally acknowledged that the use of copper started during the Middle Chalcolithic (MChal, ca. 3400–2900 B.C.) and Late Chalcolithic (LChal, ca. 2800–2400 B.C.) periods (e.g., Weinstein-Balthazar 1990, 92–95; Knapp, Held, and Manning 1994; Webb et al. 2006). Muhly (1993, 243) argues that the Chalcolithic artifacts were made of native copper. In contrast, Gale states that we cannot be certain if the objects "represent any use of Cypriot copper ores at all" (Gale 1991, 57). He concludes that since objects were not made of native metal but were smelted from copper ores, it is unlikely that islanders developed the use of copper in conjunction with the exploitation of picrolite pebbles from rivers (cf. Peltenburg 1982). In the absence of further scientific analyses, I shall try to show that a contextual approach rather than one that isolates material from its cultural associations strengthens the case for procurement from indigenous sources before the Early Bronze Age (EBA). To do so, we should distinguish material from the two Chalcolithic phases, MChal and LChal. But first, we need to delete two items from Gale's list of Chalcolithic metals.

An axe, KM 457, in fact comes from Unit 66 in the disturbed Early Bronze Age Period 5 levels over Building 86 at Kissonerga (Peltenburg 1998b, 231).

SITE CODE	SITE AND CONTEXT	CLASS	DESCRIPTION	PUBLICATION	FIGURE
			Middle Chalcolithic		
—	Erimi depth 0.4–0.6 m	Narrow band (bead?)	Bent strip, L. 2.9 cm	Gale 1991, 44–45	1.1:B
—	Erimi	Blade(?)	L. 2.3, w. 1.4, th. 0.2–0.25 cm	Gale 1991, 44–45	1.1:C
388	Erimi Layer IX	Chisel	Tip of cutting edge	Dikaios 1938, 50, 80	1.1:D
—	Erimi depth 0.0–0.20 m	"Needle"	One of two "bronze needles"	Bolger 1985, 182 (from Dikaios's daybook)	—
—	Erimi depth 0.60 m	"Needle"	One of two "bronze needles"	Bolger 1985, 182 (from Dikaios's daybook)	—
KM 2109	Kissonerga-*Mosphilia* Gr. 554	Mineral	Malachite in bivalve	Peltenburg 1998a, fig. 97:5; 1998b, 231–233	1.1:A
SL 428	Souskiou-*Laona* T. 158	Spiral bead	Flat strip coiled nine times in spiral for cylindrical bead. Traces mineralized thread, L. 2.7, dia. 0.6 cm	Crewe et al. 2005, 65, fig. 16:2	1.1:G
SL 429	Souskiou-*Laona* T. 158	Pendant(?)	six curved fragments, five with circular sections, one of these with tightly looped terminal, L. 2.5, dia. 0.2 cm	Crewe et al. 2005, 65, fig. 16:3	1.1:H
SL 554	Souskiou-*Laona* Bld. 34	Pendant(?)	Flat spiraliform band of metal, L. 2.5, w. 1.4, th. 0.4 cm	Peltenburg et al. 2006, 98, fig. 21	1.1:E
SL 569	Souskiou-*Laona* Bld. 34	Fragment	Amorphous, mineralized fragment, 0.5 x 0.3 cm	Unpublished	—
SL 570	Souskiou-*Laona* Bld. 34	Blade	Flat fragment, 1.6 x 1.8 x 0.8 cm	Unpublished	1.1:I
SL 867	Souskiou-*Laona* Bld. 34	Fragment	Amorphous lump, 0.9 x 0.8 x 0.7 cm	Unpublished	—
SV 23/34	Souskiou-*Vathrykakas* T. 23	Spiral bead	Flat strip coiled nine times in spiral for cylindrical bead. Traces mineralized thread (?), L. 3.2, dia. 0.4 cm	Peltenburg, ed., 2006, pl. 10:5	1.1:F
SV 78/19	Souskiou-*Vathrykakas* T. 78	Corrosion product	Adheres to stone pendant	Peltenburg, ed., 2006, 99	—
			Late Chalcolithic		
C 384	Kissonerga-*Mosphilia* Bld. 834	Fragments	Flakes of oxidized copper, from object?	Peltenburg 1998b, 231–232	—
KM 416	Kissonerga-*Mosphilia* above Pit 411	Awl	Square-sectioned, tapered to point, bone handle, 3.9 x 0.45 x 0.45 cm	Peltenburg 1998a, fig. 97:4; 1998b, 231	1.2:A
KM 633	Kissonerga-*Mosphilia* Unit 150	Ore	Rust-colored nugget in green envelope, 7.5 x 6 x 3.2 cm	Zwicker 1988; Gale 1991; Peltenburg 1998b, 231	1.2:F
KM 693	Kissonerga-*Mosphilia* Bld. 706	Crucible(?)	Subrectangular stone dish, L. 10.8 x h. 4.3 cm	Peltenburg 1998a, fig. 95:14; 1998b, 231–233	1.2:J
KM 694	Kissonerga-*Mosphilia* Bld. 706	Chisel	Square-sectioned, tapered above cutting edge, broken above bend, 9.8 x 1.1 x 1.1 cm	Peltenburg 1998a, fig. 97:2; 1998b, 231	1.2:B
KM 701	Kissonerga-*Mosphilia* Bld. 706	Ore	Lumps with bright green corrosion	Gale 1991; Peltenburg 1998b, 231	1.2:I
KM 986	Kissonerga-*Mosphilia* Quarry 654	Chisel	Square-sectioned, convex working edge, tapered to flat butt, 11.2 x 0.8 x 0.75 cm	Peltenburg 1998a, fig. 97:3; 1998b, 231	1.2:C
KM 1007	Kissonerga-*Mosphilia* Bld. 3	Crucible(?)	Shallow bowl, L. 17.6 x h. 6.7 cm	Peltenburg 1998a, fig. 97:15	—
KM 1182	Kissonerga-*Mosphilia* Gr. 529	Ear/hair ring	Spiral with expanded and pointed terminals	Peltenburg 1998a, fig. 97:5; 1998b, 231	1.2:H
KM 2174	Kissonerga-*Mosphilia* Bld. 834	Chisel	Tip, oval-section, beveled cutting edge, 2.5 x 0.5 cm	Peltenburg 1998a, pl. 36:2; 1998b, 231	1.2:G
LL 134	Lemba-*Lakkous* Bld. 3	Chisel	Square-sectioned, non-splayed cutting edge, L. 9.4 x w. 0.06 cm	Slater 1985, 199–200, fig. 86:5, pl. 48:11; Gale 1991	1.2:D
LL 209	Lemba-*Lakkous* Bld. 3	Blade(?)	Thin trapezoidal fragment, L. 3.5 x w. 1.5 cm	Slater 1985, 199, fig. 86:6, pls. 18:3, 48:10; Gale 1991	1.2:E

Table 1.1. List of copper and metal-related material from Chalcolithic Cyprus. KM 1182 also could belong to Kissonerga Period 5 (Philia). Bld.: building; T: tomb; Gr: grave.

It was found together with a dense spread of Philia Red Polished pottery, including two bowls characteristic of the EBA (Peltenburg 1998a, 53). Lead isotope data suggests that the axe consists of copper from northwestern Anatolia, and Gale uses this as evidence to argue that some Chalcolithic copper came from abroad (Gale 1991, 45–46, 53–54). Given the profound transformation in Cyprus at the start of the EBA, we should exclude Philia metalwork from that of the Chalcolithic period.

The second item is a hook from EChal Mylouthkia. It was one of two metal pieces found in surface scrapings of pits. The other proved to be of brass, so that leaves only the hook, which Slater (1982) regarded as a cold-worked item, and Gale (1991) thought it was smelted from ores. Its poor stratigraphic security and the absence of other evidence for metalwork at that time mean that we should treat it as unreliable evidence for metalwork on Cyprus during the EChal.

Middle Chalcolithic Copper

Turning to additional objects, there are two "needles" from Erimi, and beads, pendants, and a blade from Souskiou-*Laona* (Fig. 1.1). In his Erimi report, Dikaios (1938) only selected certain material for publication. There are no reasons for us to omit what he termed "bronze needles" in his daybook, since he also used the word "bronze" for the copper strip of Fig. 1.1:B (in Bolger 1985, 180, 182). One of the needles comes from a depth in Layer XIII, the same as the strip. Layer XIII has coherent buildings with no reported evidence of disturbance.

Spiral beads from Souskiou are made of thin strips of coiled copper (Fig. 1.1:F, G). They suggest that the Erimi strip of Fig. 1.1:B, which Porphyrios Dikaios initially described as a hook, was, or was intended as, an ornament like a spiral bead, now in an unwound state. Together with an annular pendant(?) (Fig. 1.1:H), they come from Souskiou mortuary deposits where they were probably associated on necklaces with cruciform pendant-figurines, the birthing symbols that played such a defining role in MChal society. A similar emphasis on metal for ornaments is also evident in the settlement.

Building 34 yielded a unique spiraliform pendant, a blade, and two small lumps of copper (Figs. 1.1:E, I). A more complex pendant parallel in the

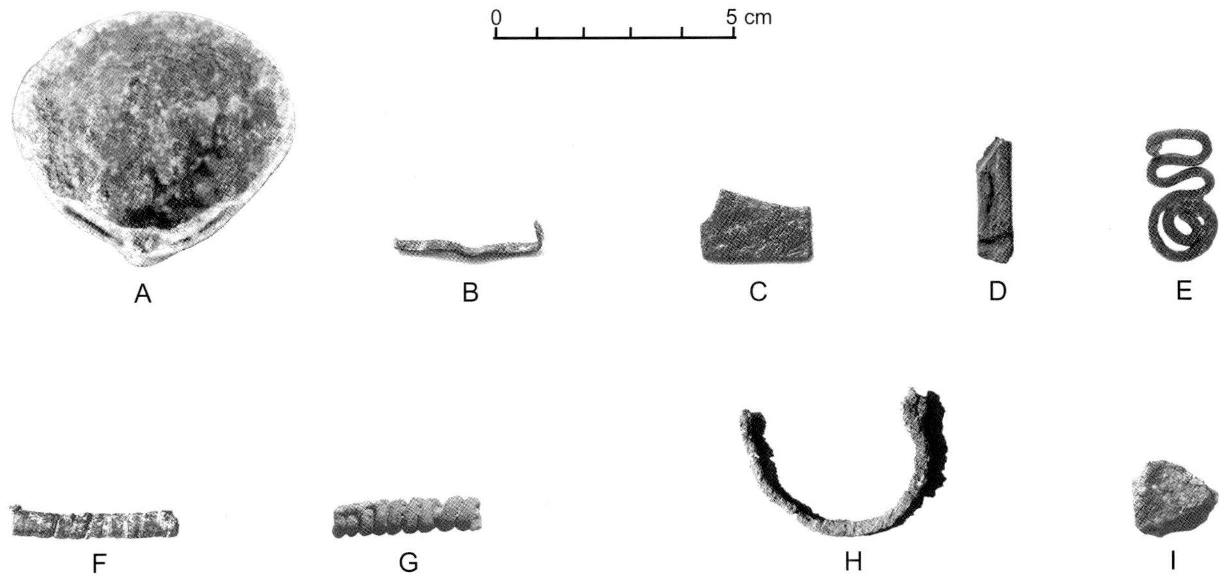

Figure 1.1. Middle Chalcolithic metal objects from Cyprus. For object descriptions, see Table 1.1. A: KM 2109; B: Erimi narrow band; C: Erimi blade; D: Erimi chisel 388; E: SL 554; F: SV 23/34; G: SL 428; H: SL 429; I: SL 570.

Museum of Cycladic Art in Athens allegedly comes from Lemba-*Lakkous* (N.P. Goulandris Foundation: Th.N. Zintilis Collection, no. 775; I owe this reference to H. Pareskeva). As its provenance is uncertain, it has not been included here. None has been analyzed yet, but we may infer that copper was more widespread than is generally believed from the occurrence of items in settlements where metal was curated. And the lumps raise the possibility of debris from manufacture for ornaments. This would not be the only instance of ornament production at Souskiou. To judge from dentalium bead cut-offs and abundant picrolite wasters, bead and pendant production was a significant activity. The finished products were very likely destined for the adjacent cemeteries, well known for yielding the richest assemblage of picrolite and dentalia objects on the island. The picrolite wasters are especially interesting since they include many quarried slabs that had to be extracted from serpentinite sources in the hills (Peltenburg et al. 2006, 85, fig. 10). Previously, scholars believed islanders obtained picrolite as pebbles from riverbeds. The unambiguous association of some of the earliest copperwork on the island with the significant intensification of picrolite bead and pendant production and evidence for active quarrying for the raw material provide a plausible context for the start of small-scale copper extraction. Clearly, incentives for the introduction of copperwork in the later 4th millennium B.C. on Cyprus included the quest for rare materials for valued personal adornments, and not just utilitarian tools as is often stated (e.g., Steel 2004, 95).

This MChal intensification also led to the introduction of segmented and other blue faience beads. Among the earliest in the Mediterranean, they come from Souskiou-*Vathyrkakas* Tombs 29, 55, and 78 (Peltenburg, ed., 2006), and Souskiou-*Laona* Tombs 138, 158 (with copper spiral bead and pendant), and 221 (Crewe et al. 2005). From their analysis of one of these faience objects, Andrew Shortland and Mike Tite (2006) conclude that the glaze was "colored by copper, perhaps introduced into the glaze as bronze" and that "if scrap bronze has been used for the colorant, the Late Bronze Age date is far more likely than the ca. 3000 B.C. date." They calculate that the bronze would have had 7% tin, much higher than is attested in Chalcolithic coppers and in the vast majority of bronzes from Early to Late Cypriot (EC–LC) Cyprus. However, since these beads occur so frequently in secure Chalcolithic tomb contexts, we must seriously consider a Chalcolithic date.

In general, tin bronzes occur early in Anatolia and adjacent regions (Thermi, Troy II/III, Amuq G, Qara Qûzâq), but even if tin production existed throughout the 3rd millennium B.C. at Göltepe and Kestel, it was not until later in the 3rd millennium that they were even moderately widespread (Muhly 1993). Alternatively, tin may have been added deliberately to the glaze recipe. Producers of faience were occasionally aware of the glossy effects of enhanced amounts of tin, as in the British Bronze Age (McKerrell 1972). Unfortunately, our knowledge of its occurrence in faience is extremely limited, due in part to the paucity of analyses of Western Asiatic faience objects. While they obviously demand further investigation, the faience beads raise the intriguing question of greater knowledge of the properties of metals from ca. 3000 B.C. than is often thought to be the case.

Further questions relating to Cyprus during the M/LChal transition arise as a result of the discovery abroad of items made of copper that is consistent with a Cypriot ore source according to lead isotope analysis. They include an axe from Pella in Jordan and two daggers, a fish hook, and an awl from Hagia Photia, Crete (Philip, Clogg, and Dungworth 2003; Stos-Gale and Gale 2003, 91–92, table 5). The axe is from a well-dated EB II hoard of ca. 3000 B.C.; the others come from the Cretan EB I cemetery of roughly the same date (Day, Wilson, and Kiriatzi 1998 for dating; Davaras and Betancourt 2004, 4). The extraction, production, and export of Cypriot copper at such an early date seems counterintuitive in view of its small-scale use on the island, let alone the lack of imports and evidence for communities geared for such endeavors. However, our information is biased to the copper-poor south and west of the island, away from the copper-rich north.

We have little idea what was happening in the north in the centuries following the abandonment of the Kythrea and Lapithos MChal settlements, and before the apparently sudden participation of Philia groups in widespread copper exchanges involving ingots and finished objects (Webb et al. 2006). That there was slightly earlier intra-island division is evident from the enormous disparity in the occurrence of iconic cruciform figurines in the north and the south, based on pre-1974 incidences. Was there a split similar to that in contemporary EB I Crete where some would argue that foreigners at the northern sites of Poros and Hagia Photia worked metals (Day, Wilson, and Kiriatzi 1998)? James Muhly pointed to a disjunction later in Cyprus, at the turn of the 3rd–2nd millennium when Minoan imports are concentrated in the north where there are sites with exceptionally rich deposits of copper and bronze objects (Muhly 1991), and where lies the likely origin for a Cypriot amphora from Knossos (Catling and MacGillivray 1983). It was probably also from the north/center that Cypriot painted pottery was imported into EB II Tarsus since that class was no longer produced in the west of the island then (Goldman 1956, fig. 283:279, 280; cf. Peltenburg, ed., 2006, pl. 36:4, 5). These are tenuous hints pointing to the existence of northern precursors for "sudden" Philia engagement in inter-regional contacts. Earlier, according to Philip (2002), contacts thrived on the transmission of consumables, raw materials, and technological innovations.

As a consequence of the spatial imbalance in current research, regional information bias constitutes a major handicap in reconstructing the early history of copper use on Cyprus. Unfortunately, that bias continues in the LChal when only the Lemba cluster of western sites furnishes information on metals.

Late Chalcolithic Copper

There is one addition to the meager inventory of LChal objects since Gale's survey: a fragmentary chisel from Building 834 at Kissonerga (Fig. 1.2:G). But, objects treated in isolation can be misleading. It is important to evaluate related material such as a possible crucible and the amorphous fragments left behind in the settlement. They imply much more extensive use. Another object (C 384) from the same Period 4a building also belongs to the first half of the 3rd millennium B.C. This is significant because it has been argued that all LChal metal objects may be related to intrusive Philia groups (Webb and Frankel 1999, 31). Kissonerga Period 4a, however, antedates the presence of Philia on the island, unlike Period 4b when indigenous and Philia groups could have co-existed. Among the other amorphous pieces is KM 701 (Fig. 1.2:I) from the slightly later Building 706. It was associated with a chisel and a copper-rich stone dish (Fig. 1.2:B, J). The dish is unique at Kissonerga by virtue of its pouring lip, which potentially allowed it to function as a crucible. Much of the material in Building 706 was derived from the immediately underlying Pithos House. Below that house was another, almost identical metal chisel (Fig. 1.2:C). The concentration of copper objects, waste(?) and possible

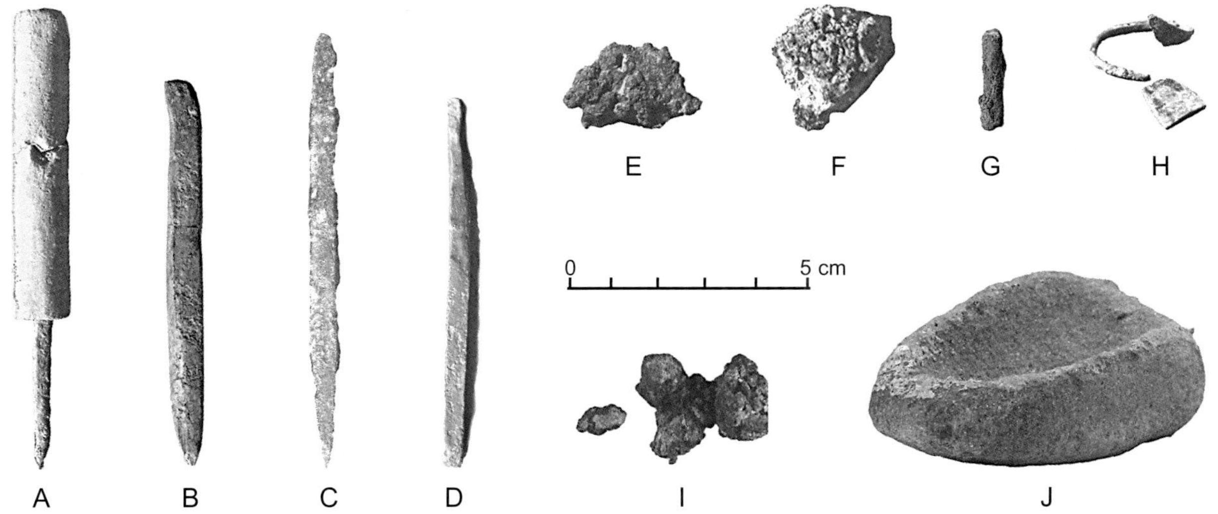

Figure 1.2. Late Chalcolithic metal objects from Cyprus. For object descriptions see Table 1.1. A: KM 416; B: KM 694; C: KM 986; D: LL 134; E: LL 209; F: KM 633; G: KM 2174; H: KM 1182; I: KM 701; J: KM 693.

crucible(s) in this location provides strong circumstantial evidence for small-scale working of copper.

Apart from the transitional LChal/Philia spiral ear/hair ring (Fig. 1.2:H), western LChal metalwork only consists of small tools, a contrast to the MChal and Philia repertoires. This may be due to our poor sample, but other changes furnish a more meaningful explanation. We have seen that many copper items in the MChal period belonged to a symbolic system in which cruciform birthing figures play a key role. The latter virtually disappear in the LChal period, as do the great variety of pendants and other symbols of identity associated with those representations. The change in metal usage, therefore, should be seen in terms of major social and ideological changes that occurred in the early 3rd millennium on the island, ones that, according to the evidence from the Pithos House at Kissonerga, were associated with an increasing emphasis on explicit control of the means of production.

Conclusion

The first secure evidence for metalwork on Cyprus occurs in the MChal period. It includes a chisel from Erimi Layer IX, the last layer before Red-on-White pottery began to dominate at that settlement. A similar ceramic sequence at Kissonerga places this in the mid–late 4th millennium B.C. Since copper was used then for ornaments on necklaces with anthropomorphic figures, we can see that its introduction was linked to the visual display of prestige goods and to the ideology expressed by birthing figures. Contextual analysis at Souskiou relates the appearance of these copper items with the floruit in the use of picrolite, some of which was quarried. While quarrying provides a conceptual and practical context for extracting copper from ores, only analysis can determine if MChal copper was obtained from Cypriot sources. There is a shift in production toward more utilitarian items in the LChal, one that emphasizes the economic and political context of the use of copper. If the few indications of exported Cypriot copper in the early 3rd millennium B.C. are confirmed, then regional exploitation patterns varied significantly, and the inventory of simple, hammered items presented here must be regarded as only valid for an understanding of the development of copperwork in the south and west of the island.

The contrast between the MChal–LChal copperwork and the weaponry that characterizes the subsequent Philia assemblage is striking. Indigenous groups encountered newcomers and allies capable of wielding metal knives, axes, metal-tipped spears, and other hook-tang weapons. When we consider that the introduction of so much weaponry occurred at a time of widespread abandonment of western settlements, and the destruction and abandonment of at least one of these, Lemba, the Philia phenomenon in the west would seem less a matter of processes of hybridization than of marginalization or encapsulation of indigenous groups.

Acknowledgments

I am grateful to Stephen Bourke and Graham Philip for information on the Pella hoard. Photos by author except: Figure 1.1:B, D courtesy of the Department of Antiquities of Cyprus; Figure 1.1:C courtesy of Diane Bolger.

References

Bolger, D. 1985. *Erimi*-Pamboula*: A Study of the Site in Light of Recent Evidence*, Ph.D. diss., University of Cincinnati.

Catling, H., and J. MacGillivray. 1983. "An Early Cypriot III Vase from the Palace at Knossos," *BSA* 78, pp. 1–8.

Crewe, L., K. Lorentz, E. Peltenburg, and S. Spanou. 2005. "Treatments of the Dead: Preliminary Report of Investigations at Souskiou-*Laona* Chalcolithic Cemetery, 2001–2004," *RDAC* 2005, pp. 41–67.

Davaras, C., and P.P. Betancourt. 2004. *The Hagia Photia Cemetery* I: *The Tomb Groups and Architecture* (*Prehistory Monographs* 14), Philadelphia

Day, P., D. Wilson, and E. Kiriatzi. 1998. "Pots, Labels and People: Burying Ethnicity in the Cemetery of Aghia Photia, Siteias," in *Cemetery and Society in the Aegean Bronze Age* (*Sheffield Studies in Aegean Archaeology* 1), K. Branigan, ed., Sheffield, pp. 133–149.

Dikaios, P. 1938. "The Excavations at Erimi, 1933–1935," *RDAC* 1936, pp. 1–81.

Gale, N. 1991. "Metals and Metallurgy in the Chalcolithic Period," *BASOR* 282 (3), pp. 37–62.

Goldman, H. 1956. *Excavations at Gözlü Kule, Tarsus* II: *From the Neolithic through the Bronze Age*, Princeton.

Knapp, A.B., S. Held, and S. Manning. 1994. "The Prehistory of Cyprus: Problems and Prospects," *Journal of World Prehistory* 8, pp. 377–453.

McKerrell, H. 1972. "On the Origins of British Faience Beads and Some Aspects of the Wessex-Mycenae Relationship," *PPS* 38, pp. 286–301.

Muhly, J. 1991. "Copper in Cyprus: The Earliest Phase," in *Découverte du Metal*, J.P. Mohen and C. Eluère, eds., Paris, pp. 357–374.

———. 1993. "Early Bronze Age Tin and the Taurus," *AJA* 97, pp. 239–253.

Peltenburg, E. 1982. "Early Copperwork in Cyprus and the Exploitation of Picrolite: Evidence from the Lemba Archaeological Project," in *Early Metallurgy in Cyprus, 4000–500 B.C.*, J. Muhly, K. Maddin, and V. Karageorghis, eds., Nicosia, pp. 41–62.

———. 1998a. *Lemba Archaeological Project, Cyprus* II.1A: *Excavations at Kissonerga*-Mosphilia, *1979–1992* (*SIMA* 70 [2]), Göteborg.

———. 1998b. *Lemba Archaeological Project, Cyprus* II.1B: *Excavations at Kissonerga*-Mosphilia, *1979–1992* (*Department of Archaeology Occasional Paper* 19), Edinburgh, http://www.arcl.ed.ac.uk/arch/publications/cyprus/kissonerga/ or http://ads.ahds.-ac.uk/catalogue/collections/blurbs/279.cfm.

Peltenburg, E., ed. 2006. *The Chalcolithic Cemetery of Souskiou*-Vathyrkakas, *Cyprus: Results of the Investigations of Four Missions, from 1950 to 1997*, Nicosia.

Peltenburg, E., D. Bolger, M. Kincey, A. McCarthy, C. McCartney, and D.A. Sewell. 2006. "Investigations at Souskiou-*Laona* Settlement, Dhiarizos Valley, 2005," *RDAC* 2002, pp. 77–105.

Philip, G. 2002. "Contacts between the 'Uruk' World and the Levant during the Fourth Millennium B.C.: Evidence and Interpretation," in *Artefacts of Complexity: Tracking the Uruk in the Near East*, J.N. Postgate, ed., Warminster, pp. 207–235.

Philip, G., P. Clogg, and D. Dungworth. 2003. "Copper Metallurgy in the Jordan Valley from the Third to the First Millennia B.C.: Chemical, Metallographic and Lead Isotope Analyses of Artefacts from Pella," *Levant* 35, pp. 71–100.

Shortland, A., and M. Tite. 2006. "Report on the Analysis of Cylindrical Bead SVP 29/32," in Peltenburg, ed., 2006, pp. 95–96.

Slater, E. 1982. "The Copper Artefacts and Their Analyses," in Peltenburg 1982, pp. 41–47.

———. 1985. "The Metal Artefacts," in *Lemba Archaeological Project* I: *Excavations at Lemba*-Lakkous *1976–1983* (*SIMA* 70 [1]), E. Peltenburg, Göteborg, pp. 198–200, 292–294.

Steel, L. 2004. *Cyprus before History: From the Earliest Settlers to the End of the Bronze Age*, London.

Stos-Gale, Z., and N. Gale. 2003. "Lead Isotopic and Other Isotopic Research in the Aegean," in *METRON: Measuring the Aegean Bronze Age* (*Aegaeum* 24), K. Foster and R. Laffineur, eds., Liège, pp. 83–100.

Webb, J., and D. Frankel. 1999. "Characterizing the Philia Facies: Material Culture, Chronology, and the Origin of the Bronze Age in Cyprus," *AJA* 103, pp. 3–43.

Webb, J., D. Frankel, Z.A. Stos, and N. Gale. 2006. "Early Bronze Age Metal Trade in the Eastern Mediterranean. New Compositional and Lead Isotope Evidence from Cyprus," *OJA* 25, pp. 261–88.

Weinstein-Balthazar, J. 1990. *Copper and Bronze Working in Early through Middle Bronze Age Cyprus* (*SIMA-PB* 84), Jonsered.

Zwicker, U. 1988. "Investigations of Material from Maa-*Palaeokastro* and Copper Ores from the Surrounding Areas," in *Excavations at Maa*-Palaeokastro *1979–1986*, by V. Karageorghis and M. Demas, Nicosia, pp. 427–448.

CHAPTER

2

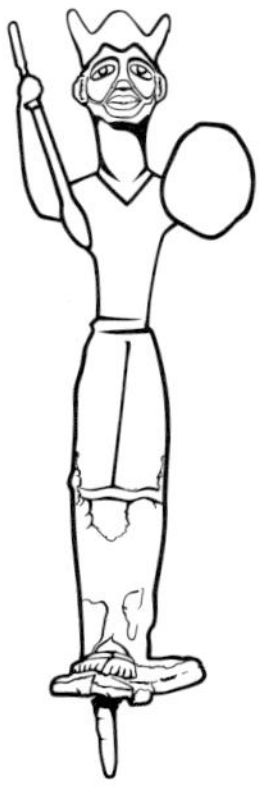

Miniature Ingots from Cyprus

Alessandra Giumlia-Mair, Vasiliki Kassianidou, and George Papasavvas

Introduction

Numerous analytical studies have focused on the main product of the Late Cypriot copper industry (i.e., the oxhide ingots), on utilitarian objects such as tools and weapons, or on the waste products of metallurgical processes (i.e., slag and remains of smelting and casting installations). Very few analysis programs, if any, however, have been published on the most important works of art produced in this period, such as the two gods from Enkomi. This is not surprising: analytical techniques such as Atomic Absorption Spectroscopy or Inductively Coupled Plasma Spectroscopy, thought to be the most appropriate for metal analysis, cannot be used without sampling, and museum curators are usually reluctant to allow sampling of such unique pieces. In order to resolve this problem we initiated a research project, funded by the University of Cyprus, which aims to analyze some of the most important metal artifacts found on the island without taking a sample. The analysis is done with a portable X-ray fluorescence (XRF) analyzer, which was brought to the Cyprus Museum, where most of these items are on display or in storage.

As a tribute to Jim Muhly we present here a small, but—we think—interesting section of the project: the analysis of seven miniature oxhide ingots dating to the Late Bronze Age. A catalog of the artifacts is presented at the end of the paper.

Miniature Ingots from Enkomi

Six of the analyzed miniature ingots come from Enkomi (Fig. 2.1). Three were excavated by Dikaios in Area I (French Mission, Quartier 4W). They were all found in the vicinity of the Ashlar Building, which incorporates the Sanctuary of the Horned God, albeit on different levels. Interestingly, none was found within the cult complex itself.

One ingot (Fig. 2.2: Enk. 1995) was found in Court 64, in Level IIIA (which, according to Dikaios [1971, 496], dates to 1220–1190 B.C.) of the Reconstructed Ashlar Building, and thus would antedate the introduction of the cult of the Horned God (Dikaios 1971, 691). The Ashlar Building was initially a residence, and only from Level IIIB was part of it used as a sanctuary, as demonstrated by the discovery of the Horned God statuette and other finds of cultic significance (Dikaios 1971, 524). A crucial question is whether any cultic activities took place in the same location before the erection of the Ashlar Building, which stands on the remains of an older structure that had been severely damaged and subsequently leveled to make ground for the new edifice (Dikaios 1971, 513). It has already been argued that the Horned God originated in an earlier period than that envisaged by Dikaios (Muhly 1980, 159–160; Knapp 1986, 87). It is, therefore, probable that his cult in this place may have had a longer history.

The second example (Fig. 2.2: Enk. 885) comes from Level IIIB (which, according to Dikaios [1971, 496], dates to 1190–1100 B.C.) and was found in Room 44 of the Ashlar Building (Dikaios 1971, 729). The room might have been associated with cult, either independently or, more possibly, in relation to the neighboring Sanctuary of the Horned God: some objects that might have been used for rituals, such as a knife and a terracotta bull rhyton, were found there (Webb 1999, 241).

The third miniature ingot (Fig. 2.2: Enk. 774; Dikaios 1971, 764) comes from Level IIIC (which, according to Dikaios [1971, 496], dates to

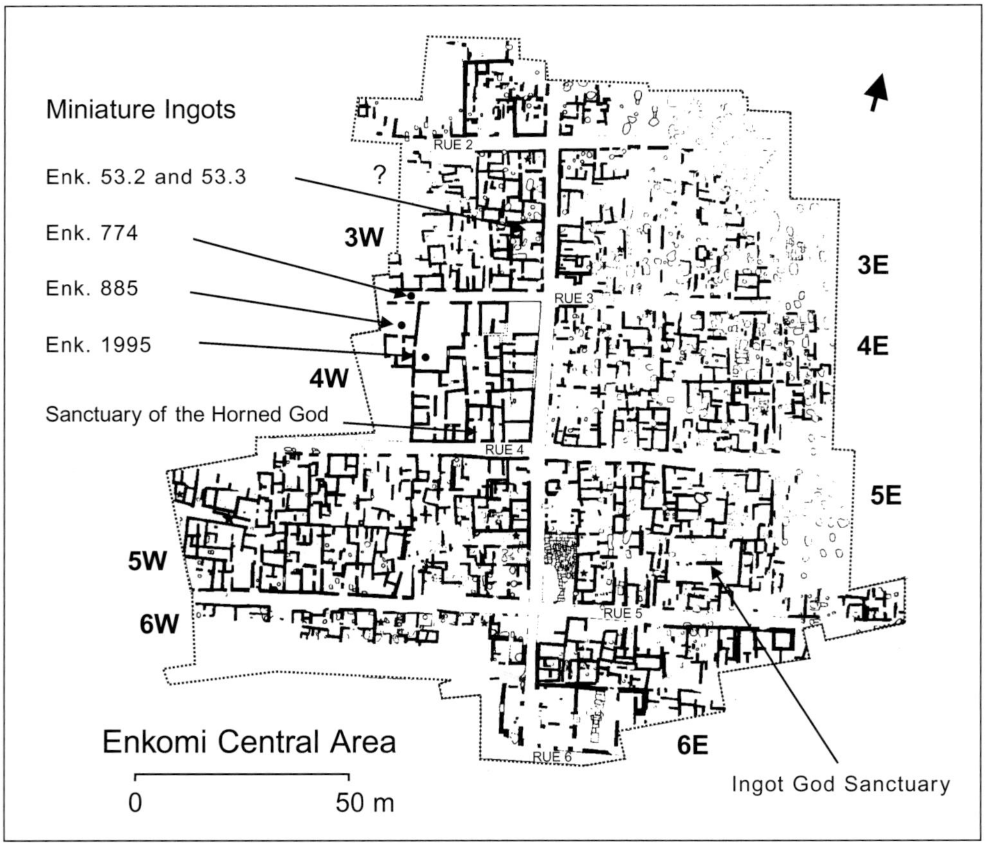

Figure 2.1. Plan of the central part of Enkomi showing the findspots of the miniature ingots under study.

1100–1075 B.C.)—a phase subsequent to the "introduction of worship" in the Ashlar Building (Dikaios 1971, 524). It was found in a street (French Mission, Rue 3) to the northwest of the Ashlar Building. However, this can hardly be its original position. Perhaps it ended up there after yet another destruction of the Ashlar Building that dates to this phase.

Another two miniature ingots were found together by the French Mission (Fig. 2.2: Enk. 53.2, Enk. 53.3). According to O. Masson (1971, 451 n. 16), they were found on the floor of a 12th century residence on Point Topographique 71 in Quartier 3W, just to the north of Dikaios's Area I. It is a challenge to ascertain the exact location of the French Mission's "Points Topographiques," as they have never been indicated on a published plan of their excavations. However, Megaw (1954, 172) states that the ingots were found in a house located on the main north–south street of Enkomi. In other words, all five excavated miniature ingots from Enkomi—to this date the most extensively excavated Late Cypriot site—were found within a radius of 50 m.

A sixth miniature ingot (Fig. 2.2: 1936-VI-19/1), acquired by the Department of Antiquities in 1936, is also believed to have come from Enkomi (Buchholz 1959, 20; O. Masson 1971, 450–451, no. 1).

A Miniature Ingot from the Mathiatis Hoard

During the study of the oxhide ingot fragments of the Mathiatis hoard, a rather interesting discovery was made (Kassianidou 2009, 52). The smallest fragment in the group (Fig. 2.2: 1936/VII-17/9i) is thinner, has a smoother surface, and is homogeneous in section, and thus it stands apart from the rest. It is, in fact, a fragmentary miniature oxhide ingot. Catling (1964, 269, pl. 52b:37) mentions one fragmentary miniature oxhide ingot among the part of the Mathiatis hoard that was exported as part of a private collection—but this is not the one now in the Cyprus Museum. This means the Mathiatis hoard included two fragmentary miniature oxhide ingots. The newly identified miniature ingot from Mathiatis is remarkable for another reason: if it had been preserved fully, it would be the largest known from Cyprus.

Besides the ingots, the hoard also included many broken or damaged bronzes and pieces of scrap metal. This led Catling (1964, 284–285) to the conclusion that it was a "melting" hoard or what Knapp, Muhly, and Muhly (1988, 237) define as a founder's hoard. The fact that miniature oxhide ingots were broken and eventually ended up in a founder's hoard is an interesting issue (see below). As the hoard was not found during an excavation, it is without context and, therefore, difficult to date. However, on the basis of the typology of the bronze artifacts, Catling (1964, 283) suggested a 12th century date. This was challenged by Knapp, Muhly, and Muhly (1988, 244–246), who argued that the Mathiatis hoard, as well as other hoards from Cyprus—all dated by Catling to the 12th century—very well could date to Late Cypriot IIC, that is, to the 13th century.

The identification of the miniature ingots as such was never challenged, because they realistically reproduce the shape (although not the texture) of oxhide ingots. Nevertheless, their function remains a matter of debate.

Function and Manufacture of the Miniature Ingots

Buchholz (1959, 19–20) was the first to argue that these objects were "votives." His suggestion was primarily based on the fact that some of them bear Cypro-Minoan inscriptions that he believed were dedications. Catling (1971, 29), having argued for a close relation between the Cypriot copper industry and cult, took the miniature ingots as further proof for this association. He proposed that they were offered to deities associated with metallurgy, most probably the ones manifested in metal by the Ingot God and the Bomford Goddess.

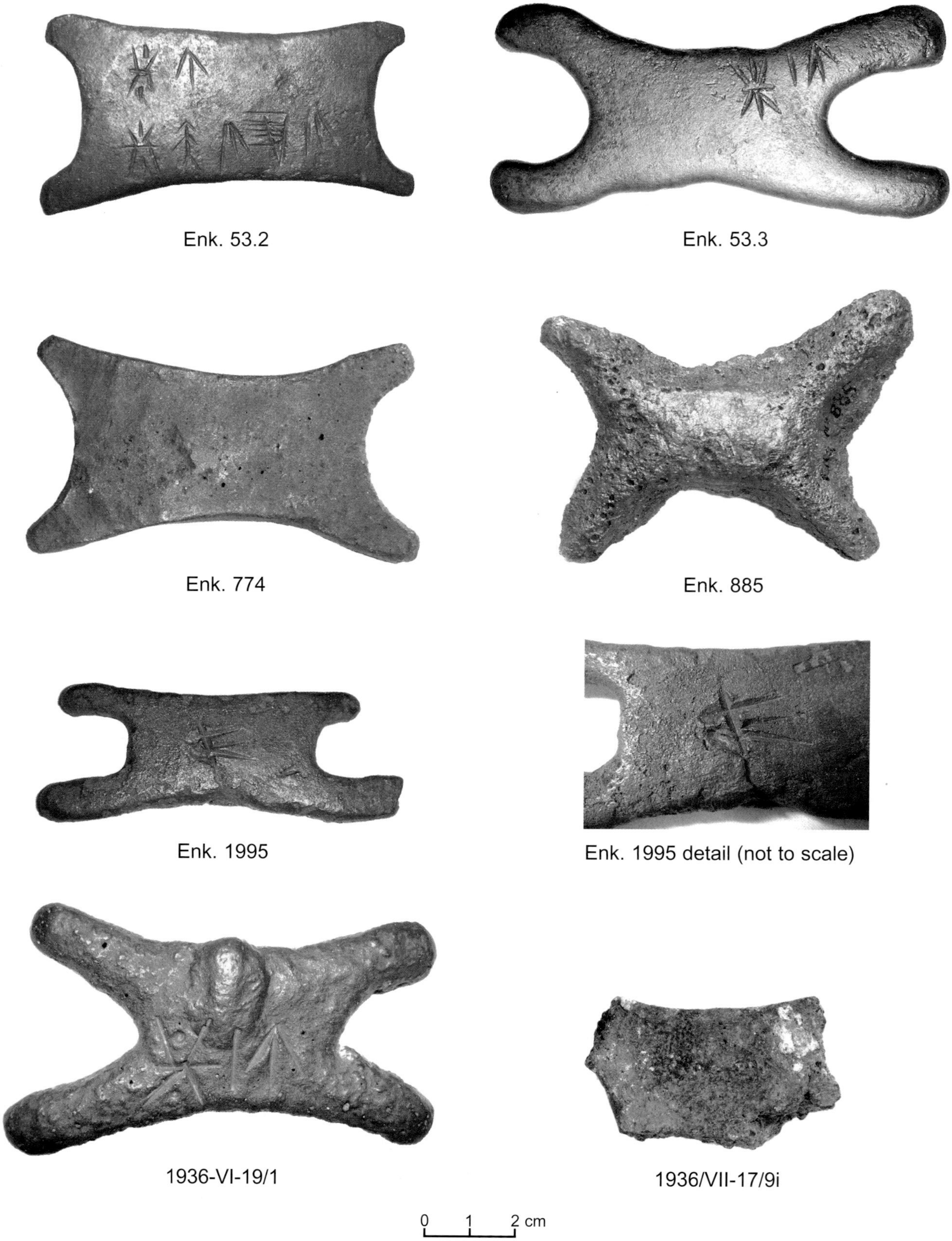

Figure 2.2. Six miniature ingots from Enkomi (inv. nos. Enk. 53.2, Enk. 53.3, Enk. 774, Enk. 885, Enk. 1995, 1936-VI-19/1) and one from Mathiatis (1936/VII-17/9i).

Buchholz's and Catling's views influenced all subsequent interpretations and discussions on these artifacts (for the most recent comprehensive review of miniature ingots, see Papasavvas 2009, 101–104). O. Masson (1971, 449, 454) remarked that the Ingot God himself offers undeniable evidence for the sacred nature of the miniature ingots and agreed that their inscriptions are dedicatory. Knapp (1986, 116) interpreted them in light of religious ideology and suggested that the miniature ingots, together with the Ingot God and other objects depicting ingots, were symbols employed by the elites in order to legitimize, even sanctify, their control over the production and distribution of copper.

However, close examination of the context of the miniature ingots revealed that none was found in association with cultic assemblages, with the exception of one example from Enkomi (inv. no. Enk. 885) and a fragmentary one recently found at Alassa (Webb 1999, 237). Webb consequently questioned their function as votives. This led to the proposition that the miniature ingots may have had the more mundane role of weights (Kassianidou 2005, 135). However, after measuring and weighing seven miniature ingots from Cyprus for this project, it is evident that this idea has to be recanted: variations in both size and weight is too great. We will return to the issue of function at the end of the paper.

The miniature ingots are produced by the lost-wax technique, and in this they differ from their full-size counterparts. Furthermore, the artisan who made them did not try to depict the rough texture of the originals, although on the wax model this would have been easy to do, as shown by the ingots carried on the shoulder of men depicted on Late Cypriot four-sided stands (see Papasavvas 2001, cat. nos. 23, 29, 64). Finally, according to Knapp (1986, 28), "[a]s if to distinguish emphatically the miniature from the oxhide ingots that they represent, those that have been analyzed are made of bronze, not copper." This observation became established in the literature, and Knapp has often been quoted, including by two of the authors here (Kassianidou 2005, 135; Papasavvas 2009, 101). Knapp based his comment on the published analysis of a miniature oxhide ingot of unknown provenance, now in the Ashmolean Museum, but allegedly found on Makarska, an island off the coast of Croatia (De Jesus 1976, 231–232). He also refers to Courtois (1982, 159) who lists the miniature ingot (inv. no. Enk. 885) among "objets de bronze" found by Dikaios in Quartier 4W. However, Courtois' comment was not based on any analysis but rather on the common practice to name all copper-based objects without any distinction "bronzes." As a result of our project, we now know that Cypriot miniature ingots are not made of bronze.

Chemical Analysis of Miniature Ingots in the Cyprus Museum

XRF spectrometry is a well known, non-destructive, analytical method that can determine the chemical composition of an object without sampling or coming into physical contact (Hall, Schweizer, and Toller 1973; Hackens, McKerrell, and Hours, eds., 1977; Helmig, Jackwerth, and Hauptmann 1989). However, if the object is covered by a rough or thick patina, the upper layer should be carefully removed in a very small area in order to guarantee reliable results. In the case of the miniature ingots this was only necessary for the example from Mathiatis, as all the rest had been cleaned and conserved in the past. The XRF analyzer can quantify over 30 elements and can rapidly determine the elements present in the object and their relative concentration. The portable device can be taken to the object—virtually anywhere—and can perform analyses in situ, even on an excavation. In sum, the method offers a fast, effective, and low-cost performance and is ideal for museum pieces that cannot be sampled.

The measurements are performed by illuminating with X-rays a small, flat, clean area on the object for a short time (typically 3–5 minutes), but the measurement also can be longer if required. The X-rays are emitted by a miniaturized X-ray tube. The size of the irradiated area has a diameter of approximately 1.5–2 mm, but the analyzed spot can be smaller or larger, as required by the size of the object, the detail to be analyzed, and the surface texture. The measurements are accomplished at a fixed angle and from a fixed distance from the

No.	Inv. No.	Part	Cu	Sn	Pb	As	Sb	Fe	Ni	Ag	Zn	Co	Hg	Bi	Mn	Au
1	Enk. 1995		99	—	—	—	—	0.3	—	—	—	tr.	—	tr.	tr.	tr.
2	Enk. 885		99	—	—	—	0.2	0.2	—	0.2	—	—	tr.	tr.	—	—
3	Enk. 774		99	—	—	0.4	—	0.2	—	—	—	—	—	—	—	tr.
4	Enk. 53.2		99	—	—	0.2	0.4	0.2	—	—	tr.	—	—	—	—	—
5	Enk. 53.3		99	—	—	0.3	0.2	0.2	—	0.2	tr.	—	—	—	—	—
6a	1936-VI-19/1	handle	99	—	—	—	0.4	0.4	—	0.3	tr.	—	—	—	—	—
6b	1936-VI-19/1	bump	98	—	—	0.4	0.3	0.3	—	0.3	tr.	—	—	tr.	tr.	tr.
7	1936/VII-17/9i		99	—	—	0.3	tr.	0.5	tr.	—	—	tr.	tr.	—	tr.	—

Table 2.1. Chemical composition of miniature ingots from Cyprus analyzed by X-ray fluorescence in weight % (tr. = trace amount).

sample. At least three readings were obtained for confirmation in case of unclear results.

Past experience has shown that a wide range of elements—particularly metals and alloys—can be simultaneously quantified with a high degree of precision if proper standards and some cautions are used (e.g., Hahn-Weinheimer, Hirner, and Weber-Diefenbach 1995; Lutz and Pernicka 1996). The different standards of various composition employed during our measurements have been expressly produced by AGM Archeoanalisi (located in Merano, Italy) for the analysis of ancient metal alloys and represent an important tool in the evaluation of the results.

The data obtained was subsequently processed in order to give the quantitative results. Interference effects—for instance, enhanced results for elements such as iron or low results for zinc in a copper matrix and similar phenomena—were taken into account while evaluating the results.

Analytical Results

The analytical results clearly demonstrate that all miniature ingots from this small but representative group are made of purified and unalloyed copper (Table 2.1). Tin was not identified in any of the analyzed pieces, not even at a trace level. Indeed, the metal used to cast, with over 99% Cu, seems to be even more completely refined than the very pure copper of oxhide ingots (for analysis of oxhide ingots, see Lo Schiavo et al., eds., 2005; Giumlia-Mair 2006, 2009; Kassianidou 2009). In fact, the bubbles and the porosity on the surface of many of the ingots are characteristic of unalloyed cast copper.

The main impurities present in the metal are iron, arsenic, antimony, and silver, but faint traces of other elements such as manganese, cobalt, and bismuth have also been determined. The low traces of zinc found on three of the miniature ingots are certainly due to the electrolytic cleaning process, a technique routinely used in the past to clean corroded bronzes. Manganese, cobalt, and the nickel determined on the ingot fragment from Mathiatis, seem to be related to the presence of iron percentages over 0.3%.

Some traces of gold were also determined in three of the ingots. This seems to be a known characteristic of the copper from Cyprus. The presence of slight traces of mercury is rather interesting and does not seem to have been noticed before, but we have to keep in mind that the method of analysis employed for this project can only collect data from the surface of the objects. More precise analyses would be required to confirm the presence of mercury in the copper of the miniature ingots.

In the case of the miniature ingot said to come from Enkomi (inv. no. 1936-VI-19/1), slight differences of composition were determined when analyzing the handle and the rather evident bump in the center of the piece. No arsenic was identified by the analysis carried out on one of the handles, but in the measurement carried out on the central bump, 0.4% of arsenic was determined. Traces of bismuth, manganese, and gold were also identified on the bump, but not on the handle. The differences are certainly very slight and might be due to simple inhomogeneity of the alloy or the presence of corrosion, but they are worth mention (for an interpretation, see Papasavvas, forthcoming).

Concluding Remarks

The lack of any evident practical use, the fact that they are so conspicuously imitating identifiable utilitarian objects (i.e., oxhide ingots), even in their chemical composition, and the fact that some of them bear inscriptions lead us to interpret the miniature ingots as cultic artifacts, even if presently their contexts do not seem to offer much support for this proposition.

The Cypro-Minoan signs inscribed on some of the miniature ingots after they were cast (Fig. 2.2: Enk. 1995) enhance the symbolic meaning of these artifacts and underline their great importance for the people who made and used them. The signs are very limited in number in all known examples except one (inv. no. Enk. 53.2), but we should bear in mind that only a very small number of Late Bronze Age objects with inscriptions have been excavated on Cyprus (Papasavvas 2003, 89–90).

It is remarkable that two of the examples bearing inscriptions (inv. nos. 1936-VI-19/1 and Enk. 53.3) have the same sequence of two signs, divided by a horizontal bar. The same two symbols are found on the ingot inscribed with seven signs arranged on two lines (inv. no. Enk. 53.2; O. Masson 1971, 451–452). It is perhaps not out of place to remark that the first sign on the three miniature ingots from Enkomi coincides with the second sign, after a vertical bar, inscribed on the two bronze ring-stands from Myrtou-*Pigades* (Catling 1957, 87, fig. 34: 419–420; Papasavvas 2001, 246, nos. 35–36, figs. 90–94). These artifacts were found in a sanctuary, so their inscriptions are most probably dedicatory. This can be taken as indirect evidence for the votive character of the miniature ingots. The same signs are adequately represented on the clay balls—which mostly also carry only two signs—(e.g., E. Masson 1971, 490, no. 27) from various Cypriot sites. Emilia Masson (1971, 504) believes them to be proper names, corresponding perhaps to names of gods. Since we are not familiar with the dedicatory practices of Late Bronze Age Cyprus, we can only conjecture as to whether the inscriptions on the miniature ingots refer to names of gods, to names of the votaries, to dedications, or to something totally different. Any suggestions can only be highly speculative and will remain so as long as the Cypro-Minoan script is undeciphered (Masson 1973, 110; Knapp 1986, 11).

More questions about the miniature ingots remain unanswered. For example, why do the miniature ingots first appear toward the end of the Late Bronze Age (the ones from Enkomi can be dated from the beginning of the 12th to the beginning of the 11th century B.C.)? There is now evidence that oxhide ingots were used on Cyprus as early as the 14th century B.C. and that Cypriot copper was exported to Crete in the shape of oxhide ingots as early as the 16th century B.C. (Kassianidou 2009). Another question is: why are they concentrated in such a limited area, and is it coincidental that this is near one of the most important sanctuaries of Enkomi—that of the Horned God?

Finally, why are we finding fragmentary miniature ingots—three are currently known, two from the Mathiatis hoard and one from Alassa (Hadjisavvas 1986, 66)? The objects are solid copper castings and therefore cannot have been accidentally broken. Indeed, the fact that two of them are broken in half seems to imply that they were deliberately broken. If they were meant to be recycled (as suggested by their presence in a founder's hoard), there was no need to break them, as they are small. Furthermore, the example from Alassa was found in a unit identified as a domestic cult place. Is the act of breaking a ritual one?

Our analysis unraveled one of the "secrets" of the miniature ingots—that of their composition—but we have a long way to go before we fully understand them.

Catalog of Analyzed Miniature Ingots

Enkomi

Enk. 1995. Miniature ingot inscribed with two signs. L. 6.95, w. 2.7, th. 1.7 cm; wt. 70 g. Excavations of the Department of Antiquities. Area I, Court 64 (Quartier 4W) southwest corner in depression filled of pebbles under floor V. Level IIB–IIIA (Dikaios 1971, 691).

Enk. 885. Miniature ingot. L. 7.6, w. 5.7, th. 1.5 cm; wt. 190 g. Excavations of the Department of Antiquities. Area I (Quartier 4W), Room 44, between floors IV and IIIA. Level Middle IIIB (Dikaios 1971, 729).

Enk. 774. Miniature ingot. L. 9.1, w. 5.1, th. 1.4 cm; wt. 240 g. Excavations of the Department of Antiquities. Area I (Quartier 4W), north street (Rue 3), M–N, 10–12 north. Level IIIC (Dikaios 1971, 764).

Enk. 53.2. Miniature ingot inscribed with seven signs organized in two lines. L. 8.5, w. 4.3–4.4, th. 1 cm; wt. 162.95 g. French Mission excavations. Quartier 3W Point Topographique 71 (O. Masson 1971).

Enk. 53.3. Miniature ingot inscribed with two signs separated by a vertical line. L. 11, w. 5, th. 1 cm; wt. 226.15 g. French Mission excavations. Quartier 3W Point Topographique 71 (O. Masson 1971).

1936-VI-19/1. Miniature ingot inscribed with two signs separated by a vertical line. Said to be from Enkomi. L. 9.5, w. 4.9, th. 2 cm; wt. 212.80 g (O. Masson 1971).

Mathiatis

1936/VII-17/9i. Fragmentary miniature ingot. Broken along the long axis of the body, as well as the handle. L. 5.5, w. 2.8, th. 1.4 cm; wt. 120 g. Part of the Mathiatis hoard stored at the Cyprus Museum (Kassianidou 2009, 101).

References

Buchholz, H.G. 1959. "Keftiubarren und Erzhandel in zweiten vorchristlichen Jahrtausend," *PZ* 37 (1), pp. 1–40.

Catling, H.W. 1957. "The Metal Objects and Coins," in *Myrtou*-Pighades: *A Late Bronze Age Sanctuary in Cyprus*, J. du Plat Taylor, ed., Oxford, pp. 86–96.

———. 1964. *Cypriot Bronzework in the Mycenaean World*, Oxford.

———. 1971. "A Cypriot Bronze Statuette in the Bomford Collection," in Schaeffer, ed., 1971, pp. 15–32.

Courtois, J.C. 1982. "L'activité métallurgique et les Bronzes d'Enkomi au Bronze Récent (1650–1100 avant J.C.)," in *Early Metallurgy in Cyprus 4000–500 B.C.*, J.D Muhly, R. Maddin, and V. Karageorghis, eds., Nicosia, pp. 155–174.

De Jesus, P.S. 1976. "Report on the Analyses of the 'Makarska' Tools and Some Implications," *SMEA* 17, pp. 221–233.

Dikaios, P. 1971. *Enkomi: Excavations 1948–1958*, vol. II, Mainz am Rhein.

Giumlia-Mair, A. 2006. "Metallurgical Analyses on the Lipari Hoard," in *Proceedings of the Workshop "Archaeometallurgy in Sardinia," September 2004, Cagliari* (*Instrumentum* 23), A. Giumlia-Mair, ed., Montagnac, pp. 14–16.

———. 2009. "The Hoard under the *alpha* II Hut on the Acropolis of Lipari. A Metallurgical Study," in Lo Schiavo et al., eds., 2009, pp. 166–214.

Hackens, T., H. McKerrell, and M. Hours, eds. 1977. *X-Ray Microfluorescence Analysis Applied to Archaeology* (*Physical and Chemical Techniques in Archaeology* 1), Strasbourg.

Hadjisavvas, S. 1986. "Alassa. A New Late Cypriot Site," *RDAC* 1986, pp. 62–67.

Hahn-Weinheimer, P., A. Hirner, and K. Weber-Diefenbach. 1995. *Röntgenfluoreszenzanalytische Methoden—Grundlagen und praktische Anwendung in den Geo-, Material- und Umweltwissenschaften*, Wiesbaden.

Hall, E.T., F. Schweizer, and P.A. Toller. 1973. "X-Ray Fluorescence Analysis of Museum Objects: A New Instrument," *Archaeometry* 15, pp. 53–78.

Helmig, D., E. Jackwerth, and A. Hauptmann. 1989. "Archaeometallurgical Fieldwork and the Use of a Portable X-Ray Spectrometer," *Archaeometry* 31, pp. 181–191.

Kassianidou, V. 2005. "Was Copper Production under Divine Protection in Late Bronze Age Cyprus? Some Thoughts on an Old Question," in *Cyprus: Religion and Society from the Late Bronze Age to the End of the Archaic Period. Proceedings of an International Symposium on Cypriote Archaeology, Erlangen, 23–24 July 2004*, V. Karageorghis, H. Matthäus, and S. Rogge, eds., Bialystock, pp. 127–141.

———. 2009. "Oxhide Ingots in Cyprus," in Lo Schiavo et al., eds., 2009, pp. 41–81.

Knapp, A.B. 1986. *Copper Production and Divine Protection: Archaeology, Ideology and Social Complexity on Bronze Age Cyprus* (*SIMA-PB* 42), Göteborg.

Knapp, A.B., J.D. Muhly, and P. Muhly. 1988. "To Hoard is Human: LBA Metal Deposits in Cyprus and the Aegean," *RDAC* 1988, pp. 233–262.

Lo Schiavo, F., A. Giumlia-Mair, U. Sanna, and R. Valera, eds. 2005. *Archaeometallurgy in Sardinia—From the Origins to the Early Iron Age* (*Monographies Instrumentum* 30), Montagnac.

Lo Schiavo, F., J. Muhly, R. Maddin, and A. Giumlia-Mair, eds. 2009. *Oxhide Ingots in the Central Mediterranean*, Rome.

Lutz, J., and E. Pernicka. 1996. "EDXRF Analysis of Ancient Copper Alloys," *Archaeometry* 38 (2), pp. 313–323.

Masson, E. 1971. "Boules d'argile inscrites trouvées à Enkomi de 1953 à 1969," in Schaeffer, ed., 1971, pp. 457–504.

Masson, O. 1971. "Deux petits lingots de cuivre inscrits d' Enkomi (1953)," in Schaeffer, ed., 1971, pp. 449–455.

———. 1973. "Remarques sur les cultes chypriotes à l'époque du Bronze Récent," in *Acts of the International Archaeological Symposium "The Mycenaeans in the Eastern Mediterranean" 27 March–2 April 1972*, Nicosia, pp. 110–121.

Megaw, A.H.S. 1954. "Archaeology in Cyprus, 1953," *JHS* 74, pp. 172–176.

Muhly, J.D. 1980. "Bronze Figurines and Near Eastern Metalwork," *IEJ* 30, pp. 148–161.

Papasavvas, G. 2001. *Χάλκινοι Υποστάτες από την Κύπρο και την Κρήτη*, Nicosia.

———. 2003. "Writing on Cyprus: Some Silent Witnesses," *RDAC* 2003, pp. 79–94.

———. 2009. "The Iconography of the Oxhide Ingots," in Lo Schiavo et al., eds., 2009, pp. 83–132.

———. Forthcoming. "A Missing Ingot Deity?" in *Acts of the Fourth International Cyprological Conference*, A. Demetriou, ed., Nicosia.

Schaeffer, C.F.A., ed. 1971. *Alasia* I. *Nouvelles Missions en Chypre 1946–1950* (*Mission Archéologique d'Alasia* IV), Paris.

Webb, J.M. 1999. *Ritual Architecture, Iconography and Practice in Late Cypriot Bronze Age* (*SIMA-PB* 75), Jonsered.

CHAPTER

3

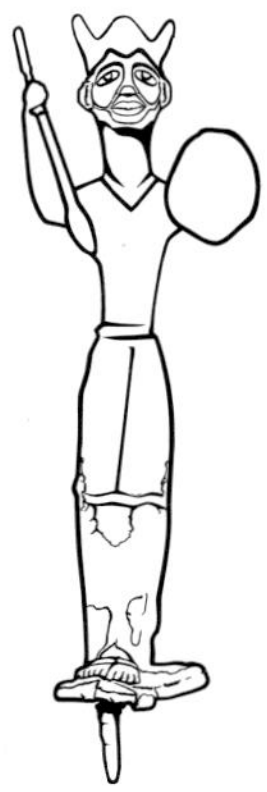

Broken Symbols: Aspects of Metallurgy at Alassa

Sophocles Hadjisavvas

A broken miniature oxhide ingot found at Alassa-*Pano Mandilaris* during the 1984 excavation season put under dispute all earlier interpretations related to the ritual character or the symbolic values of the artifact (Knapp 1986; 1996, 79).* Was it a signal of the collapse of the complex society in Cyprus, or was it merely a demonstration of the lack of copper during the final phase of the Late Cypriot (LC) period, which led the owners to re-melt even the weights, as Zwicker identified the miniature ingots? This question and others related to the bronzes from Alassa will be discussed in this chapter, always within the broader context of metallurgy on the island of Cyprus.

Introduction

Toward the end of 1987, Vassos Karageorghis asked me to organize a permanent exhibition on ancient metallurgy in the Cyprus Museum. A few months later I went to Cambridge to spend the Lent term after a fellowship with the British School at Athens, planning to complete my Ph.D. on olive oil production. Instead, I spent most of my time trying to introduce myself to the secrets of metallurgy. The bibliography on the subject was, for a newcomer in the field, quite impressive, going as far back as the 16th century when Georgius Agricola described medieval mining activities in full detail (Agricola 1950). Alas, for Cyprus the only information was a report in *The Swedish Cyprus Expedition* by a mining engineer working at the Skouriotissa mines

* Prof. A.B. Knapp is thanked for useful suggestions.

(Bruce 1937). Theoretically prepared, I returned to Cyprus and faced the real challenge of organizing the first permanent exhibition on ancient metallurgy on the island of copper. Soon after that, I realized that only some fine objects in the possession of the Cyprus Museum were related to metallurgy. An educational exhibition, however, as I had proposed, needed much more. Help was asked from the Cyprus Geological Survey Department, from mining companies, and from individuals. A number of artifacts, mainly mining tools collected from ancient galleries, were presented to the Museum. The identification and preparation of proper documentation understandable to visitors was the second target. By a happy coincidence Jim Muhly was spending the year as a Senior Fulbright Professor in Cyprus, studying the remains of metallurgical activities at Enkomi and other sites. With his constant guidance, the first-ever educational exhibition was organized in the Cyprus Museum. This article may be considered a small return to his generosity in sharing his knowledge with me and for his tremendous contribution toward the study of ancient Cypriot metallurgy and the Late Bronze Age (LBA) period in general.

Aspects of Metallurgy at Alassa

The site of Alassa–*Pano Mandilaris* produced a number of bronze artifacts of domestic, ritual, and military use (Fig. 3.1). Eleven of them, all excavated at the lower site of Alassa at the locality of *Pano Mandilaris* in 1984, have been analyzed by Ulrich Zwicker. Their composition provides some clues to the possible provenance of the ore. Despite the fact that objects related to a coppersmith's workshop were excavated in Room P, it is not certain whether the objects were locally made, as no molds have been discovered at the site. This assumption, however, is not conclusive, bearing in mind the limited exposure of the settlement and the destruction that had occurred at the site both through deep tillage and by looter's activities. Some evidence about secondary smelting, or rather re-melting, was produced in the form of scrap metal along with a handle fragment of an actual ingot (the latter was found much later at Alassa-*Palaiotaverna* and was not analyzed). Slag and copper alloys such as sulphide ore were found in insufficient quantities at four different parts of the site. This is an indication that some primary smelting on a small, perhaps family-based, scale was practiced within the houses of the settlement. Six out of the 11 objects investigated from the excavations at *Pano Mandilaris* by the University of Erlanger consist of arsenical bronze (Zwicker 1992, 166, table 2a ff 6, 7). This fact is consistent with the investigation of the slag recovered at the site.

Catalog of Bronzes

Alassa 1984/3. Fragment of miniature oxhide ingot. Pres. L. 6.5, w. 5 cm; wt. 112 gr. Cu 82.1%, As 0.5%, Pb 0.38%.

Alassa 1984/23. Fragment of bronze sickle. Pres. L. 13 cm. Cu 79.0%, Sn >3.0%, As 0.7%, Pb (2.0)%.

Alassa 1984/38. Flat axe or weight. L. 16 cm; wt. 223 g. Low-alloyed tin bronze. Widening from butt to blade. Slightly convex cutting edge with small flanges. Cu 79.9%, Sn 3.8%, Pb 2.9%.

Alassa 1984/83. Small chisel. L. 9 cm. High content of arsenic provides good strength. Cu 53.4%, As 2.8%, Pb 1.4%.

Alassa 1984/93. Sword fragment. Pres. L. 7 cm. High toughness. Cu 67.6%, As 1.9%, Pb 1.5%.

Alassa 1984/97. Fragment of miniature oxhide ingot(?). Pres. L. 3 cm. Cu 74.7%, As 0.8%, Pb 0.4%.

Alassa 1985/147. Hoe. Sides of blade widen slightly below socket. Arsenical bronze with good mechanical properties. Cu 78.0%, Sn 3.1%, As 0.73%.

Alassa Tomb 3/76. Hemispherical bowl. Dia. 13 cm. Cu 75.8%, As 0.13%, Pb (0.2)%.

Alassa Tomb 3/84. Dagger with three rivets. Cu 80.1%, Sn 3.9%, As 0.5%, Pb (0.23)%, Zn (-0.5)%.

Alassa Tomb 5/2. Dagger with three rivets. Cu 69.6%, As 1.2%, Pb (0.5)%.

Alassa 1985. Tomb 3/8. Bronze fragment. As 0.65%, Cu 78.0%, Pb (0.68%), Sn > 1.3%, Zn (1.8%).

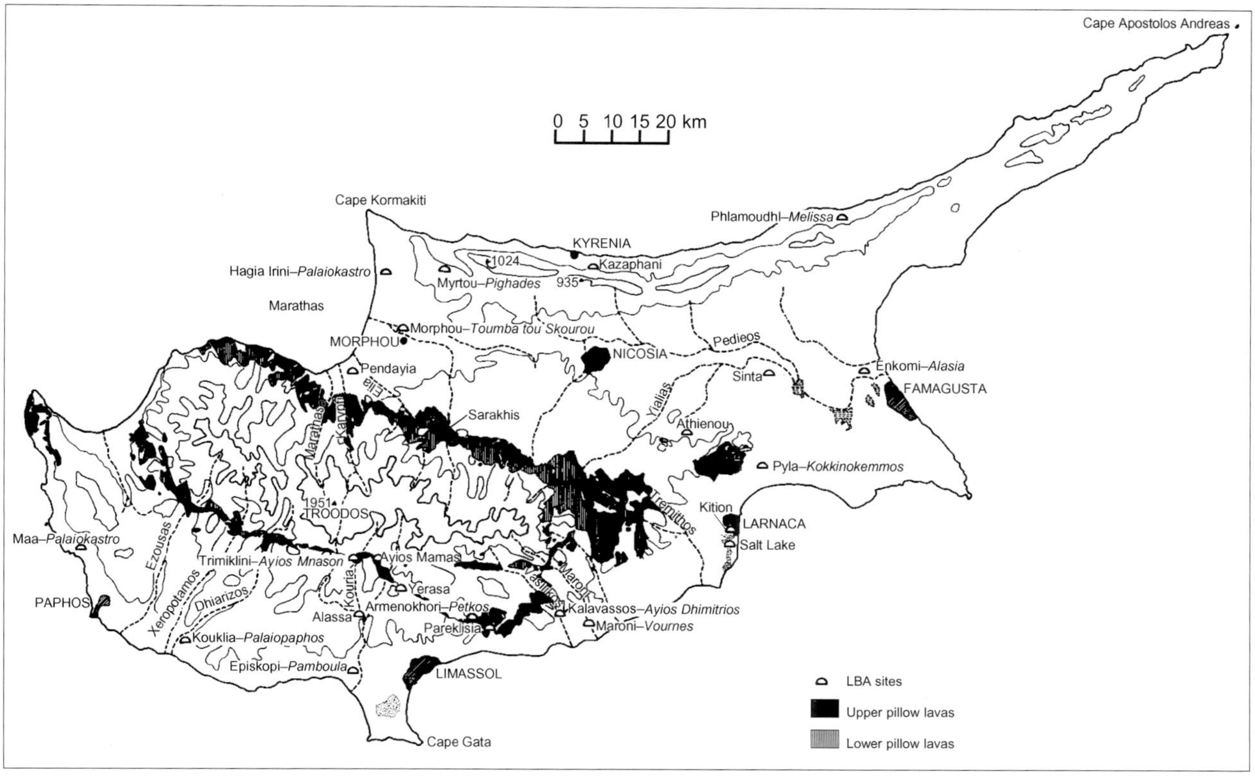

Figure 3.1. Geological map of the Troodos region showing copper-producing sites.

Discussion

During the 1991 excavation season at Alassa a small "hoard" of bronze weapons was found while weeding the site of *Pano Mandilaris*. The bronzes were most likely hidden in times of emergency within a hole opened 30 cm above the floor on the vertically cut bedrock. The "hoard" is made of a spearhead with tubular socket, a rear spike from a spear butt, and a javelin point (Hadjisavvas 1994, 112, pl. 21:2, 3). These latter bronzes have yet to be analyzed.

Although Alassa is not too far from the copper-producing regions of Ayios Mamas and Gerasa (see Fig. 3.1), the analysis of three stratified slag pieces, found at *Pano Mandilaris*, has shown that they contained a high proportion of arsenical bronze. This could probably be produced by the addition of an iron arsenide ore, which can be mined at the locality of *Pefkos*, northeast of Limassol, which in comparison to the already mentioned sites, is quite distant from Alassa (Hadjisavvas 1989, 40). This fact is either an indication of interregional exchanges, or it suggests that the Alassa regional polity exercised control over a large area extending as far as the present day village of Parekklisha.

The broken miniature (Alassa 1984/3) ingot was found in Room P, adjoining a coppersmith's workshop located in Room D, in which the best preserved, thus far, pot bellows on the island was found (Fig. 3.2). The miniature ingot (Fig. 3.3), weighing 112 g, is less than half the weight of well-known counterparts from Enkomi (Hadjisavvas 1986, 66). Another fragment from Alassa, considered by Zwicker as part of a miniature ingot, was identified during the process of analyzing the metals from the site (1992, 166, pl. 32:4). Zwicker identifies the miniature ingots as weights, basing his interpretation on the fact "that they fit well into the two systems of shekel and quedet used during the Bronze Age in the Eastern Mediterranean region" (Zwicker 1990, 10–11). In the same Room P a group of five terracotta bull figurines was found along with an incense burner, a fact that would have led some excavators of the previous generation to identify the locus as

Figure 3.2. Pot bellows from Alassa–*Pano Mandilaris*.

Figure 3.3. Miniature ingot from Alassa–*Pano Mandilaris*, almost one-half extant.

"a sanctuary dedicated to two deities, protectors of metallurgy and agriculture." In a previous article I tentatively interpreted the locus as a cult place—meaning a corner of a house where domestic rituals were practiced—equivalent to the lighting of a candle in front of an icon in today's orthodox religion (Hadjisavvas 1989, 38).

Miniature oxhide ingots are considered to be symbolic items, indicators of metallurgical sites or related to the elite engaged in copper production (Knapp 1986, 25–29). Thus far, proper miniature ingots from excavated sites are known only from Enkomi and Alassa, but the two sites are not compatible either in the scale of copper production or of metalworking. It is true that Alassa is located close to the rich cupriferous zone of the Troodos massif and occupies a strategic position on the Kouris Valley route between the mountains and the sea. However, there is no other evidence to substantiate the site as a major metallurgical center. At the same time, objects related to agriculture outnumber considerably the ones associated with metallurgy and related functions. Even if we were to accept the miniature ingot as a symbolic or ritual item related to metallurgy or copper production, we have only one-half of an object versus 30 or so terracotta bull figurines and a number of bull representations in a variety of media.

Regardless of the fact mentioned above, it is always metallurgy and related activities that are highlighted for the major urban centers that have been excavated thus far. One would assume that historians and archaeologists alike either have overestimated the role played by metallurgy in the economy of the island or confused the remains of metalworking with those of metallurgical production. Cyprus, although one of the richest regions in copper deposits in antiquity, could not have based its economy entirely on copper production, not least because agriculture was the main occupation of the Cypriots well into the last quarter of the 20th century, when it was finally replaced by tourism in economic terms.

Economics of Cypriot Metallurgy

Although it was not my intention to elaborate on the LBA economy of the island, I could not resist challenging a long-standing account that the economy of all major LBA sites was based on metal production and trade in copper alone. Kalavassos–*Ayios Dhimitrios*, for example, despite the fact that it is situated only eight kilometers from the mines, mostly produced evidence of olive oil production and storage. Maroni-*Vournes* and Maroni-*Tsaroukas* had a similar economic orientation, while Alassa also participated in wine production. Based on the available evidence only the sites of Enkomi and Kition could be justified as metallurgical centers. Apliki-*Karamallos* (Kling and Muhly 2007) and Athienou–*Pamboularin tis Koukkouninas* (Dothan and Ben-Tor 1983) are much smaller sites. The fact that the two major metallurgical centers are situated on the east coast of Cyprus could be explained on the one hand by the influx of settlers from sites on the Syrian coast, such as Ugarit, or on the other hand by the ability of the two towns (Enkomi and Ugarit) to control the trade in copper. The role played by sites such as Kalavassos and Alassa, situated close to the mines and probably the smelting sites yet to be discovered, could have been the administration of the mining and primary smelting operations and the subsistence of the people involved. One such smelting site was discovered at Politico-*Phorades*, where no remains of habitation have been revealed (Knapp, Kassianidou, and Donnely 1999). "Phorades was a smelting workshop: once it was abandoned such things as were left behind (i.e., the slag and the broken installations) no longer had any use. Other artifacts that could be curated and used again, elsewhere—such as diabase pestles, the mortars used to prepare the ores, the metal tools and the pottery—were all removed once the operation shut down" (Knapp and Kassianidou 2008). The site of Ayia Varvara–*Almyras*, although of a much later period, presents the same picture (Fasnacht et al. 1992).

It is expected that the people involved in copper mining and smelting were living in perishable huts or in the open air, a situation observed in the 1920s with the reopening of the mines on the island (see below). The products of primary smelting in whatever form, occasionally in the shape of the bottom of the furnace, were most probably sent to the urban coastal centers for further refinement (Muhly 1991, 183–184, for discussion; Kassianidou 1999). It is also possible that primary smelting sites produced the so-called bun ingots, such as the one from Delikipos, which contains 64% metallic copper (Zwicker 1992, 166, table 2a, fig. 35:5). The dating, however, of this "bun ingot" is obscure.

Certainly the introduction of metallurgy on an island so rich in copper deposits represented a decisive factor for development and accelerated the (somehow late) process of urbanization. However, even during the Roman period, when the mines of Cyprus experienced the most intensive exploitation, Strabo praised the Mediterranean agricultural triad—namely grain, wine, and olive oil—and only added that the island produced sufficient quantities of copper at Tamassos (Strabo 14.6.5). The archaeological evidence gathered thus far attests to a continuity of Cypriot society and economy with some interludes, which only briefly experienced dramatic changes. Agrarian societies are conservative by nature, and this of course is not a local Cypriot phenomenon (Muhly 1996, 46–47). Following the development of the means of production from the Early Bronze Age through the middle of the 20th century, we can hardly observe any evolution. Agricultural goods such as grain, wine, and olive oil were produced by methods described by Hesiod and the Roman writers Cato, Columella, and Pliny the Elder. Based on this conservatism, I would dare to suggest a comparison between the LBA and the first quarter of the 20th century, which is quite revealing.

In the field of crafts it was only during the 1920s A.D. that a significant improvement was observed, due mainly to the influx of refugees from the region of Smyrna. A distinct interlude in the evolution of traditional practices of Cyprus is closely related to the re-opening of the copper mines during the same period. Thousands of people engaged in agriculture were temporarily resettled closer to the mines. "The workers were staying in dry stone huts, built by themselves and were sleeping directly on the earthen floor . . . once every eight or 15 days we used to go to our villages on foot" (Varnava 1990, 107). Many others from nearby villages were traveling for hours either on a donkey's back or on foot to be at the mines by dawn and back to their villages at dusk. This process, which lasted for a period of less than two decades, on the one hand resulted in the specialization of labor and on the other hand created a new class of proletariat. The period between the mid-1920s and the beginning of World War II could easily be characterized as a period of industrial revolution on Cyprus. But even during this period agriculture was still predominant in the economy of the island. The 1931 census of population is in accord with this account. More than three-quarters of the island's population lived in the countryside (Angelides 1996, 224, table 1). From a total number of 43,837 workers, 23,000 (or 55%) were agricultural workers, not including the landowners and their families. At the peak year of 1926, the Skouriotissa mine employed 1,751 workers (Varnava 1993, 25).

The 1931 census is supportive of my earlier suggestion that mining and metallurgy in the LBA were, for the majority of the peasants, a side occupation either to satisfy their needs in copper or as a source of supplementary income (Hadjisavvas 1996, 28). Some miners' settlements could have also existed as in the case of the 1920s and 1930s. Most of them were seasonal, but through time a few became permanent. These villages are made up of people coming from settlements situated near the mines, who owned property in the region. Most of the remaining seasonal settlements fell into ruins, as their inhabitants were coming from villages some distance away. A similar situation could have existed in the LBA, judging from recent excavations at the smelting site of Politico-*Phorades* (see above).

By accepting the identification of miniature ingots as objects of symbolic character, we return to the first question set in this paper. Why would such an object on a site close to the mines have to be broken and possibly re-melted? (And, let's not forget that it was recovered in a context related to a coppersmith's workshop.) Another fact related to objects of symbolic significance is the recovery at the excavations of *Pano Mandilaris* of an incredibly large number of bull figurines dating to the final phase of the settlement. This last observation is in accord with the development of the Cypriot economy and its reflection in art. The bull, reflecting the unquestionably important role of agriculture and stock breeding down to the beginning of the LBA, is the prevailing artifact in ritual contexts of the time. Toward the end of the Bronze Age and after the establishment of a more complex society, miniature oxhide ingots appear in ritual contexts along with the bull, marking the new character of the economy. The protective deities needed to adopt another attribute to satisfy the new character of the economy. Whereas at the beginning of the LBA, artists favored the female human being and male cattle, toward the end of the period they invented another deity combining the male human and the bull. In the first instance the combination of the female human and the cattle reflects the means of production as defined in verses 405 and 406 of Hesiod's *Works and Days*: "first of all, get a house, and a woman and an ox for the plow—a slave woman and not a wife, to follow the oxen as well." In the second instance the traditional protective attributes of the bull were complimented by the male human who is associated with mines and metallurgy in general. The new image of a god standing on an ingot and wearing a horned helmet appears only in sites related to copper production, while the bull still preserves its predominant position in all other settlement sites (Hadjisavvas 2004, 116). The breakage of the ritual ingot in combination with the increased number of bull figurines at Alassa is, perhaps, a manifestation of the collapse of metallurgy soon followed by the collapse of complex society and final abandonment of the site.

The excavations at the upper site of Alassa-*Palaiotaverna*, which was expected to provide some answers because it contains the most important administrative building in the Kouris Valley polity, failed to do so. Despite its monumentality, Building II produced no evidence to explain its presence in the region, nor did it contain any finds to provide answers to the many questions raised as the result of the previous series of excavations at the lower site of *Pano Mandilaris*. What was actually confirmed was the fact that both sites were abandoned in times of emergency, leaving behind most of the occupants' holdings. At the lower site, these holdings were recovered in the excavation process. At the upper site, however, which was of much greater importance and where three ashlar buildings were found, the abandonment was accompanied by destruction, as manifested by the obvious traces left by the fire

Figure 3.4. Northern wall of the storeroom of Building II at *Palaiotaverna* showing the traces left by severe fire on the ashlar blocks.

that destroyed the storeroom of Building II (Fig. 3.4). The legacy of this building continued as far as the Cypro-Classical period when it was looted by people carrying coins of the kingdom of Paphos. The final obliteration of the evidence occurred some time during the Medieval period when squatters occupied the standing parts of the building.

The continuous use of the site is a negative factor toward our effort to investigate whether the scarcity of copper during the final years of the LC IIIA period might have led to the melting of symbolic artifacts, or whether the breakage by itself symbolizes one last episode in this period.

References

Agricola, G. 1950. *De Re Metallica*, trans. H.C. Hoover and L.H. Hoover, New York.

Angelides, S. 1996. "The Cyprus Economy under British Rule (1878–1960)," in *The Development of the Cypriot Economy: From the Prehistoric Period to the Present Day*, V. Karageorghis and D. Michaelides, eds., Nicosia, pp. 209–230.

Bruce, J.L. 1937. "Antiquities in the Mines of Cyprus," in *The Swedish Cyprus Expedition* 3, E. Gjerstad, J. Lindros, E. Sjöqvist, and A. Westholm, eds., Stockholm, pp. 639–671.

Dothan, T., and A. Ben-Tor. 1983. *Excavations at Athienou, Cyprus, 1971–1972* (*Qedem* 16), Jerusalem.

Fasnacht, W., K. Zubler, R.S. Morris, D. Gerbothé, V. Kassianidou, A. Connolly, and T. Rehren. 1992. "Excavations at Ayia Varvara-*Almyras*. Fourth Preliminary Report," *RDAC* 1992, pp. 60–74.

Hadjisavvas, S. 1986. "Alassa: A New Late Cypriote Site," *RDAC* 1992, pp. 62–67.

———. 1989. "A Late Cypriot Community at Alassa," in *Early Society in Cyprus*, E.J. Peltenburg, ed., Edinburgh, pp. 32–42.

———. 1994. "Alassa Archaeological Project 1991–1993," *RDAC* 1994, pp. 107–114.

———. 1996. "Alassa: A Regional Centre of Alasia?" in *Late Bronze Age Settlement in Cyprus: Function and Relationship* (*SIMA-PB* 126), P. Åström and E. Herscher, eds., Jonsered, pp. 23–38.

———. 2003. "The Bull in Ancient Cyprus," in *The Bull in the Mediterranean World: Myths and Cults* (*Exhibition Catalogue. Barcelona—Athens. Cultural Olympiad 2001–2004*), pp. 112–117.

Kassianidou, V. 1999. "Bronze Age Copper Smelting Technology in Cyprus—The Evidence from Politico Phorades," in *Metals in Antiquity* (*BAR-IS* 792), S.M.M. Young, A.M. Pollard, P. Budd, and R.A. Ixer eds., Oxford, pp. 91–97.

Kling, B., and J.D. Muhly. 2007. *Joan du Plat Taylor's Excavations at the Late Bronze Age Mining Settlement at Apliki Karamallos, Cyprus*, Part 1 (*SIMA* 134), Sävedalen.

Knapp, A.B. 1986. *Copper Production and Divine Protection: Archaeology, Ideology and Social Complexity on Bronze Age Cyprus* (*SIMA-PB* 42), Göteborg.

———. 1996. "Settlement and Society on Late Bronze Age Cyprus: Dynamics and Development," in *Late Bronze Age Settlement in Cyprus: Function and Relationship* (*SIMA-PB* 126), P. Åström and E. Herscher, eds., pp. 54–80, Jonsered.

Knapp, A.B., V. Kassianidou, and M. Donnely. 1999. "Excavations at Politiko-*Phorades*, 1996," *RDAC* 1999, pp. 125–146.

Knapp, A.B., and V. Kassianidou. 2008. "The Archaeology of Late Bronze Age Copper Production: Politiko *Phorades* on Cyprus," in *Anatolian Metal* IV: *Frühe Rohstoffgewinnung in Anatolien und seinen Nachbarländerz* (*Die Anschnitt, Beiheft*), Ü. Yalçin, ed., Bochum, pp. 135–147.

Muhly, J.D. 1991. "The Development of Copper Metallurgy in Late Bronze Age Cyprus," in *Bronze Age Trade in the Mediterranean* (*SIMA* 90), N.H. Gale, ed., Jonsered, pp. 180–196.

———. 1996. "The Significance of Metals in the Late Bronze Age Economy of Cyprus," in *The Development of the Cypriot Economy: From the Prehistoric Period to the Present Day*, V. Karageorghis and D. Michaelides, eds., Nicosia, pp. 45–60.

Varnava, P. 1990. *Παλεύοντας για τη ζωή (αναμνήσεις Βετεράνων)*, Nicosia.

———. 1993. *Τα μεταλλεία της Κύπρου*, Nicosia.

Zwicker, U. 1990. "Archaeometallurgical Investigations on the Copper—and Copper-Alloy—Production in the Area of the Mediterranean Sea (7000–1000 B.C.)," *Bulletin of the Metals Museum* 15, pp. 3–32.

———. 1992. "Non-Destructive and Other Investigations on Metal Objects of the Archaeological Museum," in *Studies in Honour of Vassos Karageorghis*, G.C. Ioannides, ed., Nicosia, pp. 165–178.

CHAPTER

4

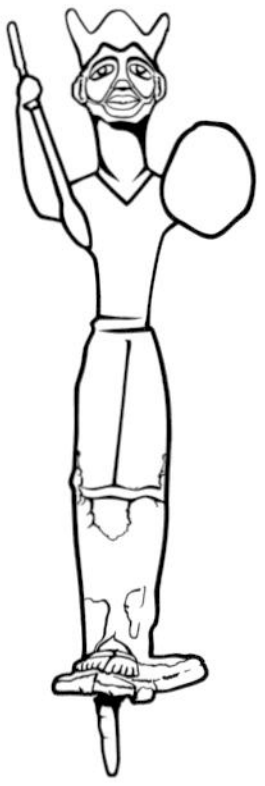

A Metallurgical Feast?

Vassos Karageorghis

The excavations of the Hebrew University of Jerusalem at Athienou–*Pamboularin tis Koukkouninas*, Cyprus, in 1971 and 1972 revealed a building with its courtyard and the area east of the building with pits (*favissae*?) in which rich collections of unusual vessels and special objects such as an ivory rhyton were found, constituting a cultic site connected with metallurgy.* Stratum III extends from the late 16th to the end of the 13th century B.C. The building founded in Stratum III continued to exist in Stratum II, dated to the 12th century B.C. (Dothan and Ben-Tor 1983).

Apart from the conical ivory rhyton, which imitates a purely Aegean shape and is not an unusual phenomenon in Cyprus and the Levant, the site yielded a variety of Mycenaean pottery; normal-size vessels were derived predominantly from pits, while miniature Mycenaean vessels, mainly juglets, were found in a group in the courtyard together with miniature votive vessels of local manufacture. The Mycenaean vessels (Figs. 4.1–4.3) included three cups, two of which dated to Mycenaean IIIB and one was a local imitation (Fig. 4.1; Dothan and Ben-Tor 1983, fig. 12); one small krater, local imitation (Fig. 4.2; Dothan and Ben-Tor 1983, fig. 13.1); two locally made shallow conical bowls (Fig. 4.2; Dothan and Ben-Tor 1983, fig. 13.2, 3); and one large Mycenaean IIIB stirrup jar (Fig. 4.2; Dothan and Ben-Tor 1983, fig. 13.4). There were also seven miniature Mycenaean IIIB juglets (Dothan and Ben-Tor 1983, fig. 14.1–6, 10), three full-size jugs (Dothan and Ben-Tor 1983, fig. 14.7, 8, 14), fragments from two three-handled piriform jars (Dothan and Ben-Tor 1983, fig. 14.11, 12), a

* About 20 years ago I met Jim Muhly in Sydney, and I asked him, “What brings you here Jim?” His answer was: “Good museums, good universities, and very good food and wine.” With this short paper on feasting, written in his honor, I would like to thank him for his friendship and collaboration over many decades and to wish him many more feasts in scholarship and real life.

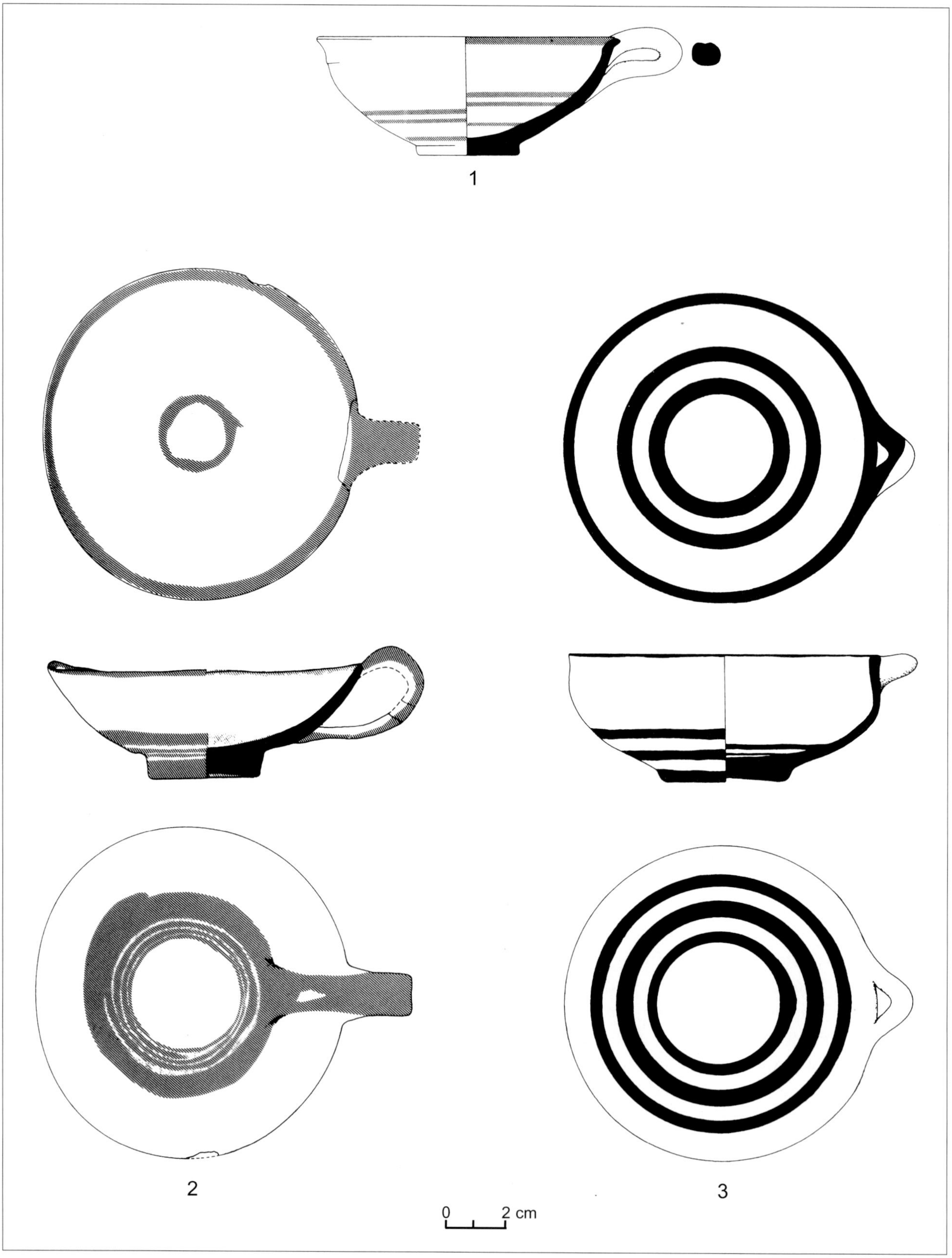

Figure 4.1. Mycenaean IIIB (nos. 1, 2) and imitation Mycenaean (no. 3) drinking cups from Athienou–*Pamboularin tis Koukkouninas* (after Dothan and Ben-Tor 1983, fig. 12, with permission from the Hebrew University of Jerusalem).

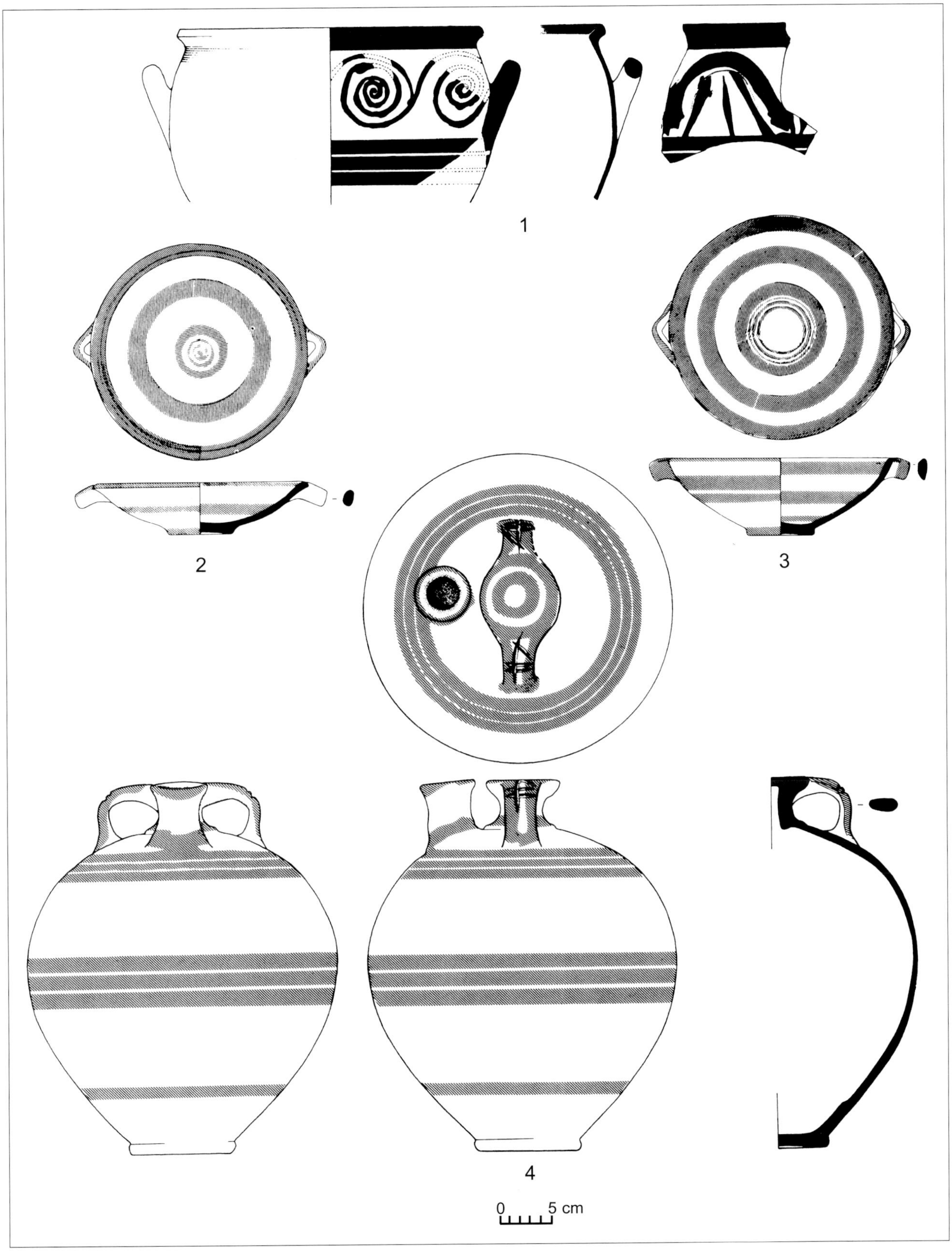

Figure 4.2. Mycenaean IIIB (nos. 2–4) and imitation Mycenaean (no. 1) vases from Athienou–*Pamboularin tis Koukkouninas* (after Dothan and Ben-Tor 1983, fig. 13, with permission from the Hebrew University of Jerusalem).

Late Minoan IIIB closed vessel (Dothan and Ben-Tor 1983, fig. 14.13) and a large Mycenaean IIIB(?) bowl (Dothan and Ben-Tor 1983, fig. 14.9). Of particular importance are two large Late Minoan IIIB stirrup jars (Fig. 4.3; Dothan and Ben-Tor 1983, fig. 15.1, 2).

The pottery of local fabrics included White Slip I and (mainly) II ware bowls and jugs (Dothan and Ben-Tor 1983, figs. 5, 6), Monochrome bowls (Dothan and Ben-Tor 1983, fig. 6.3–5), Coarse Monochrome kraters (Dothan and Ben-Tor 1983, fig. 6.7, 8), Base-Ring I and II juglets and bowls (Dothan and Ben-Tor 1983, figs. 7, 8), White Shaved juglets (Dothan and Ben-Tor 1983, fig. 9), Bucchero juglets (Dothan and Ben-Tor 1983, fig. 24), and a very large number of locally made miniature votive juglets, bowls, and cups that were mostly crudely made, which the excavators rightly believe "could not have been intended for a mundane function" (Dothan and Ben-Tor 1983, 139). They therefore assumed that "they were votive vessels connected with a cultic area," all made at the site (Dothan and Ben-Tor 1983, 139).

The pottery of Stratum II includes large kraters and pithoi, as well as skyphoi of Mycenaean IIIC:1b type, locally made (Dothan and Ben-Tor 1983, figs. 51–53). Apart from the pottery, the site yielded a small quantity of metal objects and scrap metal and half of a ton of copper ore and slag in the form of small copper nodules, found in association with the heaps of votive vessels.

Of particular note are quantities of bones, shells, and sea urchins, which the excavators thought might be "associated with the metal working process" (Dothan and Ben-Tor 1983, 140), as well as the skull and horns of a bull, which the excavators associate with the bucrania found in sanctuaries at Enkomi and Kition (Dothan and Ben-Tor 1983, 140). Some small prestige objects should also be mentioned, namely a bronze seal ring with hieroglyphs, other seals, and a scarab (Dothan and Ben-Tor 1983, pl. 38.1–9), indicating the high status of the owner.

I would like to draw particular attention to the conical ivory rhyton (Fig. 4.4), already mentioned above and discussed at some length by the excavators (Dothan and Ben-Tor 1983, 123–125). It is ca. 12.5 cm high, and its outer surface is decorated with four registers of engraved pictorial motifs that include human heads, horned animals, birds, fish, and plants.

It is not my intention to examine all of the material found at Athienou nor to discuss in any detail the character of the entire building and all its functions. I only wish to make some suggestions about a possible function, as suggested, I believe, by the archaeological finds. But first I should mention the conclusion of the excavators (Dothan and Ben-Tor 1983, 140):

> Athienou must be understood as a station on the trade route leading from the mining area to the large marketing centers on the east coast of the island. The variety of the finds, astonishing for such a small inland site, testifies to close connections with these coastal centers and, through them, with the cities of both the Aegean and the eastern Mediterranean.

Jennifer Webb (1999, 21–29) described and briefly discussed the architecture and the finds from Athienou and concluded (Webb 1999, 28–29):

> Little can be determined of the layout of the site at Athienou and precise indications for the working of the cult, apart from the accumulation of vessels and metal waste, are also lacking. The recovery of a possible bull figurine, a basin with bucrania protome and a "cult chariot" with bulls suggests that this animal was important in cult iconography. As frequently suggested, the specialized nature of the cult may have been determined by Athienou's position as an intermediary between the copper mining districts of the eastern Troodos and the workshops and trading centers on the east and southeast coasts. It is reasonable to conclude that the cult was abandoned at the end of Stratum III and the area given over entirely to the secondary processing and refining of copper ore and the storage of agricultural commodities.

In an article published in 2005 Vassiliki Kassianidou, in an attempt to question the traditional belief that metallurgy in Cyprus was associated with religion, made a brief survey of the results of the excavations at Athienou and concluded (Kassianidou 2005, 138):

> The discovery of roasted ore, which is an intermediate product in the long series of steps of turning the mineral to metal, is intriguing. As an intermediate product, roasted ore is usually a rare find even on smelting sites, unless for some reason it was discarded, which could be the case here, if for example it was deemed too poor in metal to be further processed. Is a waste product an appropriate votive for a deity, as has been suggested by Stech? And why did the intermediate product not go through the rest

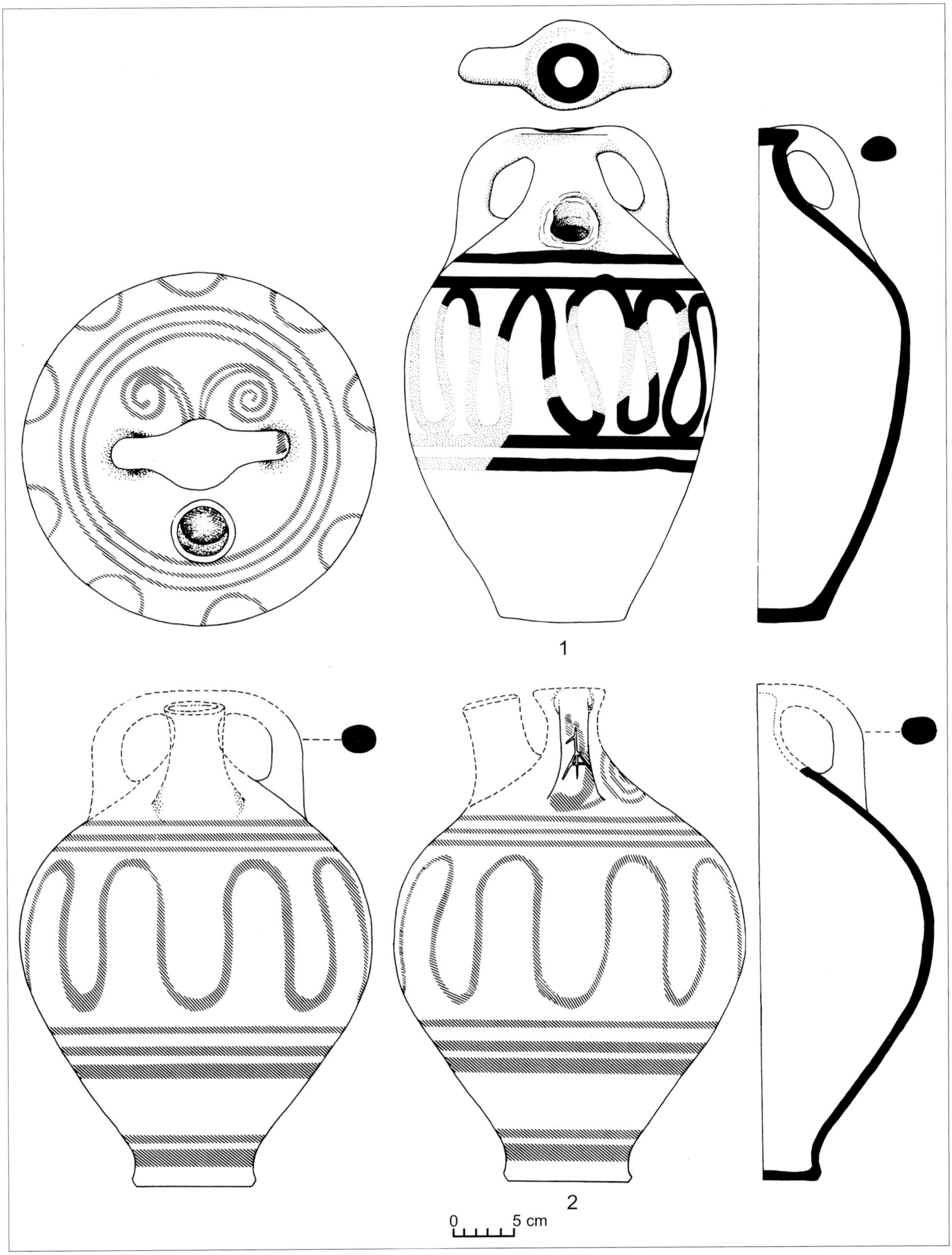

Figure 4.3. Late Minoan IIIB stirrup jars from Athienou–*Pamboularin tis Koukkouninas* (after Dothan and Ben-Tor 1983, fig. 15, with permission from the Hebrew University of Jerusalem).

of the process in order to be turned into metal? And if these are the waste products of the roasting process, where are the waste products of the smelting process? In other words where is the slag, as at least some slag would undoubtedly have been produced, if they had wanted to produce metal from these nodules. Where are the furnaces and where are the tuyères? Because of the absence of archaeometallurgical debris consistent with a smelting workshop, I find it difficult to identify Athienou as a copper production center and, therefore, it cannot be used to support the idea that the copper production was under religious control.

The location of this sanctuary, which is supposed to be involved in copper production is also intriguing. Although rationalized as an ideal site for a copper workshop, because of its proximity to the copper ores and fuel, in other words the forests, neither of these arguments are correct. The closest mine is Troulli, which lies at the distance, considerable by LBA standards, of 8 km, while trees are sparse in the Mesaoria which receives very little rainfall (although in the Bronze Age environmental conditions and vegetation may have been different from those of the present day). That the location may have been chosen because of its proximity to the route that led from the mines to the coastal towns, as suggested by the excavators, is more logical but it does not explain the nature of the finds.

If, then, the site of Athienou was not a sanctuary, and if it was not associated with a sanctuary, and it was not a copper production center, how are we to interpret the material found there? May I put forward the suggestion that part of it, if not most of it, was associated with a feast, which included eating and drinking but not without the element of a ritual ceremony? As a result of the study of the Aegean and Cypriot ceramic material of the Late Bronze Age (LBA) found at Sidon, I have come to the conclusion that this material was used during feasting that included eating and drinking. Although the publication is forthcoming, some preliminary reports have appeared (Karageorghis and Doumet-Serhal 2006; see also Karageorghis and Doumet-Serhal, 2009). This material is in compliance with practices in the LBA Aegean that spread to the Eastern Mediterranean and the Levant with which the Mycenaeans were in contact (cf. Wright 2004, 50). In LBA Cyprus there are several instances of symposia involving eating and drinking, e.g., at Kalavassos–*Ayios Dhimitrios* and Maa-*Palaiokastro* (cf. Karageorghis 2007, 257–258; see also Steel 2004a). There were many opportunities for members of the elite to offer such feasts. I quote from a very concise article by J.C. Wright (2004, 50):

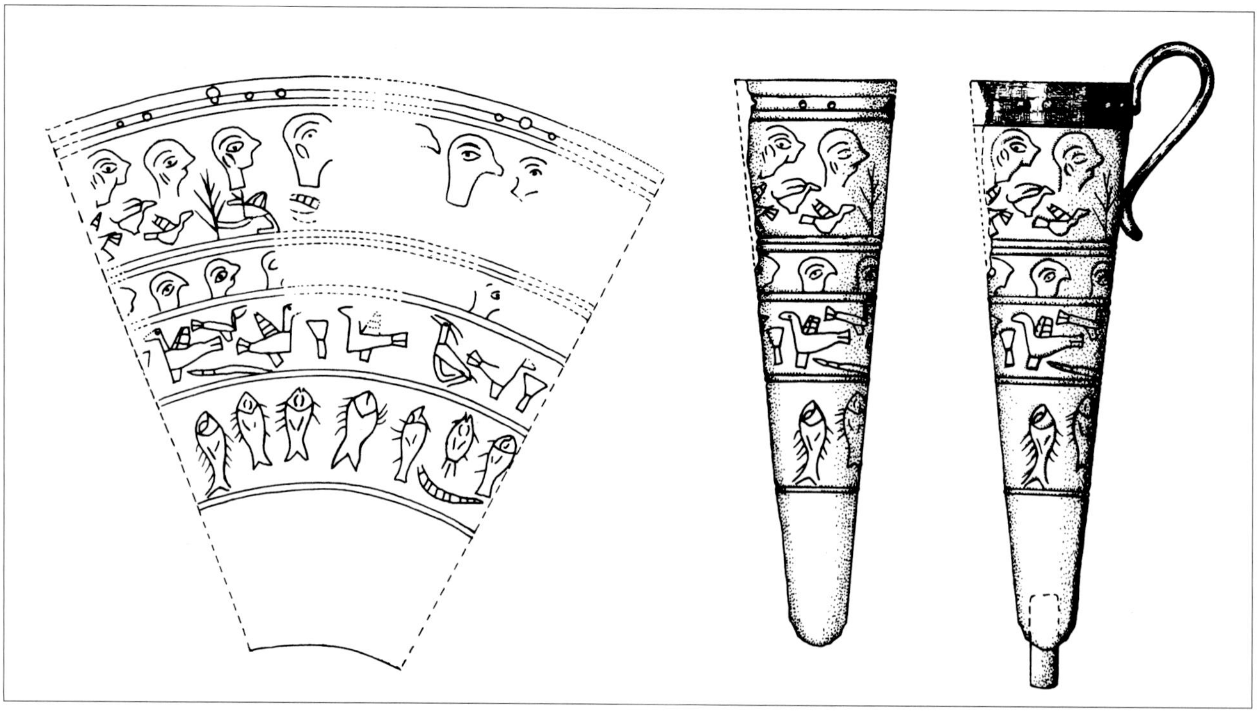

Figure 4.4. An ivory rhyton (ca. 12.5 cm tall) from Athienou–*Pamboularin tis Koukkouninas* (after Dothan and Ben-Tor 1983, with permission from the Hebrew University of Jerusalem).

> Drinking as a central activity of feasting is consistent with its historic function of aggrandizement by elites for whom rituals of drinking were associated with social strategies for consolidating their leadership and building the allegiance of a retinue. The practices of feasting and drinking gave hosts and guests alike opportunities for signaling their positions and status.
>
> Sponsored feasts were probably held exclusive for their kin, important retainers, and allied peers. The importance of this activity is registered by the appearance in tombs of the great metal vessels used for preparing and serving feasts. Over time this activity became evident in other categories—as deposits of cooking vessels in tombs and in domestic contexts, as records in Linear B, and in frescoes.

If the Athienou site was an intermediate center between the copper mines and the markets of the east and south coasts—e.g., Enkomi, Kition, and Hala Sultan Tekke—then the following passage from Wright's article, on Mycenaean feasts, is particularly relevant, as Athienou must have attracted a large number of middlemen, laborers, and others who were involved in this operation, obviously under the leadership and responsibility of a high ranking citizen or a number of them, who had both power and wealth (Wright 2004, 50):

> For the elites, however, the ability to sponsor feasts represented real economic value. The size and importance of a feast denote the amount of surplus the sponsor can draw on, which is symbolized through particular vessels, such as the bronze tripod kettles displayed in the assemblages of bronze cooking and serving vessels found in so many tombs of the elite. Since the surplus is collected from agricultural activities, its economic dimension is both geographically and demographically broad. Any substantial feast affected directly and indirectly a large and diverse population throughout the territory held by a community, as T. Palaima indicates in this volume in his study of the Linear B evidence (Palaima 2004, 106). Therefore, the sponsor of a feast demonstrates the ability to bring together large groups (through coalitions and alliances), to mobilize labor, and to command surplus and distribute it. The sponsor gains in prestige through these activities and advances his family, lineage, and allies both within and beyond the community.

In support of my suggestion, the function of most of the ceramic material and other objects may be enumerated as follows:

1. Large stirrup jars, both Late Helladic IIIB and Late Minoan IIIB, were frequently used for the transport of liquids, mainly wine and oils, to Cyprus and the Levant. We know that both Crete and the Mycenaean mainland were involved in this trade (cf. Knapp 1991, 29–30; Bounni, Lagarce, and Lagarce 1998, 33 n. 110; Yon 2000, 11; Van Wijngaarden 2002, 262; Monchambert 2004, 301–303; Steel 2004b, 73). Occasionally they are found in tombs, in which case their meaning was different from that of the original transport vessels (cf. Van Wijngaarden 2002, 269). The fact that three such vessels were found at a provincial site like Athienou is no doubt significant.
2. Drinking cups, both Mycenaean and local imitations, as well as a large number of local fabrics (White Slip, Base-Ring, Monochrome, and Plain White wares), are indicative of the large numbers of people who were using such vessels, no doubt on a special occasion. The skyphoi from Stratum II are also suitable for drinking. White Slip ware and perhaps Base-Ring ware bowls were suitable for serving hot liquids and food. It should be noted that at Kalavassos-*Ayios Dhimitrios*, the vessels which were used for a "feast" were shallow cups, as well as conical shallow bowls, both Mycenaean and local imitations (see South 1988; South and Russell 1993); there were also White Slip and Base-Ring ware bowls. At Maa-*Palaiokastro* where the population may have been Aegean immigrants, the drinking vessels were carinated footed bowls, which were prevalent in the Aegean. At Kalavasos there were also clay "wall-brackets," as at Athienou (Dothan and Ben-Tor 1983, fig. 16.2–4). Were they used for burning incense during the feast?
3. Kraters (e.g., Dothan and Ben-Tor 1983, figs. 6.7, 8, 11.19, 13.1, 14.9, 51.1, 2) as well as numerous jugs and juglets, both of Mycenaean and local fabrics, may also support the suggestion that large numbers of people were using them for their individual supply of wine (Figs. 4.5, 4.6). The number of White Shaved

ware juglets (Fig. 4.7; Dothan and Ben-Tor 1983, pl. 13) is very impressive, and these may have been used as dippers. The Late Minoan III(?) jug (Dothan and Ben-Tor 1983, pl. 34.1, 2), is particularly suitable for serving wine. We do not, however, suggest that all the jugs and juglets found at the site were used during drinking sessions.

4. The conical ivory rhyton, found in a pit, is of great significance. Conical rhyta in the Aegean were used for pouring libations during cultic ceremonies, but also for the preparation of fluids during festivities which included alcohol (Privitera 2006). Conical rhyta were imported to Cyprus, both in clay and also in faience (e.g., the well-known rhyton from Kition, which originally belonged to the furniture of a tomb). The shape was also imitated locally in Base-Ring ware, and several such rhyta were exported to Ugarit (see Yon 1980; for the Kition and Athienou rhyta, see also Koehl 2006, 245–246, with bibliography). Although of a small size, it could perform its function during the ritual part of the feast, even in a symbolic way. I do not know if the figured motifs of the engraved surface decoration are related in any way to the feasts in which the rhyton was used, namely the sacrificed animals, fish, and birds that people were consuming (see below).

5. The miniature vessels are enigmatic. They should, I believe, be associated with the copper nodules. They are crudely made, and the excavators are right in saying that they could not have been intended for a mundane function, and therefore they could have been offered during a ceremony, together with the copper nodules, as a symbolic ritual action for the increase of copper production. There may, of course, be other interpretations, but I will not expand on this problem. What I would, however, like to mention, is that miniature votive vessels (kylikes) were found, 20–22 in number, in association with a feast in the Palace of Nestor at Pylos. They were found near a heap of burned bones and may have been associated with feasting ritual (Stocker and Davis 2004, 67–69). It has been suggested that the diners did not use them at the table as drinking vessels for a liquid to drink, since they were too small, and that their content, if any, probably had a symbolic meaning, as the small copper nodules suggested above. (For a further discussion of the miniature kylikes, see Stocker and Davis 2004, 70–73.)

Figure 4.5. White Shaved ware juglets in situ (after Dothan and Ben-Tor 1983, with permission from the Hebrew University of Jerusalem).

The consumption of meat is an integral part of feasting, and this has been well demonstrated, for example, at Kalavassos–*Ayios Dhimitrios*, where, together with the discarded vessels used during the feast, there were "large quantities of bones from meat joints of sheep and goat, game birds and fish and seed remains of food plants" (South and Russell 1993, 306; for animal bones associated with feasting in the Aegean, see Stocker and Davis 2004, 63–67).

It is unfortunate that the information about material other than artifacts is very poor in the Athienou report. The faunal remains from Athienou have been studied by David Reese (Reese 2005). Of these the majority are bones of sheep and goats (Fig. 4.8), of which at least 21 examples are under 10 months old, the age when the meat is very tender and suitable for roasting on charcoal. Some of the bones were burned (Fig. 4.9) and others cut. Cut bones were observed at Maa-*Palaiokastro* outside the communal halls where feasts were taking place. Other bones from Athienou belong to cattle, pigs, fallow deer, birds, fish, and a hedgehog. There are also land snails and seashells (few, reflecting the distance of the site from the sea). The sea urchin shells mentioned in the excavation report are not in the Cyprus Museum; they were not seen by Reese and were probably left at the site owing to their fragility. We are informed that "in pit 543, bone, ash, shells, and sea urchins were found in quantity" (Dothan and Ben-Tor 1983, 140). The excavators believe that these may have been associated with the metalworking process (Dothan and Ben-Tor 1983, 140), but this is most unlikely. The same pit yielded Mycenaean IIIC–type pottery (of Stratum II) as well as the skull and horns of a bull. Although the excavators associate it with the bucrania of Kition and Enkomi, it may well be the remnant of a feast, remembering that an animal's head is a delicacy in Mediterranean feasting. Also a delicacy are the sea urchins, especially in the summer, when they are full of tasty eggs. It is tempting to associate the animals, birds, and fish depicted on the ivory rhyton with all the food consumed during a normal feast (Fig. 4.4). Sea urchins continue to be a Mediterranean delicacy, like the oysters of Britain for Roman soldiers, the shells of which are found in heaps in Roman camps. It is quite possible that this feast took place in the open air, in the courtyard of the site, most probably in the summer when it was possible to collect sea urchins in quantities, as well as fish, and when an open-air assembly of people would be possible.

I mentioned the case of Sidon where I believe the ceramic material suggests a ritual feast with eating and drinking. At this stage it is not possible to present a concrete argument, because the excavation is not yet complete, but another comparable Late Bronze Age site at Hazor in Israel recently has generated discussion and controversy. A monumental structure located in the center of the Hazor acropolis has been identified by some as a temple and by others as a "ceremonial palace." Amnon Ben-Tor, the most recent excavator of Hazor, has very generously introduced me to the gist of the discussion and, having considered the published accounts, I feel inclined to believe that at Hazor we have a phenomenon not unlike that of Sidon, comparable with that discussed

Figure 4.6. Medium-size and miniature Plain White ware juglets in situ (after Dothan and Ben-Tor 1983, with permission from the Hebrew University of Jerusalem).

Figure 4.7. White Shaved ware juglets (after Dothan and Ben-Tor 1983, pl. 13, with permission from the Hebrew University of Jerusalem).

above regarding Athienou (for a recent discussion and bibliography, see Ben-Tor 2006; also Bonfil and Zarzechi-Peleg 2007; for short annual reports on the progress of recent excavations at Hazor, see Ben-Tor 1999, 2000, 2001, 2002, 2003, 2004a, 2004b, 2005).

The possibility that there may be similar phenomena at other Levantine sites during the LBA should also be investigated. For example, the "House with Sink" at Ugarit, where a stone trough was discovered filled with tableware and identified probably as a "local tavern," very well may be part of a house for ritual feasting (see Yon 2006, 59, with bibliography).

The *marzeah* feast, well known in biblical archaeology, is also very relevant to this discussion. It had a long history in Israel from the 14th century B.C. down to the Roman period, and it involved eating, drinking, and sexual intercourse (King 1988). Manfred Bietak has tentatively identified the architectural remains of a sacred precinct at Tell el-Dab'a as a "Bêt Marzeah" (Bietak 2003).

This short re-examination of the archaeological evidence from Athienou, 35 years after the completion of the excavation, certainly does not solve all the problems connected with the site, which has remained enigmatic ever since the publication of the results in 1983, but it does propose a more realistic interpretation of the archaeological data. Hiding behind "cult" and "religion" often shows our inability to face in a rational way problems that are much simpler than the nebulous aspects of religion. This does not mean that we deny that religion often provides answers to what we are unable to understand in terms of mundane life. Athienou, however, together with the evidence from the two Levantine sites of Sidon and Hazor, in the light of what we now know in the Aegean world as Mycenaean feasting, may offer a new dimension in our understanding of some aspects of society in Late Bronze Age Cyprus.

Postscript

In a recently published article by Sharon Zuckerman (2007), very similar material from Hazor and other Near Eastern sites is presented. The author stresses the following elements that are strikingly similar to the discussion above, relating to Athienou:

1. Monumental structures, *attached courtyards*.
2. Large quantities of animal bones and ashes, especially around the installations of the countryard.
3. Large quantities of serving vessels (especially open bowls); large storage jars (pithoi) for long-term storage, craters, *votive/miniature vessels*, and cooking pots.
4. *Prestige items* (cf. ivory rhyton at Athienou) and imported vessels and other paraphernalia.

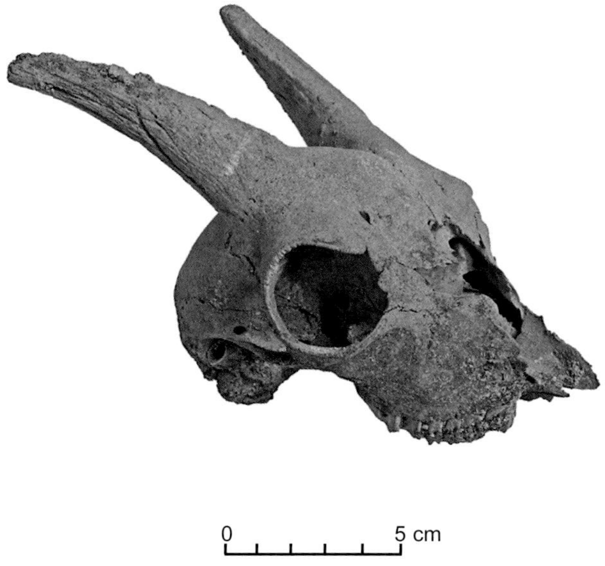

Figure 4.8. The skull of a small animal, a kid (photo courtesy of the Cyprus Museum and the excavators).

Figure 4.9. Burned animal bones (photo courtesy of the Cyprus Museum and the excavators).

References

Ben-Tor, A. 1999. "Tel Hazor, 1999," *Israel Exploration Journal* 49, pp. 269–274.

———. 2000. "Tel Hazor, 2000," *Israel Exploration Journal* 50, pp. 243–249.

———. 2001. "Tel Hazor, 2001," *Israel Exploration Journal* 51, pp. 235–238.

———. 2002. "Tel Hazor, 2002," *Israel Exploration Journal* 52, pp. 254–257.

———. 2003. "Tel Hazor, 2003," *Israel Exploration Journal* 53, pp. 218–223.

———. 2004a. "Tel Hazor, 2004," *Israel Exploration Journal* 54, pp. 230–235.

———. 2004b. "Hazor and Chronology," *Egypt and the Levant* 14, pp. 45–67.

———. 2005. "Tel Hazor, 2005," *Israel Exploration Journal* 55, pp. 209–216.

———. 2006. "Ceremonial Palace, not a Temple," *Biblical Archaeology Review* 32 (5), pp. 8, 78–79.

Bietak, M. 2003. "Temple or 'Bêt Marzeah'?" in *Symbiosis, Symbolism and the Power of the Past: Canaan, Israel, and their Neighbors from the Late Bronze Age through Roman Palaestina*, W.G. Dever and S. Gitin, eds., Winona Lake, IN, pp. 155–168.

Bonfil, R., and A. Zarzecki-Peleg. 2007. "The Palace in the Upper City of Hazor as an Expression of a Syrian Architectural Paradigm," *BASOR* 348, pp. 25–47.

Bounni, A., E. Lagarce, and J. Lagarce. 1998. *Ras Ibn Hani*. I: *Le palais nord du Bronze récent: Fouilles 1979–1995, synthèse préliminaire*, Beirut.

Dothan, T., and A. Ben-Tor. 1983. *Excavations at Athienou, Cyprus 1971–1972* (*Qedem* 16), Jerusalem.

Karageorghis, V. 2007. "Eating and Drinking in Cyprus, 13th–6th Centuries B.C.," in *Keimelion: The Formation of Elites and Elitist Lifestyles from Mycenaean Palatial Times to the Homeric Period. Akten des internationalen Kongresses vom 3. bis 5. Februar 2005 in Salzburg*, E. Alram-Stern and G. Nightingale, eds., Vienna, pp. 257–262.

Karageorghis, V., and C. Doumet-Serhal. 2006. "Sidon: Les fouilles du British Museum de 1998 à 2005," *CRAI* 2006, pp. 305–331.

———. 2009. "Sidon and the Aegean: Notes on Recent Discoveries," in *Δώρον. Τιμητικός τόμος για τον καθηγητή Σπύρο Ιακωβίδη*, D. Danielidou, ed., Athens, pp. 329–353.

Kassianidou, V. 2005. "Was Copper Production under Divine Protection in Late Bronze Age Cyprus? Some

Thoughts on an Old Question," in *Cyprus: Religion and Society from the Late Bronze Age to the End of the Archaic Period. Proceedings of an International Symposium on Cypriote Archaeology, Erlangen, 23–24 July 2004*, V. Karageorghis, H. Matthäus, and S. Rogge, eds., Möhnesee-Wamel, pp. 127–141.

King, P.J. 1988. *Amos, Hosea, Micah: An Archaeological Commentary*, Philadelphia.

Knapp, A.B. 1991. "Spice, Drugs, Grain, and Grog: Organic Goods in Bronze Age East Mediterranean Trade," in *Bronze Age Trade in the Mediterranean* (*SIMA* 90), N.H. Gale, ed., Jonsered, pp. 21–68.

Koehl, R.B. 2006. *Aegean Bronze Age Rhyta* (*Prehistory Monographs* 19), Philadelphia.

Monchambert, J.-Y. 2004. *La céramique d'Ougarit: Campagnes de fouilles 1975 et 1976* (*Ras Shamra-Ougarit* 15), Paris.

Palaima, T.G. 2004. "Sacrificial Feasting in the Linear B Documents," in Wright, ed., 2004, pp. 97–126.

Privitera, S. 2006. "The Rhyton Hoards as Evidence for Feasting in Neopalatial Crete: The Case of Gournia," *Creta Antica* 6, pp. 187–198.

Reese, D.S. 2005. "Faunal Remains from Israeli Excavations at Athienou–*Pampoulari tis Koukkouninas*," *RDAC* 2005, pp. 87–108.

South, A.K. 1988. "Kalavasos–*Ayios Dhimitrios* 1987: An Important Ceramic Group from Building X," *RDAC* 1988, pp. 223–228.

South, A.K., and P.J. Russell. 1993. "Mycenaean Pottery and Social Hierarchy at Kalavasos–*Ayios Dhimitrios*, Cyprus," in *Proceedings of the International Conference Wace and Blegen: Pottery as Evidence for Trade in the Aegean Bronze Age 1939–1989. Held at the American School of Classical Studies at Athens, December 2–3, 1989*, C. Zerner, ed., Amsterdam, pp. 303–310.

Steel, L. 2004a. "A Goodly Feast . . . A Cup of Mellow Wine: Feasting in Bronze Age Cyprus," in Wright, ed., 2004, pp. 161–180.

———. 2004b. "A Reappraisal of the Distribution, Context and Function of Mycenaean Pottery in Cyprus," in *La céramique mycénienne de l'Égée au Levant. Hommage à Vronwy Hankey*, J. Balensi, J.-Y. Monchambert, and S. Müller Celka, eds., Lyons, pp. 69–85.

Stocker, S.R., and J.L. Davis. 2004. "Animal Sacrifice, Archives, and Feasting at the Palace of Nestor," in Wright, ed., 2004, pp. 59–75.

Van Wijngaarden, G.J. 2002. *Use and Appreciation of Mycenaean Pottery in the Levant, Cyprus and Italy (1600–1200 B.C.)*, Amsterdam.

Webb, J.M. 1999. *Ritual Architecture, Iconography and Practice in the Late Cypriot Bronze Age* (*SIMA-PB* 75), Jonsered.

Wright, J.C. 2004. "The Mycenaean Feast: An Introduction," in Wright, ed., 2004, pp. 1–58.

Wright, J.C., ed. 2004. *The Mycenaean Feast*, Princeton.

Yon, M. 1980. "Rhytons chypriotes à Ougarit," *RDAC* 1980, pp. 79–83.

———. 2000. "Répartition et contextes de la céramique mycénienne d'Ougarit," in *Céramiques mycéniennes* (*Ras Shamra-Ougarit* 13), M. Yon, V. Karageorghis, and N. Hirschfeld, eds., Paris and Nicosia, pp. 1–27.

———. 2006. *The City of Ugarit at Tell Ras Shamra*, Winona Lake, IN.

Zuckerman, S. 2007. ". . . Slaying Oxen and Killing Sheep, Eating Flesh and Drinking Wine. . . : Feasting in Late Brinze Age Hazor," *PEQ* 139 (3), pp. 186–204.

CHAPTER

5

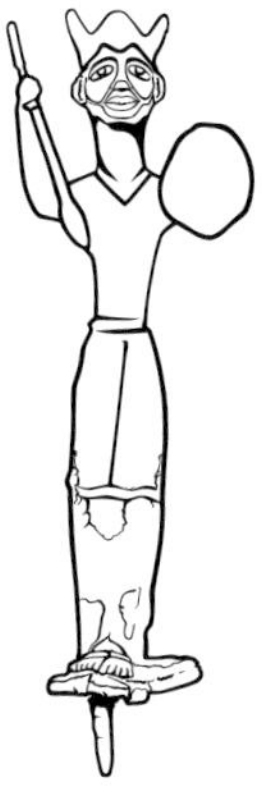

Blowing the Wind of Change: The Introduction of Bellows in Late Bronze Age Cyprus

Vasiliki Kassianidou

Jim Muhly has laid down the foundations for the study of copper production in Cyprus, and it is a great pleasure to pay tribute to his work by contributing to a festschrift in his honor.

Although it was initially difficult to decide upon a topic, the dilemma was soon overcome with the opening of a new museum in Nicosia, where among other exhibits there is one that is truly impressive for those of us who are studying ancient metallurgy. The object in question is a fragmentary pot bellows (now restored; Fig. 5.1), and it is part of the archaeological collection of the Pancyprian Gymnasium, the oldest school in Cyprus, which was founded in 1812. The school's museum was created in 1893 to house the collection of rocks and minerals that was donated by the Greek Councilor to the island (Christodoulou 2002, 5). Subsequently a variety of objects including antiquities were added to the museum. The archaeological collection was significantly expanded between 1936 and 1960—when the headmaster of the school was Constantinos Spyridakis, a respected classical scholar and ancient historian—mainly by donations from alumni. The collection was used for teaching and was stored in the school's library until the beginning of the 1990s when a decision was taken to exhibit it in the main building. More recently a series of old houses in the vicinity of the Gymnasium were bought and restored in order to create a modern museum where all of the school's collections could be put on display for the broader public. The archaeological collection was systematically recorded with the help of the Department of Antiquities, and in 2002 a catalog was published by Phryni Hadjichristophi.

More recently, Vassos Karageorghis (2007) published a paper in the *Report of the Department of*

Antiquities, Cyprus, presenting some of the most important artifacts in the collection, including the pot bellows in question. Nevertheless, I thought that the object deserved a closer look, especially since only five such pot bellows were known from Cyprus until now, but also because it can shed some light on four intriguing sherds that we uncovered during the excavation of the Late Bronze Age copper smelting workshop at the site of Politiko-*Phorades*.

The bellows (inv. no. Π.Γ.096) has a thick flat base and a bulging body that narrows slightly when it reaches the out-curved rim. The height is 29.5 cm, while the diameter is 45 cm (Fig. 5.2). In the lower part is a solidly built nozzle, trapezoidal in section (Fig. 5.3). The diameter of the aperture in the external part of the nozzle is 5.1 cm, while in the inner part of the vessel it is only 2.8 cm. The nozzle is joined to the body by a twin bridge a few centimeters below the rim. The bellows is clearly handmade, as finger impressions are still visible on the inner surface of both the base and walls (Fig. 5.4). The clay is coarse, consisting of both significant quantities of chaff and calcium inclusions. The outer surface is of a light red color while the inner surface is grayish, indicating that it was exposed to a reducing atmosphere. There are clear signs of damage due to exposure to high temperatures, especially on the bottom surface of the base, which is fractured.

The provenance of the object unfortunately is unknown, and an effort to find more information about it in the museum's archive was fruitless. Nevertheless, the fact that such pot bellows are known only from the Late Bronze Age in Cyprus enables us to relatively date it to this period.

Pot bellows were first discovered in Enkomi's Area III, the northern edge of the city excavated by P. Dikaios. A large building was uncovered there. Throughout its lifespan—from Late Cypriot I to the beginning of the Late Cypriot IIIA (in other words, ca. 1600–1200 B.C.)—it was dedicated to metallurgical activities (Dikaios 1971, 500, 516). The earliest pot bellows (inv. nos. 4543, 4544) date to the 14th century, are made of limestone, and were found together (Dikaios 1971, 637, pls. 8:3, 127:48, 48a). The better preserved example of the two bellows has a diameter of 42 cm and a height of 13 cm. Not knowing what they were, Dikaios (1971, 637) suggested that they could be molds for bun-shaped ingots. It was Christopher Davey (1979, 106) who correctly identified them as bellows and included them in his catalog of pot bellows from the Near East. Because they were found as a pair and due to their low height, Davey suggested that they were foot rather than hand operated (Davey 1979, 110).

Also from Enkomi but dating to Late Cypriot IIIA—in other words the end of the 13th to the beginning of the 12th century B.C.—is a well-preserved example of a ceramic pot bellows (inv. no. 1691/1), which, according to Dikaios (1971, 577, pl. 120:5), belongs to the Plain Wheel-Made pottery type. This example is quite large, having a diameter of 54 cm and a height of 24 cm.

Of a similar date is a large fragment of the base and nozzle of a ceramic pot bellows that was found at the site of Maa-*Palaiokastro* (Karageorghis and Demas 1988, 200, no. 256). This has a diameter of 30 cm (Zwicker 1988, 430).

Finally, a well-preserved ceramic pot bellows dating to Late Cypriot IIC—i.e., the 13th century B.C.—was uncovered during the excavations of the site of Alassa–*Pano Mandilaris* (Hadjisavvas 1986, 66, pl. XVIII:3; Karageorghis 1985, 935–936, fig. 91). It is different from the Enkomi example, as it is handmade and has a very thick rounded rim, as well as a ridge that runs parallel to it. A recessed area thus is formed between the two, which would be ideal for tying the leather covering in place on the bellows. The ridge extends to form a bridge that connects the nozzle to the main body. Interestingly, the Alassa pot bellows and the one in the Pancyprian Gymnasium Museum are the only known examples from the east Mediterranean that have such a feature: none of the examples presented by Davey (1979, 104–105) for the Near East, by Müller Karpe (2000, 117) for Anatolia, and by Evely (2000, 363), Blitzer (1995, 508–509), Dimopoulou (1997, 435), and Betancourt and Muhly (2006, 126) for Crete have a bridged nozzle. Indeed, Davey stated in a personal communication on the Alassa example, "I thought that the bellows with the 'handle' represented a third style." And since for the moment, the only known cases come from Cyprus, it may be suggested that this is a local type.

In fact, there may be another example, which has been radiocarbon dated to ca. 1650–1500 B.C., from the copper smelting workshop at Politiko-*Phorades* (for more information on the workshop, see Kassianidou 1999; Knapp, Kassianidou, and Donnelly 2001; Knapp 2003; Knapp and Kassianidou 2008). The site was excavated under

Figure 5.1. Ceramic pot bellows (h. 29.5 cm) in the Pancyprian Gymnasium collection (inv. no. Π.Γ.096). Photographed by Giorgos Papasavvas.

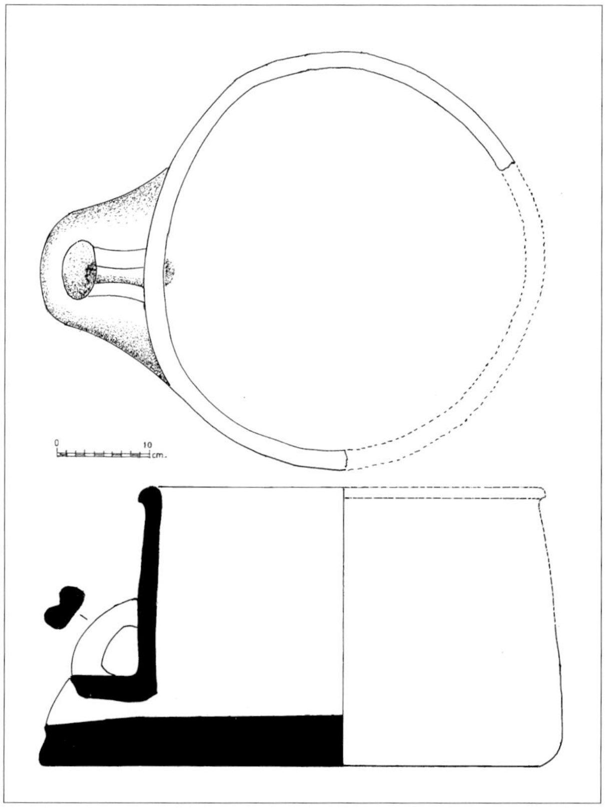

Figure 5.2. Drawing of ceramic pot bellows in the Pancyprian Gymnasium collection (inv. no. Π.Γ.096). Drawn by Elias Markou and courtesy of V. Karageorghis.

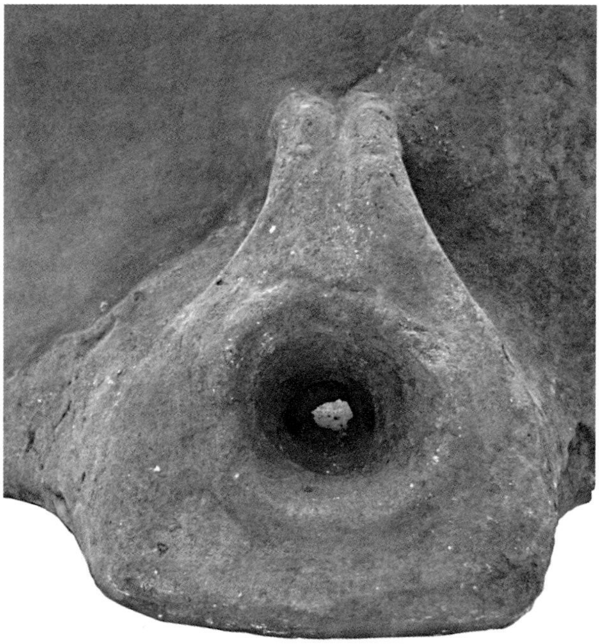

Figure 5.3. The nozzle of the pot bellows (inv. no. Π.Γ.096; outer dia. of hole 5.1 cm). Photographed by Giorgos Papasavvas.

Figure 5.4. View of the internal surface of the bellows (inv. no. Π.Γ.096; dia. 45 cm). Photographed by Giorgos Papasavvas.

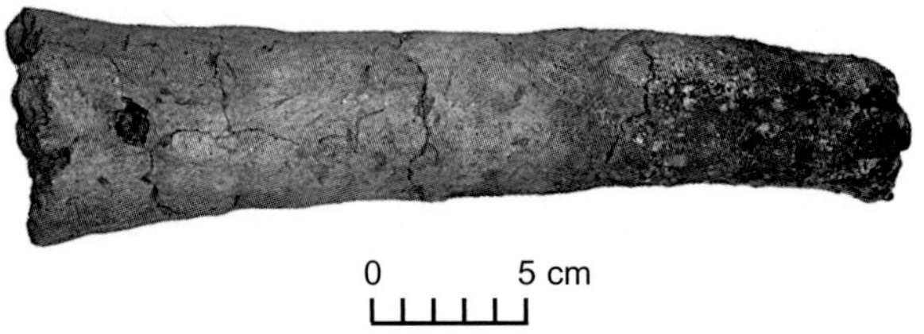

Figure 5.5. Tuyère (L. 30 cm) from Politiko-*Phorades*. Photographed by Christopher Parks.

Figure 5.6. Droplets of slag adhering to the inner surface of the airhole in a tuyère from Politiko-*Phorades*. Photographed by Christopher Parks.

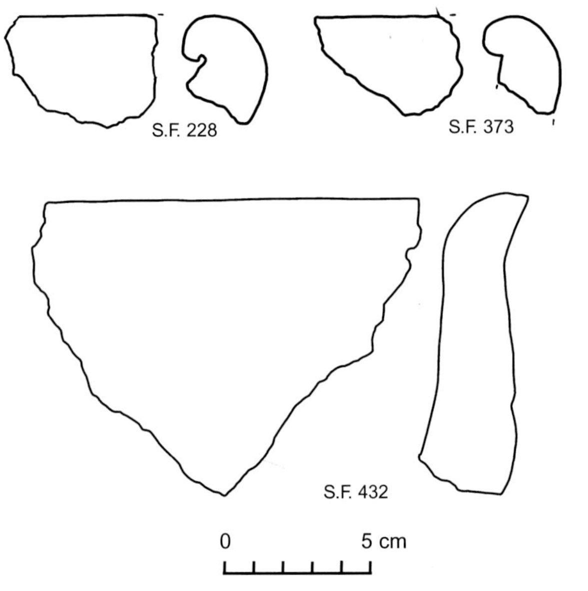

Figure 5.7. Sherds from ceramic pot bellows(?) from Politiko-*Phorades*. Drawn by Glynnis Fawkes.

the direction of A.B. Knapp and the author, and a monograph is in the final stages of preparation (Knapp et al., forthcoming).

The workshop is situated on the bank of an ancient creek where working platforms were constructed upon which the smelting furnaces were operated. When the operation closed down and the site was abandoned, thousands of furnace fragments, hundreds of fragments, and 50 almost-complete tuyères were left behind.

The most abundant find from *Phorades* is, not surprisingly, slag: over 3.5 tons were collected. The slag is homogeneous and has no metallic prills visible to the naked eye in section. It is thus completely different from the slag that has been found at Middle Cypriot sites; it is nodular and heavily encrusted with copper carbonates. The *Phorades* slag is clearly the product of a more-advanced technology, which, for the first time in Cyprus, employed tuyères and presumably bellows (Kassianidou 2007).

The use of bellows at *Phorades* was indirectly indicated by the size of the tuyères: one of the best examples has a preserved length of 30 cm (Fig. 5.5). Blowpipes are usually much smaller and were more commonly used in metallurgical processes that required a precisely directed blast of air, as is the case when working gold (Craddock 1995, 177). Furthermore, droplets of slag are commonly observed on the inner surface of the airhole in the tuyères and at some distance from the slagged tip (Fig. 5.6): they must have been deposited there after having been sucked in during the operation of the bellows. Until now, we thought that we had no finds that could clearly be identified as fragments of bellows. In all previous publications we argued that perhaps leather bellows had been used.

However, among the pottery finds from the site are four sherds (S.F. 228, S.F. 373, S.F. 432, S.F. 435), which our ceramic specialist, Ellie Ribeiro, could not classify according to the known Late Cypriot pottery typology (Fig. 5.7). Because of their fabric and appearance, she believed that they must be somehow related to the metallurgical process, although they were clearly neither furnace nor tuyère fragments. The largest of the three, S.F. 432, is a rim sherd from a large open vessel (Fig. 5.8). According to Ribeiro and Manning (in Knapp et al., forthcoming), the fabric is coarse and fairly friable and has been fired (and reheated) at a high temperature. On the interior surface of the sherd and on the top of the rim are darkened patches of scorching, indicating heat damage. A comparison of the sherd with the bellows in the Pancyprian Gymnasium reveals many similarities in the shape of the rim (which is everted and sharply "beaked"), in the fabric, and in the signs of heat damage. Indeed, when I showed the sherd to Philip Betancourt, who visited Cyprus in 2006, he immediately suggested that this very well could be a fragment from our elusive bellows. The two smaller sherds (S.F. 228, S.F. 373), which are made of the same fabric as the rim sherd and have also been damaged by exposure to high temperatures, were even harder to understand and were initially thought to be rim sherds with a hooked overhang. I now believe that these are fragments of the bridge connecting the nozzle of the bellows to the body.

If I am correct in the identification of the sherds from *Phorades*, which are securely dated to 1650–1500 B.C., as coming from bellows, then they are the earliest examples from Cyprus and are over two centuries older than the limestone ones from Enkomi. They are, therefore, extremely important as they show that pot bellows were introduced to the island, together with ceramic tuyères, at the very beginning of the Late Bronze Age. Where they originated is a question that cannot be addressed in this short paper.

Furthermore, the similarity of the *Phorades* sherds with the pot bellows from the Pancyprian Gymnasium would suggest that it, too, could be of this early date. Interestingly there is more, admittedly circumstantial, evidence that the Pancyprian bellows could be contemporaneous to *Phorades*. The school's archaeological collection includes four ceramic tuyères, three of which are slagged (Fig. 5.9; Hadjichristophi 2002, 78–79; Karageorghis 2007). The tuyères are also of types known only from the Late Cypriot period. Indeed one of them (inv. no. Π.Γ.093) is a double-walled tuyère, i.e., a tuyère that was used once and was covered with another layer of clay in order to be used again. Until now such tuyères were only known in Cyprus from Politiko-*Phorades* (Fig. 5.10). As the bellows and the tuyères came to the museum as a group, we may assume that they came from the same site. I would, therefore, argue that they all should be dated to Late Cypriot I (1600–1450 B.C.).

The introduction of tuyères and bellows to the smelting installations brought a revolution to the Cypriot copper industry. As a result, large quantities

Figure 5.8. Rim sherd (S.F. 432) of a ceramic pot bellows(?) from Politiko-*Phorades*. Photographed by Christopher Parks.

Figure 5.9. Double-walled tuyère (L. 11.2, ext. dia. 5.2 cm) in the Pancyprian Gymnasium collection (inv. no. Π.Γ.096). Photographed by Photis Mylonas and courtesy of V. Karageorghis.

Figure 5.10. Double-walled tuyère from Politiko-*Phorades*. Photographed by Christopher Parks.

of metal could be produced efficiently to satisfy both local and external demands for copper, and Cyprus could participate in the trading networks that were established in the east Mediterranean. Although copper would have been the main commodity to be exported from Cyprus, it is the pottery that elucidates this activity. Thus Levantine, Egyptian, and Aegean imports start to appear on the island, while Cypriot ceramics are found in the Levant, Egypt, Anatolia, and the Aegean, including Crete (Knapp and Cherry 1994, 43–44). Interestingly enough, it is from Crete that we now have the earliest evidence for the export of Cypriot copper in the shape of ingots: recently published lead isotope analysis of ingot fragments from Late Minoan IB deposits in Gournia and Mochlos point to a Cypriot provenance (Soles and Stos-Gale 2004, 54–58; Gale and Stos-Gale 2005, 125).

The wind of change that came with the introduction of bellows and tuyères to the island propelled Cyprus to the forefront of the metals trade in the East Mediterranean. The island would hold this leading role for centuries.

Acknowledgments

I would like to express my gratitude to the editors of the book for inviting me to participate in this festschrift. I would also like to thank the Headmistress of the Pancyprian Gymnasium, Mrs. Niki Papapetrou, for granting me the permission to study the pot bellows, as well as the curator, Mr. Michalis Phantaros, for his eagerness to facilitate my work in the museum and to search for information in the museum's archives. I am grateful to Prof. Vassos Karageorghis who allowed me to use the drawings and photographs of the objects (Figs. 5.2, 5.9), which were prepared for his own publication. Last but not least, I would like to thank Giorgos Papasavvas for providing valuable advice and moral and practical support throughout the preparation of this paper.

References

Betancourt, P.P., and J.D. Muhly. 2006. "The Pot Bellows," in *The Chrysokamino Metallurgy Workshop and its Territory*, by P.P. Betancourt, Princeton, pp. 125–132.

Blitzer, H. 1995. "Minoan Implements and Industries," in *Kommos: An Excavation on the South Coast of Crete*. I: *The Kommos Region and Houses of the Minoan Town*. Part 1: *The Kommos Region, Ecology, and Minoan Industries*, J.W. Shaw and Maria C. Shaw, eds., Princeton, pp. 403–536.

Christodoulou, G. 2002 "Ιστορία του Αρχαιολογικού Μουσείου του Παγκυπρίου Γυμνασίου," in Hadjichristophi, 2002, pp. 5–6.

Craddock, P. 1995. *Early Metal Mining and Production*, Edinburgh.

Davey, C.J. 1979. "Some Ancient Near Eastern Pot Bellows," *Levant* 11, pp. 101–111.

Dikaios, P. 1971. *Enkomi: Excavations 1948–1958* II, Mainz.

Dimopoulou, N. 1997. "Workshops and Craftsmen in the Harbour-Town of Knossos at Poros-Katsambas," in *Texnh: Craftsmen, Craftswomen, and Craftsmanship in the Aegean Bronze Age. Proceedings of the 6th International Aegean Conference, Philadelphia, Temple University, 18–21 April 1996* (*Aegaeum* 16), R. Laffineur and P.P. Betancourt, eds., Liège, pp. 433–438.

Evely, R.D.G. 2000. *Minoan Crafts: Tools and Techniques. An Introduction* (*SIMA* 92:2), Jonsered.

Gale, N.H., and Z.A. Stos-Gale. 2005. "Zur Herkunft der Kupferbarren aus dem Schiffswrack von Uluburun und der spätbronzezeitliche Metallhandel im Mittelmeerraum," in *Das Schiff von Uluburun. Welthandel vor 3000 Jahren*, Ü. Yalçin, C. Pulak, and T. Slotta, eds., Bochum, pp. 117–131.

Hadjichristophi, Ph. 2002. *Αρχαιολογικόν Μουσειόν Παγκυπρίου Γυμνασίου*. I: *Νεολιθική—Παλαιοχριστιανική περίοδος*, Nicosia.

Hadjisavvas, S. 1986. "Alassa: A New Late Cypriote Site," *RDAC* 1986, pp. 62–67.

Karageorghis, V. 1985. "Chronique des fouilles et découvertes archéologiques à Chypre en 1984," *BCH* 109:2, pp. 897–967.

———. 2005. "Kypriaka XVII," *RDAC* 2007, pp. 133–143.

Karageorghis, V., and M. Demas. 1988. *Excavations at Maa*-Palaeokastro, *1979–1986*, Nicosia.

Kassianidou, V. 1999. "Bronze Age Copper Smelting Technology in Cyprus: The Evidence from Politico-*Phorades*," in *Metals in Antiquity* (*BAR-IS* 792), S.M.M. Young, A.M. Pollard, P. Budd, and R.A. Ixer, eds., Oxford, pp. 91–97.

———. 2007. "The Formative Years of the Cypriot Copper Industry," in *Aegean Metallurgy in the Bronze Age. Proceedings of an International Symposium Held at the University of Crete, Rethymnon, Greece, on November 19–21, 2004*, I. Tzachilli, ed., Herakleion, pp. 249–267.

Knapp, A.B. 2003. "The Archaeology of Community on Bronze Age Cyprus: Politiko-*Phorades* in Context," *AJA* 107, pp. 559–580.

Knapp A.B., and J.F. Cherry. 1994. *Provenience Studies and Bronze Age Cyprus: Production, Exchange, and Politico-Economic Change* (*Monographs in World Archaeology* 21), Madison, WI.

Knapp, A.B., and V. Kassianidou. 2008. "The Archaeology of Late Bronze Age Copper Production: Politiko-*Phorades* on Cyprus," in *Anatolian Metal* IV (*Der Anschnitt Beiheft* 21), Ü. Yalçin, ed., Bochum, pp. 135–147.

Knapp, A.B., V. Kassianidou, and M. Donnelly. 2001. "Copper Smelting in Late Bronze Age Cyprus. The Excavations at Politiko-*Phorades*," *NEA* 64 (4), pp. 204–209.

Knapp, A.B., V. Kassianidou, M. Donnelly, P. Duffy, and J. Noller. Forthcoming. *The Excavations at Politiko*-Phorades *(1996–2000): The Archaeology of a Bronze Age Smelting Site.*

Müller Karpe, A. 2000. "Zur Metallverarbeitung bei den Hethitern," in *Anatolian Metal* I (*Der Anschnitt Beiheft* 13), Ü. Yalçin, ed., Bochum, pp. 113–124.

Soles, J.S., and Z.A. Stos-Gale. 2004. "The Metal Finds and Their Geological Sources," in *Mochlos* IC: *Period III. Neopalatial Settlement on the Coast: The Artisans' Quarter and the Farmhouse at Chalinomouri. The Small Finds*, J.S. Soles and C. Davaras, eds., Philadelphia, pp. 45–59.

Zwicker, U. 1988. "Appendix VIII: Investigations of Material from Maa-*Palaeokastro* and Copper Ores from the Surrounding Area," in *Excavations at Maa*-Palaeokastro, *1979–1986*, V. Karageorghis and M. Demas, Nicosia, pp. 427–448.

CHAPTER

6

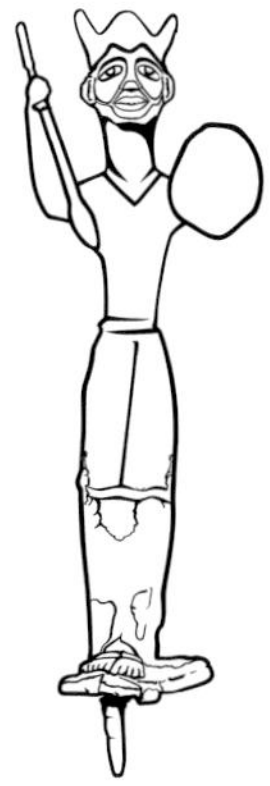

A Newly Rediscovered Cypriot Tripod-Stand in the Florence Archaeological Museum

Fulvia Lo Schiavo

A project that concerns the Cypriot collections in Italian museums is being led by Lucia Vagnetti with the assistance of Marco Bettelli and Silvana Di Paolo (all of the Istituto di studi sulle Civiltà dell'Egeo e del Vicino Oriente, Consiglio Nazionale delle Richerche [ICEVO–CNR]).* The present focus is on the Cypriot collection of more than three hundred pieces in the Florence Archaeological Museum. On the occasion of the exhibition *Egeo, Cipro, Siria, e Mesopotamia: Dal collezionismo allo scavo archeologico: In onore di Paolo Emilio Pecorella*, a sample of the objects was presented, and it included a tripod-stand that was known and published since the 1960s, but with inadequate illustrations and insufficient autoptical examination (Lo Schiavo 2008).

This short contribution in honor of Jim Muhly will highlight this very interesting document of the most typical Cypriot handicraft, whose original provenance from Cyprus is possible to determine through the evaluation of the archive documents.

Recent History

In spite of its obscure provenance, this tripod-stand is not unknown. It was acquired in 1905 by Luigi Adriano Milani, together with a hemispherical cup placed on top of the ring (Fig. 6.1; inv. nos. 82503, 82504), for the collections of the Florence Archaeological Museum, as part of a lot bought in

* To Jim, in memory of a starry night in Nicosia and a lively debate about Sardinian metallurgy—the first of many, with Robert Maddin acting as arbitrator—and also remembering many other days and nights in Cyprus, Sardinia, Athens, and New York discussing ancient metallurgy. How appropriate is this connection between Cypriot metallurgy and Florence!

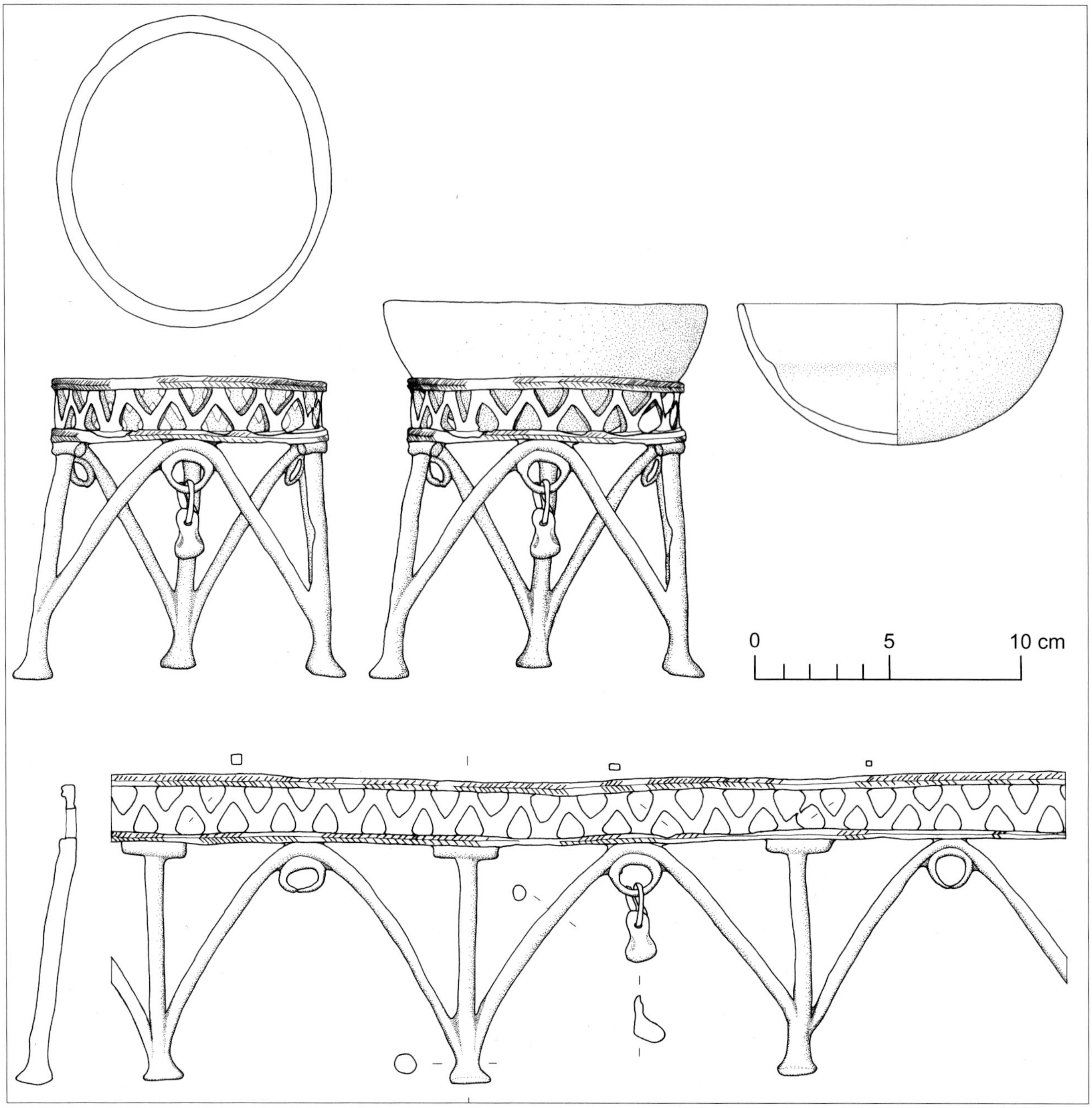

Figure 6.1. The tripod-stand (inv. no. 82503) and the cup (inv. no. 82504) in the Florence Archaeological Museum (drawn by Paola Falchi).

Greece (inv. nos. 82420–82510: "Prehellenic objects, coming from different parts of Greece, described in the enclosed to the present lading bill"). These objects were offered for sale by M. Caremfilaki (his first name is unknown, and also his stationery only has an initial), an antique dealer introduced to Milani by Giacomo Cecconi, who supplied materials to the museum. In a letter to Milani, Cecconi says: "At Vallombrosa, where I am returning in the middle of next March, I have many photographs of ancient objects *from Cyprus* that someone *from there* is offering for sale…" (28.[1].1902; emphasis added). For some months, Cecconi acted as a go-between for Milani and Caremfilaki, but from 1903 onward they established direct business relations.

Caremfilaki said, "Enclosed is a life-size photograph of a copper object that seems to me an incense burner; the upper part is removable and uncovered and the price of this object is 300 francs" (19.1.1905). Milani acknowledged the receipt of the photograph and counter-offered 250 francs (7.6.1905). The lot reached the museum in July 1905 and was cataloged and numbered 82503 (the tripod-stand) and 82504 (the cup). Since that time the two pieces have been separated both in the museum and in the bibliography (Elena Sorge recognized this and provided me with all the archival data [11.3.2008]).

According to the historical archives of the museum, it is almost certain that the nationality of M. Caremfilaki, whose correspondence in Italian is from "*Larnaca (Chypre)*," is Cypriot, and it may be presumed that the tripod-stand with the cup was found in Cyprus, because the dealer expressed his worries to Milani concerning a "new protection law" that was going to forbid archaeological excavations (17.5.1905: "To consolidate and amend the law relating to the ancient monuments and antiquities of Cyprus and to provide museums," signed C.A. King-Harman, 10th May 1905. "1. This Law may be cited for all purposes as The Antiquities Law, 1905").

Nothing else can be said, apart from the typological arguments that follow.

Documentation and Analyses

The tripod-stand, which is in very bad condition, was selected as an example of the so-called bronze disease at a time when techniques such as spectroscopy, optical microscopy, electron microscopy, X-ray diffraction, electrochemistry, and lead isotope analysis were only just beginning. "Some of these methods were used by us in order to study the cancer [bronze-disease] which had appeared on a small bronze tripod from the Antiquity Superintendence of Etruria" (Piccardi and Bordi 1961, 496). The techniques of optical microscopy, spectroscopy, and X-ray diffraction analyses were in fact applied, while those of electron microscopy and electrochemistry were not available. The elemental analyses were carried out in a "specialized laboratory" by Comm. Bruno Bearzi: "the alloy. . . was 97–98% copper, 2% tin and impurities of lead and other elements" (Piccardi and Bordi 1961, 497; these results show that the piece is almost pure copper. On the contrary, the title of the paragraph, *Piccolo tripode di bronzo* ["Small Bronze Tripod-Stand"], demonstrates that in Italian the common use of the word "bronze" is not a specific one and does not imply a reference to the elemental composition). It is worth mentioning that the same Bearzi was conducting analyses of the bronze horses from the Basilica of San Marco in Venice, and the "bronze-disease" was discussed at some length with reference to Etruscan metallurgy.

What was described as "a simple method of localized, slow and controlled cathodic or anodic cleaning" was applied to the tripod-stand (Piccardi and Bordi 1961, 498). The photographs seem to have been taken before restoration (Piccardi and Bordi 1961, figs. 1, 2), as was the one published by H.W. Catling, who was the first to link the tripod-stand to Cyprus and who acknowledged the late W.L. Brown for bringing the piece to his attention (neg. no. 7921 in Catling 1964, 193 n. 2, no. 4, table 27:f).

Unfortunately, in the Conservation Center of the Florence Superintendence there is no precise description of the treatments applied to the object. The photographic campaign carried out on materials on exhibition on the museum after the flood of November 4, 1966, shows the result of heavy cleaning through the removal of all corrosion on the tripod-stand (neg. no. 17574), while the cup is in good condition and still covered by a sound patina (neg. no. 17575).

In 1984, E. Macnamara examined the tripod-stand in Florence (Macnamara, Ridgway, and Ridgway 1984, 3, no. 18). The lustrous appearance of the foreground leg of the tripod-stand in the photograph provided by the museum to H. Matthäus shows that at that time the process of re-patination was finished (neg. no. 35951:2 in Matthäus 1985, 303, no. 692, pl. 93). The same picture was published 16 years later by G. Papasavvas (2001, 233, 234, no. 2, fig. 2).

To sum up, in the 1960s G. Piccardi, S. Bordi, and Bearzi had direct contact with the object. They were followed in the mid-1980s by E. Macnamara who took notes and measurements. Catling, Matthäus, and Papasavvas were provided with photographs by the museum.

Description

The tripod-stand (h. 10.5–11.2, dia. of the ring 10–11 cm) consists of a ring that is supported by three legs (Fig. 6.2). The ring is made from two bars (each of which is square in section) that have a zigzag openwork band between them. On their outer faces, the two bars are decorated by a horizontal groove and short diagonal incisions that form a herringbone pattern; the ring is flat on the inner side.

The legs are made from rods (round in section) with three simple cylindrical spacers set on top at a location between the legs and the ring; the feet are coarsely molded to represent bovine hooves.

Outer struts rise diagonally from each side of the legs, about 2.5 cm from the base of the feet to the ring; then they bend and join the strut of the adjacent leg; at the halfway point between the legs, irregularly modeled pendant loops are roughly set.

Only one of these loops is still holding a pendant, which was probably originally bud-shaped, but it is shapeless now, hanging down through a narrow coiled ring that is square in section.

Figure 6.2. Four views of the tripod-stand (inv. nos. 82503) in the Florence Archaeological Museum. H. 10.5–11.2, dia. of the ring 10–11 cm.

Inexperienced Craftsmanship?

This tripod is surprising from at least two main points of view: it is made of almost pure copper, and although the casting was successfully achieved, the wax model was very roughly shaped.

The first point is difficult to evaluate because only a few other stands were analyzed, but there is no doubt that while an alloy with a content of up to 10% tin and a complementary amount of lead is considered to be a good casting alloy, this object is far from it (see Table 6.1). Bearing in mind the absence of inner struts and leg volutes, this is probably best regarded as an experiment, both in its shape and in its casting alloy.

The object has not been produced by joining together different metal bars and separately modeled metal pieces. No trace of soldering, nor of any different alloy at the junctions between the spacers and the ring, has been detected. On the contrary, the use of a lost-wax technique is self-evident. Different wax bands and bars, as separately modeled pieces, have been attached together.

The wax model seems to have been the product of a very inexperienced craftsman, who was unable to conceive the shape destined to be covered by clay to be filled by the metal casting, and he did not foresee the effects that all of these imperfections would have caused to the final product. These are (from top to bottom): the irregularly traced herringbone pattern (Fig. 6.3), the variable cut of the zigzag pattern, the hasty rendering of the legs, which are different in diameter from one another, the twisted loops pasted in the upper elbows of the struts (Fig. 6.4), the pendant lump (Fig. 6.5), the shapeless feet (Fig. 6.6), and the bad closing of the

Figure 6.3. Detail of the ring and a spacer on the tripod stand.

Figure 6.4. Detail of the ring and a loop on the tripod stand.

Figure 6.5. Detail of an inner strut and a loop with the pendant on the tripod stand.

Figure 6.6. Detail of a leg and the two adjacent inner struts on the tripod stand.

ring with a bulge to give an oval profile, instead of a circular one (see Fig. 6.1). They show that the artisan was not familiar with the molding of a wax model. The ring was cut out of a flattened and incised band with a sharpened instrument that left a square cross-section to the zigzag openwork motif. The legs with their spacers, loops, and struts were joined by pressing the wax bars together and noticeably squashing them. The tripod-stand was cast in one piece by pouring the molten metal into the clay mold, with the feet acting as funnels for the pouring and the inner struts as internal channels. The last, and solely ornamental, operation was the connection with the hanging pendant.

The Hemispherical Cup

As said above, a picture of the hemispherical cup on top of the tripod stand was shown to Milani along with an offer for its sale. As the two pieces were naturally associated, they both came to the Florence Archaeological Museum, where they were registered with two consecutive inventory numbers. Until now the cup has not received any particular attention among the many rich and varied objects of the Cypriot collection. Its shape is simple: the outside is hemispherical and smooth, with a thin rounded rim h. 5.3, dia. 12 cm); inside a horizontal rib runs at ca. 2.3–2.5 cm below the rim (Fig. 6.7). The cup fits perfectly on top of the tripod-stand, so that the round bottom is entirely held in the ring (Fig. 6.8).

The cup is not made of a thin sheet, but it is quite thick (th. 0.4 cm at the bottom, 0.5 cm at the rib, 0.3 cm at the rim), so it could have been manufactured either by casting in lost-wax technique or even by the hammering of a metal disk. It is covered by a sound and stable patina, and because restoration was not necessary, no analyses were performed on it.

Figure 6.7. The cup (inv. no. 82504) in the Florence Archaeological Museum. H. 5.3, dia. 12 cm.

Figure 6.8. The tripod-stand with the cup placed on top.

Site or Museum	Cu	Sn	Pb	Ag	Publications
Enkomi Tomb 15	88.0%	10.0%	2.0%	—	Catling 1964, 194, no. 7; Macnamara and Meeks 1987, 58, no. 3; Papasavvas 2001, 43–44, no. 3
Ashmolean Museum	77.5%	6.4%	16.3%	—	Catling 1964, 201, no. 24; Craddock 1976, 102, 108; Papasavvas 2001, 45, no. 10
Santa Maria in Paulis in the British Museum	92%	7.6%	trace	trace	Craddock and Tite 1984, 20; Macnamara, Ridgway, and Ridgway 1984, 2–7, no. 2; Papasavvas 2001, 45, no. 11
Liverpool Museum	93.50%	1.89%	3.77%	—	Catling 1964, 196, no. 14; Pickles and Peltenburg 1998, 73, no. 28; Papasavvas 2001, 44, no. 5

Table 6.1. The results of metallographical analyses (in %) that were made on tripod-stands and four-wheeled-stands were collected and discussed by Papasavvas (2001, 43–45; 2003, 27).

Who, Where, and When

If this tripod-stand had been found in Sardinia, any diagnosis as an early work by an incompetent artisan undertaking a new and complex task would have been indisputable.

In Sardinia, a ring with such an openwork zigzag motif between two plain bars was a part of the Samugheo hoard (Vivanet 1890; Taramelli 1903, 490, fig. 19) and was included by E. Macnamara in her study on the tripod-stands of Cypriot provenance or inspiration found in Sardinia (Lo Schiavo, Macnamara, and Vagnetti 1985, 42, no. 3, fig. 14:3). She reported that Taramelli, although giving no dimensions, had pointed out the fine quality of the manufacture and suggested that it was cast whole by the lost-wax process and interpreted it as an arm-ring. Therefore, Macnamara attributed the ring either to a tripod-stand or to a four-sided stand, both of miniature size, and likely to be a true Late Cypriot III import.

Another tripod-stand, with the ring consisting of an openwork zigzag motif between two double bars, is known to be in a Sardinian private collection (Lo Schiavo 1983, 310–313, fig. 7; Lo Schiavo, Macnamara, and Vagnetti 1985, 36, no. 1, fig. 14:1–2, pl. IIa). The tripod-stand is complete, although broken into four fragments, and represents a more developed Late Cypriot III model with double spirals at the top of the "composite" legs and the inner ring supported by inner struts and decorated with three birds. "It is probable that this miniature tripod-stand was cast whole by the investment or lost-wax technique. . . The writer believes that the investment or lost-wax process was more widely used in the manufacture of Group A Rod Tripods, or at least those of miniature size" (Lo Schiavo, Macnamara, and Vagnetti 1985, 38).

At this point, Macnamara noted: "The writer has examined Catling 1964, no. 4, a miniature Group A Rod Tripod III type, now in Florence, and believes at least the ring and perhaps the whole, was cast by the investment or lost-wax process" (Lo Schiavo, Macnamara, and Vagnetti 1985, 38, no. 38).

Then possible answers to these questions are:

> "Who made the Florence tripod-stand?"—probably an unskilled Cypriot artisan.

"Where was it made?" — in Cyprus. Why not near Larnaca, where M. Caremfilaki lived and probably collected his merchandise, before shipping it to Italy? Let us dream of Hala Sultan Tekke, where "fragments of the clay investment used for casting a bronze rod tripod are thought to have been found in Hala Sultan Tekke by Åstrom (2000, 33–34), in an area associated with metalworking" (Papasavvas 2003, 48).

"When was it made?" — at the start of their production, while destined for success later, that is to say the Late Cypriot II period (Papasavvas 2001, 97–98; 2003, 23).

The similarity with the tripod-stand of unknown provenance in the Nicosia Museum published in 1971 is striking and did not escape Vassos Karageorgis (1971, 344, no. 17, fig. 23). G. Papasavvas, in his masterly work, rightly attributes the two pieces to the same workshop (Papasavvas 2001, 77–78, 90 no. 1, 94 no. 2, 335 no. 2). At sight, the structure of the Nicosia example is thinner and more brittle (and in fact the ring is broken), but it is impossible to proceed without metallurgical and elemental analyses.

One final question remains to be answered: "Why?" Apart from the evident function as a support, shared by the tripod and the four-wheeled stands, and apart from the observation that such a structure was not destined for anything other than ritual purposes, the scanty information concerning the conditions of the discovery prevent any conjecture (Matthäus 1985, 306–307). Moreover, in the case of stands, the question may be shifted to the containers held on top that might have received various liquids or even solids (Papasavvas 2001, 125–129).

As to the miniature tripod-stands, if their association with hemispherical cups was a general rule, one of the possible uses that might be cautiously advanced is an incense burner. This, considering its small dimensions, is the first suggestion that the Florence tripod-stand and cup bring to mind.

Acknowledgments

My gratitude is due to the entire team of the Superintendence and of the Archaeological Museum in Florence—and most of all for the kind assistance of Maria Cristina Guidotti. We worked together to set up the exhibition in honor of our greatly esteemed friend and colleague Paolo Emilio Pecorella (known as "Pamì"). This paper was conceived while choosing the pieces for the exhibition, and then taking them out of their cases.

I would like to express my gratitude to Roberto Magazzini for the splendid photographs, to Paola Falchi for the drawings, to Gabriella Campini and Miriana Ciacci for research in the archaeological deposits, to Alessandra Nistri for research in the photographic files, to Elena Sorge for investigations in the historical archives, to Giandomenico De Tommaso for translations from Greek, to Paola Rendini for translations from German, to Marcello Miccio for updates about casting, and to Alan Williams for the correction of my English.

I am, as always, indebted to Lucia Vagnetti for her invaluable critical support, and I thank her very much for her friendly help.

References

Åström, P. 2000. "A Coppersmith's Workshop at Hala Sultan Tekke," in *Periplus: Festschrift für Hans-Günther Buchholz zu seinem achtzigsten Geburtstag am 24. December 1999* (*SIMA* 127), P. Åström and S. Sürenhagen, eds., Jonsered, pp. 33–35.

Catling, H.W. 1964. *Cypriot Bronzework in the Mycenaean World*, Oxford.

Craddock, P.T. 1976. "The Composition of Copper Alloys Used by the Greek, Aegean, and Roman Civilisations.

1: The Greeks before the Archaic Period," *JAS* 3, pp. 93–113.

Craddock, P.T., and M.S. Tite. 1984. "Report on the Composition of Bronzes from the Hoard Found at Santa Maria in Paulis, Sardinia," Appendix in Macnamara, Ridgway, and Ridgway 1984, pp. 19–21.

Karageorghis, V. 1971. "Chronique des fouilles et découverte archéologiques a Chypre en 1970," *BCH* 95, pp. 335–432.

Lo Schiavo, F. 1983. "Le componenti egea e cipriota nella metallurgia della tarda età del Bronzo in Italia," in *Magna Grecia e Mondo Miceneo, Atti del XXII Convegno di Studi sulla Magna Grecia (Taranto, 7–11 ottobre 1982)*, Naples, pp. 285–320.

———. 2008. "Cenni sulla produzione e circolazione del rame e dei manufatti metallici a Cipro," in *Egeo, Cipro, Siria e Mesopotamia: Dal collezionismo allo scavo archeologico In onore di Paolo Emilio Pecorella. Catalogo della Mostra, Firenze, Museo Archeologico Nazionale, 1 dicembre 2007–4 maggio 2008*, M.C. Guidotti, F. Lo Schiavo, and R. Pierobon Benoit, eds., Florence, pp. 174–187.

Lo Schiavo, F., E. Macnamara, and L. Vagnetti. 1985. "Late Cypriot Imports to Italy and their Influence on Local Bronzework," *Papers from the British School at Rome* 53, pp. 1–71.

Macnamara, E., and N. Meeks. 1987. "The Metallurgical Examination of Four Late Cypriot III Stands now in the British Museum," *RDAC* 1987, pp. 57–59.

Macnamara, E., D. Ridgway, and F.R. Ridgway. 1984. *The Bronze Hoard from S. Maria in Paulis, Sardinia* (*British Museum Occasional Paper* 45), London.

Matthäus, H. 1985. *Metallgefässe und Gefässuntersatze der Bronzezeit, der geometrischen und archaischen Periode auf Cypern* (*Prähistorische Bronzefunde* 2.8), Munich.

Papasavvas, G. 2001. *Χάλκινοι υποστάτες από την Κύπρο και την Κρήτη*, Nicosia.

———. 2003. "Cypriot Casting Technology I: The Stands," *RDAC* 2003, pp. 25–52.

Piccardi, G., and S. Bordi. 1961. "Sul cosiddetto 'cancro' dei bronzi etruschi," *StEtr* 29, pp. 493–502.

Pickles, S., and E. Peltenburg. 1998. "Metallurgy, Society and the Bronze/Iron Transition in the Eastern Mediterranean and in the Near East," *RDAC* 1998, pp. 67–100.

Taramelli, A. 1903. "Fordongianus. Antiche terme di Forum Traiani," *Notizie degli Scavi* 1903, pp. 469–492.

Vivanet, F. 1890. "Samugheo. Vaso fittile contenete frammenti di bronzo rinvenuti nel territorio del Comune," *Notizie degli Scavi* 1890, p. 292.

CHAPTER

7

ing:
a Cypriot God

From all the places of the Ancient World, it is only on Cyprus that a metallic statuette standing on an ingot, such as the Ingot God from Enkomi (Figs. 7.1–7.3), does not come as a surprise. His discovery in 1963 and subsequent publication (Schaeffer 1965; 1971, 506–510) triggered an enduring discussion on the possible association of religion and metallurgy, as well as on his date, style, and identity. Jim Muhly, who has discussed this statuette on several occasions (e.g., Muhly 1980), was among the very first to capture the originality of Cypriot metalwork and to state that ". . . Cypriot bronzeworkers seem capable of doing whatever caught their fancy . . ." (Muhly 1988, 335).

Although the Ingot God is not as large as the Horned God that comes from the same site (Dikaios 1969–1971, I, 295, IIIa, pls. 139–143), he is still one of the largest metallic statuettes of the Bronze Age Mediterranean. His pose, with a spear in his raised right hand, recalls the "Smiting Gods" as represented in bronze statuary in the Near East, Anatolia, Egypt, the Aegean, and as far west as Sicily (Seeden 1980, 132). Only nine examples have been recovered on Cyprus to this point (Webb 1999, 223).

The Ingot God is heavily armed. Apart from the spear and the small round shield, he also bears a conical helmet with projecting horns. This equipment, however, is not complemented by any greaves, contrary to what is stated in most assessments of the statuette (see, however, Buchholz 1979, 85; Balmuth 1994). In fact, what is mistaken as greaves is a second layer of metal wrapping the lower part of the figurine in an irregular manner. This layer clearly stands out from the metal lying underneath, as it has a very distinct, albeit uneven, outline, which happens to be placed at about the height of the knees (Figs. 7.3–7.5). This explains why it was taken as the outline of a pair of greaves despite the fact that it has no consistent shape, as it is placed rather diagonally on the front, while at the back it rises up to the thighs. It also reaches down to the ingot at the base, and, in fact, as it will be shown, it is there because of the

Figure 7.1. The Ingot God from Enkomi, front view. H. 35 cm.

Figure 7.2. The Ingot God from Enkomi, side view.

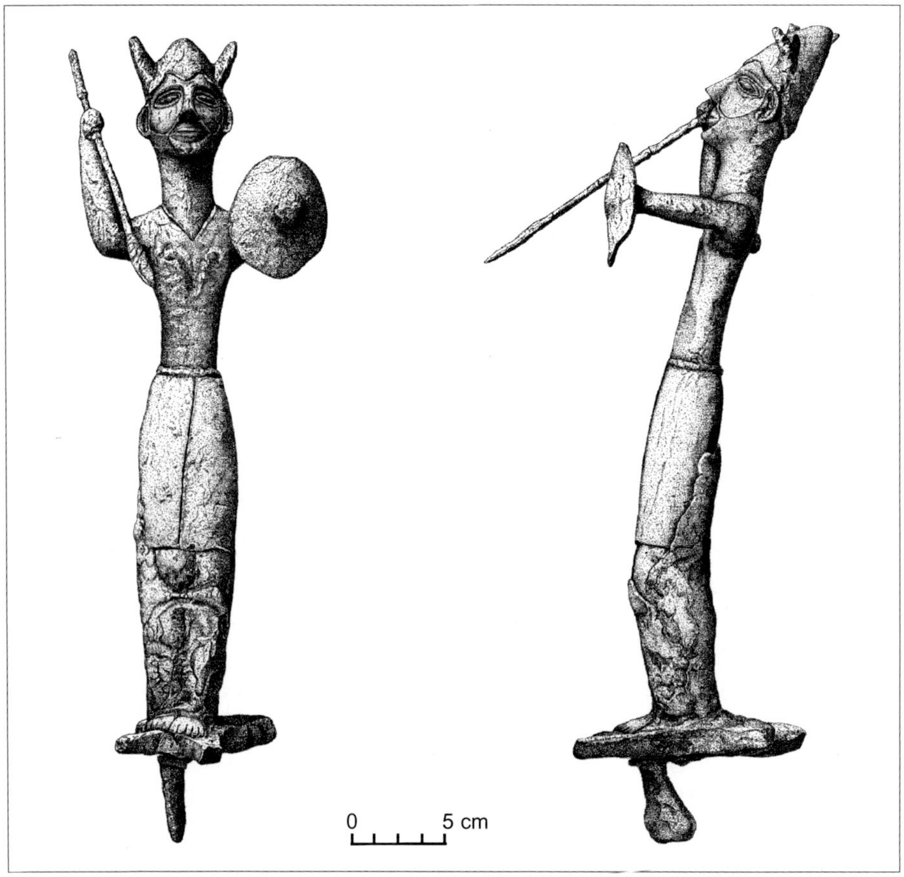

Figure 7.3. Drawings of the Ingot God, front and side views (by Clara Vasitsek).

ingot. This anomaly was the result of the use of the casting-on technique for this part of the body. This technique involves the casting of a bronze part above another cast item, and in general, albeit not exclusively, it is used for repairs or for additions of some parts upon others (Drescher 1958). The Ingot God offers one of its earliest manifestations.

There are two possible explanations for the use of the casting-on technique with this statuette: either there was a casting problem on the lower part of the figurine, such as a miscast of the legs that had to be corrected, or the lower part of the statuette was damaged in some way and had to be repaired (cf. Buchholz 1979, 84–85; Lagarce and Lagarce 1986, 74).

Although the Ingot God appears to be of an inferior artistic quality in comparison with the Horned God, he is most categorically the product of an experienced artist who was working in a different style. On the contrary, the metallic part added with the casting-on technique is the product of a craftsman who had no understanding of form or style. The face, with the finely executed and precisely rendered eyes, mouth, and nose, all enclosed within well-defined and balanced areas by the contour of the beard, is not matched by the careless, rather sloppy work done on the added layer of metal on the lower part of the body. The badly outlined and designed feet of the figure (Figs. 7.4–7.6) also seem to have been executed in a rather hasty way and are of no comparison either to the face or to the well-articulated body.

The discrepancy between the upper part of the figurine and the metallic layer covering its lower part is so sharp that it implies that the craftsman who applied the casting-on technique was a different person from the one who cast the statuette in the first place. This would suggest that the technique was not used to correct a casting problem because in this case the first artist presumably would have corrected the mistake himself as soon as he realized it. The remaining possibility is that the *casting-on* technique was used by a second, less-skilled artisan in order to correct damage suffered by the figurine at some point for an unknown reason.

Despite the apparent resemblance in posture, the Ingot God does not conform strictly to the type of the "Smiting Gods" and cannot be classified as such (cf. Webb 1999, 223). His garment is different from the short kilt worn by the majority of the statuettes belonging to this type (Seeden 1980, 133–134), while his shield is a rare attribute among the standing, armed figurines of the Near East (Seeden 1980, 139, 144). In fact, it is possible that the shield, which is held by a deficiently shaped hand at a rather badly executed joint (Fig. 7.2), also was added to the figurine with the casting-on method, perhaps after it also had been damaged. Alternatively, the figure may have been holding a different weapon, such as a sword or a dagger, as seems to be the case for the bronze statuette of a "Smiting God" found in the vicinity of the sanctuary of the Ingot God (Schaeffer 1971, 510–513). More importantly, the fact that the Ingot God is set on a base cast in one piece with the figurine (even if at a second stage) is most atypical for this group.

The outline of the element serving as a basis for the figure (Figs. 7.7, 7.8), with its curved sides and prolonged apices, allows no doubt as to the identification as an ingot (see, however, the reservations of Balmuth [1994] and the tenuous suggestion by E. Masson [1992; cf. Papasavvas 2009]). A miniature ingot is, nevertheless, an odd selection for a base, as it does not provide the figurine with more stability than any other plain orthogonal support would. In fact, the choice of the smith to give this statuette a base at all is surprising. Statuettes of this period usually were equipped with pegs extending down from their feet that were inserted into wooden or stone supports (e.g., see Negbi 1976, pls. 28–29; Seeden 1980, pls. 95–99, 117). Even the Horned God, larger than the Ingot God by 20 cm and much heavier, did not have such a base (Dikaios 1969–1971, I, 295, IIIa, pls. 139–143). The Ingot God also had a large peg projecting from underneath the ingot (Fig. 7.7; Schaeffer 1971, pls. 3, 6), and, as if this was not enough to support his weight, a second dowel projecting from his back (Fig. 7.2), now broken (Buchholz 1979, 85), possibly was meant to secure him on a vertical surface. These two elements clearly show that the ingot was not added merely to act as a base, but because it was significant. This supposition is supported further by the fact that the artist had positioned the vertical axis of the figurine and the upper surface of the ingot at a wide angle, which gave the figure a backward tilt and at the same time raised the rear part of the ingot (Figs. 7.1, 7.3), making its outlines very clear for the viewer.

Another feature that clearly differentiates the Ingot God from the armed standing figures of the

Figure 7.4. Detail of the lower part of the Ingot God, seen from the front.

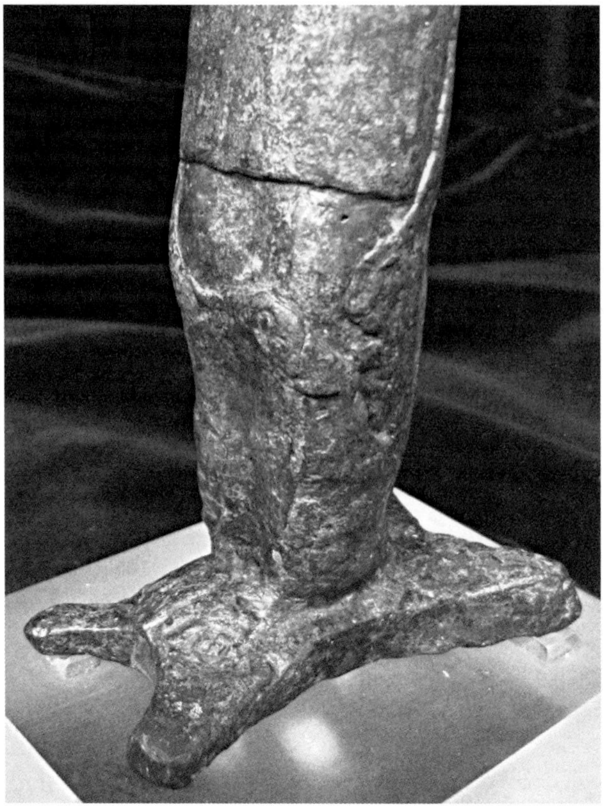

Figure 7.5. Detail of the lower part of the Ingot God, seen from the side.

Figure 7.6. Detail of the Ingot God showing the ingot base, seen from the side.

Figure 7.7. Detail of the Ingot God showing the ingot base, seen from below.

Figure 7.8. Detail of the Ingot God showing the ingot base, seen from above.

Near East is the fact that his legs and feet are kept close together. The vigorous movement forward that characterizes the "Smiting Gods" and denotes the power necessary for using the spear, is effected through a broad step seen in the majority of these figurines. The Ingot God, on the contrary, keeps his legs and feet side-by-side in a tight position, which results in a rather static posture. The figure's right leg, however, at the height of the knees, is advanced slightly, yet distinctly enough (Fig. 7.5). This feature, together with the fact that the added metallic layer conceals the lower part of the body from this point downward (Figs. 7.3), sustains the proposition that the Ingot God was originally more comparable with the "Smiting Gods" and that his legs also were shown in a striding posture. The legs perhaps were starting to extend away from each other at a lower point, probably below the knees. Although the remaining length of the legs does not seem to be enough to form a broad step, it appears as if the calves of the figure are proportionally short in relation to his slender, elongated body. It is possible, then, that the lowest part of the figurine's legs was cut off as a result of the damage suffered. A X-ray perhaps would reveal the situation under the added metal, and such X-rays apparently were carried out in the Römisch-Germanisches Zentralmuseum in Mainz in the 1960s as part of restoration operations (Buchholz 1979, 84 n. 40; Lagarce and Lagarce 1986, 74). In order to resolve these issues, the Ingot God was among the objects analyzed recently with a portable X-ray fluorescence (XRF) analyzer. The analytical data will be processed and presented elsewhere in the near future (for the project, see Giumlia-Mair, Kassianidou, and Papasavvas, this volume).

On the other hand, the ingot itself appears to have acquired its present form after two casting operations. Its outline and thickness, as well as its width, differ distinctly between its front and rear parts, as the former is more sharply defined than the latter. At the same time, the rear is broader and thicker and has an uneven surface, while the front is narrower, more flat, and smooth. Furthermore, two successive horizontal layers of metal are discernible from the side on the rear of the ingot, the upper being thinner and less sharply defined than the lower (Figs. 7.5–7.8). These features suggest that not only was the ingot added to the figurine with the casting-on technique, but also that for this addition an already cast, miniature ingot was possibly used. An older figurine was thus transformed into something new, and a warrior deity was assigned new responsibilities and became the Ingot God (cf. Buchholz 1979, 85).

The iconographic and typological details of the Ingot God, his magnitude, and the grandeur of his posture indicate that we, indeed, face a divinity. The context of the statuette also is suggestive of a deity (Courtois 1971; Knapp 1986, 18–20; Webb 1999, 102–113, 119–122; 2001), while its recovery within a small room of a monumental building that was littered with ritual paraphernalia even make the Ingot God qualify as a cult statue (Webb 1999, 223).

The identity of the Ingot God has been discussed by many scholars who have reached different conclusions. A review recently has been published by the author and does not need to be repeated here (Papasavvas 2009). The claim of the Ingot God, as well as the Bomford figurine (Catling 1971), over divine status is sustained mainly through the exceptional presence of the ingot under their feet, which, consequently, must be informative for their nature. These statuettes follow traditional types of statuary of the indigenous coroplastic art in the former case (Catling 1971, 24–29), and of the Near Eastern bronze statuary in the latter (Webb 1999, 225), while the addition of the ingot comes as a novelty.

Two interdependent questions need to be asked at this point. In what respect were copper production, as expressed by the ingot, and the martial character of the Ingot God linked to each other? And what was the factor that allowed the fusion of the pronounced female nature of the Bomford figurine with the metallurgical qualities imparted in the ingot under her feet (Catling 1971, 19–23)? In the latter case, the ingot probably was intended to attach a metallurgical notion to the aspect of productivity thought to be expressed by nude female figurines (Catling 1971, 30–32). On the other hand, the ingot under the feet of the male warrior probably communicates a concept of command and subordination, enhanced by the power of weapons. At the same time, the unequivocal, visual correlation of weapons and metallurgy and the inauguration of a warrior god into copper affairs imply that the production and distribution of the metal occasionally had to be maintained by warlike attitudes against internal or external agents, such as those mentioned by the king of Alasia in one of the Amarna Letters (Moran 1992, 111–112 [EA 38]).The combination of metallurgy, warfare, and female nudity under the common factor of the ingot seems to demonstrate the way that Cypriots perceived their world, with a focus on copper production and a concern for its security and prosperity (cf. Seeden 1980, 148–155).

If two different deities suddenly appeared with such an imposing feature that is alien in their proper spheres of interaction, this suggests that the ingot had acquired the power of a symbol—one that probably declared the subordination of copper production under divine authorities who, however, did not dispose of their other established attributes and roles. When the artists (and the commissioners) of these two statuettes perceived them as standing on an ingot, they were not following an iconographic tradition: rather, they were creating a new image out of an idea that was eminent among Cypriots. Moreover, since there was no iconographic type for a deity associated with metallurgy on Cyprus, they created one from a different iconographic type, the warrior god, and from a symbol, the miniature ingot, thereby blending together two religious entities (on the possible religious function of the miniature ingots as votives, see Buchholz 1959, 19–20). If the choice to use this latter symbol as a support seems to demean it, note that there were not many options for such a combination: the figure could not be made to hold the ingot, as not only were both his hands occupied with the weapons, but also because he would then have been misunderstood as a common ingot-bearer like those shown on the four-sided bronze stands (Papasavvas 2009, figs. 2, 5, 6, 11).

It is not an unexpected development that, in a world so closely dependent on metal production and distribution, some established gods could be provided with newly acquired metallurgical facets or charged with the protection of metal production. Such a dual nature is best exemplified in the case of Hathor, named the protector of the copper mines at Timna. This, however, does not make her any exclusive or universal patron deity of copper (for a discussion and references, see Kassianidou 2005, 132–133). The inception of the ingot in Cypriot statuary, thus, should not be taken as symptomatic of new gods (cf. Kassianidou 2005, 130–132). Conversely, the deities represented by the Ingot God and the Bomford figurine did not suffer from a split personality. They still could be represented without the ingot, and the emphasis then could be placed upon their other abilities. Thus, not every bronze smiting figure found on Cyprus had to step on an ingot: a small bronze statuette of this type with no base at all was found in the vicinity of the sanctuary of the Ingot God (Schaeffer 1971, 510–513). Similarly, none of the terracottas that are typologically associated with the Bomford figurine present such a feature.

The circumstances that led to the deposition of the Ingot God prior to the abandonment of the sanctuary are very intriguing (see the exemplary discussion of the depositional sequence in Webb 2001). The statuette was found resting on Sol III in an upright position, within a compact mudbrick fill, which was sealed by Sol II. This hints at a deliberate removal of the statuette from its cultic environment. The room that contained the statuette continued to be used for some time but was abandoned shortly thereafter. Although the statuette was not destroyed but only withdrawn and, in fact, had not abandoned its primary location, this state of affairs is reminiscent of similar, more or less synchronous behavioral patterns in other areas of Enkomi. The Horned God himself also seems to have been cached in the room that once housed him, marking the termination of the use of his sanctuary (Webb 1999, 66–69). In both cases the circumstances of deposition show no disrespect but rather a decision to formally terminate the function of these sanctuaries.

The exact date of the abandonment of Enkomi and of the sanctuary of the Ingot God depends on the chronology of the latest deposits of the site and on the classification of their ceramic material. The main phase of use of the sanctuary and its abandonment both fall within Sols III–I that provided a wide range of Proto White Painted ware shapes and are accordingly assigned to Late Cypriot IIIB. Iacovou (1988, 8–11) argues that no comparable ceramics were found elsewhere in the town, or at least none are present among the published finds, and concludes that the sanctuary of the Ingot God continued in use at a time when the rest of the settlement had been deserted. Kling (1989, 174), on the other hand, recognizes parallels of the wares found in Sols III–I of the sanctuary in other areas of the site and argues that the use of the sanctuary overlapped the final occupation of the rest of the town. Webb (2001) builds on the idea that the settlement seems to have been abandoned gradually and presumes that some parts of the town might have been inhabited as long as the sanctuary of the Ingot God was used, while other areas were already left by their inhabitants.

As to the date of the statuette itself, Muhly (1980, 156–161) has questioned the absolute reliance on ceramics to date such an exceptional metallic find and argued for a dating at least a century earlier than its ceramic context indicates, while others have suggested an even earlier date (Knapp 1986, 87; Carless Hulin 1989, 133–134). Sol III, the main phase of use for the sanctuary, only gives us the date of the use of the Ingot God as such, that is of the statuette that was transformed with the addition of the ingot. Since, however, the statuette does not seem to have been repaired due to a miscast immediately after the casting operation, its manufacture and first use must predate Late Cypriot IIIB. According to the excavator (Schaeffer 1971, 525–526), the sanctuary of the Ingot was erected upon an earlier structure, possibly also serving cult purposes (Courtois 1971, 198–211; Webb 1999, 102, 119–123; 2001, 69–72), which was ruined during the 12th century B.C. Its remains were found in two earlier floors (Sols VI–V), which correspond to thin layers of gray clay covered with ash and charcoal. The statuette very well might have originated in this earlier building, which, however, was only partially excavated and very briefly reported. This suggestion is supported further by the observation that some vases of earlier periods, possibly originating in the earlier building, succeeded by the sanctuary of the Ingot God were found together with the Ingot God (Iacovou 1988, 9). In any case, Sols VI–V are assigned by the excavator to Late Cypriot IIC (Courtois 1971, 198–211), although Kling (1989, 37–8) has associated Sol V with Late Cypriot IIIA. Late Cypriot IIC thus can be taken as the latest possible date for the inception of the Ingot God at Enkomi, although at that stage he was probably not an Ingot God at all.

The decision of the community of Enkomi to modify an existing, most probably damaged statuette of a warrior god into a new icon that emphasized a fabricated metallurgical aspect of a deity seems to indicate that this was done at a time of crisis, if not for the entire island than at least for the town of Enkomi, whose abandonment coincides with extensive religious performances practiced in the very same sanctuary of the Ingot God (Webb 1999, 102–113). Most importantly, the poor quality of the work done for this transformation and the unskillful application of the *casting-on* technique suggest that no skilled craftsmen were around at that particular time in order to work competently on the statuette. Given the importance of this sanctuary that lies in the heart of the site and is equipped with a very large cultic assemblage, such a scenario would have been inconceivable in a period other than the one that preceded the abandonment of the town. Furthermore, although the sanctuary of the Ingot God was located in the vicinity of metallurgical workshops, these, according to Muhly (1985, 34 n. 95), were not in use after Late Cypriot IIIA.

On the basis of some topographical observations made for Enkomi, Pickles and Peltenburg (1998; see the discussion in Webb 2001, 79–80) have described a process from centralized to decentralized authority at that site starting from the 14th century B.C. This process is made obvious in the wide distribution of copper-working residues across the town, in contrast to earlier periods when such installations were located in a single place. This situation seems to reflect a decline of centralized authority and the establishment of competing social groups who would have set up independent metalworking facilities in different sectors of the town. This process continued during Late Cypriot III when an even greater degree of local functionalism can be observed. The proximity of the sanctuaries of the Ingot God and of the Horned God to each other, and even their diverse character, could also be indicative of the existence of different social groups operating at the same site, who were appealing to different deities (Webb 2001, 79–80). Under this light, the transformation of the Ingot God with the addition of the ingot was perhaps an effort to fabricate a new godly dimension associated with copper—one that would provide the instigators with more authority and power in their social environment (cf. Knapp 1986). The importance of this invention was lost when the community that introduced the Ingot God eventually had to abandon the town. Even then, the Ingot God was cached in a respectful way, probably because, even with the ingot, he still was something more, the original god before he was given the ingot.

This tranformation is no less dramatic than the deliberate deposition of the statuette, which possibly indicates the awareness that new times had come, leading to the final abandonment of Enkomi and to the establishment of a new coastal site at nearby Salamis. This new site, perhaps unsurprisingly, was inaugurated with a new sanctuary, possibly meant to replace the sanctuary of the Ingot God (Iacovou 1989, 55).

References

Balmuth, M. 1994. "Reconsideration of the Bronze 'Ingot God' from Enkomi," *AJA* 98, p. 289.

Buchholz, H.G. 1959. "Keftiubarren und Erzhandel in zweiten vorchristlichen Jahrtausend," *PZ* 37 (1), pp. 1–40.

———. 1979. "Beobachtungen zum prähistorischen Bronzeguss in Zypern und der Ägäis," in *Acts of the International Symposium "The Relations between Cyprus and Crete, ca. 2000–500 B.C.," Nicosia, 16th April–22nd April 1978*, Nicosia, pp. 76–86.

Carless Hulin, L. 1989. "The Identification of Cypriot Cult Figures through Cross-cultural Comparison: Some Problems," in *Early Society in Cyprus*, E. Peltenburg, ed., Edinburg, pp. 127–139.

Catling, H.W. 1971. "A Cypriot Bronze Statuette in the Bomford Collection," in Schaeffer, ed., 1971, pp. 15–32.

Courtois, J.-C. 1971. "Le sanctuaire du Dieu au Lingot d'Enkomi-Alasia," in Schaeffer, ed., 1971, pp. 151–362.

Dikaios, P. 1969–1971. *Enkomi: Excavations 1948–1958* I–III, Mainz am Rhein.

Drescher, H. 1958. *Der Überfangguss. Ein Beitrag zur vorgeschichtlichen Metalltechnik*, Mainz am Rhein.

Iacovou, M. 1988. *The Pictorial Pottery of Eleventh Century B.C. Cyprus* (*SIMA* 79), Göteborg.

———. 1989. "Society and Settlement in Late Cypriot III," in *Early Society in Cyprus*, E. Peltenburg, ed., Edinburg, pp. 52–59.

Kassianidou, V. 2005. "Was Copper Production under Divine Protection in Late Bronze Age Cyprus? Some Thoughts on an Old Question," in *Cyprus: Religion and Society from the Late Bronze Age to the End of the Archaic Period. Proceedings of an International Symposium on Cypriote Archaeology, Erlangen, 23–24 July 2004*, V. Karageorghis, H. Matthäus, and S. Rogge, eds., Möhnesee-Wamel, pp. 127–141.

Kling, B. 1989. *Mycenaean IIIC:1b and Related Pottery in Cyprus* (*SIMA* 87), Göteborg.

Knapp, A.B. 1986. *Copper Production and Divine Protection: Archaeology, Ideology and Social Complexity on Bronze Age Cyprus* (*SIMA-PB* 42), Göteborg.

Lagarce, J., and E. Lagarce. 1986. "La métallurgie," in *Enkomi et le bronze récent à Chypre*, J.-C. Courtois, J. Lagarce, and E. Lagarce, eds., Nicosia, pp. 60–99.

Masson, E. 1992. "Le dieu guerrier d'Enkomi: Est-il debout sur un lingot?" in *Studies in Honor of Vassos Karageorghis*, G.K. Ioannidis, ed., Nicosia, pp. 155–156.

Moran, W. 1992. *The Amarna Letters*, Baltimore.

Muhly, J.D. 1980. "Bronze Figurines and Near Eastern Metalwork," *IEJ* 30, pp. 148–161.

———. 1985. "The Late Bronze Age in Cyprus: A 25 Years Retrospect," in *Archaeology in Cyprus 1960–1985*, V. Karageorghis, ed., Nicosia, pp. 20–46.

———. 1988. "Concluding Remarks," in *Bronzeworking Centres of Western Asia, c. 1000–539 B.C. Colloquium held in the British Museum, July 1986*, J. Curtis, ed., London, pp. 329–342.

Negbi, O. 1976. *Canaanite Gods in Metal: An Archaeological Study of Ancient Syro-Palestinian Figurines* (*Publications of the Institute of Archaeology, Tel Aviv University* 5), Tel Aviv.

Papasavvas, G. 2009. "The Iconography of the Oxhide Ingots," in *Oxhide Ingots in the Central Mediterranean*, F. Lo Schiavo, J.D. Muhly, R. Maddin, and A. Giumlia-Mair, eds., Rome, pp. 83–132.

Pickles, S., and E. Peltenburg. 1998. "Metallurgy, Society and the Bronze/Iron Transition in the East Mediterranean and the Near East," *RDAC* 1998, pp. 67–100.

Schaeffer, C.F.-A. 1965. "An Ingot God from Cyprus," *Antiquity* 39, pp. 56–57.

———. 1971. "Les Peuples de la Mer et leur sanctuaires à Enkomi-Alasia aux 12e–11e s. av.n.è," in Schaeffer, ed., 1971, pp. 505–566.

Schaeffer, C.F.-A., ed. 1971. *Alasia: Première série* (*Mission Archéologique d'Alasia* 4), Paris.

Seeden, H. 1980. *The Standing Armed Figurines in the Levant* (*Prähistorische Bronzefunde* 1.1), Munich.

Webb, J.M. 1999. *Ritual Architecture, Iconography and Practice in the Late Cypriot Bronze Age* (*SIMA-PB* 75), Jonsered.

———. 2001. "The Sanctuary of the Ingot God at Enkomi: A New Reading of its Construction, Use and Abandonment," in *Contributions to the Archaeology and History of the Bronze and Iron Ages in the Eastern Mediterranean. Studies in Honor of Paul Åström*, P.M. Fischer, ed., Vienna, pp. 69–82.

PART II

Metallurgy of Crete

CHAPTER

8

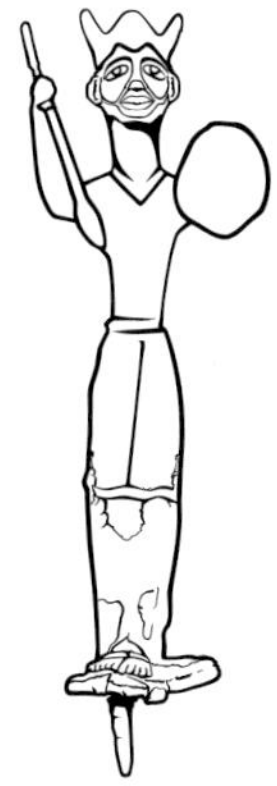

Reconstructing Early Cretan Metallurgy: Analytical Evidence from Kephala Petras, Siteia

Mihalis Catapotis, Yannis Bassiakos, and Yiannis Papadatos

Introduction

Recent excavations at Kephala Petras in eastern Crete (Papadatos 2008; Papadatos et al., forthcoming) brought to light new evidence for copper-smelting activities in Crete, dating at least to the earliest part of the Early Minoan (EM) I period or possibly to the end of the Final Neolithic (FN) period (Papadatos 2007a). The finds, typical of early copper-smelting processes, include two pieces of copper ore, six slag fragments, and four pieces of iron ore. To these probably are to be added a small number of deformed clay fragments clearly subjected to high temperatures, which might represent refractory material used in the metallurgical process.

The importance of these finds, despite their small quantity, lies in the fact that they are among the earliest direct evidence for copper-smelting activities on Crete. Evidence of similar date is extremely rare both within Crete (Betancourt 2006, 2007) and in the rest of the Aegean (Nakou 1995; Muhly 2002, 2006; Papadatos 2007a). For this reason, an analytical project was undertaken by the Laboratory of Archaeometry of the National Center for Scientific Research (NCSR) "Demokritos" for the technological study of the metallurgical activities evidenced at Kephala Petras. The project combined microscopic analysis, chemical analysis, and phase microanalysis of ore and slag samples in an effort to elucidate the major technological aspects of the smelting process. This paper presents the analytical results from the study of the metallurgical finds and discusses their implications for our understanding of the early metallurgical technology in this area.

Raw Materials

Direct evidence for the nature of the raw materials used in the smelting process at Kephala Petras is provided by the two small pieces of copper ore found at the site. Microscopic examination of the ore samples indicated that they consist of secondary minerals of copper and iron oxides in a siliceous matrix, containing only a small amount of residual pyrite and chalcopyrite (Fig. 8.1), which is also reflected in the low levels of sulfur in their bulk composition (Table 8.1). That the smelted ores were of the oxidized type is further corroborated by the low levels of sulfur and the limited presence of matte inclusions in all slag samples.

The smelting of oxidized ores for the production of copper was standard throughout the southern Aegean during the EBA, as shown by several studies of smelting sites on Seriphos, Keros, Kea, Kythnos, and Crete (Gale et al. 1985; Bassiakos and Doumas 1998; Papastamataki 1998; Bassiakos and Catapotis 2006; Bassiakos and Philaniotou 2007; Catapotis and Bassiakos 2007; Georgakopoulou 2007). The principle of the process was the conversion of copper oxides to metallic copper by creating a reducing atmosphere in the furnace (Newton and Wilson 1942, 151). Copper is a fairly noble metal, and its oxides could be relatively easily reduced. The major challenge was rather the removal of the "earthy" component of the ore (i.e., the gangue minerals) by creating a fluid slag which could allow the separation of the reduced metal. This involved firstly the manipulation of the composition of the charge and secondly the attainment of sufficiently high temperatures during the process.

Unfortunately, the evidence is too limited to provide a quantifiable picture of the gangue minerals of the ores smelted at the site. Bulk analysis of the two copper ore samples proved one of them to be iron rich and the other rather siliceous (Table 8.1). This simply could reflect minor variations among different pieces collected from the same ore deposit, given that such deposits are always heterogeneous bodies. It is equally possible, however, that copper ores brought to Kephala Petras for smelting (if indeed more than one smelting event took place there) did not always have the same mineralogical composition. Whether this implies the use of ores from different sources is open to further analysis. It remains, however, an intriguing possibility, given that lead isotope analysis of slags from the other copper-smelting site on Crete, namely Chrysokamino, revealed that ores from various copper deposits were probably smelted at the site (Stos-Gale and Gale 2006, 313–316).

Whatever the case, the chemical composition of the slag from the site reveals that some care was probably taken to ensure that the composition of the charge would enable the formation of a liquid slag during the smelt. Indeed, as seen in Figure 8.2, the chemical composition of all five slags falls into an area of the FeO_x-SiO_2-CaO-Al_2O_3 phase diagram characterized by relatively low liquidus temperatures in the range of 1250°C. Moreover, it is notable that the samples form a fairly tight compositional cluster, which contrasts with the variability reflected in the two copper ores. Although we should not dismiss the possibility that all slag samples come from a single smelt, this consistency might suggest that smelters at Kephala Petras exerted some control over the composition of the smelting charge.

This control could have been achieved through a range of possible beneficiation and fluxing techniques. As far as beneficiation is concerned, the small size of the copper ore fragments (1.6–2.8 cm) and the large concentrations of copper oxides clearly visible in their surface (see also Table 8.1) may suggest that some sort of beneficiation process was taking place prior to smelting, involving the crushing of the ore and the selection of the richest pieces for the smelt. As for the possibility of fluxing, the pieces of iron ore found at the site may be illuminating. Of the four iron ores discovered during the excavations, only one has been analyzed so far. Arguably, its chemical composition (Table 8.1) does not seem ideal for the fluxing of either siliceous or ferrous copper-ores. Nonetheless, the fact that it contains no traces of copper, coupled with its distinctively high levels of alumina, suggests that this piece of ore probably came from a different deposit from that of the copper ores found at the site. Moreover, that the iron ore was found in the same context with a piece of siliceous copper ore suggests that it might have been somehow related to the metallurgical activities at the site. Put together, this evidence may raise the possibility that this piece of iron ore is an example

Find number	Type	Percent (%)									
		FeO	CuO	SiO_2	Al_2O_3	MgO	CaO	K_2O	TiO_2	BaO	SO_3
KP 03/473	copper ore	48	39	6.5	0.4	0.5	0.4	0.1	—	—	5.9
KP 03/1156	copper ore	16	48	34	—	0.3	0.4	0.1	—	—	1.7
KP 03/855	iron ore	36	—	35	27	—	0.4	0.3	1.4	—	—
KP 03/1119	slag	48	2.2	32	5.4	1.3	7.5	1.2	—	3.2	0.3
KP 03/230	slag	50	2.9	35	4.2	1.2	6.4	0.7	—	—	0.3
KP 03/244	slag	50	1.3	32	4.6	1.3	9.0	1.1	0.2	—	—
KP 03/759	slag	27	3.8	45	9.4	1.8	9.0	1.7	—	1.8	—
KP 03/194	slag	43	1.6	37	5.7	1.8	9.0	1.2	0.5	—	—

Table 8.1. Chemical composition from area scans of ore and slag samples from Kephala Petras using scanning electron microscopy–energy dispersive spectroscopy (SEM-EDS) (multiple scans; normalized data).

Figure 8.1. Copper-ore sample KP 03/1156. Note the residual sulphide minerals seen as white in the micro-photograph (optical microscope; cross-polarized light [XPL]).

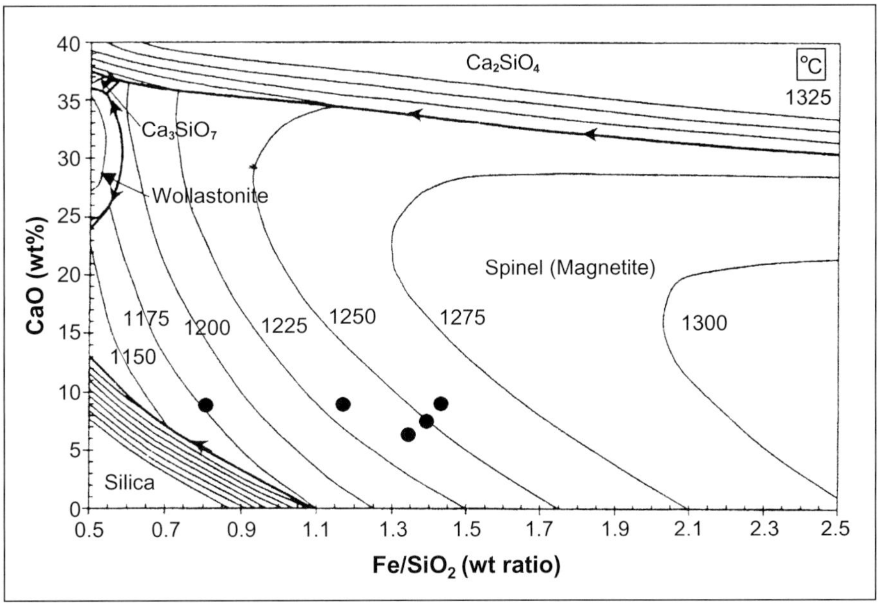

Figure 8.2. Reduced chemical composition of slag samples from Kephala Petras plotted on the Fe/SiO_2–CaO (+7% Al_2O_3) phase diagram (diagram after Kongoli and Yazawa 2001, fig. 11).

(though not representative) of minerals that were added to the charge, when necessary, to adjust its composition in order to ensure the formation of a liquid slag. In other words, although further research is necessary, it seems possible that fluxes were used in the smelting process at Kephala Petras.

Smelting Conditions

To seek evidence for the temperature and redox conditions that prevailed during the smelting process at Kephala Petras, the mineralogy and texture of the slag were examined. Macroscopic and microscopic examination of five samples indicated that although the slag is not fully liquified it contains only a few pieces of unreacted raw materials, such as copper ore and, more commonly, silica (Fig. 8.3). This suggests that the maximum temperatures attained during the smelt were close to, though not necessarily above, the liquidus temperature of the slag, which has been estimated to be about 1250°C.

That the maximum smelting temperature was close to the liquidus temperature of the slag is also demonstrated by the notable levels of copper in the slag. More specifically, the Cu weight-% in all five samples is above 1%, reaching a maximum of 3% in the more siliceous slag (Table 8.1). This suggests that the smelting temperature was not high enough to enable the production of a fluid slag. The slag remained instead fairly viscous, leading to the entrainment of metallic prills and thus increasing copper losses.

Evidence for the redox conditions during the smelt is offered by the oxidation state of copper and iron as well as the composition of metal-rich prills in the slag. The virtual absence of primary (i.e., not post-depositional) cuprite and the copper content of the slag (1%–3%) suggests that the partial pressure of oxygen (pO_2) did not exceed the 10^{-5}atm. On the other hand, the low levels of iron in the copper prills (0.4%–3.0%) and matte inclusions (1.1%–4.0%) (Table 8.2), coupled with the predominance of magnetite (against wustite) in all slag samples, argue against pO_2 values below the 10^{-11}atm. Overall, the mineralogy of the slag samples from Kephala Petras points to moderate redox conditions, with the pO_2 ranging from 10^{-6} to 10^{-11}atm.

What is, perhaps, more important is that the mineralogy of the slag samples reflects significant variability in the redox conditions prevailing during the smelt. Although the small size of the sample does not allow firm conclusions, the co-existence of delafossite, magnetite, and wustite in a single slag sample (Fig. 8.4) strongly suggests that the smelting conditions were far from stable. Similar conclusions can be drawn from the highly variable levels of iron in copper prills (Table 8.2). Considering also the small number and size of the slag pieces found at the site and the absence of any evidence for a major metallurgical installation, this variability could be taken to imply that the smelting process was taking place inside a small ceramic container, possibly a bowl-shaped crucible, or inside a small hearth, lined with refractory clay.

We could consider, at this point, the deformed clay fragments found at the site, whose state of vitrification suggests that they have been subjected to high temperatures (Papadatos 2007a, figs. 10.7, 10.8). Although none of the sherds bears traces of copper or slag, given the absence of any fire-destruction layers at the site, it is very likely that these burned ceramics may relate to a pyrotechnological process. If this is indeed the case, it is possible that they are the remains of small ceramic vessels or hearths used as crucibles for copper smelting. This is reinforced further by petrographic analysis, which shows that they are made of a semicoarse fabric containing much organic temper and rare inclusions of quartz and quartzite (E. Nodarou, pers. comm.)—a fabric not used for any other classes of pottery at the site. Moreover, fabrics of this type with organic temper were very common for the manufacture of refractory material used for metallurgical purposes in Early Minoan Crete (Doonan, Day, and Dimopoulou-Rethemiotaki 2007, 104–105).

Find number	Phase	Percent (%)		
		Cu	Fe	S
KP 03/1119	matte inclusion	77	4.0	19
KP 03/1119	matte inclusion	81	1.1	18
KP 03/1119	copper prill	99	0.5	0.3
KP 03/1119	copper prill	98	1.9	0.2
KP 03/230	copper prill	97	3.0	0.4
KP 03/244	copper prill	98	1.6	0.2
KP 03/759	copper prill	99	0.5	0.6
KP 03/194	copper prill	99	0.4	0.2

Table 8.2. Chemical composition of metallic inclusions in slag samples from Kephala Petras determined by SEM-EDS (normalized data).

Figure 8.3. Slag sample KP 03/230 containing piece of unreacted copper-ore (dark inclusion at the center) surrounded by magnetite skeletons (optical microscope with plain-polarized light [PPL]).

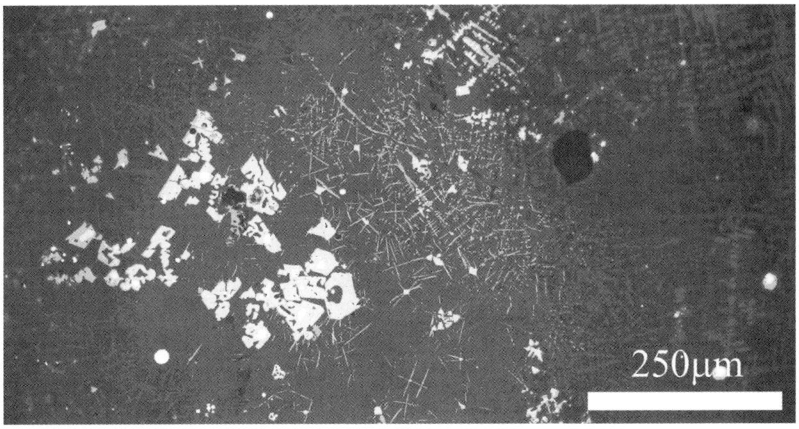

Figure 8.4. Slag sample KP 03/1119 that features the co-presence of delafossite laths (center), magnetite skeletons (left), and wustite dendrites (right) (optical microscope with PPL).

Product

Analysis of the metallic phases embedded in the slag samples suggests that the product of the smelting activities evidenced at Kephala Petras was metallic copper containing low levels of iron and possibly some copper-rich matte inclusions. Arsenic and other impurities were below the detection limit of the analytical instrument, as was the case with the two copper ore samples from the site. Such a product is compatible with the technology reconstructed at the site, i.e., with the use of oxidized copper ores containing only small amounts of residual sulfides and the attainment of moderate redox conditions during the smelt. A final refining stage would have been necessary in order to remove iron and other impurities before using the metal for the manufacture of artifacts.

This new evidence strengthens the view that the predominance of arsenical copper in the metalwork of the Early Bronze Age (EBA) southern Aegean should not be attributed to a widespread practice of smelting arsenic-rich raw materials. Indeed, as previous studies have shown, most copper-smelting sites in that area produced arsenic-free copper (Bassiakos and Doumas 1998; Papastamataki 1998; Bassiakos and Philaniotou 2007). Arsenical-copper smelting is only evidenced at the sites of Chrysokamino on Crete (Catapotis and Bassiakos 2007); Daskaleio-Kavos on Keros, where arsenic-free copper and arsenical copper were produced by two distinct processes (Georgakopoulou 2007); and Skouries on Kythnos, where a small number of slag samples were found to contain arsenic-rich copper prills (Gale et al. 1985; contra Bassiakos and Philaniotou 2007, 51–52).

Kephala Petras in the Context of Early Cretan Metallurgy

Kephala Petras and Chrysokamino

Despite the limited available evidence, it would be worth putting the results of our analysis within the wider context of early Cretan metallurgy. Interesting conclusions can be drawn from a technological comparison of the evidence from Kephala Petras with its sole counterpart on Crete, the copper-smelting site of Chrysokamino. As demonstrated in Table 8.3, the copper-smelting technologies employed at Kephala Petras and Chrysokamino bear similarities but also significant differences.

At both sites, oxidized copper ores were used, and the composition of the charge seems to have been controlled, to a lesser or greater extent, through some combination of ore selection, beneficiation, and/or fluxing. What differentiated Chrysokamino, however, was the use of arsenic-rich raw materials, which enabled the repeated production of arsenical copper (Catapotis and Bassiakos 2007).

The smelting process at Chrysokamino was conducted in furnaces consisting of a perforated conical shaft (a technology also evidenced in the western Cyclades and Attica) laid on top of a clay-lined pit, which probably enabled the smelting of tens of kilograms of ore in every single smelt. At Kephala Petras there is no direct evidence for the type of vessel used in the smelting process. However, the size, texture and mineralogy of the analyzed slag samples seem to suggest the use of a small clay-lined hearth or a crucible, which would have enabled the smelting of some hundreds of grams of ore in each smelting episode. The difference in the capacity of the smelting installations and the quantity of the metallurgical debris from the two sites may reflect the difference between regular processing of copper ores at an activity-specific location (Chrysokamino) on the one hand, and occasional small-scale smelts conducted very close to or within the borders of a settlement (Kephala Petras) on the other.

At Chrysokamino, oxygen was supplied into the system in two ways. First, wind penetrated through the numerous perforations on the shaft, enabling the early combustion of the fuel and the efficient preheating of the charge (Catapotis, Pryce, and Bassiakos 2008). At the lower levels of the furnace air was supplied by pot-bellows, evidence of which has been reported from the site (Betancourt 2006,

Smelting Processes	Chrysokamino	Kephala Petras
Copper ores	Oxidized	Oxidized
Beneficiation	Yes?	Yes?
Fluxing	Iron ore?	Iron-ore?
Arsenical ingredient	Yes	—
Smelting vessel	Perforated shaft furnace	Crucible?
Air supply	Wind + pot-bellows	?
Furnace temperature	≈1300°C	<1250°C
Redox conditions	Moderate	Variable
Slag processing	Crushing, pulverizing(?)	?
Copper losses	1%	1%–3%

Table 8.3. Comparison of technical aspects of the smelting processes at Kephala Petras and Chrysokamino (Catapotis and Bassiakos 2007).

125–132). The temperature and redox atmosphere thus created inside the furnace were sufficient to ensure the reduction of most copper-oxide to its metallic state, the formation of a fluid slag, and a good separation of the two formed phases. In the case of Kephala Petras, the manner of air-supply in the process remains uncertain, but smelting conditions appear to have been variable, and the temperatures attained were not sufficient to lower the viscosity of the slag, resulting in a less successful slag-metal separation compared to Chrysokamino. Nonetheless, it should be noted that copper losses in the slag from Kephala Petras (1%–3% in analyzed samples) are still of the same order of magnitude as those of Chrysokamino and other smelting sites in the Aegean (Catapotis and Bassiakos 2007, fig. 4.7).

Assigning a chronological significance to the differences evidenced at the two sites, though tempting, is not without problems. Direct metallurgical evidence at Kephala Petras (ores and slag) comes from secondary deposits, containing FN and EM IA pottery, which were sealed by the floors of an EM IA building. Only the burned clay fragments come from undisturbed pure FN deposits, but their association with metallurgy, though possible, has not yet been firmly established. Thus, the current evidence cannot provide a precise dating for metallurgy at Kephala Petras; it can only set a terminus ante quem to the EM IA period (Papadatos 2007a).

Establishing the chronology of the metallurgical activities at Chrysokamino presents greater difficulties. The metallurgical evidence derives from an extensive slag heap, which constituted a disturbed, unstratified deposit, containing a mixture of FN, EM I–II, and mostly EM III–MM IA pottery (Betancourt 2006, 68–71). Only a small area of this extensive deposit was found sealed by the successive floors of an EM III–MM IA hut. Therefore, the EM III–MM IA floor is the secure terminus ante quem for the dating of the slag heap. Based on this evidence, it is possible that metallurgical activity started as early as the FN (Betancourt 2006, 179), but the main period of use seems to be EM III (Muhly 2006, 155). Further support comes also from the pot bellows, which can be securely dated solely to the EM III–MM IA period (Betancourt 2006, 126).

Whatever the case, the fact that two copper-smelting sites located on the northern shores of eastern Crete demonstrate such different smelting techniques is indicative of the technological variability that characterized early metallurgical practices in the southern Aegean during the 3rd millennium B.C. Similar variability has been also noticed in the Cyclades, sometimes within the same island, e.g., Kythnos (Bassiakos and Philaniotou 2007) and Seriphos (Georgakopoulou 2005) or even the same settlement, e.g., Kavos North on Keros (Georgakopoulou 2004, 2007).

Kephala Petras and the Kampos Group Metallurgical Sites

One of the most important features of the early part of the EBA in Crete is the appearance on the northern shore of sites with strong Cycladic affiliations, namely Poros-Katsambas, Hagia Photia, and Gournes (Davaras and Betancourt 2004; Galanaki 2006; Wilson, Day, and Dimopoulou-Rethemiotaki 2008). Known also as the Kampos Group in the EBA Cyclades, they date to the EM IB–IIA period. Recent approaches associate these sites with the emergence of nodal points controlling the inflow of exotic materials and techniques of Cycladic origin, including metals and metallurgy, to Crete (Day, Wilson, and Kiriatzi 1998; Broodbank 2000, 300–304; Papadatos 2007b; Wilson, Day, and Dimopoulou-Rethemiotaki 2008).

Two sites, Poros-Katsambas and Hagia Photia, have yielded metallurgical evidence. At Poros-Katsambas, a rich metalworking assemblage has been discovered, including crucibles, molds, tuyères, copper ingots, arsenic-bearing minerals, casting spills, melting slag, and dross (Doonan, Day, and Dimopoulou-Rethemiotaki 2007). At Hagia Photia the material comes from two different contexts. The cemetery has yielded copper-based artifacts and two complete crucibles, which, according to preliminary analyses, were used for the melting of copper (Betancourt and Muhly 2007, 150). At the adjacent settlement, nine chisel molds have been discovered in an EM IB–IIA deposit underlying the Middle Minoan strata (Tsipopoulou 2007).

What is important to note is that at both sites the evidence relates to metalworking, not to the production of metal from its ores. In other words, the evidence suggests that metal from the Cyclades and/or other sources, reached the "Kampos Group" sites in Crete in its ready-made form. The evidence from Kephala Petras shows, instead, that copper ore was brought to the settlement in order to be smelted for the extraction of the metal. Whether this should be taken to reflect a different mode of metal procurement (Papadatos 2007a, 163–165) in action on Crete before the Kampos Group phase remains to be addressed by future research. In any case, this contrast reinforces the picture of variability in the spatial arrangement of metallurgical activities evidenced in the archaeological record of the Early Bronze Age in the southern Aegean (Catapotis 2007).

Conclusions

Despite the limited amount of metallurgical evidence from Kephala Petras, it was possible to reconstruct a copper-smelting technology that is compatible with what is known so far about early smelting techniques in the southern Aegean. The process involved the smelting of oxidized ores, possibly mixed with a flux, inside a small clay container where fairly high temperatures and moderate redox conditions were attained. This enabled the production of copper with low levels of iron that were probably removed in a final refining stage, not attested so far at Kephala.

Although the evidence from the site seems to point to small-scale smelting activities, it proves beyond doubt that copper production was taking place on Crete before the "Kampos Group" period. Therefore, the paucity of metal objects and their absence from tombs of the FN–EM IA period, such as Pseira (Betancourt and Davaras, eds., 2003), Partira (Mortzos 1972), and Hagios Nikolaos–Palaikastro (Tod 1902–1903), should not be viewed as evidence for the non-practicing of metallurgy on Crete during that time. Rather, it is indicative of different depositional practices, which have been convincingly associated with the changing role of metal in the southern Aegean societies during the EB I–II period (Nakou 1995; Papadatos 2007a).

Acknowledgments

The authors would like to thank the excavator of Petras, Dr. Metaxia Tsipopoulou, for granting us permission to study and publish the material from Kephala. We would also like to thank the director of the 24th Ephorate of Prehistoric and Classical Antiquities, Vili Apostolakou, for permissions to sample and study the metallurgical evidence. Finally, we thank the anonymous reviewer for useful comments on an earlier version of this paper. The study related to this paper took place as part of the project "Study and Publication of the FN–EM I Settlement at Kephala Petras, Siteia," funded by the Institute for Aegean Prehistory (INSTAP).

References

Bassiakos, Y., and M. Catapotis. 2006. "Reconstruction of the Copper Smelting Process at Chrysokamino Based on the Analysis of Ore and Slag Samples," in Betancourt, 2006, pp. 329–353.

Bassiakos, Y., and C. Doumas. 1998. "The Island of Keros and Its Enigmatic Role in the Aegean EBA: A Geoarchaeological Approach," in *Αργυρίτις Γη*, A. Fragiskos, ed., Athens, pp. 55–64.

Bassiakos, Y., and O. Philaniotou. 2007. "Early Copper Production on Kythnos: Archaeological Evidence and Analytical Approaches to the Reconstruction of Metallurgical Process," in Day and Doonan, eds., 2007, pp. 19–56.

Betancourt, P.P. 2006. *The Chrysokamino Metallurgy Workshop and Its Territory* (*Hesperia Suppl.* 36), Princeton.

———. 2007. "The Final Neolithic to Early Minoan III Metallurgy Site at Chrysokamino, Crete," in Day and Doonan, eds., 2007, pp. 57–67.

Betancourt, P.P., and J.D. Muhly. 2007. "The Crucibles from the Aghia Photia Cemetery," in Day and Doonan, eds., 2007, pp. 146–153.

Betancourt, P.P., and C. Davaras, eds. 2003. *Pseira* VII: *The Pseira Cemetery* 2. *Excavation of the Tombs* (*Prehistory Monographs* 6), Philadelphia.

Broodbank, C. 2000. *An Island Archaeology of the Early Cyclades*, Cambridge.

Catapotis, M. 2007. "On the Spatial Organisation of Copper Smelting Activities in the Southern Aegean during the Early Bronze Age," in Day and Doonan, eds., 2007, pp. 207–223.

Catapotis, M., and Y. Bassiakos. 2007. "Copper Smelting at the Early Minoan Site of Chrysokamino on Crete," in Day and Doonan, eds., 2007, pp. 68–83.

Catapotis, M., O. Pryce, and Y. Bassiakos. 2008. "Preliminary Results from an Experimental Study of Perforated Copper-Smelting Shaft Furnaces from Chrysokamino (Eastern Crete)," in *Aegean Metallurgy in the Bronze Age: Proceedings of an International Symposium Held at the University of Crete, Rethymnon, Greece on November 19–21, 2004*, I. Tzachili, ed., Athens, pp. 113–122.

Davaras, C., and P.P. Betancourt. 2004. *The Hagia Photia Cemetery* I: *The Tomb Groups and Architecture* (*Prehistory Monographs* 14), Philadelphia.

Day, P.M., and R.C.P. Doonan, eds. 2007. *Metallurgy in the Early Bronze Age Aegean* (*Sheffield Studies in Aegean Archaeology* 7), Oxford.

Day, P.M., D.E. Wilson, and E. Kiriatzi. 1998. "Pots, Labels and People: Burying Ethnicity in the Cemetery of Aghia Photia, Siteias," in *Cemetery and Society in the Bronze Age* (*Sheffield Studies in Aegean Archaeology* 1), K. Branigan, ed., Sheffield, pp. 133–149.

Doonan, R.C.P., P.M. Day, and N. Dimopoulou-Rethemiotaki. 2007. "Lame Excuses for Emerging Complexity in Early Bronze Age Crete," in Day and Doonan, eds., 2007, pp. 98–122.

Galanaki, K. 2006. "Πρωτομινωικό ταφικό σύνολο στην πρώην Αμερικανική Βάση Γουρνών Πεδιάδος," in

Πεπραγμένα Θ΄ Διεθνούς Κρητολογικού Συνεδρίου A (2), Herakleion, pp. 227–242.

Gale, N.H., A. Papastamataki, Z.A. Stos-Gale, and K. Leonis. 1985. "Copper Sources and Copper Metallurgy in the Aegean Bronze Age," in *Furnaces and Smelting Technology in Antiquity* (*British Museum Occasional Paper* 48), P. Craddock and M. Hughes, eds., London, pp. 81–93.

Georgakopoulou, M. 2004. "Metallurgy at the Early Bronze Age Site of Daskaleio-Kavos on the island of Keros (Cyclades, Greece)," *Institute of Archaeometallurgical Studies Newsletter* 24, pp. 3–12.

———. 2005. *Technology and Organization of Early Cycladic Metallurgy: Copper on Seriphos and Keros, Greece*, Ph.D. diss., University College London.

———. 2007. "Metallurgical Activities within Early Cycladic Settlements: The Case of Daskaleio-Kavos," in Day and Doonan, eds., 2007, pp. 123–134.

Kongoli, F., and A. Yazawa. 2001. "Liquidus Surface of FeO-Fe_2O_3-SiO_2-CaO Slag Containing Al_2O_3, MgO, and Cu_2O at Intermediate Oxygen Partial Pressures," *Metallurgical and Materials Transactions* 32B, pp. 583–592.

Mortzos, C. 1972. "Πάρτιρα, μία πρώϊμος μινωϊκὴ ὁμάς," *Ἐπετηρίς Ἐπιστημονικῶν Ἐρευνῶν* 3, pp. 386–421.

Muhly, J.D. 2002. "Early Metallurgy in Greece and Cyprus," in *Anatolian Metal* II (*Der Anschnitt* 15), Ü. Yalçin, ed., Bochum, pp. 77–82.

———. 2006. "Chrysokamino in the History of Early Metallurgy," in Betancourt 2006, pp. 155–177.

Nakou, G. 1995. "The Cutting Edge: A New Look at Early Aegean Metallurgy," *JMA* 8 (2), pp. 1–32.

Newton, J., and C.L. Wilson. 1942. *Metallurgy of Copper*, New York.

Papadatos, Y. 2007a. "The Beginning of Metallurgy in Crete: New Evidence from the FN–EM I Settlement at Kephala Petras, Siteia," in Day and Doonan, eds., 2007, pp. 154–167.

———. 2007b. "Beyond Cultures and Ethnicity: A New Look at Material Culture Distribution and Inter-regional Interaction in the Early Bronze Age Southern Aegean," in *Mediterranean Crossroads*, S. Antoniadou and A. Pace, eds., Athens, pp. 419–451.

———. 2008. "The Neolithic–Early Bronze Age Transition in Crete: New Evidence from the Settlement at Petras Kephala, Siteia," in *Rethinking the Cretan Neolithic* (*Sheffield Studies in Aegean Archaeology* 8), V. Isaakidou and P.D. Tomkins, eds., Oxford, pp. 261–275.

Papadatos, Y., P. Tomkins, E. Nodarou, and I. Iliopoulos. Forthcoming. "The Beginning of Early Bronze Age in Crete: Continuities and Discontinuities in the Ceramic Assemblage at Kephala Petras, Siteia," in *The Aegean Early Bronze Age: New Evidence*, C. Doumas, A. Gannikouri, and O. Kouka, eds., Athens.

Papastamataki, A. 1998. "Μεταλλουργικές δραστηριότητες στην Κέα κατά την Αρχαιότητα, νεότερα στοιχεία," in *Κέα-Κύθνος: Ιστορία και Αρχαιολογία* (*Meletemata* 27), L.G. Mendoni and A. Mazarakis-Ainian, eds., Athens, pp. 759–764.

Stos-Gale, Z.A., and N.H. Gale. 2006. "Lead Isotope Analyses of Slags from Chrysokamino," in Betancourt 2006, pp. 299–319.

Tod, M.N. 1902–1903. "Excavations at Palaikastro, II. Hagios Nikolaos," *BSA* 9, pp. 336–343.

Tsipopoulou, M. 2007. "Aghia Photia–Kouphota: A Centre for Metallurgy in the Early Minoan Period," in Day and Doonan, eds., 2007, pp. 135–145.

Wilson, D.E., P.M. Day, and N. Dimopoulou-Rethemiotaki. 2008. "The Gateway Port of Poros-Katsambas: Trade and Exchange between North-Central Crete and the Cyclades," in *Horizon: A Colloquium on the Prehistory of the Cyclades*, N. Brodie, J. Doole, G. Gavalas, and C. Renfrew, eds., Cambridge, pp. 261–270.

CHAPTER

9

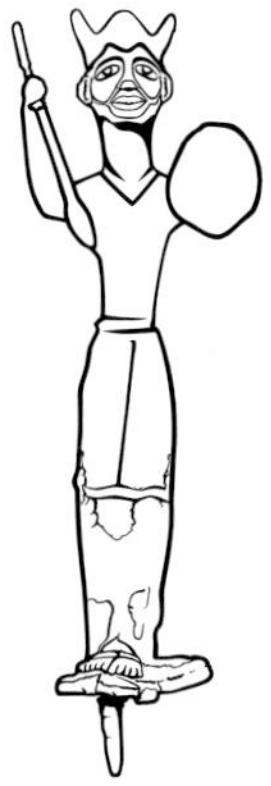

Silver and Bronze Artifacts from the EM I Necropolis at Gournes, Pediada

Calliope E. Galanaki, Yannis Bassiakos, and Vassilis Perdikatsis

Excavation Data

A rescue excavation carried out in 1999 by the 23rd Ephorate of Prehistoric and Classical Antiquities on the north coast of Central Crete (ca. 8 km east of Herakleion in the area of the former American military base) revealed an Early Minoan (EM) I necropolis and a building dated to the Middle Minoan (MM) period (Fig. 9.1; Galanaki 1999; 2006; Blackman 2001, 129). The well known EM I–II cave tomb excavated by Stephanos Xanthoudides in 1918 lies in the adjacent modern village of Kokkini Chani, at a distance of approximately 1.5 km (Xanthoudides 1918). Architectural remains of the Early Minoan period were uncovered in 1979 on the east side of this tomb, thus on the west side similar relics are clearly visible but not yet excavated.

The necropolis consists of 36 tombs in all, which can be classified into two sections lying 65 m apart. The first section extends along the southeastern side of the plot. It contains nine rock-cut chamber tombs of various sizes that are composed of the main burial chamber and a smaller antechamber. Furthermore, seven shallow pit graves are oriented south or southeast in this section of the cemetery (Galanaki 2006, 227, 236, pl. I:4.1).

The second section extended northwest from the first and is located on the center and west side of the plot. This area contained 20 similar rock-cut chamber tombs facing south (Galanaki 2006, 236, pl. I:4.2). Three of these were found within the western boundary of a rectangular three-roomed Minoan building (Galanaki et al., forthcoming).

The burial architecture finds close parallels both within Crete, at the cemetery of Hagia Photia in Siteia (Davaras and Betancourt 2004), and far beyond, at the cemeteries of Ano Kouphonisi in the Cyclades (Zapheiropoulou 1984) and Manika in Euboea (Sampson 1988; Sapouna-Sakellaraki 1987).

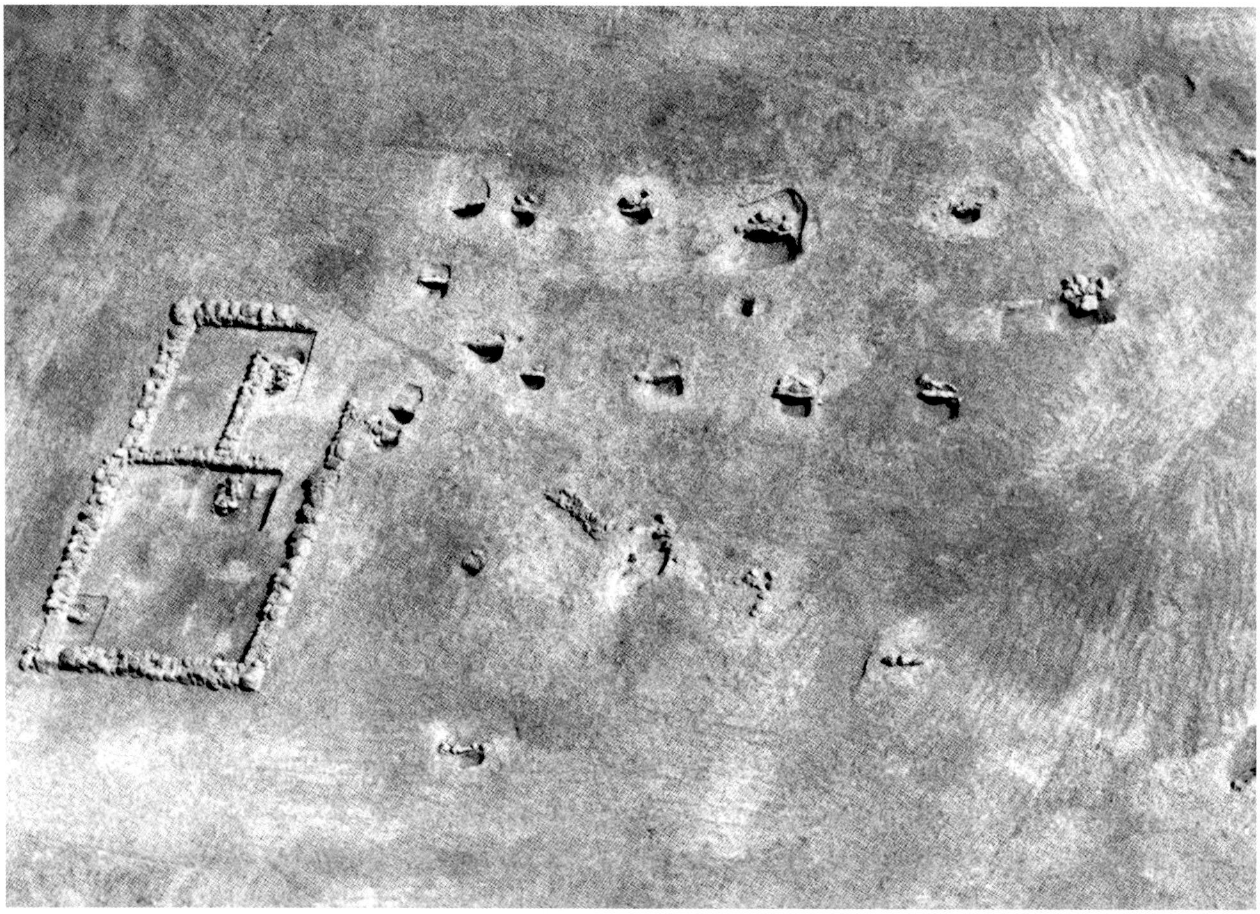

Figure 9.1. Aerial photograph of the EM I cemetery at Gournes, Pediada, and the MM building (Section 2).

The tombs contained the following categories of grave offerings: 130 clay vessels, 32 obsidian blades, 17 stone objects, numerous groups of seashells, three silver objects, and six bronze objects (Galanaki 2006, 229–232, 239–241, pls. IV–VI).

It was not possible to correlate the Early Minoan necropolis with the nearby settlement, despite an extensive surface survey carried out in the surrounding area; the only evidence was in the form of vague indications found on the coastline, approximately 300 m north of the necropolis.

The Early Minoan burial assemblage forms a link in the chain of early Prepalatial burial and other sites on the north coast of eastern and Central Crete (Hagia Photia and Kephala Petras near Siteia, Krassi, Gournes, Pyrgos, the Eileithyia Cave and the smaller burial cave at Amnissos, Poros, Knossos, Archanes, Kyparissi, Moni Odigitria, and Krotos near Tripiti) (Marinatos 1929; Alexiou 1951; Dimopoulou 1997; Galanaki 2006, 231; Wilson, Day, and Dimopoulou-Rethemiotaki 2008).

Crete's location, at a considerable distance from the Cyclades, allowed for geographical and cultural independence alongside equal participation in the ring of Early Bronze Age Aegean culture. Thus, the relationships between the social groups that created the EM I–III settlements and cemeteries with common traditions in architecture, pottery, and technology are being re-examined in terms of the constant movement of raw materials, commodities, artifacts, and ideas (Papadatos 2007a; Muhly 2008).

Silver Jewelry

Thirty-one silver necklace beads come from Prepalatial sites on Crete (Hagia Triada, Mochlos, Krassi, Kousses, Gournia, and Zakros) in compositions of varying types and materials (Vasilakis 1996, 180, 183–188, 190–192, 194; 2008, 75–78). One isolated disk-shaped bead derives from a Prepalatial burial site at Zakros (Vasilakis 1996, 188).

A unique EM I silver necklace from Tomb 2 (HM X-A 1707α) consists of 458 beads, 455 of which are solid-cast disk-shaped beads, while three are compressed spherical ones (Fig. 9.2). The necklace also has cylindrical beads consisting of blue stone (Fig. 9.3) that are barely visible to the naked eye; their diameter is equal to that of the stringing hole in each disk-shaped bead.

The Gournes silver necklace directly points to its counterpart in Tomb 26 of the Early Cycladic (EC) I cemetery at Louros on Naxos (Fig. 9.4; Papathanassopoulos 1961–1962, 135, pl. 67c, NM 6205[1], 8826), which preserves approximately 200 disk-shaped beads with a compressed spherical bead in the center. It further points to the smaller sized Late Neolithic necklace found at the entrance to the Alepotrypa Cave at Diros Mani (Fig. 9.5; Papathanassopoulos 1971; 1996, 227, no. 41; 1998, 65, no. 65, NMD, D918a–c), which preserves 168 disk-shaped beads, one ring-shaped bead, and an almond-shaped bead in the center. The Gournes necklace differs apart from the technique from these two parallels due to the addition of the miniature cylindrical blue stone beads and the number of compressed spherical beads, which are three in this particular case.

Furthermore, three slightly compressed spherical beads of differing sizes (Fig. 9.6; HM X-A 1707β, HM X-A 1708, and HM X-A 1709) from three necklaces in Tombs 2, 27, and 33 point to similar EM II–III silver beads from Gournia (Davaras 1973a; 1973b; Vasilakis 1996, 188, no. 62–70).

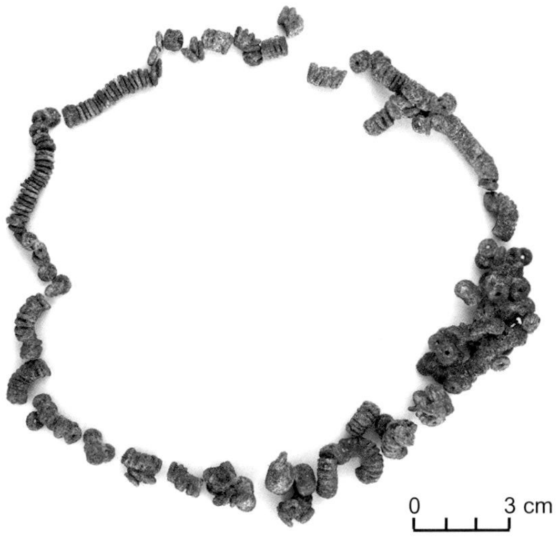

Figure 9.2. Silver necklace from the EM I cemetery at Gournes, Pediada (Tomb 2, HM X-A 1707α). Photo by Y. Papadakis-Ploumidis.

Figure 9.3. Cylindrical blue stone bead from the silver necklace of the EM I cemetery at Gournes, Pediada (Tomb 2, HM X-A 1707β). Photo by Y. Papadakis-Ploumidis.

Figure 9.4. Silver necklace from the EC I cemetery at Louros on Naxos (Papathanassopoulos 1961–1962, 135, pl. 67c:NM 6205 [1], 8826; Tomb 26).

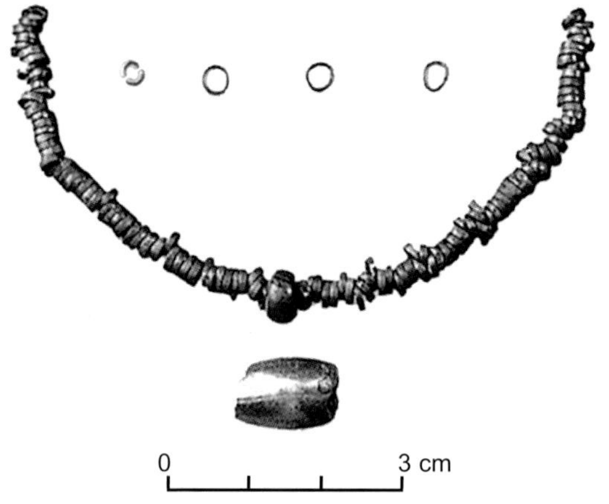

Figure 9.5. Silver necklace from the possible burial site at Alepotrypa, Diros Mani, Laconia (Papathanassopoulos 1998, 65, no. 65:NMD 918a–c).

Figure 9.6. Silver beads from the EM I cemetery at Gournes, Pediada (HM X-A 1707β, HM X-A 1708, HM X-A 1709; Tombs 2, 27, 33). Photo by Y. Papadakis-Ploumidis.

The beads from the Gournes silver necklace allow comparison with similar beads made of different materials (steatite, faience, schist, and serpentine) that come from other Prepalatial sites on Crete such as Archanes (Sakellarakis and Sapouna-Sakellaraki 1997, 614–615, 619, 624–627, pls. 643, 650, 657–658, 669, 673–674), Koumasa (Xanthoudides 1924, pls. XXVIa:809, 811, XXXII:548–549, 554), Porti (Xanthoudides 1924, pl. XXXIXb:672), Kalathiana (Xanthoudides 1924, pl. XLII:828–829), Platanos (Xanthoudides 1924, pl. LVIII:1152, 1165, 1169, 1176, 1186), Moni Odigitria "Του Σκανιάρη ο Λάμμος," Krotos, Trypiti (under publication), Lebena (Alexiou 1961–1962, 88–91; Alexiou and Warren 2004, pl. 114A:A, C, D), and Mochlos (Seager 1912, 48, 55, 66, 75, pls. 20:I.vii, 25:vi.35, 36:xvi, 36:xx.8). The Gournes necklace resembles four necklaces of steatite and faience beads in the Metaxas Collection (HM 593, HM 2214–HM 2216), as well as those from earlier sites outside Crete, such as Neolithic Dimini (Papathanassopoulos 1961–1962, 135, NM 6205 [2], pl. 67:ε; 1996, 72, 73, nos. 95, 96; Kyparissi-Apostolika 2001, 509, 515, 519, pls. 12, 13 [discoid beads], 402, 405–414, 501, 510, 520–521, 541–543, 521–552, 555, 558, 561–562, pls. 12–16 [compressed spherical beads]), which yielded a small Late Neolithic necklace with 46 disk-shaped stone beads from Alepotrypa at Diros Mani. Similar disk-shaped beads derive from Early Bronze Age sites in the Cyclades (Rambach 2000, pls. 178, 179, 180).

The production of compressed spherical beads from other materials (faience and steatite) was already common in EM I, as is evident from those found at Platanos, Koumasa, Porti, Lebena, and Mochlos. This clearly points to the three similar beads in the Gournes silver necklace.

Bronze Tools and Jewelry

Three EM I bronze borers with quadrilateral profile (Fig. 9.7:HM X 5798, HM X 5799, HM X 5802) derive from Tombs 2 and 21 and clearly point to similar bronze tools from the Early Minoan cemeteries at Krassi, Kanli Kasteli, and Platanos in the Mesara (Branigan 1974, 26, 171, nos. 958, 961, 965–966), and from EC I Tomb 26 at Louros, Naxos (Papathanassopoulos 1961–1962, 135, pl. 67c:NM 6205 [1], NM 8826), and from Chalandriani on Syros (Rambach 2000, pl. 170). It should be noted that the silver necklace was found together with two of the three bronze borers in Tomb 2 at the Gournes necropolis, just as in Tomb 26 at Louros on Naxos, where three similar bronze borers were found together with the necklace (Papathanasopoulos 1961–1962, 134, pl. 68c:NM 8824).

The isolated bronze crescent-shaped earring (Fig. 9.8:HM X 5803) with pointed endings was found south of Tomb 21 in the EM I necropolis at Gournes, Pediada and points clearly to similar gold and silver earrings found at Hagios Onouphrios, Platanos, and Archanes (Branigan 1974, 45–46,

Figure 9.7. Bronze borers from the EM I cemetery at Gournes, Pediada (HM X 5798, HM X 5899, HM X 5802; Tombs 2, 21).

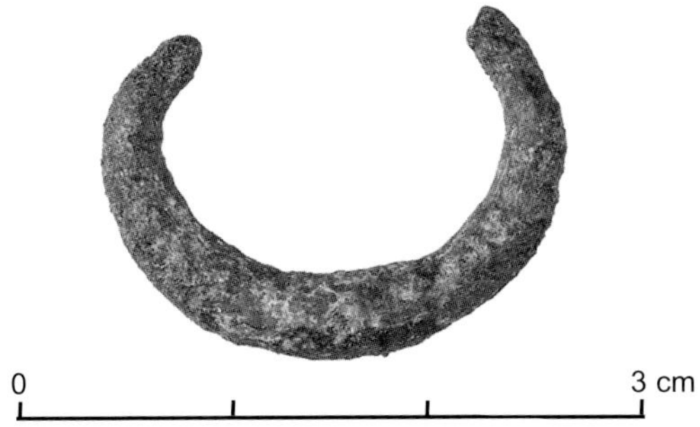

Figure 9.8. Bronze crescent-shaped earring from the EM I cemetery at Gournes, Pediada (HM X 5803; Tomb 21).

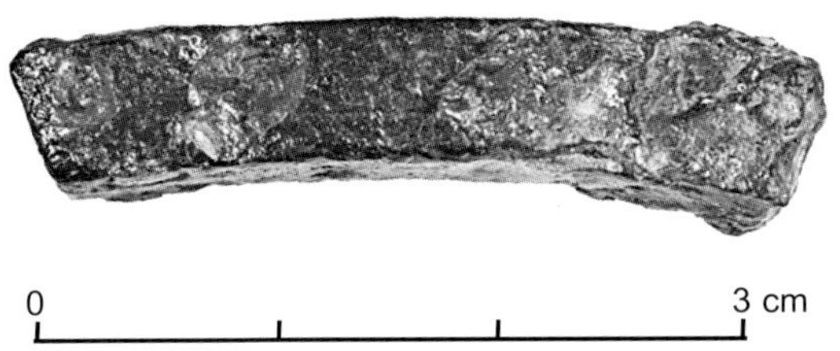

Figure 9.9. Bronze shank from the EM I cemetery at Gournes, Pediada (HM X 5801; Section 2).

Figure 9.10. Irregular corroded ferrous mass from the EM I cemetery at Gournes, Pediada (HM X 5800; Tomb 9).

189, nos. 25, 27, pl. 23; Vasilakis 1996, 135–137, nos. 1–2, 5; Sakellarakis and Sapouna-Sakellaraki 1997, 641–642, pls. 703–704). This particular type survived on Crete until the Late Minoan period.

Also from section 2 is a bronze shank (Fig. 9.9:HM X 5801) with quadrilateral profile, and from section 1 (Tomb 9) is an irregular corroded ferrous mass (Fig. 9.10:HM X 5800).

Analytical and Technological Approaches

General

During the last seven years, the Group of Paleoenvironmental and Ancient Metals Studies (GP-AMS) at the NCSR "Demokritos" Center in Athens has launched a new inter-disciplinary archeometric project entitled "Investigation on the Beginnings of Minoan Metal Production," in collaboration with the 23rd and 24th Ephorates of Prehistoric and Classical Antiquities, the Archaeological Department of the University of Crete, the INSTAP Study Center for East Crete, and excavator-archaeologists of American universities. In the frame of this program, two members of GP-AMS, Yannis Bassiakos and Constantine Athanassas have undertaken since 2005 analytical and technological studies on ca. 15 metallic finds from the excavation at Gournes.

The analytical works are still in progress; the results presented here therefore should be regarded as preliminary. Among the main goals of the ongoing analytical approaches are four points: (1) the correlation of the Early Minoan metallurgical practices, currently studied at eastern Cretan sites such as Chrysokamino (Bassiakos and Catapotis 2006), Poros-Katsambas (Doonan, Day, and Dimopoulou-Rethemiotaki 2007), Kephala Petras in Siteia

(Papadatos 2007b), Hagia Photia (Betancourt and Muhly 2007; Tsipopoulou 2007), and elsewhere, with contemporaneous metallic objects; (2) the investigation of the particular Minoan techniques of transformation of metallic crude materials (i.e., copper ingots) to finished objects; (3) the search for possible links between the Early Minoan metal-working technologies (Tselios 2006) and the corresponding practices employed on other Aegean islands, in the Greek mainland, and around the wider Mediterranean region; and (4) the examination of issues regarding provenancing, metallic materials transmission, and technological interactions between early workshops or production centers within Crete and beyond the island.

The relatively limited number of samples that have been analyzed until now seems insufficient to accommodate adequately these rather long-range goals; some of the results, however, as stressed below, correlate with data coming from other similar ongoing studies and provide the basis for some enlightening evaluations.

A new, portable unit performing X-ray fluorescence (XRF) analyses was used for in-situ elemental determination of the museum's metallic intact objects excavated at Gournes. For a small number of worn finds laboratory analyses were allowed: the techniques of optical microscopy, scanning electron microscopy and energy dispersive microanalyses (SEM/EDX). In one case, when mineralogical characterization was desired, X-ray diffractometry (XRD) was employed. The laboratory analyses were conducted at the Institute of Material Sciences in the NCSR "Demokritos" Research Center, Athens, and in the Department of Mineral Resources Engineering at the Technical University of Crete, Chania.

Results

Table 9.1 presents the results of point analyses (2.5 mm width) on metallic objects (e.g., borers, handle, earring, shank, and silver inlays) from Gournes that were performed at the Archaeological Museum of Herakleion in September 2006. The type of in-situ XRF analytical technique (for major and minor elements) holds the privilege of obtaining the so-called non-destructive analyses very quickly and, more importantly, it can be applied on integral or unique finds, from which extraction of a sample for laboratory studies is not feasible. Elements of atomic number 13 (aluminum) and above can be adequately detected and measured with this new analytical unit.

In an attempt to retrieve as much technological information as possible from the objects (most of them being badly corroded), we decided to apply laboratory techniques to three samples: (1) one fragment of a silver inlay to be examined under optical and scanning microscopes; (2) two tiny blue stone beads, one broken and one intact and initially thought to consist of lapis lazuli, for analysis by SEM/EDX and XRD, respectively; and (3) one tiny fragment of the ferrous mass to be analyzed by laboratory XRF and XRD.

The study of the silver inlay from the silver necklace (HM X-A 1707α) under the polarizing microscope showed, unfortunately, a very advanced stage of corrosion throughout the silver metallic mass (Fig. 9.11), resulting from millennia of burial as a whole object in an oxidizing environment near the coastline. The once shiny surface of the inlay is now covered by a layer of corrosion products that is ca. 100 μm thick, and the extended dislocation of the metallic bonds in the internally retained silver "islets" makes any attempt for metallographic study impossible. Some notes indicated on Figure 9.11 have to do with the constituents of the corrosion products and are based on data given in Table 9.2, which also contains data attributed to the broken blue stone bead. The analyses for both materials are discussed in the next section.

The above mentioned irregular corroded ferrous mass (HM X 5800) that had shown iron and arsenic content from the in-situ museum analysis (Table 9.1) had no flat or smooth surface to allow quantitative chemical analysis. Laboratory preparation of a flat-surface-offering-pellet, after grinding and homogenizing, allowed us to perform a more accurate chemical analysis as well as a mineralogical XRD analysis. That sample contained quite noticeable amounts of arsenic (among other elements), as seen in Table 9.3. The subsequent XRD analysis of the same sample revealed as the main mineralogical phases hematite, limonite, and lollingite, the latter being an iron arsenide ($FeAs_2$). An evaluation of the above analytical data and their possible relevance to technological and archaeo-metallurgical issues follow.

Herakleion Museum Catalog Number	Object Type	Fe	Cu	Sn	As	Br	Ag
HM X-A 1707α	silver necklace (mass)	3.3	0.6	ND	ND	10.9	85.2
HM X-A 1707α	silver inlay (single)	0.6	ND	ND	0.3	ND	99.2
HM X-A 1707β	silver inlay (single)	0.6	ND	ND	0.2	ND	99.3
HM X-A 1708	silver inlay (single)	1.4	15.4	ND	ND	6.9	76.2
HM X-A 1709	silver inlay (single)	+	+	ND	ND	+	+
HM X 5798	bronze borer	0.4	98.3	ND	1.3	ND	ND
HM X 5799	bronze borer	0.3	96.4	ND	3.3	ND	ND
HM X 5800	irregular corroded "bronze"	+	+	+	+	ND	ND
HM X 5801	"bronze" shank	1.9	1.4(?)	84.6(?)	11.9(?)	ND	ND
HM X 5802	bronze borer	0.3	96.8	ND	2.9	ND	ND
HM X 5803	earring (crescent-shaped)	0.7	90.1(?)	8.3(?)	0.9(?)	ND	ND
HM X 5800	irregular corroded ferrous mass	+	ND	ND	+	ND	ND

Table 9.1. XRF surface analysis of metallic finds from the excavation at Gournes. The analyses are given by elements. Total error for the major elements: <10%. Data are normalized by weight and %. ND = not detected. +: element detected but not quantitatively analyzed. The analyzed surface was ca. 3 mm². Concentrations ≤0.2% should be considered indicative. Some values of the objects HM X 5801 and HM X 5803 are indicated with a question mark. Very probably they do not correspond to the initial alloys and only reflect the analysis of the surface, the latter very often affected by taphonomic effects and diagenetic processes over millennia of burial.

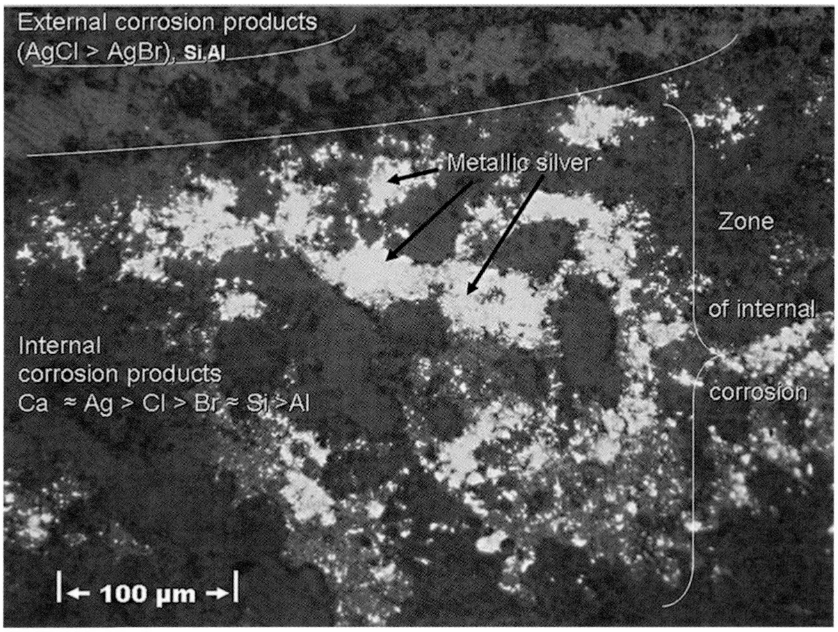

Figure 9.11. Characteristic microstructure of the badly corroded silver inlay coming from the silver necklace (HM X-A 1707α), exhibiting external corrosion layer and one weathered "body," which retains few islets of metallic silver, unfortunately not allowing etching and metallographic analysis. Examined under the polarizing microscope with parallel Nicol prisms.

Herakleion Museum Catalog Number and Description	Na	Mg	Al	Si	P	S	Cl	Ag	K	Ca	Br	Cu
HM X-A 1707α, external corrosion layer*	2.6	1.5	10.2	ND	0.3	ND	12.5	60.2	0.2	6.1	6.5	ND
HM X-A 1707α, retained metallic silver**	ND	0.9	0.7	ND	ND	ND	2.9	91.9	ND	ND	2.1	1.5
HM X-A 1707α, internal corroded area (a)*	ND.	1.6	6.5	1.9	ND	ND	11.6	52.2	ND	21.6	4.4	0.7
HM X-A 1707a, internal corroded area (b)**	ND	1.7	2.8	7.8	ND	ND	4.3	30.3	ND	47.4	1.5	4.2
Blue stone bead, "window" analysis 1**	0.5	29.6	2.6	62.9	0.9	1.2	0.1	ND	0.2	0.6	ND	ND
Blue stone bead, "window" analysis 2**	0.8	31.8	2.6	62.0	0.6	0.8	<0.1	ND	<0.1	0.5	ND	ND

Table 9.2. SEM/EDX analyses ("window") on the corroded silver inlay (HM X-A 1707α) and on broken blue stone bead. Analyzed areas on the polished section of HM X-A 1707α: 1–1.5 mm^2. ND = not detected. Total error for the major elements: <10%. * Analysis by elements. ** Analysis by oxides.

Herakleion Museum Catalog Number and Description	Fe	Ca	Ti	Cl	Cu	As	Mn	Cr
HM X 5800, irregular ferrous object, 1st analysis	62.5	6.4	0.5	19.3	2.7	8.3	0.2	0.1
HM X 5800, irregular ferrous object, 2nd analysis	59.9	2.4	0.3	12.4	0.6	24.1	<0.1	<0.1

Table 9.3. Laboratory XRF analyses of the irregular ferrous sample (HM X 5800). Analysis in weight percent, by oxides.

Evaluation of Data and Concluding Remarks

Most of the results obtained so far were produced by employing the non-destructive XRF technique. The respective chemical analyses, therefore, correspond to the surface of the examined findings, not to their internal mass. Special attention was paid to select clear, flat surfaces that were as free of patina as possible in order to obtain both qualitative and quantitative results. This was not always possible, however. For example, the analysis of the bronze shank (HM X 5801, indicated with question marks in Table 9.1) is technically correct and scientifically incorrect, as can be deduced even from the first data evaluation. Also, for the findings of HM X-A 1709 and HM X 5800 (Table 9.1) that missed appropriate flat and clear surfaces, only their qualitative analysis is given. Further issues on the advantages and disadvantages regarding the use of non-destructive XRF analyses on archaeological metallic finds—including the phenomenon of metallic component migration in an ancient alloy during burial, the role of the external metallic surface as "interface," etc.—are now examined, based on new analytical data, by the members of GP-AMS at "Demokritos." In the set of results presented here, one can note that the analyses provided by the portable XRF unit are repeatedly verified by other laboratory techniques as is the case of the silver inlay (HM X-A 1707α) for the elements Fe, Cu, Br, and Ag. Although it is impossible to conduct tests of the analytical reliability for all results obtained by the portable unit in museums (mainly due to the fact that some objects are complete), such cross-checking, when feasible, increases the validity and the usefulness of the portable unit.

Based on the above results, the analyses of the three arsenical-bronze borers (HM X 5798, HM X 5799, HM X 5802) seem correct for technological evaluations; their arsenic content (1.3%, 3.3%, and 2.9%) falls near the most prevalent As content of early Aegean arsenic bronzes (3%) (Papadimitriou 2001; however, it is not always detected, see Table 9.1). Also, their Fe content is relatively low (0.4%, 0.3%, and 0.3%, respectively). Low iron content in ancient bronzes, up to 0.4%, comprises a technological merit (Papadimitriou 2001), as allowing desirable formation of the objects by hammering without beating-cracks caustation. The fact that some metallic finds (albeit an inadequate number) have an iron content of up to 0.4% suggests that EM metalsmiths were aware of the technological demand of removing the iron, which is rather systematically present in levels well above the limit of 0.4% as indicated from analyses on copper prills trapped in EM copper smelting slags (e.g., at Chrysokamino; see Bassiakos and Catapotis 2006). The removal of excess iron contained in the ingots or nuggets of copper that have been produced by smelting was achieved by melting the copper in crucibles and then pouring the molten copper into molds. A number of used crucibles and molds are already known from various Early Minoan sites on Crete. New crucibles and molds were recently published from Hagia Photia (Betancourt and Muhly 2007; Tsipopoulou 2007) and Poros-Katsambas (Doonan, Day, and Dimopoulou-Rethemiotaki 2007). All these indicate that the Early Minoan smiths were quite capable in manufacturing arsenical copper.

Little has been achieved until now with the laboratory study of the silver inlay in terms of better understanding early silver manufacturing. The noticeable bromium concentration in the corrosion products raises new questions about the initial bromium content of the original metallic substance and its provenance. On the other hand, the extremely fine structure of the blue stone beads, with absolute symmetry in cutting and micro-piercing, generates further unanswered questions.

Of particular importance was the identification of the blue stone beads of the necklace, initially taken for lapis lazuli during the conservation stage of this valuable object. The hypothesis of lapis lazuli was

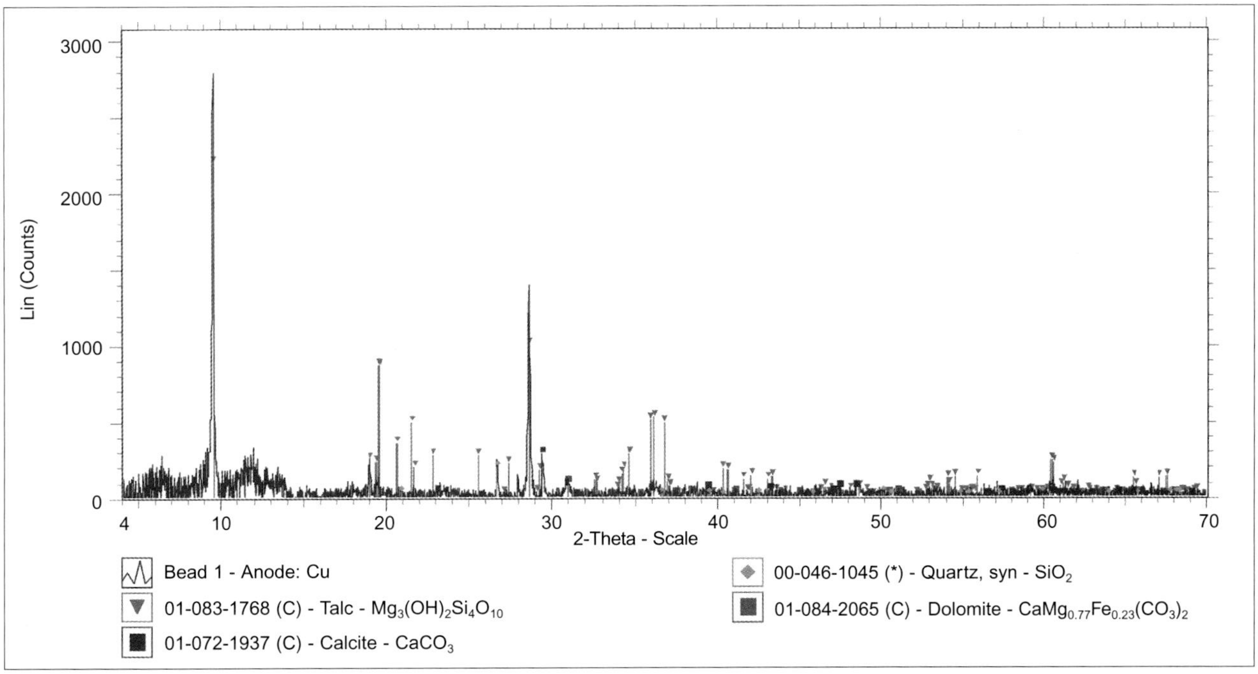

Figure 9.12. The XRD spectrum of a stone bead from the silver necklace, indicating that the main mineralogical component of the stone is talc, the basic mineral of steatite.

supported by the fact that the blue color of the beads was observed not only on their surface but also in their interior mass, as noted on fractured surfaces of the beads under the stereoscope and microscope. Lapis lazuli is a semiprecious stone of blue color, often associated with marble and/or dolomitic marble and geologically created during contact metamorphic processes. Its main mineralogical component is lazurite, a complex silicate with a hardness of 5–5.5 on the Mohs scale. Quantitative EDX chemical analyses of one broken bead is given in Table 9.2. Some elements such as Na, Al, Si, S, and Ca, part of the chemical formula of lazurite, are detected in the given analyses of Table 9.2, albeit stoichiometrically not consistent with lazurite. Bassiakos initially adopted the opinion that the blue bead consisted of an idiomorphic lapis lazuli. The element of magnesium, found to be abundant in the analysis but not present in lazurite's chemical formula was believed to originate from any dolomite (a rock rich in magnesium), the latter genetically associated with lapis lazuli, as stated above. However, the mineralogical analysis, undertaken at the Department of Mineral Resources Engineering (Chania), resulted in undoubted characterization of the examined beads by employing the XRD technique. The analysis was performed by using a Siemens A D8000, X-Ray powder Diffractometer with copper tube. The crystalline phases were identified using DIFFRACPlus software (Brucker AXS) and JCPDS Powder Diffraction File. Figure 9.12 presents the diffractogram of the studied blue bead. The main component of the bead is talc, $[Mg_6(OH)_4Si_8O_{20}]$, a soft mineral with hardness of 1 on the Mohs scale, with a minor content of dolomite, calcite, and quartz. An additional feature, which sustains the validity of talc as the main mineralogical component of the blue bead, is its remarkable softness noted under the microscope; if slightly scraped by a thin microscopic pin, the bead is incised quite obviously.

Thus, careful evaluation of data coming from both mineralogical determination (XRD) and quantitative analysis (SEM/EDX) led us to conclude that the raw material from which the beads were constructed was steatite, which is a fine-grained talc. According to the EDX analysis, the quantitatively prevailing elements of Si and Mg belong to talc, while few portions of these elements along with the remaining lesser elements might belong to the other identified minor minerals of dolomite, calcite, and quartz, or even to lesser (and not detected) clay minerals and other phases that derive from contaminations in the burial environment.

The incompatibility between the blue color of the beads and their mineralogical characterization is serious. To our knowledge, steatite of blue color does not exist; even blue-dyed steatites (dyed throughout their mass as the beads examined here) without copper glazing are not known from any archaeological site. Therefore, the possibilities of dealing with blue steatites or with artificially (and successfully) dyed ones, require much more study, archaeological and scientific.

Last—but not least—is the identification of lollingite ($FeAs_2$), an iron arsenide, among the other minerals of a ferruginous piece found in a tomb at Gournes. The same mineral was recently reported (Doonan, Day, and Dimopoulou-Rethemiotaki 2007) in the metallurgical assemblage of Poros-Katsambas. Moreover, at Chrysokamino the copper prills entrapped in the metallurgical slags appear to contain arsenic in noticeable levels (3%–8%; Bassiakos and Catapotis 2006). Hence, the logically emerging question—whether these iron arsenides are indigenous or not—as well as technological issues related to the manufacturing of the arsenical-bronzes by the early Minoans comprise challenges for more archaeological and analytical studies.

Acknowledgments

Part of this project was financed by the Institute for Aegean Prehistory (INSTAP), Philadelphia, PA. Dr. Bassiakos additionally wishes to acknowledge the generous funding by the Stavros Niarchos Foundation that made possible the acquisition and evaluation of the XRF unit for cultural purposes.

References

Alexiou, S. 1951. "Πρωτομινωικαί ταφαί παρά το Κανλί Καστέλλι," *CretChron* 1951, pp. 257–294.

———. 1961–1962. "Οι Πρωτομινωικοί τάφοι στη Λεβήνα και η εξέλιξη των πρωτομινωικών ρυθμών," *CretChron* 1961–1962, pp. 15–16, 88–91.

Alexiou, S., and P. Warren. 2004. *The Early Minoan Tombs of Lebena, Southern Crete* (*SIMA* 30), Savedälen.

Bassiakos, Y., and M. Catapotis. 2006. "Reconstruction of the Copper Smelting Process, Based on the Analyses of Ore and Slag Samples," in *The Chrysokamino Metallurgy Workshop and Its Territory* (*Hesperia Suppl.* 36), P.P. Betancourt, Princeton, pp. 329–353.

Betancourt, P.P., and J.D. Muhly. 2007. "The Crucibles from the Hagia Photia Cemetery," in Day and Doonan, eds., 2007, pp. 146–153.

Blackman, D. 2001. "Archaeology in Greece. 2000–2001," *Archaeological Reports for 2000–2001*, pp. 1–144.

Branigan, K. 1974. *Aegean Metalwork of the Early and Middle Bronze Age*, Oxford.

Davaras, C. 1973a. "'Αρχαιότητες καὶ μνημεῖα 'ανατολικῆς Κρήτης," *Amaltheia* 1973, pp. 40–44, 48–49.

———. 1973b. "'Αρχαιότητες καὶ μνημεῖα 'ανατολικῆς Κρήτης," *ArchDelt* 28, pp. 585–589.

Davaras, C., and P.P. Betancourt. 2004. *The Hagia Photia Cemetery* I: *The Tomb Groups and Architecture* (*Prehistory Monographs* 14), Philadelphia.

Day, P.M., and R.C.P. Doonan, eds. 2007. *Metallurgy in the Early Bronze Age Aegean* (*Sheffield Studies in Aegean Archaeology* 7), Oxford.

Dimopoulou, N. 1997. "Workshops and Craftsmen in the Harbour-Town of Knossos at Poros-Katsambas," in *Τεχνη: Craftsmen, Craftswomen and Craftsmanship in the Aegean Bronze Age. Proceedings of the 6th International Aegean Conference, Philadelphia, Temple University, 18–21 April 1996* (*Aegaeum* 16), R. Laffineur and P.P. Betancourt, eds., Liège and Austin, pp. 433–438.

Doonan, R.C.P., P.M. Day, and N. Dimopoulou-Rethemiotaki. 2007. "Lame Excuses for Emerging Complexity in the Early Bronze Age Crete: The Metallurgical Finds from Poros Katsambas and Their Context," in Day and Doonan, eds., 2007, pp. 98–122.

Galanaki, K. 1999. "Γούρνες. Οιμόπεδο ΙΘΑΒΙΚ," *ArchDelt* 54, pp. 853–856.

———. 2006. "Πρωτομινωικό ταφικό σύνολο στην πρώην Αμερικανική Βάση Γουρνών Πεδιάδος," in *Πεπραγμένα Η΄ Διεθνούς Κρητολογικού Συνεδρίου*, Herakleion, A2, pp. 227–241.

Galanaki, K., E. Nodarou, C. Papadaki, and J. Triantafillidi. Forthcoming. "Μινωικό κτίριο στο Πρωτομινωικό νεκροταφείο Γουρνών Πεδιάδος," in *Πεπραγμένα Ί Διεθνούς Κρητολογικού Συνεδρίου*, Chania.

Kyparissi-Apostolika, N. 2001. *Τα προϊστορικά κοσμήματα της Θεσσαλίας*, Ministry of Culture edition, Athens.

Marinatos, S.P. 1929. "Πρωτομινωικὸς θολωτὸς τάφος στὸ Κράσι Πεδιάδος," *ArchDelt* 12 (Chronika), pp. 102–141.

Muhly, J.D. 2008. "Ayia Photia and the Cycladic Elements in Early Aegean Metallurgy," in *Aegean Metallurgy in the Bronze Age. Proceedings of an International Symposium Held in Rethymnon, Crete, Greece, November 19–21, 2004*, I. Tzachili, ed., Athens, pp. 69–74.

Papadatos, Y. 2007a. "Beyond Cultures and Ethnicity: A New Look on Material Culture Distribution and Inter Regional Interaction in the Early Bronze Age Southern Aegean," in *Mediterranean Crossroads*, S. Antoniadou and A. Pace, eds., Athens, pp. 419–451.

———. 2007b. "The Beginnings of Metallurgy in Crete: New Evidence from the FN–EM I Settlement at Kephala-Petras, Siteia," in Day and Doonan, eds., 2007, pp. 154–167.

Papadimitriou, G.D. 2001. "Η εξέλιξη των κραμάτων χαλκού στον Ελλαδικό χώρο μέχρι το τέλος της γεωμετρικής εποχής: κραματικές προσμείξεις και τεχνολογικήν εξέλιξη," in *Archaeometry Issues in Greek Prehistory and Antiquity*, Y. Bassiakos, E. Aloupi, and G. Facorellis, eds., Athens, pp. 587–608.

Papathanassopoulos, G.A. 1961–1962. "Κηκλαδικά Νάξου," *ArchDelt* 17, pp. 104–151.

———. 1971. "Σπήλαια Διρού," *AAA* 4, pp. 12–26, 149–154, 289–304.

———. 1996. *Ο νεολιθικός πολιτισμός στην Ελλάδα*, Athens.

———. 1998. Catalog entry nos. 64–66 in *Κοσμήματα της ελληνικής προϊστορίας. Ο νεολιθικός θησαυρός*, K. Demakopoulou, ed., Athens, p. 65.

Rambach, J. 2000. *Kykladen* I: *Die frühe Bronzezeit Grab- und Siedlungsbefunde*, Bonn.

Sakellarakis, I., and E. Sapouna-Sakellaraki. 1997. *Αρχάνες. Μια νέα ματιά στη Μινωική Κρήτη*, Athens.

Sampson, A. 1988. *Μάνικα* II: *Ο πρωτοελλαδικός οικισμός και το νεκροταφείο*, Athens.

Sapouna-Sakellaraki, E. 1987. "New Evidence from the Early Bronze Age Cemetery of Manika, Chalkis," *BSA* 82, pp. 233–264.

Seager, P. 1912. *Exploration in the Island of Mochlos*, Boston and New York.

Tselios, T. 2006. *Η μεταλλουργία του χαλκού στην Προανακτορική Κρήτη: Τεχνολογικές εξελίξεις και κοινωικές*, Ph.D. diss., University of Crete, Rethymnon.

Tsipopoulou, M. 2007. "Aghia-Photia-Kouphota: A Center for Metallurgy in the Early Minoan Period," in Day and Doonan, eds., 2007, pp. 135–145.

Vasilakis, A. 1996. *Ο χρυσός και ο άργυρος στην Κρήτη κατά την πρώιμη εποχή του Χαλκού*, Herakleion.

———. 2008. "Silver Metalworking in Prehistoric Crete. A Historical Survey," in *Aegean Metallurgy in the Bronze Age. Proceedings of an International Symposium Held at the University of Crete, Rethymnon, Greece, November 19–21, 2004*, I. Tzachili, ed., Athens, pp. 75–85.

Xanthoudides, S. 1918. "Μέγας πρωτομινωικὸς τάφος Πύργου," *ArchDelt* 4 (Chronika), pp. 136–170.

———. 1924. *The Vaulted Tombs of Messara*, London.

Wilson, D.E., P.M. Day, and N. Dimopoulou-Rethemiotaki. 2008. "The Gateway Port of Poros-Katsambas: Trade and Exchange between North-central Crete and the Cyclades in EB I–II," in Ορίζων: *A Colloquium in the Prehistory of the Cyclades*, N. Brodie, J. Doole, G. Gavalas, and C. Renfrew, eds., Cambridge, pp. 261–270.

Zapheiropoulou, P. 1984. "The Chronology of the Kampos Group," in *The Prehistoric Cyclades: Contributions to a Workshop on Cycladic Chronology*, J.A. MacGillivray and R.L.N. Barber, eds., Edinburgh, pp. 31–40.

CHAPTER

10

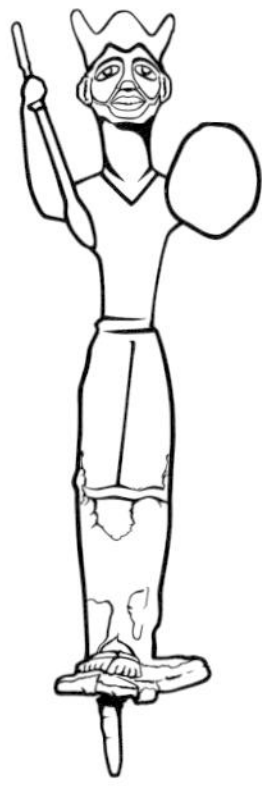

The Dog Diadem from Mochlos

Jane Hickman

In the 1985 article "Beyond Typology: Aegean Metallurgy in Its Historical Context," James D. Muhly argues for a comprehensive examination of metal objects associated with early periods in the Aegean.* While it is essential to organize and categorize material culture, one must move beyond the classification of objects in order to place them into a broader historical and cultural framework. Objects should be evaluated in relation to their archaeological context, rather than as static pieces in museum cases (Muhly 1985, 134). Quoting Professor Muhly, "typology is quite different from explanation" (1985, 109).

Although the scope of this paper is limited to the study of just one golden diadem from Mochlos, its intent is to go beyond a description of the object in order to explain its function in Prepalatial Minoan society. A close examination of the manufacturing, use, and deposition of the diadem reveals decisions made by artisans and patrons as well as cultural practices present at the site.

Find Context and Description

In 1908, during the excavation of the cemetery at Mochlos in Crete, Richard Seager discovered a rich cache of jewelry. This hoard of gold, perhaps originally contained in a cloth bag or wooden box, had been deliberately placed into a deep pit at the far end of Tomb II, part of the Tomb I/II/III complex. Among the diadems, pendants, flower and leaf pins, beads, and other ornaments was a long gold band

* It is a pleasure to offer this contribution to a volume honoring James D. Muhly.

decorated with four dogs in dot repoussé (Figs. 10.1, 10.2; Seager 1912, 26–27, figs. 8–9:II.4). The Dog Diadem was recovered in two pieces; perforations on either side of the break indicated that the band was repaired in antiquity. Objects in the cache have been dated to Early Minoan (EM) II–III (ca. 2500/2400–2000 B.C.). Vasiliki Ware (EM IIB) and one EM III jug, described by Seager (1912, 23) as "light-on-dark geometric ware," were recovered from the soil above the deposit (Seager 1912, 23–24; see also Soles 1992, 49–50). A Prepalatial date for the stratum that contained the jewelry was also indicated by the presence of short triangular daggers, ivory seals (Seager 1912, 24), and two "Koumasa-style vases" (Soles 1992, 50).

Stylistic comparisons to similar objects of gold confirm an EM date for the Dog Diadem. Costis Davaras (1975, 101–102) stated that a diadem decorated with abstract animals, recovered from a silver cup found in Mochlos Tomb VI (see p. 97), was associated with EM II–III pottery sherds. Additionally, a fragment of a gold strip or diadem, also decorated in dot repoussé, was recovered from Tomb VII at nearby Pseira; Vasiliki Ware sherds provided an EM IIB date for the last use of the tomb (Betancourt and Davaras, eds., 2003, 66). It should be noted, however, that the Dog Diadem may actually be older than the deposit where it was found. Since the diadem showed clear signs of wear and was part of a secondary deposit, it was likely manufactured many years before it was placed in the pit at the back of the tomb.

The Dog Diadem (Herakleion Museum number [hereafter HM] 269) is made from thin sheet gold, with a length of 32.3 cm and a maximum width of 2.9 cm. The weight is 5.1 g. Four dogs are incised, the outlines filled in with dot repoussé. At the center of the diadem, between two opposing dogs, is a uniform pattern of dots (Figs. 10.3, 10.4). A border

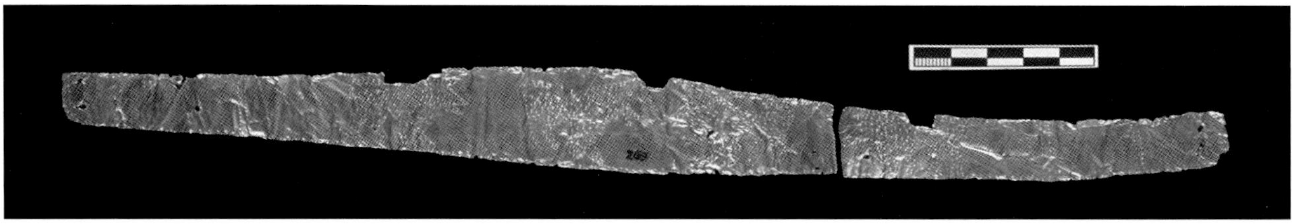

Figure 10.1. Dog Diadem (HM 269), as recovered in two pieces from Mochlos Tomb II.

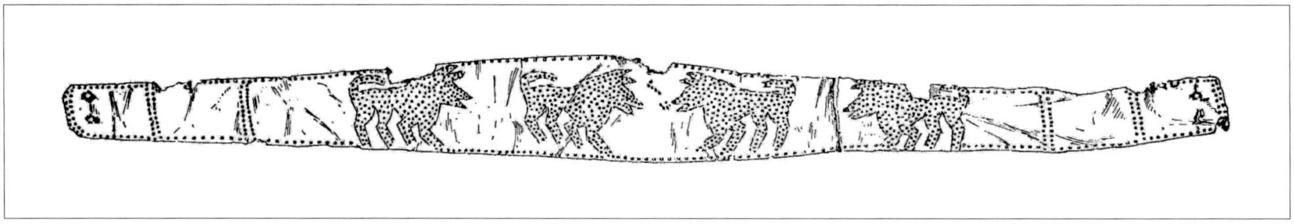

Figure 10.2. Dog Diadem (HM 269; after Seager 1912, fig. 9:II.4).

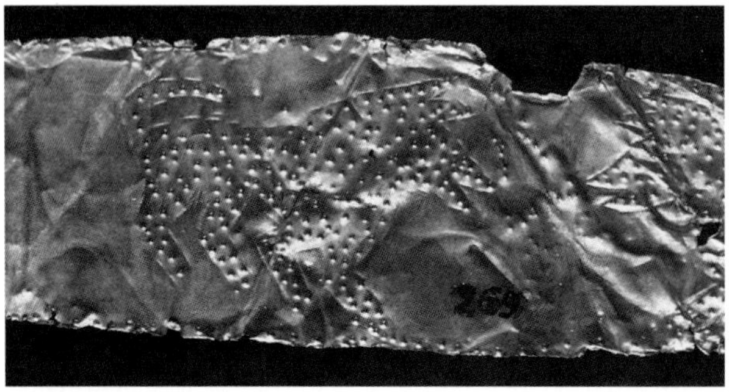

Figure 10.3. Detail near center of Dog Diadem (HM 269).

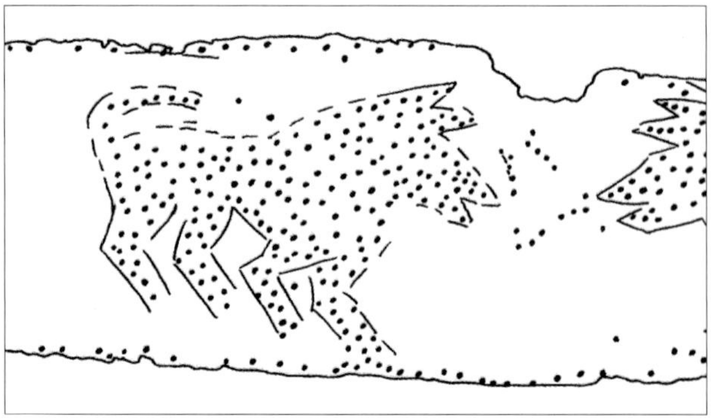

Figure 10.4. Detail near center of Dog Diadem (HM 269).

and six sets of vertical lines are also created in dots. Two perforations at each end of the band are crudely punched, with ragged edges; these holes were used to attach the diadem around the head or to a material that has perished. A pin or pendant may have been stuck through two additional perforations on the left side. Along the top edge of the diadem are three irregular notches where the metal has been broken away.

The Choice of Gold

The decision to use gold for the diadem was likely based on the metal's shiny appearance, association with foreign lands, ease in working, and availability. One does not know whether the patron or the artisan provided the gold. In Egypt, Mesopotamia, and later Classical Greece, patrons were known to supply raw materials to artisans (Williams 1998, 99; Moorey 1999, 13–14). Silver was not used for diadems in Prepalatial Crete, at least insofar as those objects that have survived. Fragments of copper or bronze strips, one of which had a pierced hole near one end, were recovered from the Hagios Charalambos Cave (P.P. Betancourt, pers. comm.); this may indicate that copper or bronze was used for thin diadems during early periods in Crete.

No sources of gold to Early Bronze Age (EBA) Mochlos can be securely identified. However, northern Greece and Anatolia are frequently proposed as sources to the Aegean during this period (Branigan 1974, 63, 90, 98 [Macedonia and northwestern Anatolia]; Lehrberger 1995, 137; Higgins and Higgins 1996, 108 [Macedonia and Thrace]; McGeehan-Liritzis 1996, 234 [Saratse and Vardarophtsa in northern Greece]; Yener 2000, 77, 80–81 [central Taurus Mountains of Anatolia]; Muhly 2001, 730 [Kestel in the Taurus Mountains of Anatolia]). Other possible sources include Syria (Maxwell-Hyslop 1971, 230; Higgins 1980, 54), the Eastern Desert of Egypt, and Nubia (Wilkinson 1971, 7; O'Connor 1993, 62; Ogden 2000, 161). The Cycladic islands of Melos (Lehrberger 1995, 137) and Siphnos (Weisgerber and Pernicka 1995, 174, 177; Higgins and Higgins 1996, 177)—the last location, a known source of ancient silver (Gale and Stos-Gale 1981, 216–217)—are also described as potential sources of gold.

Gold used for jewelry in Crete also may have originated in the Balkans. A late fifth millennium B.C. source in the Balkans is suggested by substantial finds of gold jewelry at Varna, Bulgaria (Gimbutas 1977; Renfrew 1978; Ivanov 1991; Misch-Brandl 1994; Bailey 2004). Also, gold ornaments dated to the Final Neolithic (FN) (ca. 4500–3300 B.C.), recovered from mainland Greece (e.g., the Aravissos hoard in Demakopoulou 1998, 63, nos. 58–60) and the Cyclades (e.g., the strip from the Zas Cave in Zachos 1990, 30), show a strong similarity to objects from the Balkans. Muhly (2002, 78; 2006, 157) argues for a Balkan origin of Greek Neolithic metallurgy, supported by the "dominant Balkan character" of FN metal finds from the Aegean. Unfortunately, the trail grows cold before the third millennium B.C. No evidence exists to support the movement of gold from the Balkans to Crete during the EBA, either overland or by sea. If gold had been transported south through the Cyclades during the third millennium B.C., one would expect to find more than a single EBA gold bead in that region (from Phyrrhoyes on Naxos in Papathanassopoulos 1981, 136–137, fig. 63).

The circumstances surrounding the arrival of gold in Crete are unknown. Gold may have been brought to Mochlos as a traded commodity or offered as a gift to local elites. It is also possible that it was obtained abroad by Minoan traders. Gold probably arrived in Crete as gold dust or nuggets, although it may have been transported as sheet metal or ingots (for evidence of gold ingots, see Archi 2002, 181 on ingots referenced in third millennium B.C. texts from Ebla; Gopher et al. 1990, 441 on possible ring ingots from fourth millennium B.C. Israel; Ogden 2000, 162 on ingots from the early second millennium B.C. Tôd Treasure in Egypt).

Gold used in the manufacture of Minoan jewelry may have come from various sources, with these sources changing through time. Exotic raw materials like gold and perhaps hippopotamus ivory, which arrived in Crete around the same time (EM IIA), may have originated elsewhere and subsequently been reshipped through intermediary port cities such as Byblos or Ugarit on the Levantine coast (cf. Bevan 2004, 109 on objects moving from Egypt to Byblos to Crete; Muhly 1973, 337 on the movement of copper and tin through Syria, with a reference to the importance of Byblos and Ugarit; Muhly 1985, 132–133 on Ugarit's role in the early second millennium B.C. tin trade; Soles 2005, 429 on later connections between Ugarit and Mochlos). A direct or indirect connection between Mochlos and the Levant is supported by the find of a silver cylinder seal, probably of Syrian origin and dated to the early to mid-third millennium B.C., recovered from Mochlos Tomb I (Seager 1912, 22, fig. 36:I.n; Aruz 1984; 2008, 40–41, no. 106, fig. 59; Soles 2005, 429–430).

Form

The earliest extant gold diadems in Crete may be from the Pyrgos Cave, located on the north coast of the island (Xanthoudides 1918). Two undecorated fragments of gold bands, each with perforations along one long edge, are associated with EM I–II burial deposits at the cave (Hickman 2008, 18–20, 333–334, pl. 3A–B). By the EM II–III period, gold diadems or diadem fragments are found at various sites in Crete, including Lebena, Koumasa, Archanes, Mochlos, and Pseira; diadems are identified from other sites in the Mesara, but these objects may date to MM IA or later (Hickman 2008, charts 1, 4). With the exception of Mochlos, where 19 diadems were recovered from three tombs, only one or two complete or fragmentary diadems were recovered from each site (Hickman 2008, chart 1).

Early Minoan diadems were manufactured in various sizes and shapes. Some were plain bands, and others included attachments in the form of vertical strips or perhaps flowers or leaves. Dot repoussé was used to decorate many diadems, with borders, lines, geometric motifs, animals, and anthropomorphic images all executed in dots (Hickman 2008, 137–143; Sakellarakis and Sapouna-Sakellaraki 1997, 643–645). Diadems crafted from sheet gold were also found in graves and hoards dated to the third millennium B.C. in the northeastern Aegean, western Asia, Mesopotamia, and Egypt (Hickman 2008, 243–284, 305–307). The custom of adornment with diadems, connected with the first widespread use of gold, was associated with sites throughout the eastern Mediterranean.

As evidenced by an anthropomorphic pouring vessel found at Mochlos (Fig. 10.5; Seager 1912, 64, fig. 34:XIII.g), which probably dates somewhat later than the Dog Diadem, a tradition of wearing leather or cloth headbands may have

Figure 10.5. Clay pouring vessel (after Seager 1912, fig. 34).

existed at this site during the Prepalatial period. Creating *gold* headbands would have provided maximum visual impact for this striking and rare material. Decorated headbands were also one of the easiest forms for a craftsman to produce from sheet metal, requiring nothing more than cutting and punching. Perhaps the production of gold diadems was stimulated by a developing market for prestige goods, driven by increasing social complexity and evolving social systems (cf. Sbonias 2000 on Prepalatial seal production). The desire for objects such as the Dog Diadem, likely used as adornment in public ceremonies, may have existed on an individual level, or it may have related more to group cohesion or competition.

The Dog Motif

The diadem-maker used four heraldic dogs for decoration. The spacing of the figures was decided ahead of time, and the animals were sketched in proportion. The outline of each dog was incised without the use of a template. The hind legs of the center dogs are higher than the front legs, perhaps to indicate movement. Between the two center dogs is a curious pattern of dots, which Andonis Vasilakis (1996, 99) suggests may represent *θήραμα*—quarry or game (Figs. 10.3, 10.4). Since the mouths of the dogs are open, their ears are forward, and they are portrayed in opposition to one another, the dots may also represent the sound of barking or perhaps barking at winged prey.

Dogs decorate other EBA objects from Crete and the Cyclades. A dog with a curving tail stretches out on the serpentinite (steatite) cover of a vessel, also recovered from Tomb II at Mochlos (Fig. 10.6; Seager 1912, fig. 5:II.i). A stone pyxis from Zakros includes a long-legged dog virtually identical to that on the Mochlos lid (Sakellarakis 1993, 14–15, no. 2719).

Two quadrupeds that resemble dogs decorate a poorly preserved EBA diadem from the settlement of Kastri on the Cycladic island of Syros (Fig. 10.7; Papathanassopoulos 1981, 132–133, fig. 61; Tsountas 1899, pl. 10.1). Compared to the dogs on the Mochlos diadem, the Syros dogs are depicted with more features: large eyes, collars, and male genitalia. In addition to the dogs, images of bird-like figures and ornamented rosettes adorn the Cycladic diadem. The Syros diadem is also larger, with a length of approximately 47 cm and a maximum width of 5 cm. Crafted from silver rather than gold, the diadem appears more carefully manufactured. However, distinct similarities are revealed in a close look at the diadems and the rendering of the dogs. Both diadems utilize dot repoussé as the primary decorative technique, with dot borders and pairs of vertical rows of dots.

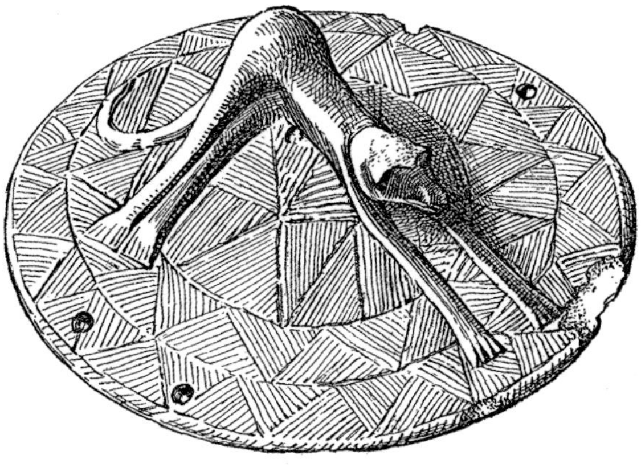

Figure 10.6. Stone cover (after Seager 1912, fig. 5).

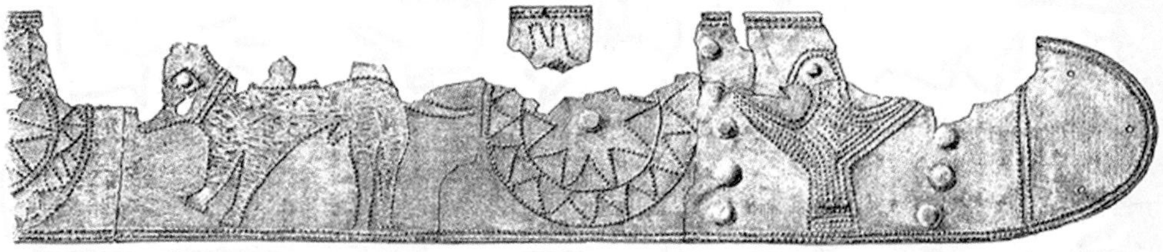

Figure 10.7. Silver diadem from Kastri, Syros (after Tsountas 1899, pl. 10.1). A later reconstruction of the full diadem by Papathanassopoulos (1981, 132–133, fig. 61) includes fragments illustrating the erect ears of the dogs.

Each diadem is wider in the center, with tapered ends. The bilaterally symmetrical dogs are comparable in appearance, with pointed ears, open mouths, and textured coats. While the center dogs on the Mochlos diadem have raised hind legs, the front legs of the Syros dogs are elevated off the ground. Similarities between the diadems suggest a cultural connection between Mochlos and Kastri, perhaps illustrative of the "international spirit" (Renfrew 1967, 15) associated with the Aegean Early Bronze II period.

Dogs were part of the Minoan artistic repertoire, perhaps due to their place in early Cretan social and economic life. Working dogs must have served as hunters of prey (cf. Andrews 1994, 64 on Old Kingdom Egypt; Immerwahr 1990, pls. 68, 70, 80 on later wall paintings from Tiryns and Pylos) and as companions to shepherds (Lonsdale 1979, 149). Dogs may have also represented the work or the attributes of a family or clan buried in Mochlos Tomb I/II/III.

Manufacturing Techniques

No jewelry-making workshops have been identified from Prepalatial Crete. Therefore, it cannot be stated with certainty that the Dog Diadem was made at Mochlos by local artisans. Gold jewelry may have been manufactured elsewhere, on or off the island, and arrived in Mochlos as trade goods or gifts. Except for the use of a similar dog motif at Syros, however, diadems at Mochlos do not resemble those from elsewhere in Crete, the Aegean, or the Eastern Mediterranean. Some Mochlos diadems are crudely manufactured. Unfinished edges, ragged perforations, and lines designed to aid in cutting remain on the sheet metal; dot repoussé borders, on the Dog Diadem and other Mochlos diadems, are uneven (e.g., Seager 1912, fig. 9.II.1, 9.II.7). A study of metal jewelry from Prepalatial Crete indicates no mass production or standardization of design (Hickman 2008, 162). The early use of gold jewelry in Crete appeared to be an indigenous development with regional and site-specific variation (Hickman 2002; 2008, 313–316). If this was the case, the Dog Diadem may have been manufactured at or near Mochlos.

Techniques employed in the construction of the Dog Diadem were not complex. No specialist knowledge of jewelry-making was evident. No special tools were required. Fabrication methods included hammering (if sheet metal had not been previously manufactured), cutting, chasing, embossing, perforating, manipulating the sheet metal by hand, and perhaps burnishing. Since the metal was about as thick as heavy foil, at 0.1 mm (Vasilakis 1996, 99), great strength was not required. The artisan used a knife or an obsidian blade to cut the band from sheet metal. The blade would have been extremely sharp to avoid any pulling or tearing of the thin metal. If any burnishing was required to remove fine scratches or smooth out wrinkles in the metal, it was probably done at this point. Subsequent decoration of the band from the back of the piece would have been best undertaken on a hard surface covered with a pliable material like leather.

In her examination of a diadem from Mochlos Tomb IV/V/VI (Fig. 10.8; Davaras 1975, 103–104, fig. 3), Corinne de Checchi (2006, 20–21) described the difficulty that would have been encountered by an artisan as he attempted to incise a continuous line, as required to outline what appeared to be abstract animals. She noted an accidental double line near the top of one of the animals where the artisan's tool must have slipped (de Checchi 2006, fig. 7). Unintentional tool marks are also noticeable on the Dog Diadem, near the tail of one dog and in two places—the ear and the neck—on the second dog (Fig. 10.9). In each case, the artisan stopped, moved his hand, and changed the direction of his outline. This problem was avoided somewhat when the backs and tails of the dogs were incised with short strokes that formed broken rather than continuous lines.

The artisan created an outline of each dog. The outlines were then filled with dots, with some dots spilling into the border (Figs. 10.3, 10.4). Two types of awls were required: a scratch awl with a sharp tip for chasing or incising, and a rounded-tip or dull awl for creating the embossed dot repoussé. The diadem-maker was familiar enough with sheet gold to know that if he used a sharp tool to create the dots, he would have cut through the thin metal. Since the perforations for attachment at the ends of the band are larger than those used for the later repair, two differently sized punches or perhaps needles were used as well. A hammer would not have been necessary to

Figure 10.8. Diadem from Mochlos Tomb IV/V/VI, restored by Costis Davaras (Hagios Nikolaos Museum 4313).

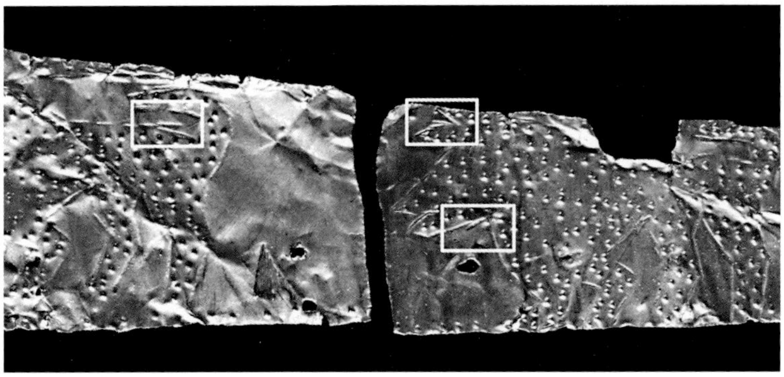

Figure 10.9. Dog Diadem (HM 269). Unintentional tool marks in white boxes; also note three perforations indicating repaired break.

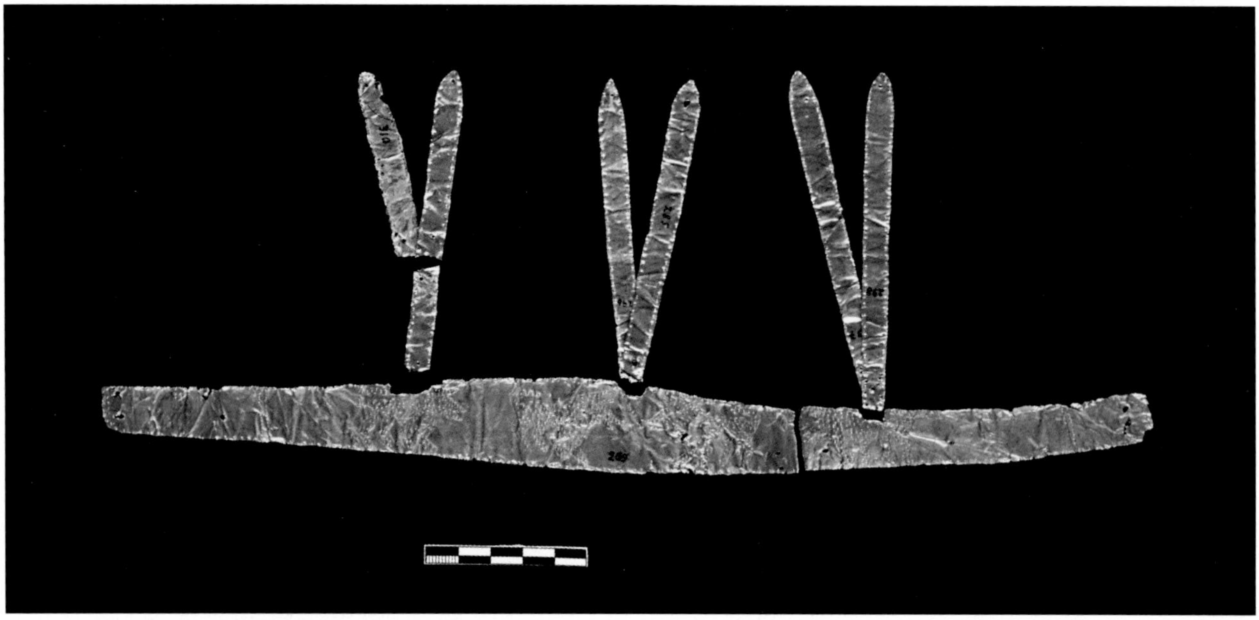

Figure 10.10. Dog Diadem (HM 269), with vertical extensions (HM 285, HM 296–HM 298, HM 310a–c) recovered from Tomb II, as reconstructed by author.

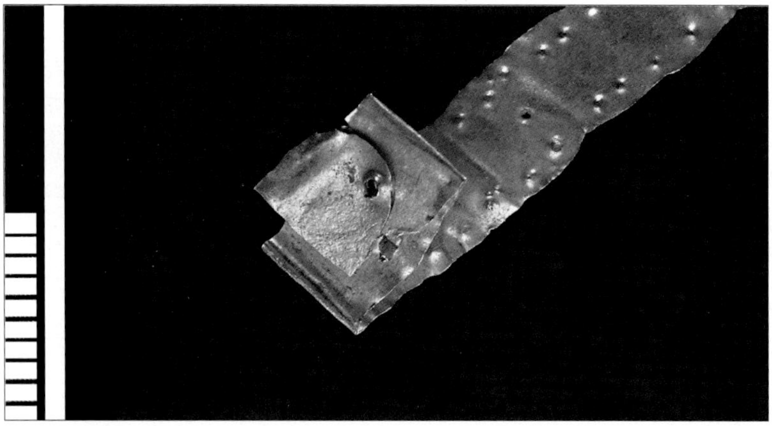

Figure 10.11. Attachment mechanism on antenna-like extension from Mochlos Tomb XIX (HM 295).

drive the awls and punches, as these tools could be pushed into or through the thin metal with ease. If workshops dedicated to the production of gold jewelry had not yet been established at Mochlos, all tools needed to make the diadem—a knife or obsidian blade, awls, punches or needles, and burnishers—would have been available in a leatherworking or woodworking shop. However, it is quite possible that metal tools were not used at all; sharpened sticks, worked animal bone, pieces of shell, and animal teeth could have created the same effects.

At the top of the Dog Diadem, the gold has been significantly broken away in three places (Seager [1912, 26–27] identified broken metal at five places). When this writer began her study of Prepalatial jewelry, Costis Davaras (pers. comm.) suggested that vertical extensions may have originally been attached to the diadem (see also Davaras 1975, 109). In the early 1970s, Davaras restored what he called "antennae" to the Mochlos diadem decorated with abstract horned animals, perhaps *agrimia* or Cretan wild goats, from Tomb IV/V/VI (Fig. 10.8; Davaras 1975, 103–104, fig. 3).

At least 14 antenna-like extensions were recovered from Tomb II. Six of these (HM 285, 296–298, 310a–c [310 included three fragments, comprising two extensions]) included a rounded tip with a single perforation and a border in dot repoussé; some were broken with no visible means of attachment. Others contained single or multiple perforations at the attachment end, and in the case of HM 310, a few small holes along the length of the extension. The preserved length of the most complete antennae in this group ranged from 9.4 to 10.5 cm (Vasilakis 1996, 109–110). Figure 10.10 illustrates how the Dog Diadem may have appeared with antennae in place. Of course, single rather than double extensions may have projected up from each notch. Apparently, the artisan slipped the extensions through horizontal slits he had cut at the top of the diadem. He then folded the end of each extension over several times in the back before perforation and further fastening (Fig. 10.11; intact example from Mochlos Tomb XIX, HM 295). When the antennae were later torn off, notches of a roughly rectangular shape remained.

Perforations at the top of the extensions indicated that these strips were likely affixed to a tall leather or cloth headdress, perhaps sewn on by another artisan. Since the extensions were too flimsy to have stood up on their own, the diadem-maker must have known that they would be attached to another material. However, extensions may not have been part of the original diadem design; the notch on the right was cut deeply into the back of a dog. Antenna-like extensions, without associated diadems, were recovered from several tombs at Mochlos. This indicates that headdresses with gold bands and tall vertical extensions may have been a tradition at this site.

Use and Deposition

The Dog Diadem was surely used in a ceremonial context, perhaps worn repeatedly as part of public display and ritual. This may have taken place near the tomb where the diadem was found. Cemeteries with paved outdoor spaces appeared to be the focal point for community ritual during early periods in Crete (Branigan 1993; Marinatos 1993; Dickinson 1994; Branigan, ed., 1998), with tombs revisited for ceremonies after initial burials took place (Betancourt 2007, 37). Such a space existed in front of Mochlos Tomb IV/V/VI, cut into the highest cliff, overlooking the sea (Seager 1912, fig. 15; Soles 1992, plan 3). Ceremonies may have marked stages in the transition from corpse to the realm of ancestor, as the dead became part of the collective identity of the community (cf. Keswani 2004 on Bronze Age Cyprus; Metcalf and Huntington 1991; Murphy 1998 on the Mesara). Annual celebrations such as harvest festivals may have also been associated with these spaces (Branigan 1993, 130–136). Public gatherings would be appropriate occasions for adornment with gold jewelry, especially for pieces as dramatic as the Dog Diadem.

The Dog Diadem was recovered in poor condition. However, instead of being recycled for its gold content, it was retained and later buried deep in the back of a tomb, possibly many years after its use as adornment and as a grave offering. This piece must have been highly valued. Perhaps it was an heirloom. During the period of its use, the band would have deteriorated even more if it had not been attached to a backing. The Dog Diadem was worn, broken, repaired, and most likely worn again, all before its final deposition. The repair is confirmed by three perforations, indicating someone mended the band, perhaps securing it with wire or thread, as it was reattached to its backing (Fig. 10.9).

The circumstances of the diadem's burial are unknown. Did it decorate a corpse when the body was first interred? Were the extensions removed at the time of initial burial or years later when bones and grave goods were moved to the back of the tomb? In the last role it played in Minoan life, did the Dog Diadem—as part of a buried cache of gold jewelry—represent something other than personal adornment?

The antenna-like extensions were not carefully unfolded and removed from the diadem. They were ripped off, either by someone who did not understand the construction of the piece or in a deliberate forceful act. Evidence from the Aegean Bronze Age confirms the deliberate destruction of grave goods such as pottery, bronze vessels, daggers, and marble objects (Åström 1987, 215; Hamilakis 1998, 122; Soles 1999, 2001; Broodbank 2000, 268; Betancourt and Davaras, eds., 2003, 136–137). Several explanations have been offered to better understand this custom. Grinsell (1961, 476; 1973) compares the ceremonial "killing" of objects to the killing of humans and animals during funerary ritual; when an object is broken, its spirit is released, and it is then able to accompany the dead to the afterlife. Grave goods of a personal nature may also be destroyed in order to prevent them from being used again, either in deference to the deceased or because of a fear of objects associated with the dead (Fossey 1985, 23; Grinsell 1961, 477). Some objects, especially ceramic drinking vessels, may be broken as a symbolic act after use in burial rituals (see Betancourt and Davaras, eds., 2003, 136–137 for a discussion of ritual "killing" at Pseira). Intentional destruction of grave goods also illustrates the high rank of those able to destroy or remove valuable objects from circulation; in this process, the status of the deceased and his descendents may be elevated (see Broodbank 2000, 267–268 on the EBA Cyclades) or lowered.

Other gold ornaments from Mochlos were ripped or cut (e.g., Seager 1912, fig. 8:II.7, II.8, II.18). The bottom edge of the diadem reconstructed by Costis Davaras (Fig. 10.8) was purposely cut off. Perhaps families associated with specific pieces of jewelry kept small parts of objects because of the value of gold (Davaras 1975, 110) or for sentimental or other reasons (cf. Carter 2007, 100 on long obsidian blades). Gold fragments held back from final burial may have been melted down and used to create new objects. It is also possible that, years after burial, the objects had lost their human association and were thought of strictly as bullion; parts of objects could be removed from tombs with some gold left behind as gifts to ancestors or to the gods.

Conclusions

A detailed study of the manufacturing, use, and deposition of the Dog Diadem provides information on the crafting of jewelry and insight on cultural practices present at Mochlos. No specialized knowledge of jewelry-making or metallurgy was required to create the diadem. Although the sheet metal was carefully cut and the dogs were sketched accurately, the spacing of the dots along the border was irregular, with some dots haphazardly placed (Figs. 10.3, 10.4). Also, the attachment holes were crudely punched and unfinished. This may indicate that craftsmen of different abilities contributed to the manufacture of the piece, signifying collaboration on the production of early gold jewelry before the advent of full-scale workshop production. When the diadem became part of a headdress, another artisan with different skills, such as a seamstress or leatherworker, may have created the head covering and completed the construction of the piece.

The function and form of the Dog Diadem changed through time. The diadem was originally designed as an object of personal adornment, likely worn on the head in public ceremonies. Decisions related to the choice of gold, form, and decoration may have been made by the artisan or the patron, or may have been a collaborative effort. At the time of initial manufacture or at some point later in the diadem's use, vertical extensions were added, and a headdress was created. Evidence of additional antenna-like extensions without associated diadems from Mochlos Tombs I/II/III, IV/V/VI, and XIX, indicated that a tradition of wearing gold diadems with tall vertical extensions existed at this site. The repair and subsequent reuse of the Dog Diadem suggested that it was highly valued, perhaps an heirloom worn by multiple generations.

As social groups move toward greater complexity, as was the case during the Prepalatial period in Crete, public display and ritual may take on added importance. Ceremonies are a way to demonstrate collective identity and shared cultural beliefs and are a means to maintain social relations during times of transition. Gold jewelry found at Mochlos—especially objects that adorned the head such as the Dog Diadem—was surely used in a public manner. The active nature of gold jewelry, with its connection to distant, foreign lands, must have elevated the status of those who commissioned it, made it, wore it, and were buried with it. The Dog Diadem may have been associated with emerging elites at a time of increasing individualism (cf. Day and Wilson 2002 on EM Knossos) or with community leaders focused on demonstrations of shared ideology.

The diadem was taken out of circulation and buried in Tomb I/II/III. The deliberate destruction of the object, which occurred when the extensions were ripped off, suggested that the diadem was ritually killed at either the time of initial burial or later when grave goods and bones were moved within the tomb. The presence in several tombs of ripped or cut jewelry indicates that this practice was a custom at Mochlos, perhaps associated with community rituals practiced at the cemetery. The inclusion of the diadem in a cache of gold jewelry buried deliberately in a deep pit suggests a ritual or votive deposit. This last deliberate act associated with the diadem indicates that, at this time, it may have been valued more as bullion than as an object of adornment.

Acknowledgments

The author wishes to acknowledge the Greek Ministry of Culture and the archaeological museums in Athens, Herakleion, and Hagios Nikolaos for permission to study the objects discussed here. Photographs of the jewelry were taken by Chronis Papanikolopolous of the INSTAP Study Center publication team. Permission to publish the photos was granted by the archaeological museums in Herakleion and Hagios Nikolaos. Funding for research was provided by the Fulbright Foundation/Institute for International Education and the Louis J. Kolb Foundation of the University of Pennsylvania Museum of Archaeology and Anthropology.

References

Andrews, C. 1994. *Amulets of Ancient Egypt*, Austin.

Archi, A. 2002. "Jewels for the Ladies of Ebla," *ZA* 92, pp. 161–199.

Aruz, J. 1984. "The Silver Cylinder Seal from Mochlos," *Kadmos* 23, pp. 186–188.

———. 2008. *Marks of Distinction: Seals and Cultural Exchange between the Aegean and the Orient (ca. 2600–1360 B.C.)*, Mainz.

Åström, P. 1987. "Intentional Destruction of Grave Goods," in *THANATOS: Les coutumes funéraires en Egée à l'âge du Bronze (Aegaeum* 1), R. Laffineur, ed., Liège, pp. 213–217.

Bailey, D.W. 2004. "Varna," in *Ancient Europe 8000 B.C.–A.D. 1000: Encyclopedia of the Barbarian World.* I: *The Mesolithic to Copper Age (c. 8000–2000 B.C.)*, P. Bogucki, and P.J. Crabtree, eds., New York, pp. 341–344.

Betancourt, P.P. 2007. *Introduction to Aegean Art*, Philadelphia.

Betancourt, P.P., and C. Davaras, eds. 2003. *Pseira* VII: *The Pseira Cemetery* 2. *Excavation of the Tombs*, Philadelphia.

Bevan, A. 2004. "Emerging Civilized Values? The Consumption and Imitation of Egyptian Stone Vessels in EM II–MM I Crete and its Wider Eastern Mediterranean Context," in *The Emergence of Civilisation Revisted*, J.C. Barrett and P. Halstead, eds., Oxford, pp. 107–126.

Branigan, K. 1974. *Aegean Metalwork of the Early and Middle Bronze Age*, Oxford.

———. 1993. *Dancing with Death: Life and Death in Southern Crete c. 3000–2000 B.C.*, Amsterdam.

Branigan, K., ed. 1998. *Cemetery and Society in the Aegean Bronze Age*, Sheffield.

Broodbank, C. 2000. *An Island Archaeology of the Early Cyclades*, Cambridge.

Carter, T. 2007. "The Theatrics of Technology: Consuming Obsidian in the Early Cycladic Burial Arena," in *Rethinking Craft Specialization in Complex Societies: Analyses of the Social Meaning of Production (Archaeological Papers of the American Anthropological Association* 17 [1]), R. Flad and Z. Hruby, eds., Berkeley, pp. 88–107.

Davaras, C. 1975. "Early Minoan Jewellery from Mochlos," *BSA* 70, pp. 101–114.

Day, P.M., and D.E. Wilson. 2002. "Landscapes of Memory, Craft and Power in Prepalatial and Protopalatial Knossos," in *Labyrinth Revisited: Rethinking 'Minoan' Archaeology*, Y. Hamilakis, ed., Oxford, pp. 143–166.

de Checchi, C. 2006. "Les techniques d'orfèvrerie minoenne. L'apport des objets de Mochlos," in *Les Ors des Mondes Grec et "Barbare." Actes du colloque de la Société d'Archéologie Classique du 18 novembre 2000*, G. Nicolini, ed., Paris, pp. 17–33.

Demakopoulou, K. 1998. *Κοσμήματα της Ελληνικής Προϊστορίας: Ο Νεολιθικός Θησαυρός*, Athens.

Dickinson, O. 1994. *The Aegean Bronze Age*, Cambridge.

Fossey, J.M. 1985. "The Ritual Breaking of Objects in Greek Funerary Contexts: A Note," *Folklore* 96, pp. 21–23.

Gale, N.H., and Z.A. Stos-Gale. 1981. "Cycladic Lead and Silver Metallurgy," *BSA* 76, pp. 169–224.

Gimbutas, M. 1977. "Varna: A Sensationally Rich Cemetery of the Karanovo Civilization, about 4500 B.C.," *Expedition* 19, pp. 39–47.

Gopher, A., T. Tsuk, S. Shalev, and R. Gophna. 1990. "Earliest Gold Artifacts in the Levant," *CurrAnthr* 31, pp. 436–443.

Grinsell, L.V. 1961. "The Breaking of Objects as a Funerary Rite," *Folklore* 72, pp. 475–491.

———. 1973. "The Breaking of Objects as a Funerary Rite: Supplementary Notes," *Folklore* 84, pp. 111–114.

Hamilakis, Y. 1998. "Eating the Dead: Mortuary Feasting and the Politics of Memory in the Aegean Bronze Age Societies," in Branigan, ed., 1998, pp. 115–132.

Hickman, J. 2002. "Regional Diversity in Early Bronze Age Aegean Diadems," *AJA* 106, pp. 297–298.

———. 2008. *Gold before the Palaces: Crafting Jewelry and Social Identity in Minoan Crete*, Ph.D. diss., University of Pennsylvania, Philadelphia.

Higgins, M.D., and R. Higgins. 1996. *A Geological Companion to Greece and the Aegean*, Ithaca.

Higgins, R.A. 1980. *Greek and Roman Jewellery*, Berkeley.

Immerwahr, S.A. 1990. *Aegean Painting in the Bronze Age*, University Park, PA, and London.

Ivanov, I.S. 1991. "Les objets métalliques de la nécropole chalcolithique de Varna," in *Découverte du Métal*, J.-P. Mohen, ed., Paris, pp. 9–11.

Keswani, P. 2004. *Mortuary Ritual and Society in Bronze Age Cyprus*, London.

Lehrberger, G. 1995. "The Gold Deposits of Europe: An Overview of the Possible Metal Sources for Prehistoric Gold Objects," in *Prehistoric Gold in Europe: Mines, Metallurgy and Manufacture*, G. Morteani and J.P. Northover, eds., Dordrecht, pp. 115–144.

Lonsdale, S.H. 1979. "Attitudes towards Animals in Ancient Greece," *GaR* 26, pp. 146–159.

Marinatos, N. 1993. *Minoan Religion*, Columbia.

Maxwell-Hyslop, K.R. 1971. *Western Asiatic Jewellery c. 3000–612 B.C.*, London.

McGeehan-Liritzis, V. 1996. *The Role and Development of Metallurgy in the Late Neolithic and Early Bronze Age of Greece* (*SIMA 122*), Jonsered.

Metcalf, P., and R. Huntington. 1991. *Celebrations of Death: The Anthropology of Mortuary Ritual*, Cambridge.

Misch-Brandl, O. 1994. *The Oldest Gold in the World: Varna, Bulgaria*, Jerusalem.

Moorey, P.R.S. 1999. *Ancient Mesopotamian Materials and Industries: The Archaeological Evidence*, Oxford.

Muhly, J.D. 1973. *Copper and Tin: The Distribution of Mineral Resources and the Nature of the Metals Trade in the Bronze Age*, New Haven.

———. 1985. "Beyond Typology: Aegean Metallurgy in Its Historical Context," in *Contributions to Aegean Archaeology: Studies in Honor of William A. McDonald*, N.C. Wilkie and W.D.E. Coulson, eds., Minneapolis, pp. 109–141.

———. 2001. "Review of *The Domestication of Metals: The Rise of Complex Metal Industries in Anatolia* by K. Aslihan Yener," *AJA* 105, pp. 729–730.

———. 2002. "Early Metallurgy in Greece and Cyprus," in *Anatolian Metal* II, Ü. Yalçin, ed., Bochum, pp. 77–82.

———. 2006. "Chrysokamino in the History of Early Metallurgy," in *The Chrysokamino Metallurgy Workshop and Its Territory* (*Hesperia Suppl.* 36), P.P. Betancourt, ed., Princeton, pp. 155–177.

Murphy, J.M. 1998. "Ideology, Rites and Rituals: A View of Prepalatial Minoan Tholoi," in Branigan, ed., 1998, pp. 27–40.

O'Connor, D. 1993. *Ancient Nubia: Egypt's Rival in Africa*, Philadelphia

Ogden, J. 2000. "Metals," in *Ancient Egyptian Materials and Technology*, P.T. Nicholson and I. Shaw, eds., Cambridge, pp. 148–176.

Papathanassopoulos, G. 1981. *Neolithic and Cycladic Civilization*, Athens.

Renfrew, C. 1967. "Cycladic Metallurgy and the Aegean Early Bronze Age," *AJA* 71, pp. 1–20.

———. 1978. "Varna and the Social Context of Early Metallurgy," *Antiquity* 52, pp. 199–203.

Sakellarakis, J.A. 1993. *Herakleion Museum: Illustrated Guide*, Athens.

Sakellarakis, Y., and E. Sapouna-Sakellaraki. 1997. *Archanes: Minoan Crete in a New Light* I–II, Athens.

Sbonias, K. 2000. "Specialization in the Early Minoan Seal Manufacture: Craftsmen, Settlements and the Organization of Production," in *Minoisch-Mykenische Glyptik: Stil, Ikonographie, Funktion* (*CMS* 6), Berlin, pp. 277–293.

Seager, R.B. 1912. *Explorations in the Island of Mochlos*, Boston and New York.

Soles, J.S. 1992. *The Prepalatial Cemeteries at Mochlos and Gournia and the House Tombs of Bronze Age Crete* (*Hesperia Suppl.* 24), Princeton.

———. 1999. "The Ritual 'Killing' of Pottery and the Discovery of a Mycenaean Telestas at Mochlos," in *Meletemata: Studies in Aegean Archaeology Presented to Malcolm H. Wiener as He Enters His 65th Year* (*Aegaeum* 20), vol. III, P.P. Betancourt, V. Karageorghis, R. Laffineur, and W.-D. Niemeier, eds., Liège and Austin, pp. 787–791.

———. 2001. "Reverence for Dead Ancestors in Prehistoric Crete," in *Potnia: Deities and Religion in the Aegean Bronze Age. Proceedings of the 8th International Aegean Conference/8e Rencontre égéenne internationale. Göteborg, Göteborg University, 12–15 April 2000* (*Aegaeum* 22), R. Laffineur and R. Hägg, eds., Liège and Austin, pp. 229–236.

———. 2005. "From Ugarit to Mochlos—Remnants of an Ancient Voyage," in *Emporia: Aegeans in the Central and Eastern Mediterranean* (*Aegaeum* 25), R. Laffineur and E. Greco, eds., Liège and Austin, pp. 429–439.

Tsountas, C. 1899. "Kykladika II," in *ArchEph* 1899, pp. 73–134.

Vasilakis, A.S. 1996. *Ο Χρυσός και ο Άργυρος στην Κρήτη κατά την Πρώιμη Περίοδο του Χαλκού*, Herakleion.

Weisgerber, G., and E. Pernicka. 1995. "Ore Mining in Prehistoric Europe: An Overview," in *Prehistoric Gold in Europe: Mines, Metallurgy and Manufacture*, G. Morteani and J.P. Northover, eds., Dordrecht, pp. 159–182.

Wilkinson, A. 1971. *Ancient Egyptian Jewellery*, London.

Williams, D. 1998. "Identifying Greek Jewellers and Goldsmiths," in *The Art of the Greek Goldsmith*, D. Williams, ed., London, pp. 99–104.

Xanthoudides, S.A. 1918. "Μέγας πρωτομινωικὸς τάφος Πύργου," *ArchDelt* 4, pp. 136–170.

Yener, K.A. 2000. *The Domestication of Metals: The Rise of Complex Metal Industries in Anatolia*, Leiden.

Zachos, K. 1990. "The Neolithic Period in Naxos," in *Cycladic Culture: Naxos in the 3rd Millennium* B.C., L. Marangou, ed., Athens, pp. 29–32.

CHAPTER

11

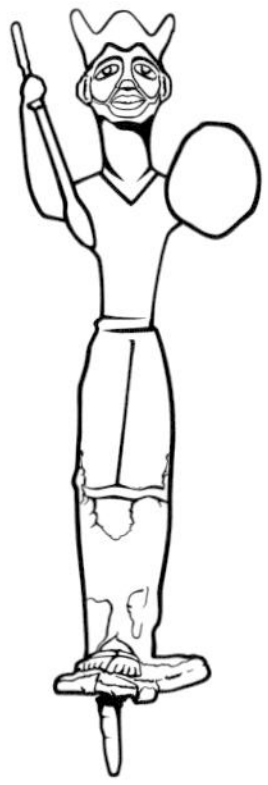

The Triangular "Daggers" of Prepalatial Crete

Keith Branigan

Forty years ago I published a paper on the Early Bronze Age (EBA) daggers of Crete (Branigan 1967), which divided them into two very distinct groups, the long daggers and the so-called triangular daggers (although with their convexly curved edges they are not, strictly, triangular). While the long daggers proved to have a long history in Crete, surviving not only throughout the Bronze Age but into later periods too, the triangular daggers seem to have gone out of production no later than Early Minoan (EM) III. Unlike the ubiquitous long daggers, they were also unique to Crete with only one possible example found elsewhere in the Aegean.

The principal conclusions of the 1967 paper can be briefly summarized. The triangular daggers probably appeared in EM I, although there were no examples from pure EM I deposits; but there was no reason to think that they were in any way ancestral to, or earlier than, the first Cretan long daggers (pace Xanthoudides 1924, 26; Pendlebury 1939, 70; Alexiou 1969, 21). The bulk of the daggers were in use and deposited in tombs in EM II, and a few notably large examples may have survived in use into EM III. They can be divided into seven typological groups, with sub-groups based on riveting variations, of which Types 1–3 accounted for over 75% of the extant blades. The dominance of the two-rivet system of hafting (86% of examples) supported the view that they were probably mostly of EM I–II date. It was suggested that they were probably made of copper from local sources and that some of them were deliberate arsenical alloys. Their function was discussed at some length, and it was noted that they were too short, broad, and weak to have made satisfactory daggers. A purely funerary or votive purpose was rejected because of the clear signs of use and repair on many daggers. Use as an all-purpose household knife, particularly in the period before long single-edged knives were produced, seemed plausible. It was, however, noted that many of the earliest Cretan long daggers then known were no longer or stronger than many of the triangular daggers. The possibilty that they were, therefore, in some sense rivals of the long daggers was allowed.

Forty years on, it seems worth revisiting the Prepalatial triangular daggers (which have not been discussed since Branigan 1974) and re-assessing their history, manufacture, and usage in the light of new research. I am delighted to have this opportunity to offer this paper in honor of Jim Muhly, whose researches in the field of prehistoric metallurgy in the Mediterranean I have read and admired for almost as long as my acquaintance with the triangular daggers of Prepalatial Crete!

The Corpus

In 1967 there were 91 triangular daggers recorded, and my 1974 catalog included 94 examples, including two said to be from Amorgos. One of those (cat. no. 90) was the sole example of an eighth "type" and on reflection should not be classed as a triangular dagger at all. On the other hand, a much damaged fragmentary dagger from Hagia Triada included in the unclassified long daggers (cat. no. 385) was probably a triangular dagger, so the 1974 total of 94 examples stands. To these we may now add four more examples, one of which had actually been discovered in 1971 and two more c. 1974–1975, bringing the total to 98. They are inserted into the 1974 catalog sequence at the end of the appropriate type/subtype list with an "A" added to the preceding catalog number.

12A. Type 1. Mochlos Tomb VI cleaning. Context: EM IIB–III. L. 10.9; max. w. 4.1 cm. Soles 1992, 61, pl. 2.3.

15A. Type 1B. Moni Odigitria? Mitsotakis Collection. L. 11.7; max. w. 5.0 cm. Marangou 1992, 258, no. 321.

73A. Type IIIA. Moni Odigitria, Tholos B, surface soil. L. 6.5; max. w. 3.8 cm. Branigan 2010, 147, no. M1, fig. 57.

80A. Type VA. Moni Odigitria, Mitsotakis Collection. L. 14.5; max. w. 6.0 cm. Xenaki-Sakellariou 1986.

Dagger 15A can be ascribed to the Moni Odigitria tholos tombs with near certainty. Almost all of the EM material in the Mitsotakis Collection came from these tombs. We might also provide here a new entry for dagger cat. no. 89 with corrected information (with thanks to Alexiou and Warren).

89. Type VII. Lebena Tomb I. Context: EM IIB? L. 8.1; max. w. 4.5 cm. Alexiou and Warren 2004, 39, no. 109, fig. 9, pl. 15.

As blades, none of the four new daggers add anything to the typological groups identified in 1967/1974, although, as we shall see, dagger 80A provides extremely important new information on the hafting of these daggers and by implication on their function or usage.

Chronology

The detailed publication of the Lebena tholos tombs provides two of the best attested dates for triangular daggers that we have so far. A dagger (cat. no. 55) from Lebena Tomb IIA was found in the lower EM II level of the deposits (Alexiou and Warren 2004, 147). A dagger (cat. no. 89) from Tomb I was found with an EM IIB Vasiliki cup and other EM II pottery (Alexiou and Warren 2004, 39). A third small dagger was found in the EM I deposits in Lebena Tomb II, but sadly it is not a triangular dagger (with convex edges) but a small long dagger. Thus it is still the case that we have no triangular daggers with pure EM I associations, and the date of their appearance can still not be fixed by stratified examples.

The four new daggers provide no significant new information about the chronology of the triangular daggers as a whole. The dagger found in the cleaning of Mochlos Tomb VI was part of a small cache or hoard of material that seems to have been deposited (re-deposited?) *outside* the tomb and then to have fallen in when a wall collapsed. The pottery associated with this material and event was 85% EM IIB–III (Soles 1992, 58), so that the dagger probably had been made no later than EM IIB. The two smaller daggers from Moni Odigitria have no context, one being from a collection and the other from surface soil. This is true of the third dagger from this site, but despite the lack of a context it does raise important chronological questions due to the decoration of the gold sheathing of its haft. The "looped rope" decoration is a motif normally found on Kamares and later pottery in Crete. Agnes Xenaki-Sakellariou (1986, 240–242, fig. 7) illustrated a particularly good example on a cup from Protopalatial phase IB at Phaistos, and concluded that the dagger from the Mitsotakis Collection was to be dated to the beginning of the Protopalatial period. This is of course much later than the latest dates for production previously proposed (Branigan 1967). We shall return to this dagger and its dating below. The present contextual evidence for the chronology of the triangular daggers can be summarized as follows:

EM II	19
EM II–III	4
EM III	2
EM I–MM I/II	63

These contextual dates all refer, of course, to the date of deposition rather than manufacture. The biggest single group of daggers comes from Hagia Triada Tholos A, which unfortunately does not allow them to be closely dated. The EM II examples come mostly from the lower stratum of Platanos Tholos A but include examples from Mochlos and Lebena. The two EM III daggers are from Tomb XXI at Mochlos, which Warren (1969, 196) and I both date to EM III. On present evidence, therefore, we have to say that we cannot be sure that the first triangular daggers were made in EM I, but they were certainly made in EM II. There is at present no contextual evidence to suggest that they were still being made in EM III, although some were still in use at that time.

Manufacturing Processes

Since 1967 there have been several technical research programs that have thrown light on Early Minoan copper- and bronzeworking in general, and together with information available in 1967, allow us to say more about the manufacture of triangular daggers.

In 1967 I suggested that the Early Minoans relied on local copper sources for their raw materials, though this has been challenged by some (e.g., Nakou 1995, 17). Since then some extensive and important copper sources have been discovered in the Cyclades and Attica (e.g., Gale et al. 1985; Stos-Gale 1994; Bassiakos and Philaniotou 2007). These, together with copper ores from Cyprus, Turkey, and Crete have been "finger-printed" by lead isotope analysis and compared with lead-isotope analyses of Minoan copper/bronze artifacts (e.g., Stos-Gale 1993). Among the analyzed items were 18 triangular daggers. Ten of these matched the Kythnos fingerprint, two were tenatively identified with either an Anatolian or Greek mainland source, and six matched the fingerprint of the Chrysostomos copper source near Kali Limenes on the south coast of Crete. The metalworkers who produced triangular daggers therefore seem to have used both imported and local copper to make these blades.

The Chrysostomos copper ore deposit is close to, and visible from, two tholos cemeteries, each with two tombs (Branigan 1993, 145, nos. 29–32). The two Lasaia tombs may flank a contemporary settlement. It seems unlikely, therefore, that the acquisition and transformation of metals was "spatially separated from the consuming population" and regarded as a secret and exclusive activity as Georgia Nakou has suggested (1995, 18). Indeed, the distribution of metalworking debris in the EBA Aegean as a whole suggests that metalworkers worked in or immediately alongside settlements rather than in remote and secret locations (Branigan 1974, 130–131).

The copper used to make triangular daggers, as well as other metal artifacts, was sometimes alloyed with other metals. In 1974 I tabulated the analyses of dozens of copper and bronze artifacts analyzed (and generously supplied to me) by S. Junghans, but I did not highlight the results from triangular daggers for special attention. I now present these results for the 19 triangular daggers analyzed by Junghans in Table 11.1 below (cat. nos. and type refer to Branigan 1974). Thresholds for each element are: Sn 1.0%, Pb 2%, As 2%, Ni 0.5%.

While it is widely accepted that tin and lead levels above 1.0% almost certainly indicate a deliberate alloy, there has been much debate about the level at which arsenical content might indicate a deliberate alloy. In 1968, based on Junghans's analyses, I suggested that in Crete, an arsenical content above 2% indicated a deliberately produced alloy. As can be seen in Table 11.1, triangular daggers made of a tin alloy are almost mutually exclusive with those containing 2% or more arsenic. Caley (1949) had previously pointed out that where metallurgists produced an arsenical bronze, they did not usually include any tin in the alloy. Colin Renfrew (1972) and Jim Muhly (1973) also came to the conclusion that many of the products of EBA metallurgists in the Aegean were deliberate arsenical alloys.

That one third of the analyzed triangular daggers prove to be deliberate tin alloys is surprising, given that tin alloys are generally regarded as something of a scarcity outside of the Troad in the Aegean EBA (Doonan, Day, and Dimopoulou-Rethemiotaki 2007, 103). The metal used to make these triangular daggers is, of course, well below the level of an "ideal" tin-bronze with 10% tin, but these triangular daggers nevertheless indicate that Crete must have been importing tin—albeit in small quantities—by EM II at the latest. The source of the tin used in the Bronze Age Aegean has been the source of much debate (e.g., Hall and Steadman 1991), in which Jim Muhly has played a major part (e.g., Muhly 1973, 1985). There is still no solution to the problem, but given the use of tin-bronze in the Troad from EB I, it seems likely that Crete must have obtained its supplies from there, whatever the ultimate source of the raw material.

Site	Catalog Number	Type	Context	%			
				Sn	Pb	As	Ni
Hagia Triada	1	I	EM I–MM II	—	2.4	3.8	0.84
Hagia Triada	27	II	EM I–MM II	—	—	2.3	0.52
Hagia Triada	30	II	EM I–MM II	—	—	3.2	1.15
Hagia Triada	34	II	EM I–MM II	—	—	—	0.92
Hagia Triada	63	III	EM I–MM II	—	2.2	4.0	1.05
Hagia Triada	78	Va	EM I–MM II	—	—	—	0.99
Hagia Triada	79	Va	EM I–MM II	—	—	2.1	—
Hagia Triada	81	Vb	EM I–MM II	—	—	2.7	—
Hagia Triada	84	VII	EM I–MM II	—	—	2.3	—
Marathokephalo	49	II	EM I–MM I	1.5	—	—	—
Marathokephalo	50	II	EM I–MM I	—	—	4.1	1.15
Marathokephalo	64	III	EM I–MM I	1.25	—	—	—
Marathokephalo	65	III	EM I–MM I	—	—	2.2	0.74
Marathokephalo	66	III	EM I–MM I	1.8	—	—	—
Mochlos Tomb XXI	16	Ic	EM III	—	—	3.5	—
Mochlos Tomb II	73	IIIa	EM I–MM II	4.8	—	—	—
Mochlos Tomb XI	74	IV	EM I–MM II	—	—	3.0	—
Platanos A,1	45	II	EM II	1.75	2.2	3.2	0.57
Platanos A,1	93	fragment	EM II	2.1	—	—	—

Table 11.1. Analyses of 19 triangular daggers from Junghans (1969) showing results for tin, lead, arsenic, and nickel. Catalog numbers and types refer to Branigan 1974.

The nickel content of the analyzed triangular daggers has been included in Table 11.1 because in 1968 I noted that many EBA copper artifacts from Crete proved to have higher nickel contents than artifacts found elsewhere in the Aegean. It was partly on this basis that I suggested that Crete used its own sources of copper. Recent work on the metallurgical materials from Chrysokamino (Bassiakos and Catapotis 2006; Catapotis and Bassiakos 2007) has thrown some light on this anomaly. Copper prills embedded in slag from Chrysokamino revealed high levels of arsenic and significant levels of nickel (Bassiakos and Catapotis 2006, 340–341, 346, table F.5; Catapotis and Bassiakos 2007, 72, fig. 4.3). Mihalis Catapotis and Yannis Bassiakos argue that these concentrations were not derived from the copper ores smelted at the site, but rather from arsenides deliberately added to the smelting charge to produce an arsenical copper. In this connection it is worth noting that of the nine triangular daggers with more than 0.5% nickel in Table 11.1, seven were found to be arsenical coppers with arsenic contents ranging from 2.2 to 4.1% (the remaining two daggers had arsenic levels at 1.5%). Thus, there is good reason to think that the high nickel content noted in EM copper artifacts, including triangular daggers, was not the result of using a particularly distinctive (and local) copper ore, but rather of the chosen alloying techniques and raw materials employed by EM metallurgists.

In the case of the triangular daggers, the smelted copper was apparently poured into open moulds, suitable for the casting of blades with the flat profile seen on 90% of the triangular daggers. Working these into finished products would have been a straightforward process. One example exists that had been more extensively worked to produce a mid-ridge (Branigan 1974, cat. no. 82). There is a small group of seven triangular daggers, however, which feature a mid-rib (Branigan Type VII), and these were presumably cast in two-part molds. Although the earliest mid-rib daggers known in Crete in 1974 were long daggers of Types V and VI from Pyrgos, I suggested a production date of EM IIB for the Type VII triangular daggers. The only example for which a reasonably firm contextual date can be provided is cat. no. 89 from Lebena Tomb I (see above), and that date is EM IIB. Nevertheless, the new discoveries at Poros-Katsambas reveal EM I–IIA two-part molds for daggers with mid-ribs (Doonan, Day, and Dimopoulou-Rethemiotaki 2007, 107–109), so that an earlier date for the introduction of the Type VII triangular dagger is likely. The fragmentary molds so far recovered from Poros-Katsambas appear to be for the manufacture of long daggers, but since not a single triangular dagger has yet been found in North-Central Crete, this is hardly surprising.

Function

Now that we have good evidence for the production of daggers with cast mid-ribs in EM I–IIA, the overwhelming preference for the adoption of a "flat" profile among the triangular daggers must be seen as a matter of choice rather than a limitation imposed by early, "primitive" casting techniques. We pointed out (Branigan 1967, 237) that triangular daggers were seemingly too broad, too weak, and too short to be convincingly identified as daggers. The average effective length of the triangular daggers (that is, the exposed blade) is only 7.0 cm—less than three inches. They are broad in relation to their length—the length to maximum width ratio, even in the slimmest examples, is never greater than 3:1. The vast majority have no strengthening mid-ridge or mid-rib, and 86% of them are attached to their haft by only two small rivets. Triangular daggers are therefore totally unsuited for one of the principal functions of a dagger, which is to stab. They are difficult to envisage being used as an offensive weapon of any sort. Their broad, short blades with convex edges seem best equipped for scraping, spreading, and cutting.

Sinclair Hood (1971, 118) suggested that they might have been used as spearheads, and I had briefly considered this possibility myself. The leaf-shaped blade with a "flat" profile is not uncommon for spearheads. But the base of the Minoan triangular blades is so wide (80% are between 4 cm and 6 cm), and the haft/shaft can be seen to be the same width/diameter, indicating that they could never have penetrated deeper than the length of the exposed blade, which, as we noted above, averages less than 7 cm (3 inches). Furthermore, the two flimsy rivets that held the blade in place would have easily and repeatedly broken under the stress of thudding into its target or, worse, rocks or broken ground, if it had missed!

In 1967 I therefore considered two possible alternative uses for triangular daggers. One was that they were manufactured mainly for deposition in tombs, and were therefore symbolic rather than practical implements. The fact that all 92 daggers with a known provenance came from tombs might be thought to support that notion, but the vast majority of all EM artifacts have come from burial places. Emphatic evidence against the production of triangular daggers as symbolic or token grave goods, however, is provided by the evidence for considerable wear and tear and, in some cases, repair, before deposition. Many triangular daggers reveal a lengthy biography. Around 25% of them have significant damage to their cutting edges, and 15% show evidence for having suffered heel damage from pressure on the rivets, or have been re-riveted at some time. The triangular daggers were clearly used in a way that put stress on both their cutting edges and their rivets.

The other suggested use for the triangular daggers was the obvious one, that they were basically domestic knives. They would have been suitable for most household tasks connected with preparing food and cleaning and cutting animal skins. Long daggers were not very suitable for some of these activities, and in this sense long and triangular "daggers" may have been complementary rather than rivals. It might also mean they were associated with different groups in society. Renfrew (1972, 394) saw daggers as male status symbols, and Todd Whitelaw (1983, 343) believed they were buried with the heads of nuclear families. Nakou (1995, 9–13) examined in some detail the evidence that daggers in the Aegean EBA were used as status symbols and concluded "that their symbolic associations were masculine." Some of the finely cast daggers with "omega" hafts from Platanos and Koumasa, and the superb ribbed dagger with a "maltese cross" (or crossed double axes?) motif from Mochlos (Branigan 1974, pl. 7) certainly

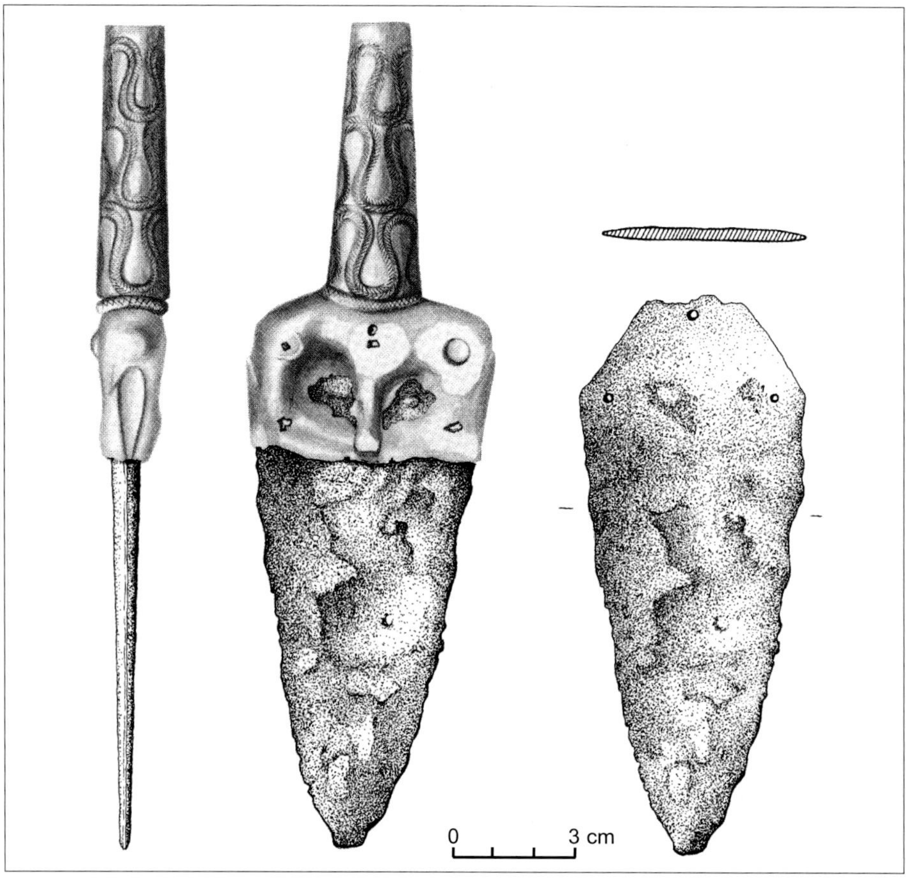

Figure 11.1. The gold-hafted dagger from Moni Odigitria in the Mitsotakis Collection (after Xenaki-Sakellariou 1986, drawing by Mme. Dringopoulou-Faklari).

look like they were produced to display status. Protopalatial Malia has yielded a descendant of these ribbed long daggers, resplendent with its gold haft sheathing preserved, which hints at what might have been in the EM examples. The contents of Levkas tomb R17A (Renfew 1972, 382), dated EB II–III, including two ribbed long daggers and a gold haft sheathing, confirm that such items of display were to be found in the Aegean EBA.

Now, of course, we have an Early Minoan dagger with a gold haft sheathing, but it is not a long dagger with fine mid-ribs but a simple "flat" triangular dagger (Fig. 11.1). If we are right in regarding triangular daggers not as offensive weapons but as domestic knives, then the choice of such a dagger for elaborate and expensive enhancement as a status symbol is difficult to understand. The gold hafted dagger from Moni Odigitria (above cat. no. 80A) requires us to think again about the function and symbolism of the triangular daggers. One is tempted to regard it as a "one off," though this leaves unexplained the choice of such a humble domestic item for such flamboyant treatment. But there is evidence to suggest that other triangular daggers may have been hafted in a similar way.

When I was recording the triangular daggers in the collections of the Herakleion Museum back in 1964, I noticed and recorded traces of the hafting systems that were used on both triangular and long daggers. In the case of the triangular daggers, I classified these into six different types (Branigan 1967, 237, fig. 9). There was a small group of daggers with corrosion marks on their heels, which did not fit into any of these six types, and at the time I could not understand. I recorded them, however, and showed these marks on the drawings of the daggers in my monograph on *Aegean Bronzework* (1974); I reproduce them here (Fig. 11.2). In each case, the daggers had a pair of round or oval corrosion marks set within the area of the haft. With the evidence of the Moni Odigitria dagger now available, we can see that these marks were created by oval or round sockets left in the haft to take a pair of inset "eyes" made of some other material (rock crystal perhaps?). Two other daggers showed variations of this treatment. One had a single circular corrosion mark rather than a pair, and the second had a heart-shaped mark. Finally, I recorded one dagger as the single example of my Type E haft with a "double-omega" form. We can now see that

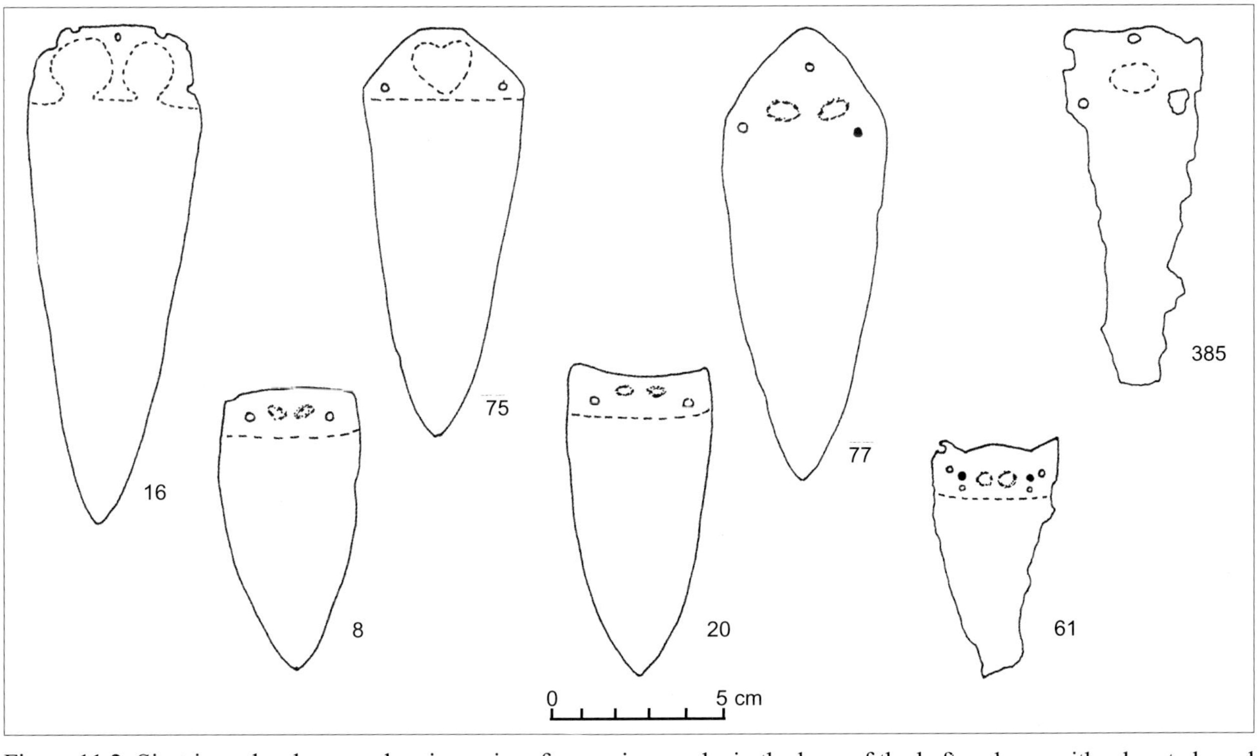

Figure 11.2. Six triangular daggers showing pairs of corrosion marks in the base of the haft and one with a heart-shaped mark (cat. nos. refer to Branigan 1974).

Catalog Number	Type	Context	Site	Length (cm)
8	I	EM I–IIB	Platanos A, 1	8.4
16	I	EM III	Mochlos Tomb XXI	14.8
20	II	EM I–MM I	Koumasa B	9.0
61	III	EM I–MM I	Hagia Triada A	7.9
77	V	EM I–MM I	Hagia Triada A	13.6
75	V	EM I–MM I	Hagia Triada A	12.1
385	Uncertain	EM I–MM I	Hagia Triada A	10.8x (incomplete)
80A	V	EM I–MM II	Moni Odigitria	14.5

Table 11.2. Seven triangular daggers with socket corrosion marks in their haft areas. Catalog numbers refer to Branigan 1974 (cat. no. 385 is damaged and of uncertain type and length). The Odigitria dagger (cat. no. 80A) is added for comparison.

it is, in fact, very similar to the Moni Odigitria dagger with two sockets flanking a flaring "nose." The details of these seven daggers are briefly set out in Table 11.2.

It is not suggested that these daggers necessarily had gold-sheathed hafts, although all four tombs involved produced considerable quantities of sheet goldwork. With the exception of perhaps Mochlos Tomb XXI, from which came a small "mask" in gold sheet, none of the published goldwork can have come from haft sheathings. But all of these daggers were apparently hafted with inset features of another material, five of them with pairs of "eyes." The elaborated inset hafts are spread across four dagger types (though three of them are Type V), and are unique to triangular daggers. Not one of the 146 long daggers I examined in the Herakleion Museum showed similar corrosion marks. This again suggests that the triangular and long daggers were regarded as quite different implements. The seven examples were recovered from five different sites (though not surprisingly all but one come from the Mesara), suggesting that they were in widespread usage and not confined to a single small locality.

The example from the lower stratum of Platanos Tholos A confirms the type was in use by EM IIB, while that from Mochlos Tomb XXI shows that it continued in use at least until EM III. This brings us back to the date of the Moni Odigitria example. Xenaki-Sakellariou concluded (1986, 242), on the evidence of the "looped rope" motif on the haft, that the dagger could be dated to the early Protopalatial period. This is significantly later than the latest context-dated triangular daggers, from Mochlos Tomb XXI, of EM III. We may note that some meander and interlace motifs on EM III–MM IA sealstones (Yule 1980, pl. 21, motif 32; pl. 22, motif 32) have similarities to the "looped rope" pattern of the Odigitria haft sheathing, with multiple sinuous loops, albeit in this case interlaced. It is possible that the dagger, as a very special item of regalia, was retained as an heirloom for several generations, but this would only be relevant to the date of manufacture if the dagger was re-hafted or its haft re-sheathed late in life. There are no signs of re-rivetting to suggest re-hafting, but it is possible that a new sheath was fitted over the existing haft. The Odigitria cemetery remained in use until MM IB, so deposition at that time is certainly possible.

The history of the Odigitria cemetery (Vasilakis and Branigan 2010) provides several possible contexts for the use of the gold-hafted dagger. The cemetery began in EM I with the building of Tholos A, followed by the construction of Tholos B and an enclosure wall. A rectangular suite of rooms was added to Tholos B in EM II, and at the end of EM II Tholos A was partially cleared and thereafter saw only sporadic use. Tholos B continued in use, although it too saw an episode of clearance. Its suite of rectangular rooms was elaborated, and meanwhile open rituals or ceremonies involving eating and drinking took place in the outer courtyard area. Eventually, in MM IB Tholos B was extensively cleared, its human remains being put into an ossuary pit. A fumigation fire was lit, and a drinking ceremony took place at the entrance to the tomb, and the tomb went out of use. There are a number of episodes in this history in which a ceremonial triangular dagger made in EM III or later may have featured. The closure of Tholos A with its attendant ceremonies took place at the end of EM IIB and is probably too early to have seen the involvement of the dagger. In EM III–MM I feasting and drinking rituals in the Outer Courtyard seem to have involved large numbers of people and would provide a fitting setting for the display and ceremonial use of the dagger. The clearance episode in Tholos B may well have been accompanied by feasting. Finally, the closure of Tholos B with the careful removal of human remains to the ossuary pit, the drinking ceremony, and the fumigation process would have provided further opportunities for communal activities where the dagger could have been displayed and used.

How might the dagger have been used in any or all of these ceremonies? Bearing in mind the discussion above of the triangular daggers' particular characteristics, there are two possibilities. One is that it was used, if only symbolically, in the preparation of food for the feasting that took place in the cemetery. This appears to have been a communal activity involving dozens, even hundreds, of people. It would have provided an ideal setting in which to flourish and display this precious item. Alternatively, the dagger may have been used in funerary or post-funerary rituals. Some tholos cemeteries have yielded evidence that human bones were manipulated and displayed in various ways (Branigan 1987). One of these involved the cutting and scraping of bones, which could have been done with the flat blade of a triangular dagger. Careful examination of the bones from Odigitria, however, provided no evidence for the practice at this cemetery (Triantaphyllou 2010). On balance, therefore, we prefer the explanation that the Odigitria dagger was ceremonially used in the preparation of communal feasts that took place in the cemetery. Such feasts were not exclusively involved with funerary occasions, and we have argued elsewhere that they involved other events in the seasonal routine of the community (Branigan 1993, 127–136). Among the deities honored in these events was the Snake Goddess, and one wonders if the "looped rope" motif on the haft of the dagger might just possibly represent a serpent rather than "rope."

Conclusion

The 98 triangular daggers now known were probably first manufactured in EM I, and were certainly in use and being deposited in EM II. Some continued in use into EM III, and the gold hafted example from Odigitria suggests that some may have had even longer lives. Various archeometallurgical researches over the last 25 years have thrown new light on the manufacture of these artifacts. Some triangular daggers prove to have been made of local copper from the southern Asterousia and others from imported copper from the Cyclades and Greek mainland. We now know that the high nickel content of some of the daggers reflects not a unique ore source but the particular alloying strategies and raw material used by Early Minoan coppersmiths in the Mesara. We can now be sure that by EM II, if not before, triangular daggers were being made of deliberate arsenical, tin, and lead alloys, and that some were cast in two-part molds. The triangular daggers reveal the same levels and range of technical skills as contemporary long daggers, and there is no reason to regard them as in any sense more "primitive." They were produced for a different function compared to long daggers, and the time has perhaps come when we should stop calling them "daggers," and re-label them as knives. Their most plausible function was as domestic knives, but the gold hafted example from Moni Odigitria was clearly elaborated for ceremonial or display purposes. Its survival has allowed us to identify five, perhaps six, other triangular daggers hafted in a similarly elaborate way with inset "eyes" in the base of the haft. These, too, were presumably used in ceremonies or rituals, very probably in the tholos cemeteries, which seem to have been the main focus for communal ritual in the Prepalatial Mesara.

References

Alexiou, S. 1969. *Minoan Civilisation*, Herakleion.

Alexiou, S., and P.M. Warren. 2004. *The Early Minoan Tombs of Lebena, Southern Crete* (*SIMA* 30), Savesdalen.

Bassiakos, Y., and M. Catapotis. 2006. "Reconstruction of the Copper Smelting Process Based on the Analysis of Ore and Slag Samples," in *The Chrysokamino Metallurgy Workshop and Its Territory* (*Hesperia Supple.* 36), P.P. Betancourt, Princeton, pp. 329–354.

Bassiakos Y., and O. Philaniotou. 2007. "Early Copper Production on Kythnos: Archaeological Evidence and Analytical Approaches to the Reconstruction of Metallurgical Process," in Day and Doonan, eds., 2007, pp. 19–56.

Branigan, K. 1967. "The Early Bronze Age Daggers of Crete," *BSA* 62, pp. 211–239.

———. 1968. *Copper and Bronzeworking in Early Bronze Age Crete*, Lund.

———. 1974. *Aegean Metalwork of the Early and Middle Bronze Age*, Oxford.

———. 1987. "Ritual Interference with Human Bones in the Mesara Tholoi," in *Thanatos: Les coutumes funéraires en Egée à l'âge du Bronze* (*Aegaeum* 1), R. Laffineur, ed., Liège, pp. 43–50.

———. 1993. *Dancing with Death*, Amsterdam.

———. 2010. "Metalwork," in Vasilakis and Branigan 2010, pp. 147–150.

Caley, E.R. 1949. "On the Prehistoric Use of Arsenical Copper in the Aegean Region," in *Commemorative Studies in Honor of Theodore Leslie Shear* (*Hesp. Suppl.* 8), Princeton, pp. 60–63.

Catapotis, M., and Y. Bassiakos. 2007. "Copper Smelting at the Early Minoan Site of Chrysokamino on Crete," in Day and Doonan, eds., 2007, pp. 68–83.

Day, P.M., and R.C. Doonan, eds. 2007. *Metallurgy in the Early Bronze Age Aegean* (*Sheffield Studies in Aegean Archaeology* 7), Oxford.

Doonan, R.C., P.M. Day, and N. Dimopoulou-Rethemiotaki. 2007. "Lame Excuses for Emerging Complexity in Early Bronze Age Crete: The Metallurgical Finds from Poros-Katsambas and Their Context," in Day and Doonan, eds., 2007, pp. 98–122.

Gale, N.H., A. Papastamataki, Z.A. Stos-Gale, and K. Leonis. 1985. "Copper Sources and Copper Metallurgy in the Aegean Bronze Age," in *Furnaces and Smelting Technology in Antiquity* (*BMOP* 48), P. Craddock and M. Hughes, eds., London, pp. 81–101.

Hall, M.E., and S.R. Steadman. 1991. "Tin and Anatolia: Another Look," *Journal of Mediterranean Studies* 4 (1), pp. 217–234.

Hood, M.S.F. 1971. *The Minoans*, London.

Junghans, S. 1969. *Studien zu den Anfängen der Metallurgíe* II, Stultgart.

Marangou, L. 1992. *Minoan and Greek Civilisation from the Mitsotakis Collection*, Athens.

Muhly, J.D. 1973. *Copper and Tin: The Distribution of Mineral Resources and the Nature of the Metals Trade in the Bronze Age* (*Transactions of the Connecticut Academy of Arts and Sciences* 43 [4]), pp. 155–535.

———. 1985. "Sources of Tin and the Beginnings of Bronze Metallurgy," *AJA* 89, pp. 275–291.

Nakou, G. 1995. "The Cutting Edge: A New Look at Early Aegean Metallurgy," *JMA* 8 (2), pp. 1–32.

Pendlebury. J.D.S. 1939. *The Archaeology of Crete: An Introduction*, London.

Renfrew, A.C. 1972. *The Emergence of Civilisation: The Cyclades and the Aegean in the Third Millennium B.C.*, London.

Soles, J.S. 1992. *The Prepalatial Cemeteries at Mochlos and Gournia and the House Tombs of Bronze Age Crete* (*Hesperia Suppl.* 24), Princeton.

Stos-Gale, Z. 1993. "The Origin of Metal Used for Making Weapons in Early Minoan and Middle Minoan Crete," in *Trade and Exchange in the Prehistoric Mediterranean*, C. Scarr and F. Healey, eds., London, pp. 115–129.

———. 1994. "The Role of Kythnos and Other Cycladic Islands in the Origins of Early Minoan Metallurgy," in *Kea-Kythnos: Historical and Archaeological Researches*, L. Mendoni and A. Mazarakis-Ainian, eds., Athens, pp. 1–12.

Triantaphyllou, S. 2010. "The Human Remains," in Vasilakis and Branigan 2010, pp. 229–248.

Vasilakis, A., and K. Branigan. 2010. *Moni Odigitria: A Prepalatial Cemetery and Its Environs in the Asterousia, Southern Crete* (*Prehistory Monographs* 30), Philadelphia.

Warren, P.M. 1969. *Minoan Stone Vases*, Cambridge.

Whitelaw, T. 1983. "The Settlement at Fournou Korifi, Myrtos, and Aspects of Early Minoan Social Organisation," in *Minoan Society*, L. Nixon and O. Krzyskowska, eds., Bristol, pp. 323–340.

Xanthoudides, S. 1924. *The Vaulted Tombs of Mesara*, London.

Xenaki-Sakellariou, A. 1986. "Poignard minoen de la collection Mitsotakis avec poignée en or ouvragée," *RA* 1986, pp. 235–244.

Yule, P. 1980. *Early Cretan Seals: A Study of Chronology*, Mainz on Rhine.

CHAPTER

12

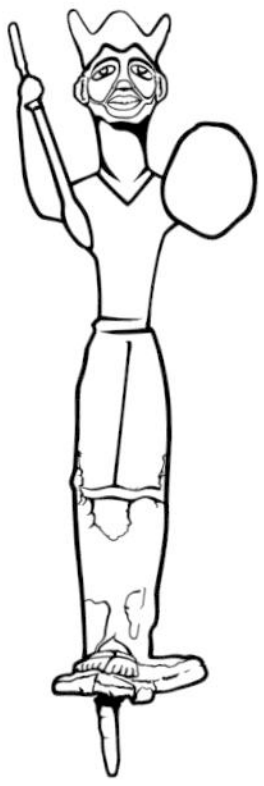

A Marine Style Gold Ring from the Hagios Charalambos Ossuary: Symbolic Use of Cockle Shells in Minoan Crete

Philip P. Betancourt

An interesting gold ring was discovered during the excavation of the Hagios Charalambos cave by Costis Davaras (for the excavation, see Davaras 1976, 1982, 1983, 1986).* It consists of a circular gold band with the addition of a decorative plate bearing marine style ornament on its upper surface (Fig. 12.1). The band was made from a thin strip whose long edges were rolled to form a double roll before it was bent to create a circular form. The decorative plate was made separately and then fused to the circular ring. Three tiny cockle shells and irregular raised areas representing rocks decorate the plate.

The plate's decoration is in the repoussé technique. Because repoussé ornament must be created by pressing the soft gold from the back to expand it and then outlining the elements and adding details by working with a tool on the front, the plate must have been created before the two parts of the ring were attached together. The decorative plate has an irregular contour; its maximum length is 1.3 cm. The ring that supports the plate has an exterior diameter of 1.57 cm and an interior diameter of 1.50 cm.

No solder can be seen at the join between ring and plate. The absence of visible solder indicates that the

* It is a pleasure to present this article to my good friend and colleague Jim Muhly. His many contributions to our knowledge of ancient metallurgy have helped define the field. I would also like to thank the many people who have worked with me on the material from Hagios Charalambos, especially Costis Davaras who made it all possible.

components were probably attached using the copper-diffusion bonding technique because fusing gold without the use of copper is a very difficult process (Carroll 1974, 1983). Copper-diffusion bonding is a jewelry-making technology that uses small amounts of Cu to lower the melting point sufficiently at the point of attachment to fuse two pieces of gold together. It has been documented for Etruscan and earlier jewelry using filigree and granulation as ornament on gold plates (Parrini, Formigli, and Mello 1982; Mello, Parrini, and Formigli 1983; Duval et al. 1989). The process has also been securely identified for northwest Anatolian jewelry of the third millennium B.C. (Betancourt 2006). It is essential for successful gold granulation, a process known from the Aegean as early as the Early Bronze Age (Bernabò-Brea 1976, pls. 241–244, 251). The finest example of the use of the technique in Crete is on the well-known bee pendant from Middle Minoan (MM) Malia (Demargne 1930, 419).

The three tiny shells have almost circular shapes with radiating incised lines, details that suggest they are intended to represent bivalve mollusks with radiating ribs. The most likely identification would be a member of the family Cardiidae, which has several species that are common in Cretan waters. The mollusks live buried in sand near the surface. They are edible, and they would be well known to the Minoans because the shells often wash up on beaches. The family Arcidae is a less likely possibility because its shells are more elongated than the circular images on the ring.

The context for this object is a Minoan ossuary in the Hagios Charalambos cave. The cavern is located on the western side of the Lasithi Plain in the interior of East-Central Crete. The condition of the finds indicates that the contents were moved to the cavern in a mixed condition. One of the lower rooms (Room 5) has a steeply inclined floor, so it was prepared by building two retaining walls inside it. After the retaining walls were constructed, long bones were laid at right angles across one another to form a grid over part of the floor, and a deposit of disarticulated human bones, animal bones, pottery, stone tools, and other artifacts that had been used as burial offerings in the original tombs was placed over the grid. Piles of human skulls were left on the upper surface of the deposit both in Room 5 and in other rooms in the cave. Pottery ranging from Neolithic to MM IIB was found randomly scattered within the deposit.

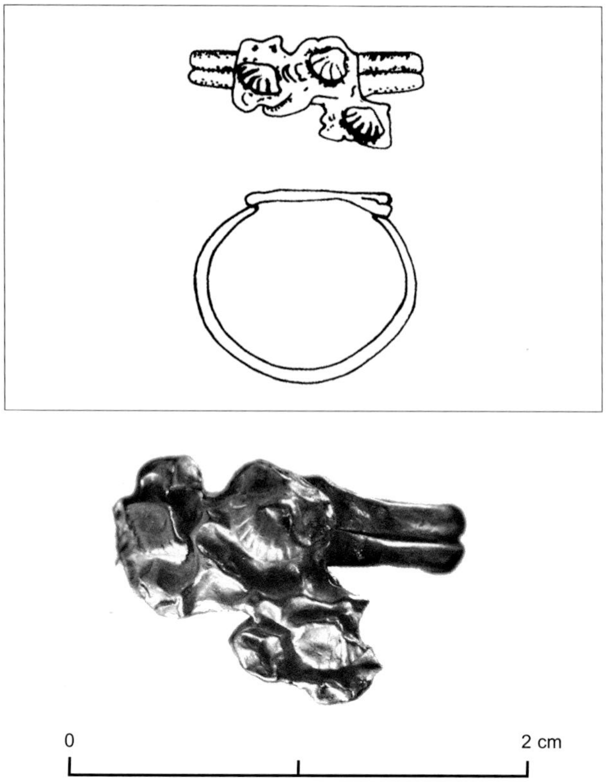

Figure 12.1. Gold ring (HNM 11,868) from the Hagios Charalambos Cave decorated with three cockle shells and irregular rocks. Drawing is 2:1 scale.

The selective placement of specific classes of human bones that had been previously sorted indicates that the deposit was not moved or disturbed after it was deposited. The random mixing of the other human bones, the pottery, and the other objects proves that the deposit was placed in the cave in a mixed secondary deposit within a relatively short period. The latest date for the majority of the deposit is MM IIB. The mouth of the cave was closed and buried by the end of the Bronze Age.

The ring would have been an import into the Lasithi Plain because both its style and its technology are foreign to this part of Crete. It must be dated stylistically because of the absence of stratigraphy in the cave. The most likely date is MM IIB, the last period for the majority of the material in the cavern (and the date of the parallels discussed below; for a later suggestion, see Pini 2010, 17).

Marine style decoration begins in Crete during MM IIB. It is present among the motifs in the MM IIB corpus of about 6,500 sealings found in the *archivio di cretule* in Vano 25 at Phaistos (Levi 1957–1958; Fiandra 1968, 1975). Shells are used on a series of horizontal bands on a MM IIB cup from Vano XIX at Phaistos (Pernier and Banti 1935–1951, color pl. 30b), and octopus motifs appear in Kamares Ware pottery found at Phaistos and elsewhere in South Crete during the same period (Pelagatti 1961–1962, pp. 101–103). The South Cretan examples of marine style pottery are much more abstract than the rather naturalistic representations on the ring, and the seal designs from Phaistos depict the triton shell with a branch or a framing line rather than placing the cockle within a rocky setting (Fiandra 1968, nos. 674, 937, 1489). Three-dimensional naturalistic representations of cockle shells set within a rocky marine setting like the image on the ring are first known from South Crete in MM III, the phase that follows this period (Levi 1976, vol. I*, pl. 219). Associations between this MM III marine style object and the gold ring from Hagios Charalambos are not possible because the ossuary was no longer being used for secondary burial at this time.

A very close stylistic correspondence comes from the northern coast of Crete from the same MM IIB period as the last period of the ossuary. Marine decoration using irregular rocks as the background for various sea creatures including crabs and shells is used to decorate the exterior of several vases found in deposits from the MM IIB destruction level in Quartier Mu at Malia (Poursat 1980; Poursat and Knappett 2005, pl. 37d, e). The naturalism of these motifs (Fig. 12.2) is much closer to the style of the ring than the more abstract style used in southern Crete at this time. A similar style occurs on a class of spouted cups with interior handles (called scoops) found at a number of sites in eastern and east-central Crete (Floyd 1999). Examples of this shape with marine style ornament in three dimensions are illustrated in Figures 12.3 and 12.4. The scoop from Malia (Fig. 12.3) closely resembles the other three-dimensional marine style appliqués from that site, and the class may have originated there. The ornament on all of these pieces is very naturalistic. The cockle shells have radiating incised lines and a life-like form, and they

Figure 12.2. Clay jug (HM 19,814) with three-dimensional marine style decoration on the exterior from Quartier Mu at Malia, MM IIB (courtesy of J.-C. Poursat). Restored ht. ca. 15 cm.

Figure 12.3. Scoop (HM 21,008) from Quartier Mu at Malia with cockle shells and irregular rocks added at the rim, MM IIB (courtesy of J.-C. Poursat). H. 4.5 cm.

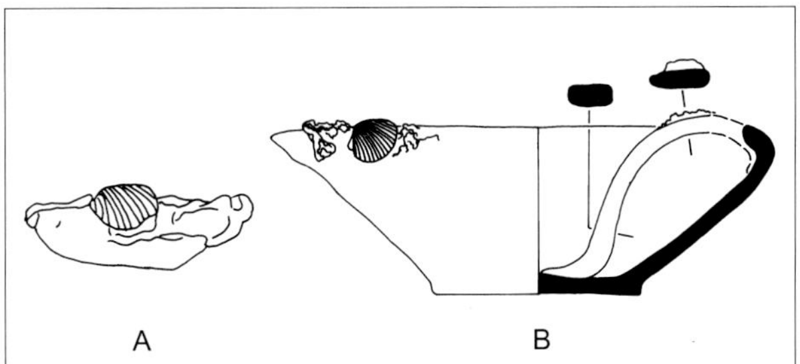

Figure 12.4. Scoops from Pseira with cockle shells and irregular rocks added at the rims, MM IIB. A: sherd from Building AA, Room AA 7, INSTAP Study Center for East Crete no. PS 799. B: complete scoop from an unknown find spot (SM 7140). Scale 1:3.

are placed within the irregular three-dimensional rocks in the same manner as on the gold ring. They suggest that the ring also may be associated with this MM IIB tradition.

The techniques of goldworking on the miniature ring also suggest an association with the region of Malia. The minute details and the use of a bonding technique that does not incorporate solder recall the style of the pendant with two bees from the Chrysolakkos tomb at Malia (Demargne 1930, 419). Such high skills and complex technology are rare in Crete from the Middle Minoan period, and nothing else like this piece is known from Lasithi.

Cockle Shells in Symbolic Contexts

These elaborate objects suggest that the motif of the cockle shell may have more than simple decorative value. In the absence of written records, the meaning of a motif can best be learned by examining the contexts in which is occurs. If a specific pattern emerges within the contexts for a specific element, this pattern can shed light on how the motif was regarded within a society. Bivalve shells occur in Minoan Crete within a limited set of contexts. The shells are sometimes used as food, but several of the contexts can only be regarded as symbolic.

One of the most interesting examples of the bivalve shell in a cultic context comes from Knossos. Hundreds of shells, some of them with painted lines, were present in the two rectangular pits called the Temple Repositories (Evans 1921–1935, I, 463–523). Their inclusion along with many other exotic items, including faience female figurines holding snakes, suggests they had a symbolic meaning. Many of the shells had been given concentric painted lines that imitate the patterns on some of the local cockle shell species, like *Acanthocardia tuberculata* (Linnaeus, 1758) and *Cerastoderma edule* (Linnaeus, 1758). The purpose of the shells in the Temple Repositories is unknown, though they have sometimes been regarded as individual offerings to a deity (Gesell 1985, 65). They must, however, have had symbolic meaning of some type.

The presence of bivalve shells in symbolic contexts continues into the Late Bronze Age. Two shrine contexts are known to have had marine shells as well as miniature vessels made of clay. A find at Phaistos that was regarded as a cultic context included examples of the dog cockle (*Glycymeris* sp.) along with miniature clay vessels (Mosso 1910, 71–73). The *Glycymeris* had been collected dead, and one of them had been modified by grinding a flat place on the shell so it would sit like a small vessel. At Kommos, 12 *Glycymeris* were found in situ on a stone table in House X, along with miniature clay vases and a complete triton shell, a context regarded as a household shrine by the excavator (Shaw 1995, 877–878, fig. 4; discussed by Ruscillo 2006, 803–805; see also Ruscillo, forthcoming).

The most remarkable context for bivalve shells also comes from Kommos (noted by both of the scholars who studied the shells: Reese 1995, 252–255; Ruscillo 2006, 803–805). Over 6,500 examples of *Glycymeris* sp. were found at the site, mostly in levels from LM IIIA:2 to LM IIIB. They had been collected dead, so they were not used as food, and some of them were so eroded from the beach sand that they were just flat remnants, so they could not have been used as spoons or small containers. They had no marks from use as scrapers. Over 1,200 specimens were in and around House X. How were these shells used? Could the Minoans have had a ceremony in which each participant brought a shell, as has been suggested for the shells in the Temple Repositories (Gesell 1985, 65)? Some type of symbolic use seems the only explanation for the presence of so many examples of just one species, all of them collected dead.

Many excavations in Crete, particularly the ones from several years ago, did not pay much attention to the presence of marine shells, but *Glycymeris* that were collected dead have been reported from as early as EM II Myrtos where they were regarded as possibly spoons or scrapers (Shackleton 1972, 325). Reese gives a long list of other sites where the same shell was collected dead and brought to the settlement for unknown reasons (Reese 1995, 256): Akrotiri, Chania, the Idean Cave, Juktas, Knossos, Malia, Palaikastro, Phaistos, Pyrgos, Syme, the

Trapeza Cave, and Tylissos. These examples suggest a widely held Minoan practice involving cockles that required dead shells collected on beaches and had no useful purpose that can be recognized in the archaeological record.

It is not surprising that people who lived on an island and believed in supernatural powers would associate some of them with the sea and its unpredictable forces. It is also an easy extension to use a seashell washed up on the beach during a violent storm as a symbol of the storm's power. The details of the symbolism cannot be understood without written records, but its general outlines were probably fairly simple. The Minoans were a seafaring people, and ceremonies involving marine associations are not only highly likely, but almost a certainty. The visual symbols of the sea might be a powerful reminder of the dangers as well as the opportunities provided by the Aegean.

Conclusions

The tiny gold ring from Hagios Charalambos is a unique product of the jeweler's art. It uses sophisticated jewelry-making techniques and a miniaturist style of three-dimensional relief in repoussé to form an object that is beautiful as well as interesting. Considering the evidence for the symbolic use of the cockle shell in Minoan society and the great distance from the sea at which the item was found, it was probably also charged with symbolic meaning. Probably the ring must also be regarded as an elite object that was intended to provide its owner with increased prestige.

Links to the palace of Malia exist both with the techniques of manufacture and with the motif of the marine shell contained within a setting of irregular rocks. It is likely that the ring was a Minoan palatial product manufactured at Malia or at some center under its direct influence and sent to Lasithi in MM IIB, before the destructions that brought this period to a close. The MM IIB destructions at Malia can be securely placed by their stratigraphic position between MM IB and MM IIIA levels in Quartier Mu (Poursat 1990; Knappett 1997, 306; for the pottery assemblage, see Poursat and Knappett 2005). Several other deposits at Malia suggest a general destruction of the palace and its environs at this time (van Effenterre and van Effenterre 1969, pl. 48; 1976, 81–82; Pelon 1970, 20–21, nos. 1–3). These destructions ended a number of practices at the site, including the tradition of art objects depicting naturalistic marine scenes within a seascape defined by irregular, knobby rocks.

The ring provides evidence in support of the theory that Lasithi was a part of a Malia state during MM II that reached all the way from Crete's north coast to Pyrgos Myrtos on the south coast. The theory has been suggested on the basis of several parallels between sites, especially pottery (Poursat 1988; Cadogan 1990, 1995; Knappett 1997, 1999; Knappett and Schoep 2000). It is also supported by the nature of the seals and other elite goods from the Lasithi Plain at this time, because they also have a close relation to Malia (Betancourt 2007). The marine style ring fits well within the picture of an influential palatial state centered at Malia during the period that ended in MM IIB.

The symbolic aspects of the motif suggest that the relationship between Lasithi and Malia may have involved belief-systems as well as economic or political alignments. The Lasithi Plain is far from the sea, and artistic motifs involving marine elements or seafaring are not likely to evolve at inland sites. The gold ring is the only object with a marine design found at the cave. If the owner of the ring held any views that associated the item's ornament with symbolic overtones, they were surely introduced from elsewhere.

References

Bernabò-Brea, L. 1976. *Poliochni: Città prehistorica nell'isola di Lemnos* II. Rome.

Betancourt, P.P. 2006. "Joining Techniques of Early Bronze Age Trojan Jewelry," *Studia Troica* 16, pp. 89–95.

———. 2007. "Lasithi and the Malia-Lasithi State," in *Krinoi kai Limenes: Studies in Honor of Joseph and Maria Shaw* (*Prehistory Monographs* 22), P.P. Betancourt, M.C. Nelson, and H. Williams, eds., Philadelphia, pp. 209–219.

Cadogan, G. 1990. "Lasithi in the Old Palace Period," *BICS* 37, pp. 172–174.

———. 1995. "Mallia and Lasithi: A Palace-State," in *Πεπραγμένα του Ζ΄ Διεθούς Κρητολογικού Συνεδρίου* A'(1), Rethymnon, pp. 97–104.

Carroll, D.L. 1974. "A Classification for Granulation in Ancient Metalwork," *AJA* 78, pp. 33–39.

———. 1983. "On Granulation in Ancient Metalwork," *AJA* 87, pp. 551–554.

Davaras, C. 1976. "῞Αγιος Χαράλαμπος," *ArchDelt* 31 (B', 2 Chronika), p. 379.

———. 1982. "Σπήλαιο Αγίου Χαραλάμπους," *ArchDelt* 37 (B', 2 Chronika), pp. 387–388.

———. 1983. "Σπήλαιο Γεροντομουρί," *ArchDelt* 38 (B', 2 Chronika), p. 375.

———. 1986. "Πρώϊμες μινωϊκὲς σφραγίδες καὶ σφραγιστοκοὶ δακτύλιοι ἀπὸ τὸ σπήλαιο Γεροντομουρὶ Λασιθίου," *ArchEph* 1986, pp. 9–48.

Demargne, P. 1930. "Bijoux minoens de Mallia," *BCH* 54, pp. 404–421.

Duval, A.R., C. Eluère, L.P. Hurtel, and M. Menu. 1989. "The Use of Scanning Electron Microscopy in the Study of Gold Granulation," in *Archaeometry: Proceedings of the 25th International Symposium*, Amsterdam, pp. 325–333.

Evans, A.J. 1921–1935. *The Palace of Minos at Knossos* I–IV, London.

Fiandra, E. 1968. "A che cosa servivano le cretule di Festòs," in *Πεπραγμένα του Β΄ Διεθνούς Κρητολογικού Συνεδρίου* A', Athens, pp. 383–397.

———. 1975. "Ancora a proposito delle cretule di Festòs: connesione tra i sistemi amministrativi centralizzati e l'uso delle cretule nell'età del bronzo," *Bolletino d'Arte* 1975, pp. 1–25.

Floyd, C.R. 1999. "The Minoan Scoop," in *Meletemata: Studies Presented to Malcolm H. Wiener as He Enters His 65th Year* (*Aegaeum* 20), P.P. Betancourt, V. Karageorghis, R. Laffineur, and W.-D. Niemeier, eds., Liège, pp. 249–257.

Gesell, G.C. 1985. *Town, Palace, and House Cult in Minoan Crete* (*SIMA* 67), Göteborg.

Knappett, C. 1997. "Ceramic Production in the Protopalatial Mallia 'State': Evidence from Quartier Mu and Myrtos Pyrgos," in *Texnh: Craftsmen, Craftswomen and Craftsmanship in the Aegean Bronze Age* (*Aegaeum* 16), R. Laffineur and P.P. Betancourt, eds., Liège and Austin, pp. 305–311.

———. 1999. "Assessing a Polity in Protopalatial Crete: The Malia-Lasithi State," *AJA* 103, pp. 615–639.

Knappett, C., and I. Schoep. 2000. "Continuity and Change in Minoan Palatial Power," *Antiquity* 74, pp. 365–371.

Levi, D. 1957–1958. "L'Archivo de Cretule a Festòs," *ASAtene* 19/20, pp. 7–192.

———. 1976. *Festòs e la civiltà minoica* I, Rome.

Mello, B., P. Parrini, and E. Formigli. 1983. "Etruscan Filigree: Welding Techniques of Two Gold Bracelets from Vetulonia," *AJA* 87, pp. 548–551.

Mosso, A. 1910. *The Dawn of Mediterranean Civilisation*, London and Leipsig.

Parrini, P., E. Formigli, and E. Mello. 1982. "Etruscan Granulation: Analysis of Orientalizing Jewelry from Marsiliana d'Albegna," *AJA* 86, pp. 118–121.

Pelagatti, P. 1961–1962. "Osservazioni sui ceramisti del I palazzo di Festòs," *CretChron* 15–16, pt. 1, pp. 99–111.

Pelon, O. 1970. *Fouilles exécutées à Mallia. Exploration des maisons et quartiers d'habitation (1963–1966), troisième fascicule* (*ÉtCrét* 16), Paris.

Pernier, L., and L. Banti. 1935–1951. *Il palazzo minoico di Festòs* I–II, Rome.

Pini, I. 2010. *Aegean and Cypro-Aegean Non-Sphragistic Decorated Gold Finger Rings of the Bronze Age* (*Aegaeum* 31), Liège and Austin.

Poursat, J.-C. 1980. "Reliefs d'applique moulés," in *Fouilles exécutées à Mallia: Le quartier Mu.* II: *Vases de pierre et de metal, vannerie, figurines et reliefs d'applique, elements de parure et de decoration, armes, sceaux et empreintes*, by B. Detournay, J.C. Poursat, and F. Vandenabeele (*ÉtCrét* 26), Paris, pp. 116–132.

———. 1988. "La ville minoenne de Malia: Recherches et publications récentes," *RA* 1988, pp. 61–82.

———. 1990. "Le début de l'époque protopalatiale à Malia," in *Ειλαπίνη: Τόμος τιμητικός για τον καθηγητή Νικόλαο Πλάτωνα*, A. Vlachopoulos and K. Birtacha, eds., Athens, pp. 461–466.

Poursat, J.-C., and C. Knappett. 2005. *Fouilles exécutées à Mallia: Le quartier Mu*. IV: *La poterie du minoen moyen II. Production et utilisation* (*ÉtCrét* 33), Paris.

Reese, D.S. 1995. "5. The Minoan Fauna: The Marine Invertebrates," in *Kommos* I: *The Kommos Region and Houses of the Minoan Town*. Part 1*: The Kommos Region, Ecology, and Minoan Industries*, Princeton, pp. 240–271.

Ruscillo, D. 2006. "Faunal Remains and *Murex* Dye Production," in *Kommos* V: *The Monumental Minoan Buildings at Kommos*, J.W. Shaw and M.C. Shaw, eds., Princeton and Toronto, pp. 776–844.

———. Forthcoming. "The Faunal Remains," in *House X at Kommos: A Minoan Mansion near the Sea*. Part 1: *Architecture, Stratigraphy, and Selected Finds* (*Prehistory Monographs*), Philadelphia.

Shackleton, N.J. 1972. "The Shells," in *Myrtos: An Early Bronze Age Settlement in Crete*, by P.M. Warren, London, pp. 321–325.

Shaw, J.W. 1995. "Brief Summary of Excavation at Kommos, Crete: Season XVI (1991)," in *Πεπραγμένα του Ζ' Διεθούς Κρητολογικού Συνεδρίου*, Rethymno A'(2), pp. 875–880.

van Effenterre, H., and M. van Effenterre. 1969. *Fouilles exécutées à Mallia: Le centre politique*. I: *L'agora (1960–1966)* (*ÉtCrét* 17), Paris.

———. 1976. *Fouilles exécutées à Mallia: Exploration des maisons et quartiers d'habitation*. IV: *Le quartier O (1956–1960), quatrieme fascicule* (*ÉtCrét* 22), Paris.

CHAPTER

13

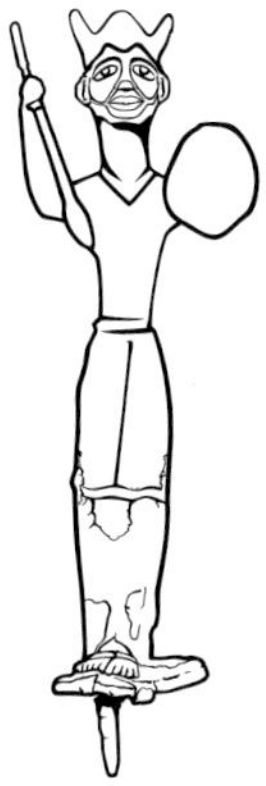

Metalworking at Malia, Quartier Mu: High or Low Technology?

Jean-Claude Poursat and Cécile Oberweiler

Bronze working is well attested at Malia during the Protopalatial period.* In Quartier Mu, three workshop houses near the two large elite residences A and B have preserved a few tiny blocks of copper ores, many smith's tools (stone molds, clay crucibles, and tuyères), as well as some remains from copper melting processes (Poursat 1996). About 100 copper or bronze tools and weapons have been found in the various buildings (Fig. 13.1). Among them, two so-called "carpenter's hoards"—axes, chisels, and saws from Buildings A and B (Poursat 1985) and a deposit of three bronze vases hidden under the floor of Building C (Vandenabeele 1980)—had most probably been produced by the craftsmen of Quartier Mu.

Most of these objects have already been published (Poursat 1996), as well as the results of some chemical and isotopic analyses (Éluère 1996; Poursat and Loubet 2005). In the meantime, however, some fragments of unpublished clay crucibles have been identified among the ceramic material of Quartier Mu. Various analyses have been conducted concerning their fabric and traces due to the melting process, which remain on the surface as well as on other tools (stone molds and clay tuyères). From these new data, additional conclusions can be drawn about bronze working at Protopalatial Malia.

---

* All the analyses concerning the tools (crucibles, tuyères, and molds) mentioned in this paper are described in a Ph.D. dissertation by C. Oberweiler (2005). Some of them have been carried out in collaboration with Y. Maniatis at NCSR "Demokritos," Hagia Paraskevi, Greece.

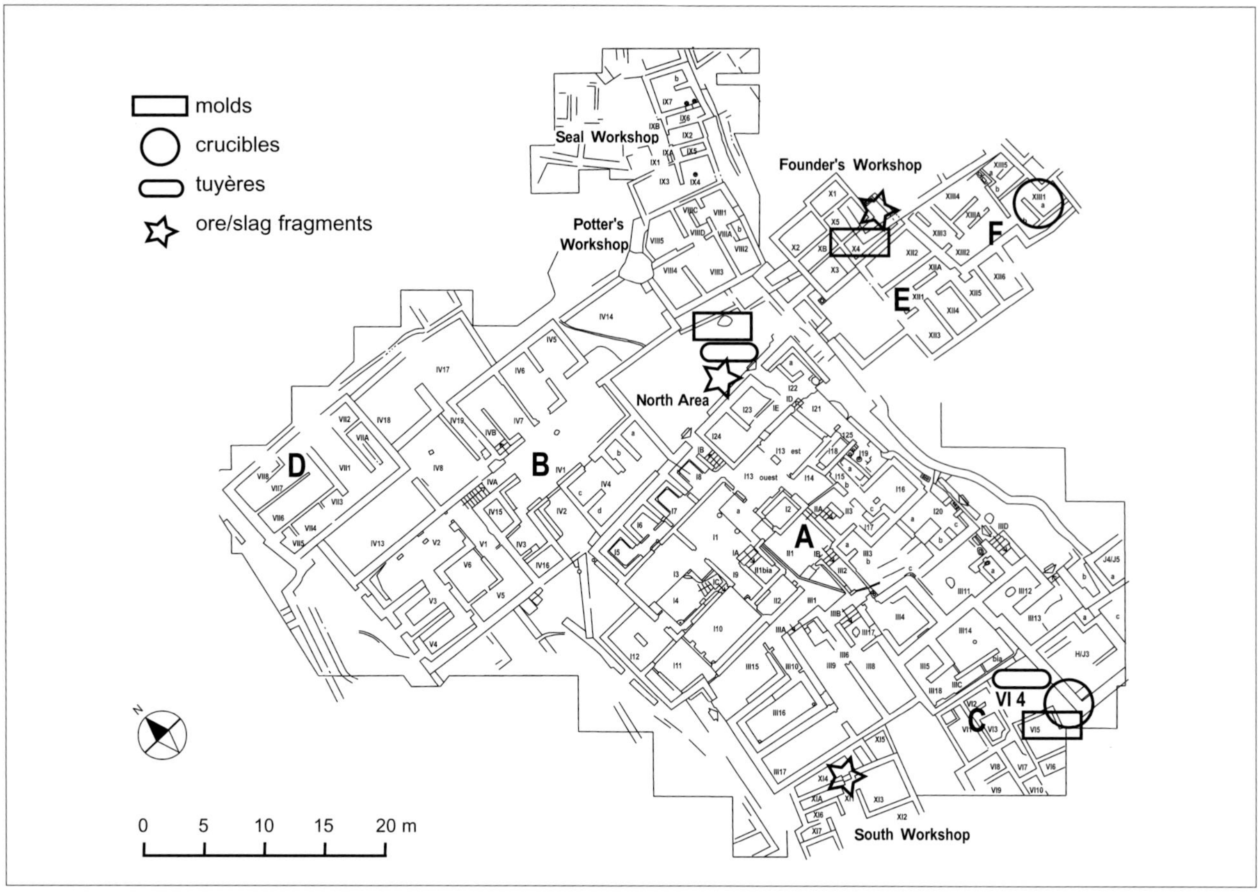

Figure 13.1. Plan of Quartier Mu, showing the places where smithing tools were found. Plan by M. Schmid and N. Sigalas.

The Smithing Tools

Casting as a distinct process in the *chaîne opératoire* includes two steps in which tools such as crucibles, bellows, tuyères, and molds are involved: (1) melting copper and/or preparing alloys as bronze, and (2) pouring the metal into the molds. During these operations there are two key constraints: the firing temperature must rise to approximately 1200°C, the efficient temperature to pour copper and bronze (whereas the melting temperature for copper is 1084°C), and the tools must be able to withstand it. Practically, the bronzesmith must melt copper or bronze without allowing the crucible, the tuyère nozzle, and the mold to melt or break during the process. In order to meet these specifications, several complementary technical solutions must be used that are related to the specific fabric, shape, size, and construction technique of the tools.

Crucibles

The crucible is the most important piece because it is exposed to such strong thermal and mechanical stresses. Its production must account for not only the technical criteria mentioned above, but also the type and amount of metal to melt (quantity related to the size and number of objects to be produced), whereas fabric and morphology are conditioned by its handling during the process. All these criteria are completely dependent.

Four fragments of clay crucibles for copper or bronze melting have been identified in Quartier Mu. Vitrified areas or alterations of the clay material due to high firing temperature are visible on their inner surface. Two of them, probably parts of the same crucible, have been found above Room XIII 1a in Building F. As no other material related

Figure 13.2. Clay crucible (M 91/3107–09a) from Quartier Mu, Malia.

to metalworking comes from this building, we may assume that this crucible was initially kept in the nearby Founder's Workshop. The bigger piece (M 91/3107–09a; Fig. 13.2) preserves the base of the crucible with a part of its pedestal; a tiny layer of bronze slag still remains on its inner surface. The other fragment (M 91/3107–09b) is part of a crucible wall; traces of bronze slag are still visible on the inner surface. X-ray fluorescence (XRF) analyses conducted on this slag have shown the presence of copper and tin and very tiny traces of arsenic.

The two other fragments of crucibles come from a destruction deposit in Building C, south of the open Area VI 4. They were found together with a stone mold (Poursat 1996, 69), and three tuyères were discovered nearby in Area VI 4. The larger fragment (M 89/2202–04; Fig. 13.3) preserves part of its pedestal, on which the hole for handling is still visible; a strong layer of bronze slag remains on the inner surface (XRF analyses have shown the presence of copper and tin, but no traces of arsenic have been noted). The smaller fragment (M 89/2202–02) is part of the wall; a thin layer of bronze slag also is present on its inner surface.

These crucibles show the well-known, Cretan, open-bowl shape with a characteristic hole through the pedestal for introducing a wooden stick as a handle (Evely 2000). This kind of crucible is known in Crete since the Early Bronze Age (EBA), as shown by two examples from the Hagia Photia cemetery (Davaras and Betancourt 2004). For the Protopalatial period, other crucibles of this kind are known from Palaikastro (unpublished), Poros (Dimopoulou 1997, 434), and Myrtos Pyrgos (Cadogan 1978, 76, fig. 17). The estimated capacity of one of the crucibles from Quartier Mu probably did not exceed 1 liter (ca. 8 kg) of copper or bronze. Clay fabric analyses have revealed a specific mixture consisting of a fine but common non-calcareous and non-refractory clay and a large amount of organic fibers, which were specifically selected and calibrated. The result is a porous and fibrous ceramic composite with good thermal insulation and resistance to thermal shocks and mechanical stresses.

Analyses of the use-wear patterns have indicated the position of the crucible in the combustion area during the melting process. The higher temperatures as estimated by scanning electron microscopy (SEM) (ca. 1000°C–1200°C) occur only inside the bowl, whereas the lower temperatures (less than 800°C) are located on the outer surface. These

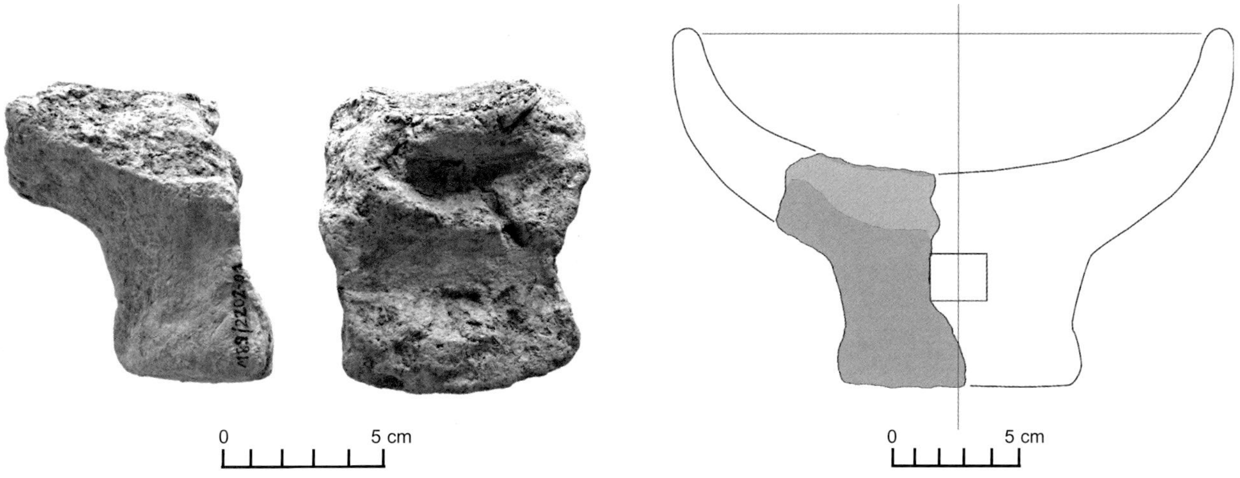

Figure 13.3. Photo and reconstruction drawing of clay crucible (M 89/2202–04) from Quartier Mu, Malia.

Figure 13.4. Tuyère (HNM 13413 [69 M 1392]) from Quartier Mu, Malia.

observations, as well as the open bowl shape of the crucibles, indicate that the metal was heated from above and not through the wall, as is the case with modern crucibles. The crucibles thus could withstand the casting processes and ensure an efficient melting of the metal.

Bellows and Tuyères

During melting processes, air supply is necessary to produce the constantly high temperature required to melt copper or bronze. It is provided by bellows that are linked to tuyères. In Quartier Mu, no bellows have been recovered, but four clay tuyères attest that they were in use. Three of them have been found in the open Area VI 4 near Building C; the fourth one comes from the North Area, not far from the Founder's Workshop. Their size is small (ca. 7 cm), and they have a straight shape with a rounded end (Poursat 1996, 117). Alterations due to firing are still visible, especially on 69 M 1392 (Fig. 13.4), which also bears tiny pieces of slag (XRF analyses have shown copper and traces of tin). Analyses of the clay fabric indicate the use of a non-calcareous clay rich in various inclusions (quartz, hematite, and schist), but no organic temper was added. This fabric is not truly refractory (as compared to the crucible fabric) because a tuyère is less exposed to high firing temperatures due to cold airflow keeping the tuyère cool and preventing the melting of the nozzle.

Molds

The main smithing tools recovered in Quartier Mu are the stone molds. At least eight pieces are large enough to be securely identified. Five of them come from the Founder's Workshop, one from the North Area, one from the Seal Workshop, and the last one from Building C (Poursat 1996, 116). All are made of schist, and they belong to the permanent mold type, which means that they could be reused several times. The complete example (Fig. 13.5) belongs to the open type and bears three matrices on the upper side and one on the lower side. The other pieces are very fragmentary; four of them, however, probably belong to the closed type. They generally comprise two or three stone elements attached together, with each of them bearing a part of the matrix (Fig. 13.6). Only five matrices can be identified; they were intended for tools or weapons of a very simple shape (chisel, knife, and double axe), as well as for an ingot. Their mean capacity is less than 60 cubic cm. Only the double-axe matrix on mold M 89/2202–01 (Poursat 1996, 69), according to other complete examples from the palace at Malia (Chapouthier and Demargne 1942, 53–66), could have a capacity between 100 and 140 cm^3. This means that pouring objects of this kind required less than about 500 g of copper or bronze for each; only the double-axes had a weight between 0.9 and 1.3 kg.

Casting Processes and Working Areas

Even if a refractory fabric and a suitable shape improve the tools' resistance, the temperature range for casting operations is always very limited: above 1200°C, the crucible begins to melt, and below 1100°C, the copper as well as the bronze are not hot enough to be poured. In order to achieve better melting conditions, the bronzesmith must attend carefully to the respective positions of the crucible and the tuyère inside the combustion structure. Experimental melting conducted with this kind of pedestalled crucible and the associated tuyère has shown the importance of their placement for a

Figure 13.5. Stone mold (B 81/C 15) from Quartier Mu, Malia.

Figure 13.6. Mold fragments from Quartier Mu, Malia (B 81/C 17, upper row; B 81/C 16, lower row).

successful outcome. Correct positioning protects the crucible and the tuyère as much as possible from heat damage and allows them to be reused.

The characteristics of the tools as well as their position inside the combustion structure also give some indications about the size and the organization of the working area. The shape, small capacity, and size of the crucibles and tuyères recovered in Quartier Mu suggest the probable use of an open ephemeral furnace of quite a small size. The low capacity of the crucibles (ca. 8 kg) corresponds to that of the molds; with one crucible of this kind, the bronzesmith could produce between six or eight double-axes, or about 15–20 small objects. The size of the working structure, however, would have depended on the number of crucibles and tuyères that could be set in the hearth. If the number of bellows, and consequently the number of blowers involved, were more important, the size of the working area would of course have been larger. The metal quantities in use during the Protopalatial period, however, do not seem to justify such an installation. The majority of the 100 copper or bronze objects found in Quartier Mu (Poursat 1996, 141) do not exceed 500 g each, which makes a total weight of 50 kg (a single oxhide copper ingot weighs ca. 30 kg; Buchholz 1959, 32).

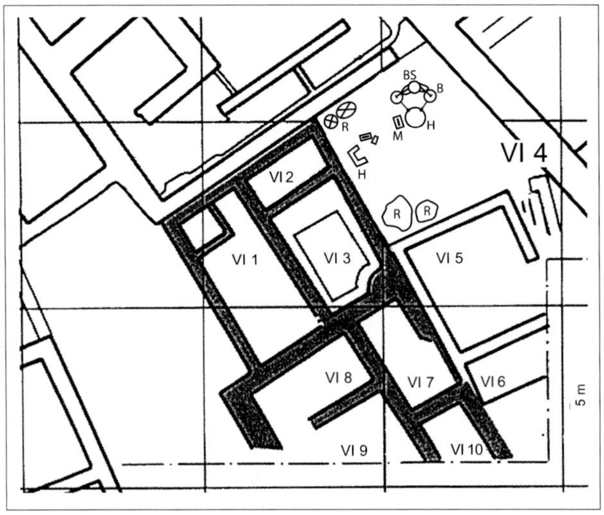

Figure 13.7. Reconstructed casting installation in Area VI 4. BS = bronzesmith, B = bellow, H = hearth, M = mold, R = raw materials (clay, sand, etc.).

Where could these small structures have been situated? Melting activities were probably conducted in the North Area next to the Founder's Workshop where some slag droplets, a mold fragment, and a tuyère have been found. The open Area VI 4, adjacent to Building C, was certainly another working space for casting activities, as indicated by the discovery of three tuyères and a tiny piece of metal slag near a deposit with a crucible and a mold. It is then possible to imagine the installation of a light structure in this small (ca. 16 m^2) courtyard (Fig. 13.7). Unfortunately, no traces of firing structures have been recovered during the excavations in these two very disturbed areas. We must assume, however, that these small melting installations were situated quite near the main buildings of Quartier Mu.

Concluding Remarks

The previous metal analyses (Éluère 1996) have shown that the bronzesmiths of Quartier Mu produced bronze objects, weapons (daggers and spearheads), and tools (axes, chisels, and drills) in which the percentage of tin varied widely and was not correlated to their nature or function. The results display a great variability, for example the spearheads have a percentage of tin that ranges from 2.7% to 9.8%. These variations can be explained by the diversity or the irregularity of the metal supplies. Some isotope analyses made on copper and bronze objects from Quartier Mu (Poursat and Loubet 2005) indicate that, during the Protopalatial period, copper probably came to Crete in small quantities from various regions of the Aegean or the East Mediterranean (Anatolia, Cyprus, and Levant). The same may be said about tin, imported from Mari via Ugarit (Muhly 1985, 282) or from Anatolia (Dercksen 2005). These variations may also occur because bronze production in Quartier Mu took place in a context of small production units that were linked with workshop houses and not strictly specialized: the South Workshop probably produced stone vases as well as bronze and perhaps bone objects. The artisan's polyvalence is one of the main characteristics of the craft activity in Quartier Mu (Poursat 1996, 150).

The recent analyses of the smithing tools and the study of the production processes reveal the technical abilities of the craftsmen. Metalworking requires a lot of complementary knowledge and *savoir-faire*, the first acquired by transmission and apprenticeship, the second by experience and practice. Bronzesmiths had to know how they could obtain an efficient melting temperature, and what the copper or bronze properties were under strong thermal and mechanical stresses. They also had to know the properties and behavior of the materials involved in the fabrication of their tools, as well as the right capacity and the suitable shape required for the crucibles according to the type and amount of metal objects to produce. In this way, tool studies shed new light on the competence of the Minoan bronzesmiths during the Protopalatial period.

References

Buchholz, H.-G. 1959. "Keftiubarren und Erzhandel im zweiten vorchristlichen Jahrtausend," *Prähistorische Zeitschrift* 37, pp. 1–40.

Cadogan, G. 1978. "Pyrgos, Crete, 1970–1977," *Archaeological Reports* 24, pp. 70–84.

Chapouthier, F., and P. Demargne. 1942. *Fouilles exécutées à Malia. Troisième rapport: exploration du palais* (*ÉtCrét* 6), Paris.

Davaras, C., and P.P. Betancourt. 2004. *The Hagia Photia Cemetery* I: *The Tomb Groups and Architecture* (*Prehistory Monographs* 14), Philadelphia.

Dercksen, J.G. 2005. "Metals According to Documents from Kültepe-Kanish Dating to the Old Assyrian Colony Period," in *Anatolian Metal* III (*Der Anschnitt* 18), Ü. Yalçin, ed., Bochum, pp. 17–34.

Dimopoulou, N. 1997. "Workshops and Craftsmen in the Harbour-town of Knossos at Poros-Katsambas," in *TEXNH: Craftsmen, Craftswomen and Craftsmanship in the Aegean Bronze Age* (*Aegaeum* 16), R. Laffineur and P.P. Betancourt, eds., Liège, pp. 433–438.

Éluère, C. 1996. "Étude en laboratoire de quelques objets métalliques du Quartier Mu de Malia (Crète)," in Poursat 1996, pp. 141–148.

Evely, D. 2000. *Minoan Crafts: Tools and Techniques. An Introduction*, vol. 2 (*SIMA* 92 [2]), Jonsered.

Muhly, J.D. 1985. "Sources of Tin and the Beginning of Bronze Metallurgy," *AJA* 89, pp. 275–291.

Oberweiler, C. 2005. *La métallurgie du cuivre et du bronze dans le monde égéen du BA au début du BR (IIIe mill. – début du IIe mill. av. J.-C.): Les techniques de fonderie*, Ph.D. diss., University of Paris I–Sorbonne.

Poursat, J.-C. 1985. "Outils de charpentier et de menuisier au Quartier Mu de Malia (Crète)," in *De l'Indus aux Balkans*, J.-L. Huot, M. Yon, and Y. Calvet, eds., Paris, pp. 119–125.

———. 1996. *Artisans minoens. Les maisons-ateliers du Quartier Mu* (*ÉtCrét* 32), Athens.

Poursat, J.-C., and M. Loubet. 2005. "Métallurgie et contacts extérieurs à Malia (Crète) au Minoen Moyen II. Remarques sur une série d'analyses isotopiques du plomb," in *EMPORIA: Aegeans in the Central and Eastern Mediterranean* (*Aegaeum* 25), R. Laffineur, and E. Greco, eds., Liège, pp. 117–122.

Vandenabeele, F. 1980. "Vases de métal," in *Le Quartier Mu II* (*ÉtCrét* 26), B. Detournay, J.-C. Poursat, and F. Vandenabeele, eds., Paris, pp. 71–89.

CHAPTER

14

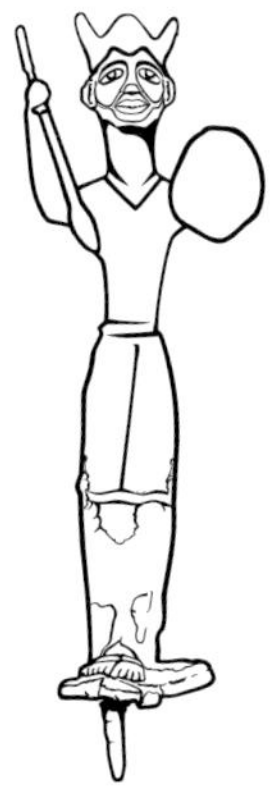

The Mochlos Sistrum and Its Origins

Jeffrey S. Soles

Four varieties of loop sistra have been reported in Crete.* The first and oldest type is a simple clay sistrum with an oval loop that dates to the Middle Minoan (MM) I–II period (ca. 2100–1700 B.C.). Iannis and Efi Sakellarakis, Phil Betancourt, and Jim Muhly have described this class in some detail (Sakellarakis and Sakellaraki 1991, 184–187, figs. 15–18; 1997, 351–356; Betancourt 2005, pl. 102c; Betancourt and Muhly 2006, figs. 1, 2). The oldest example comes from Tomb 9 at Archanes and dates to MM IA; five more come from the Hagios Charalambos burial cave and date to the MM I–II period. Two clay disks that probably belong to such sistra have also been found, one in a tomb context at Krasi, also dated MM I (Marinatos 1929, 122, fig. 15), and another from Building C.7 at Mochlos, as yet unpublished, which may belong to a MM level beneath the Late Minoan (LM) I building. The Archanes, Hagios Charalambos, and Krasi examples were found in their final resting places and had presumably been used in funerary ceremonies, whereas the Mochlos piece was located in a manufacturing building where a sistrum was probably being made, but was not yet finished. Sakellarakis and Betancourt have pointed out that these sistra are all handmade of local Cretan clay. The handles and loops are made together in one piece; two to three holes are pierced opposite each other on each side of the loop frame for the insertion of a horizontal wooden bar on which two to three disks hang. They vary in height from approximately 0.16 to 0.18 m.

The second variety of loop sistrum is a bronze instrument with a separate handle and more elongated loop that dates to the LM I period (ca. 1540–1440 B.C.). Only two examples have been found, one at Hagia Triada and the other at Mochlos (Halbherr, Stefani, and Banti 1977, 74–75, fig. 43;

* I will always be grateful to Jim Muhly for publishing my first article on Mochlos in *Expedition* (Soles 1978) and so encouraging me to write many more Mochlos articles.

Soles 2005, 433, pl. 99f; 2008), but this type is also pictured on the Hagia Triada Harvester's Vase and on Linear A tablets (Betancourt and Muhly 2006, figs. 3–7). The Mochlos example (HNM 14398) is complete and illustrates the method of manufacture most clearly (Figs. 14.1–14.3). Its handle was cast in two parts. The lower part, which the user grasped with his hand, is hollow and cylindrical and tapers toward the top. It has a slightly projecting ring base that would allow it to sit upright. The upper part or capital is U-shaped and attached to the top of the cylindrical handle; it has a hole in its center where it joins the cylindrical handle, and its two ends are flattened and provided with two rivet holes for the attachment of the loop. Two small and shapeless lumps of metal are also attached to the outer sides of the U-shaped capital. The loop is a continuous band of hammered sheet metal ca. 0.033 m wide. Its two ends are each attached by two rivets to the inside of the U-shaped capital. The frame was pierced at two points on each side to support two cross pins, the upper of which was provided with three small disks, the lower with two. Unlike the pins in the clay sistra, the ends of these pins project through the holes of the frame and are twisted at their ends to keep them in place. Both cross pins and disks have a relatively high tin content in comparison to the other parts of the sistrum; they are the only parts that were also alloyed with silver (see Giumlia-Mair, Appendix). The total length of the sistrum from top to bottom is 0.285 m, and the maximum width across its oval frame (from side to side) is 0.083 m. When the instrument was shaken, these disks struck each other and the sides of the frame, and the cross pins also moved back and forth in the frame. It made a louder sound than its clay predecessors, and the tin and silver content of the pins and disks created a fuller timber and a bell-like sustain that enabled the sound to last longer before becoming inaudible. A later bronze example from Egypt has been recorded and may be heard on the internet at http://egyptsound.free.fr/sounds.htm.

The examples from Hagia Triada and those depicted on the Harvester's Vase and Linear B tablets show similar details of construction and clearly belong to the Mochlos type. The Hagia Triada sistrum, which was only identified by Tom Brogan and Malcolm Wiener in the summer of 2006 although published in 1980 and on display in the Herakleion Museum since its discovery (HM 1234), preserves only the two parts of the handle; its top is broken and its loop is missing, but, like the Mochlos sistrum, both ends of the upper U-shaped capital are narrowed and pierced with two rivet holes for the attachment of the loop. Pieces of the loop, described as flat bands of metal in the publication (Halbherr, Stefani, and Banti 1977, 75), were apparently found with it, but were not identified.

Figure 14.1. The Mochlos sistrum in situ (HNM 14,398; length 28.5 cm) in Room 2.2 of House C.3.

The sistrum on the Harvester's Vase is depicted in some detail, and its handle is also constructed like the handle of the two surviving examples with a lower part that the singer grasps in his hand and an upper U-shaped capital to which the loop frame is attached. The Linear A sistrum sign also shows the two part handle. It is also provided with a flat base somewhat wider than the Mochlos base, but designed to show that it can also stand upright. Like the Mochlos sistrum it has two pins that project through the loop and terminate in bent or twisted ends. Like the clay sistra, the Hagia Triada sistrum was also found in its final resting place. It was being stored with the Harvester's Vase in a room that probably served as a sacristy in the Hagia Triada villa and may be the very same sistrum that was depicted on the Harvester's Vase. The Mochlos sistrum, on the other hand, was being stored in a merchant's hoard and had not yet reached its final destination.

A third variety of loop sistrum may be depicted on the famous sarcophagus from Episkopi in the LM IIIA period (ca. 1400–1300 B.C.). Emily Vermeule has suggested that the object carried in the chariot scene on the sarcophagus is a "rattle" (Vermeule 1979, 67–68). If so, it resembles the earlier Middle Minoan examples, but with a round instead of oval loop, rather than the Mochlos type since its handle supports the loop directly and lacks a U-shaped capital. As Vermeule and others have noted, the chariot scene seems to depict the journey of the dead over the seas to Elysion, and the funerary context would be appropriate for the display of a sistrum. No actual example has survived from the Mycenaean period, however.

The fourth variety of loop sistrum is much later in date, and only one example survives. No longer in situ, it was found broken and fragmentary in a mixed context with 2nd and 1st millennium B.C. finds in the sanctuary of Syme where it was probably once used in the religious rites practiced in the sanctuary. It is a Hathor-headed sistrum that preserves only the top of the cylindrical handle, and its capital, which, quite unlike the U-shaped capital of the Mochlos sistrum, consists of a two-faced head of Hathor, one on the front and one on the back, with a naked Astarte figure standing on each side of the capital. It has been carefully described by Anastasia Lambropoulou who convincingly places its origin on the Levantine coast in a Syro-Palestinian workshop dating from the 9th to the 7th century B.C. (Lambropoulou 1999).

Figure 14.2. The sistrum (HNM 14,398).

The origins of the earliest and latest sistra on Crete can be safely located then, in the first case by the material, which is local Cretan clay, and in the second stylistically by the iconography of the capital, which has many Syro-Palestinian parallels. The origins of the Mochlos type, on the other hand, are more problematic. The context of the Mochlos sistrum suggests that it was an import. It was being stored in a pit underneath the floor of a room in an ordinary house, House C.3, which was being used to shelter a bronze hoard (Fig. 14.1). The room, identified as Room 2.2, was located on the ground floor in the center of the house (Soles and Davaras 1996, 194–198, fig. 11; Soles 2008). It was easily reached, either from the vestibule of the house, where the main entrance from the street was located, through a passage that ran underneath the return of the staircase along the north side of the room, or via an internal staircase that led up from the basement of the house along the south side of the room. The room seems to have been designed originally as a living room, but at the end of the LM IB period when the pit was dug several large storage pithoi were located here. The pit, which was located against the south wall of the room, was dug into the earth floor to a depth of ca. 0.50 m. It was semicircular in shape, measuring ca. 0.50 m along its rear and 0.50 m from front to back. The south wall, which dated to the time of the house's construction in the LM IA period and extended beneath the floor, formed a straight line and solid stone wall for the rear of the pit, while loose stones were placed around the outer, semicircular side of the pit after it was dug to prevent the earth floor of the room from falling into the pit. Some traces of plaster remained to suggest that it might also have been partly plastered; it was not a particularly well-built pit, however, and it appears to have been constructed hastily. The bronze objects were laid inside, and then the pit was concealed with earth and a stone quern that was placed upside down on top of the pit.

In addition to the sistrum, the hoard contained a large number of copper ingots and other finished objects. At the top immediately beneath the quern lay one-half of an oxhide ingot (HNM 14389; Soles 2005, 434, pl. 100b). It was set at a slight angle against the south wall and sheltered the bronze sistrum that was placed immediately beneath it. As time passed, the weight of the ingot probably caused the upper part of the sistrum's loop to break. The sistrum, in turn, sat on top of a pair of tongs and a group of chisels. Beneath lay an assortment of

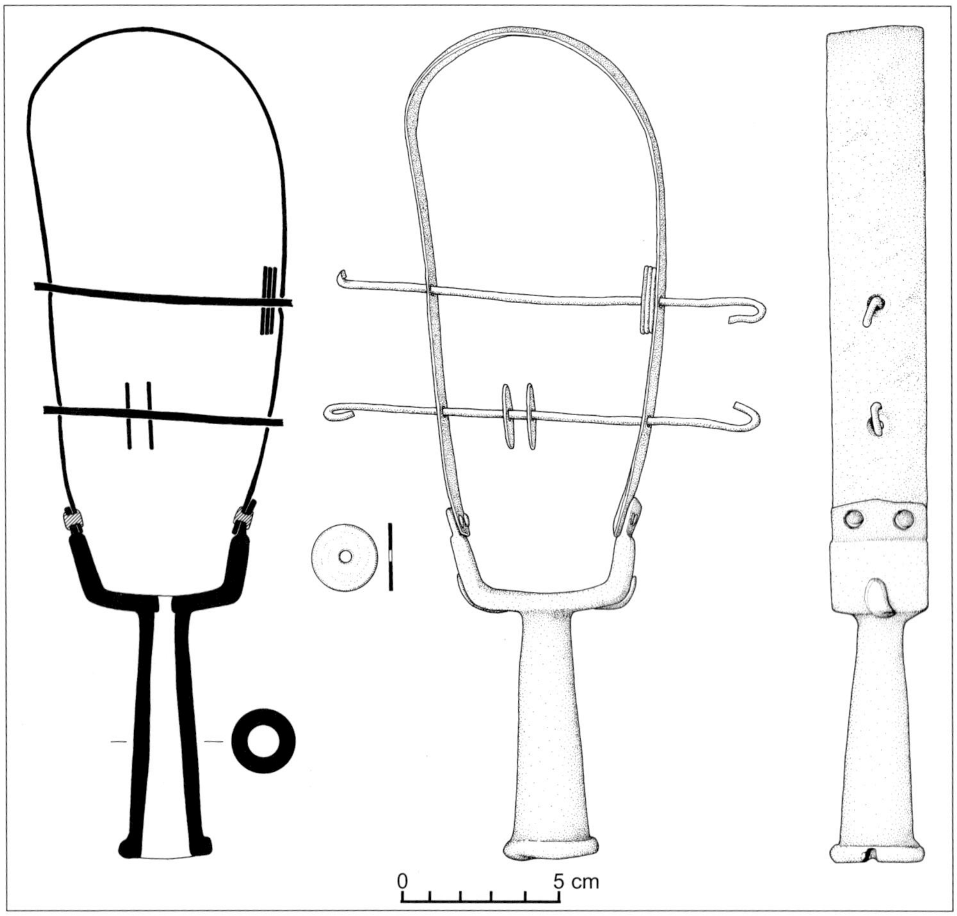

Figure 14.3. Drawing of the sistrum (D. Faulmann).

objects, mostly tools and weapons, which were grouped in clusters, nearly all intact. The tools included nine double axes, seven chisels, three tongs, three knives and a sickle, including two with rivets that once held handles and two without (and presumably without handles), three awls, a pair of balance pans, one saw, and one adze. The weapons included eight daggers, all but one with rivets that once held handles. Miscellaneous objects included two bowls, one with silver hammered on to the rivets of its handle, and one pin. Nearly 60 copper ingot fragments, many preserving traces of calcified linen threads from the bags that held them, also lay in the pit. They varied in size from as little as 0.02 m to as much as 0.16 m in length. Broken from larger oxhide ingots, many preserved the horns or edges of these ingots so the original shapes could be identified. Numerous copper strips, which were looped around themselves in circles, also lay in the pit. They formed a kind of strip ingot and were probably shipped in this form.

The hoard is one of two "utilitarian" hoards found in the house, each of which was made up of very different kinds of objects. Hoard A, which was located in the basement of the house, Room 1.3, consisted largely of scrap-metal that was intended to be melted down for recasting (Soles and Davaras 1996, 194–196, pl. 56). It was being stored with agricultural commodities at the rear of a basement storeroom and formed a metal resource for the house. It has a good parallel in a smaller but contemporary hoard from the Artisans' Quarter (Soles 2003, 20, fig. 17, pl. 12D; Soles et al. 2004, 46–52), which was sitting in a metal workshop on the verge of being melted down when it was abandoned. Both belonged to foundry hoards. Hoard B on the other hand consisted mostly of intact artifacts, many like the sistrum in pristine condition, and others like the rivetless daggers and knives unused. It could be easily reached on the ground floor of the house, and its objects were commodities of the sort that a merchant would assemble for trade. No spill or other evidence for actual casting was found in the house, but the individual who lived there was apparently a major dealer in the town's metal commerce.

The different objects in the hoard suggest that he was engaged in three different enterprises, as an importer, producer, and distributor. As an importer, he was apparently in contact with suppliers to the north in Attica and to the east in the Levant. Lead isotope (LI) analyses of the copper strip ingots in Hoard A suggest that they come from mines near Laurion (see Ch. 22 by Stos-Gale in this volume), and this may also be the case with those in Hoard B, which have not been analyzed yet. Laurion is also likely to have been the source of the silver that was hammered onto the rivets of the bowl in Hoard B. LI analyses of the copper ingot fragments in Hoard A suggest that they came from mines in northwest and eastern Cyprus (see Ch. 22). Those from eastern Cyprus probably came via an intermediate stop on the Levantine coast, and this was certainly the case with the half ingot (Fig. 14.4: HNM 14389) that lay on top of the hoard. Flat on top and bottom, with a woolly or "bubbly" top and smooth bottom, and incurved sides, this ingot belongs to Buchholz's Type 1 (Buchholz 1959). It weighs exactly 15 kilos, so whoever cut it in half knew exactly what it should weigh and employed a standardized unit of weight in order to facilitate a balanced transaction. Its standardized weight implies commodity exchange between merchants (Bradley 1985, 696). The ingot is one of only two half-ingots ever found in Crete. The other is a Type 2 half-ingot with long "ears" or handles from Hagia Triada, which appears to have come from the Apliki mine in Cyprus (Gale 1991, 1999). Such ingots also appear on the Uluburun ship, and it is clear that they were shipped to Crete in this form. Like many of the Uluburun ingots, the Mochlos ingot is also incised with an identification mark. The same mark was found on one of the Uluburun bun ingots and identified as a "quarter rudder" by Cemal

Figure 14.4. Half ingot with linear sign (HNM 14,389).

Pulak (2000, 146, fig. 4d), who argues that the mark was probably incised at "the point of receipt and export to the west, rather than at the production end of the metal" (Pulak 1988, 36). (The same marks are found on tin and copper ingots, so they could not have been incised at the mines themselves.) The mark is thought to identify ownership and suggests that the Mochlos ingot came from the same location in Syria as the Uluburun ingot, even though it predates it by many years. Contemporary ingots from Hagia Triada also bear some of the same signs found on the Uluburun ingots, and the same Syrian merchant families who were shipping ingots to Crete in the second half of the 14th century appear to have been already established 100–200 years earlier. The merchant who shipped the half-ingot to Mochlos obtained it in Cyprus, returned it to his base in Syria, incised it with his mark of ownership, then shipped it west to Mochlos.

The sistrum is anomalous. Alone among the utilitarian objects in the hoard, it was a religious object, intended like the one pictured on the Hagia Triada vase, to be used only for ceremonial purposes. It was not found in a religious context, however. Located in the hoard, it was probably part of the hoard, and like every other object it was meant to be traded and had not yet reached its final destination. Eventually it would end up in a religious context. It might be destined for the ceremonial building at Mochlos to be used in an ancestor cult, or for some wealthy villa in the neighboring countryside for use in an agricultural ceremony, or for the Gournia palace to be used in rites involving the goddess, or even farther afield at Malia or Knossos itself. But where did it come from? Its context suggests that it came with the half ingot and other copper in the hoard on a ship that sailed from the Levant along the Cypriot coast and into the Aegean. Like the three Syrian ivory tusks found with six ingots at Zakros (Platon 1971, 116, 244–245), it may have arrived at the site as part of the same cargo. Indeed, since both the Mochlos and Zakros deposits date to the same time, it is possible that the same ship delivered the sistrum, the ivory tusks, and the six and one-half ingots on the same voyage, stopping first at Zakros and continuing to Mochlos. In any case, the sistrum had only two possible sources: either it came from overseas, or it came from a manufacturing center in Crete that produced elite objects for ceremonial purposes.

Although the sistrum is at home in Egypt, it is unlikely that the Mochlos type came directly from Egypt any more than the clay MM sistra did. Egyptian loop sistra were its source of inspiration as they were for the MM sistra, but the type probably developed outside of Egypt for the simple reason that none like it has been found in Egypt. Its antecedents are found in Egypt, and the earliest surviving parallel for the type dates to the first half of the 12th Dynasty (ca. 1991–1843 B.C.). A simple, two-part sistrum with loop and handle is depicted in the tomb of Khesu-wer at Kom el-Hisn in the northwest Delta (Edgar 1915, fig. 5, pl. 25), the ancient site of Amou, whose chief deity was the goddess Hathor. The owner of the tomb was a priest of Hathor and was in charge of her temple. He superintended the women who were employed in the worship of the goddess, and in his tomb he is shown presiding over two bands of female musicians, continuing his duties in death. In the upper band he plays two loop sistra, while the women respond with similar instruments. In the lower band he claps his hands while the women respond in kind. His sistra resemble the MM clay examples very closely; like them they have only a handle connected directly to a simple elliptical loop, which is provided with three pins. Vertical lines appear to suggest the clappers on these pins as they do in depictions of naos sistra from the end of the Old Kingdom (Fischer 1962, fig. 6d). They probably served as the models for the MM Cretan examples, and the location of the site in the northwest Delta where they were used in the temple of Hathor would have made them all the more accessible to Cretan visitors. They could not have served as the models for the Mochlos type, however, since they lack its U-shaped capital.

At this point the simple loop sistrum seems to disappear in Egypt. It may have survived in perishable materials like the papyrus sistrum from Thebes preserved in the Munich Museum (Mikrakis 2000), one of which may have made its way to Crete, but it appears to have been displaced by the older and more lavish naos sistrum, which certainly becomes more popular (Davies 1920; Klebs 1931; Bonnet 1952, 716–720; Nagy 1977; Ziegler 1979, 31–62). When the loop sistrum reappears in the 18th Dynasty, it is invariably decorated with a double-faced Hathor head, which sits as a capital atop the handle, with the loop attached to the goddess's crown and usually provided with snake-like pins. It is this type that is found in contemporary Theban tombs, including the tomb of Menkheperrasonb where a Minoan embassy is also depicted (Davies and Davies 1933, 13–14, pl. 17) and a tomb belonging to wives of Tuthmosis III (Lilyquist 2003, 112, 160, fig. 93a). The simple variety without the Hathor head capital reappears in the Amarna Period when the little daughters of Akhenaten are shown playing them before the Aten in the tomb of Meryra (Davies 1903, pls. 25, 26). The Hathor-headed variety would be inappropriate in this context, so the Hathor head is eliminated, and the handle is shaped once again as it was in the Old Kingdom in the form of a papyrus stalk crowned with a papyrus bloom to which the loop is attached directly. Tutankhamun's own sistra are perhaps the closest Egyptian parallels to the Mochlos sistrum that survive from the Bronze Age (Hickman 1949, 81–82, no. 69317, pl. 53; Manniche 1976, 5–6, pls. 3, 4). They are both quite plain, and the handle, which is shaped like an octagonal column enlarged at its base, supports a simple rectangular block, both made of gilt wood, which serves as the capital in which the gilt bronze loop is inserted. By the time the small gilt shrine of Tutankhamun was made, the type was abandoned, and Queen Ankhesenamun holds the traditional naos and loop sistra with Hathor-head capitals in relief scenes on the shrine (Eaton-Krauss and Graefe 1985, 11, 13, 18, pls. 8, 9, 16). Not until the Late Period and the Greco-Roman Period does a simple loop sistrum without a Hathor-head capital become popular again (Ziegler 1979, 34, Louvre B44). These resemble the Mochlos type closely, with a rectangular capital supporting the loop; sometimes the loop is still cut from sheet metal and separately attached, but often all parts of the sistrum are now cast, including the loop (Bonnet 1952, 717, fig. 171). The capital and loop are cast as one piece, with the capital now becoming the base of the loop, the handle is cast as a separate piece, and then the two are joined. In the Greco-Roman Period the whole sistrum is sometimes cast as a single piece (Ziegler 1979, 58–62, nos. 76, 77, 81–83).

The chief problem with an Egyptian identification is the absence of contemporary parallels in Egypt. There are similarities in the details of construction between the Egyptian and Mochlos types, but these belong to later examples from the Late Period and Greco-Roman Period and are probably not sufficient to suggest an Egyptian manufacture in any case. The similarities include:

1. The handles of the two are cylindrical and taper toward the top in contemporary 18th Dynasty examples (Hickmann 1949, 90, no. 51174; Quibell 1908, 63, pl. 50) as well as later examples.
2. They are often provided with a ring base, and both are designed to stand upright, as they do in a relief scene at Dendarah where naos and loop sistra stand upright side by side on offering tables (Daumas 1970, pl. 5B).
3. The handle and loop are often made of separate pieces; if metal, the handle is usually cast solid while the loop is cut from sheet metal (Hickman 1949, 77–78, no. 69313; Ziegler 1979, 21–26, nos. 66, 68, 69).
4. If sheet metal, the loop is usually attached to the capital by rivets, and rivet holes are provided on both ends of the loop and both sides of the capital (Hickman 1949, 87, no. 69307, fig. 33).
5. The horizontal pins are often simply bent at the ends to keep them in place.

If not Egypt, then a Levantine or Cretan source is a likelier possibility. Cities on the Levantine coast were also centers with rich musical traditions of their own. They borrowed instruments from Egypt and Mesopotamia and produced their own loop and naos sistra to meet the demands of a local Hathor cult (Braun 1999, 88–90, pls. III:3–2, 3–3). They also created hybrids like the Syme sistrum and may have been influenced by an ancient type of Mesopotamian sistrum that resembles the handle and U-shaped capital of the Mochlos sistrum, without the loop. These consisted of a handle and a U-shaped capital, which also served as the frame into which the cross bars were inserted directly. They were apparently the very first sistra appearing in Mesopotamia before Egypt. One is depicted together with a zoomorphic lyre on a plaque adorning the side of a lyre from the Royal Cemetery at Ur, dating to ca. 2600 B.C. (Duchesne-Guillemin 1981, 289, pl. 34), and another is depicted with a zoomorphic lyre and a hour-glass drum on an Akkadian cylinder seal dating from ca. 2334–2193 B.C. (Dumbrill 1998, 432, pl. 88; Schmidt-Colinet 1981, 6, pl. 11). The type appears to have been imitated in Anatolia as well (Stoddert 1968, ed., 5, no. 4). There is no hard evidence however for the Mochlos type anywhere in the Levant.

This leaves Crete itself as the most likely source. In his capacity as a producer, the Mochlos merchant who owned the sistrum had close contacts with the metal workers in the Mochlos Artisans' Quarter on the coast opposite his house. Clearly he provided raw material to them. One of the awls in the hoard looks as if it were used to break ingots into small pieces, and this would have been one of his activities. He provided only small pieces of copper, as well as scrap metal from the hoard in his basement (Hoard A), to the artisans who produced the final objects. Like his counterpart in early Medieval Europe (Braudel 1982, 316–322), he may have practiced a "putting-out" system, whereby he provided the artisans with the raw material and allowed them to keep part of the finished product while he kept the other part (Soles 2003, 97). The clay molds that survive in the Artisans' Quarter indicate that the artisans were capable of producing daggers and bowls like those in Hoard B (Soles et al. 2004, 22–25, 49–50), and energy dispersive X-ray fluorescense (ED-XRF) analyses indicate that they were even able to differentiate alloys for functional and ideological reasons (Soles 2004, 141; Soles and Stos-Gale 2004, 58). They also were making objects for ritual use. Only a few unfinished or fragmentary examples survived in the workshop: three stone vases, including an unfinished offering table (Soles et al. 2004, 36, no. IC.170), a chalice (Soles et al. 2004, 38, no. IC.187), and a block vase or kernos (Soles et al. 2004, 38, no. IC.193); a clay foot that may have stood in a shrine in Building A; and a one-handled bronze lekane (Soles et al. 2004, 50, no. IC.279) thought to have been made from a rare copper in the Taurus Mountains. Most of these are modest objects that might be used locally in tombs or household shrines, like another lekane that was placed in Prepalatial Tomb VII as an offering (Seager 1912, 56, fig. 26). There is no reason the Mochlos artisans could not have made the bronze sistrum, especially if one or more of their members had traveled to the east as seems very likely (Soles et al. 2004, 141–142). Indeed, the clay sistrum disk from Building C.7 in the main settlement indicates that at one time Mochlos artisans were making sistra.

The finished metal objects from the two different locations, the Merchant's House and the Artisans' Quarter, are very different. Those that survive from the Artisans' Quarter are much more modest, utilitarian objects like fishhooks, tweezers, pins, chisels, and simple blades, while those from the merchant's hoard include weapons, more substantial tools, and even one object with precious silver metal. This difference probably reflects the way goods produced in the Artisans' Quarter were marketed. The artisans targeted a local and regional market of potential customers who lived in the small settlements and farms scattered around the Bay of Mirabello that did not have their own production centers. They lived in their shops and their shops were always open for business, so anyone could have come to the shops from a neighboring farm to acquire a fishhook, needle, or other object needed for everyday life. The metal merchant, on the other hand, was targeting a more elite market. The finished objects in Hoard B indicate that he was meeting a demand for weapons, metal-working equipment, and also axes and chisels that might be needed for building and/or maintaining ships. These objects reflect the dangers of the period in which the Minoans lived when Crete was under attack, and there was a demand for such objects to provide for their defense. His customers may have been located in larger towns or elite centers, nearby or farther afield, and he appears to have been ready to meet the demand himself. As a resident of Mochlos, he may have been a ship owner himself, who assembled this hoard as part of a cargo that he was preparing to ship to market somewhere along the north coast of Crete.

Wherever the Mochlos sistrum comes from, whether the Artisans' Quarter at Mochlos, Zakros, or some other palace workshop producing elite ceremonial goods, the ideology of the sistrum originates in Egypt. The Minoans may have associated the goddess Hathor with their own fertility goddess, but they certainly associated the sistrum's music with concepts of renewal and rebirth just as the Egyptians did. In Egypt the music of the sistrum brought about renewal in many different contexts, in tombs, temples, and pharaoh's court. One of the best illustrations of this fact, as Derchain (1970) and Nagy (1977, 54) have pointed out, is found in the story of Sinuhe, who, upon returning from exile to Egypt, was received at court by royal princes carrying sistra. While the sistra played, the king, in the guise of Atoum, restored life to Sinuhe and allowed him to become Egyptian again. The episode also alludes to the cosmic union of Hathor and Atoum and the creation of the world. Inscriptions from the tomb of Menkheperrasonb, contemporary with the Mochlos sistrum, illustrate the same point. Here a choir of female musicians greet the immortal *ka* of Menkheperrasonb with sistra enabling him to receive the "rejuvenating breath" of Amun (Davies and Davies 1933, 14, pl. 17). In an earlier tomb for Menkheperrasonb and his mother, the musicians greet the two of them with these words, "For thy ka! The sistrums and menyets of Amun. . . . Receive them presented to thy nostrils, thou who art of the millions near him and may he give freshness of breath to thy nostrils daily" (Davies and Davies 1933, 21, pl. 24). This association of the sistrum and its music with creation and rebirth traveled to Crete at a time when many Egyptian contacts were being established (Watrous 1998) and explains its appearance in its different contexts. In tombs at Archanes, Hagios Charalambos, and Krasi, it was used in funerary ceremonies in order to renew life after death and afterward deposited in the tombs. In the scene on the Episkopi sarcophagus it is shown actually performing its miracle as the deceased shakes it while he traverses the sea in his journey to the new Elysion. The Harvester's Vase is probably misnamed, then, since the men who sing to the accompaniment of the sistrum are probably planting rather than harvesting, and the music of the sistrum promotes the fertility of the land and guarantees the success of their labors. The idea, then, for generating and using the instrument arrived via diffusion even if its manufacture was local. Alfred Kroeber has described this kind of diffusion as "idea" or "stimulus diffusion" that might occur without the export of an actual artifact (Kroeber 1940). As Colin Renfrew observes, in stimulus diffusion, "it is not the details or the full complexity and elaboration of a new product or new process which is transmitted, but rather the idea, the realization that such a process is possible, and some understanding of how to bring it about" (Renfrew 1973, 110–111). It is nonetheless further indication that Minoan artisans, whether traveling on their own ships or on Syrian vessels, visited Egypt and witnessed the playing of sistra in Egyptian temples. Its association with an Egyptian fertility goddess must have appealed to Minoans who worshipped a fertility goddess of their own, and so it came to be used in Crete in the same contexts and with the same meaning as in Egypt.

Appendix: Composition of the Mochlos Sistrum

by Alessandra Giumlia-Mair

The sistrum is in good condition, only the top part of the loop having been mended from several fragments. The heavy cast handle shows a compact, smooth and dark patina, while the other parts of the object show a lighter, green patina with cuprite spots, under which the metal is still visible in some areas. The preservation is generally good and the metal is still sound, with corroded areas in the more heavily worked parts (i.e., the loop, cross pins, and disks). It is important to underline that the preservation of the two pins and of the disks is quite remarkable for such heavily worked parts, particularly when compared with the poorer condition of the upper part of the loop.

Method of Analysis

As it was not possible to obtain destructive samples by drilling (drilled samples are taken for more precise Atomic Absorption Spectrometry [AAS] or Inductively Coupled Plasma Spectrometry [ICPS] analyses), non-destructive analysis with X-Ray Fluorescence Spectroscopy (XRF) was carried out by using portable equipment—consisting of a X-ray source, a transformer, a tripod, and a laptop computer, expressly developed for analyses of cultural heritage objects—and with a dedicated program for the analysis of ancient metals. The dedicated software and several other useful characteristics make this equipment much more versatile and precise than the current XRF versions that are available on the market and have been developed for the needs of modern industry.

The X-rays are emitted by a miniaturized X-ray tube provided with a collimator that can change the size of the irradiated area as needed. The beam usually has a diameter of around 1.5–2 mm, but the analyzed spot can be smaller (down to less than 1 mm in diameter) or larger (up to over 4 mm), as dictated by the size of the object, the detail to be analyzed, and the characteristics of the patina. This important feature of the equipment allows the choice of the most suitable spots for analysis, even if the corrosion-free area is rather small, as was the case with some parts of the sistrum.

Before analysis, all parts of the object are carefully examined under the microscope both to assess their condition and to locate and select the best spots for analytical measurement. In general, metallic-looking areas are selected, or, when this is not possible, areas with a thin, preferably cupritic patina are measured. Areas with thicker rough patina or with visible corrosion products need to be avoided; otherwise the data can be considered qualitative only.

A laser pointer indicates the exact spot of the measurement on the object, and the equipment gives an audible signal when the distance from the surface to be measured is correct and within the acceptable span of +/- 0.2 mm. This ensures that the angle and the distance from the object are always the same for all measurements and allows a more precise calculation of the composition. In this way also the comparison with the known standards is much more reliable.

The standards represent the most important tool for obtaining reasonable quantitative results. They are routinely measured, first at the beginning of each day, then whenever the equipment is switched off and on, and also every couple of hours during analyses, to allow for the calculation of possible drifts due, for example, to a change of temperature, humidity, or other parameters that can affect the performance of the scientific equipment.

Several different standards of various composition, such as copper containing single (5% Sn; 1% of As, Sn, Ag, Fe, Pb; 0.5% As, Ag, Fe, respectively) and composite alloying elements (e.g., 1% Sn, As, Fe, Pb in Cu; 1% As, Sn, Ag, Sb, Fe in Cu; 1% Sn, Ag, Au, As in Cu, and various silver containing alloys) have been employed during the measurements. These have been specifically produced by AGM Archeoanalisi (laboratory located in Merano, Italy) for the analysis of ancient metals of different composition and represent a fundamental tool for a precise calculation of the results. They are also particularly useful in the case of interference effects, such as enhanced results for iron in a copper matrix or similar phenomena, which can be monitored and taken into account while evaluating the results.

The measurements take typically 10–15 minutes, but can be longer, if necessary. Longer measurement times were chosen to collect more signals when the analyzed area was small. The analysis is routinely repeated in case of particular problems or anomalous results. In the case of the sistrum, the analyses of the cross pins and the disks were repeated on different areas of the analyzed parts to maximize the reliability of the results. The results given in Table 1 are the average of several measurements.

Discussion of Results

The analyses carried out on the various parts of the sistrum have shown that the artisans were able to differentiate their alloys, depending on the properties and functions required for the various components of the object (Table 1).

The alloy employed for the cast handle contains about 2 % each of tin and arsenic and traces of antimony, while the loop is made of nearly pure copper with a low arsenic content (under 1%). The four rivets (two on each side) are rather corroded, but the results obtained on the two less altered specimens seem to indicate that they were made of a more impure copper, containing 1% of arsenic and traces of antimony and iron, i.e., the metal employed for these small functional parts was of lesser quality. The most interesting results were obtained from the mobile cross pins inserted in the sides of the loop. The alloys employed for these parts are more unusual since they contain a noticeable amount of tin (around 5%) and silver (around 8%) in addition to the arsenic content determined in most elements of the sistrum. The two disks that were analyzed also contain around 5% each of tin and silver. Regrettably, the remaining disks could not be sampled because they are stuck together and could not be moved from their position on the side of the loop.

The addition of both tin and silver very much improves the ringing quality or sound of metal

(Giumlia-Mair 2001a, 2005), and metal workers have been aware of this for some time. For example, in Iron Age and Roman times, even small bells were cast by using alloys containing higher tin percentages (Giumlia-Mair 2001b, 25). The tin amount in alloys for large bells was later much increased and went up to 17%–38% in the Middle Ages (Theobald 1933, 152–160; Brepohl 2002, 640; Tiussi and Giumlia-Mair 2002). These alloys were called "bell metal" and were in use until modern times when new alloys containing aluminum and magnesium were introduced (Elphick 1988, 79). Silver bells are still appreciated because of their especially pleasant sound. As higher tin percentages combined with silver are not present in the other parts of the sistrum, it is clear that they were added on purpose to the alloy used for the mobile and vibrating parts of the sistrum to obtain a pleasant sound from the musical instrument. The pins moved to and fro in the perforations and produced a clinking and vibrating sound when their curved ends hit the metal of the loop. This means that the most important elements for the sound of the sistrum were the vibrating pins, which in this case also have the highest tin and silver contents, while the disks were produced with a slightly lower silver percentage. The presence of both tin and silver is also almost certainly the reason for the better preservation of the cross pins and the disks, while the heavily worked loop, made of almost pure copper, and the hammered flanges of the U-shaped support on the handle have been attacked by corrosion.

Not many sistra have been analyzed, and only very few are chronologically comparable to the example from Mochlos. Among the many thousands of analyses carried out by Riederer on Egyptian material, only three are sistra (Riederer 2002, 171, 176, 193). Two belong to the Ägyptisches Museum der Staatlichen Museen zu Berlin (inventory number 2768: 8.45% Sn, 5.8% Pb; inventory number 9710: 7.77% Sn, 16.5% Pb, 0.2% As), and one belongs to the Hamburg Museum (inventory number 1917/711: 8.2% Sn, 6.36% Pb). All three are of unknown date; however, their compositions suggest that they might have been produced in Ptolemaic or Roman times. The sistrum in the Louvre (inventory number E 11201), analyzed by Mathis (2005) and dated to the 3rd Intermediate Period (1070–712 B.C.) is made of black copper (Giumlia-Mair and Quirke 1997) and has therefore a very distinctive composition (i.e., copper with approximately 1%–3% of gold to intentionally induce a black patina), while the stylistically very different sistra from Anatolia, Horoztepe, and Alaca Höyük, which would be of an earlier date, are reported to be made of arsenical copper (Zimmermann 2005). None of the analyzed pieces contain silver or have a composition similar to that of the example from Mochlos; however, it is also very difficult to draw comparisons because only one set of analysis is published for each object

Location of Analysis on Sistrum	Cu %	Sn %	Pb %	As %	Sb %	Fe %	Ag %	Mn %
Handle	95	2.2	—	1.8	traces	0.9	—	—
Loop	97	—	—	0.8	—	1.5	—	—
Rivet	97	—	—	0.7	traces	1.8	—	traces
Rivet	98	—	—	0.9	traces	traces	—	—
Rivet	97	—	traces	1	traces	1.5	—	—
Rivet	97	—	—	1.3	traces	1.2	—	—
Pin 1	86	4.8	traces	0.5	—	traces	8	—
Pin 2	87	4.9	—	traces	—	0.8	7.9	—
Disk 1	89	5.1	—	traces	—	traces	4.8	—
Disk 2	90	4.9	—	traces	—	1	4.9	—

Table 14.1. Average values of major elements that resulted from analysis on different parts of the Mochlos sistrum as determined by portable XRF.

(with the exception of the sistrum in the Louvre, where two non-sounding elements, the handle and the *uraeus*, but not the cross pieces, were examined). From the publications it is clear that only one part of the other sistra was analyzed, but we do not know which one. The analysis data suggests that only the handles (i.e., the largest parts) were sampled, but we cannot be certain.

Nevertheless, even without appropriate terms of comparison, the skill and the empirical knowledge of the artisans who produced the sistrum found in Mochlos are incontestable and quite striking. The choice of special alloys for the ringing elements of the sistrum characterize it as an object of special function and exceptional significance.

References

Betancourt, P.P. 2005. "Egyptian Connections at Hagios Charalambos," in Laffineur and Greco, eds., pp. 449–454.

Betancourt, P.P., and J. Muhly. 2006. "The Sistra from the Minoan Burial Cave at Hagios Charalambos," in *Timelines: Studies in Honor of Manfred Bietak*, vol. 2, E. Czerny, I. Hein, H. Hunger, D. Melman, and A. Schwab, eds., Leuven, pp. 429–435.

Bonnet, H. 1952. *Reallexikon der ägyptischen Religionsgeschicht*, Berlin.

Bradley, R. 1985. "Exchange and Social Distance—The Structure of Bronze Artefact Distributions," *Man* 20, pp. 692–704.

Braudel, F. 1982. *Wheels of Commerce, Civilization, and Capitalism, 15th–18th Century*, vol. 2, New York.

Braun, J. 1999. *Die Musikkultur Altisraels/Palästinas: Studien zu archäologischen, schriftlichen und vergleichenden Quellen*, Göttingen.

Brepohl, E. 2002. "Copper and Its Alloys in Theophilus Presbiters 'De Diversis Artibus' (1100–1125)," in *I Bronzi Antichi: Produzione e tecnologia. Proceedings of the XV Interational Conference on Ancient Bronzes, Grado and Aquileia, 22–26 May 2001*, A. Giumlia-Mair, ed., pp. 639–643.

Buchholz, H.G. 1959. "Keftiubarren und Erzhandel in zweiten vorchristlichen Jahrtausend," *PZ* 37, pp. 1–40.

Daumas, F. 1970. "Les objets sacrés de la désse Hathor à Dendara," *Revue d'Egyptologie* 22, pp. 63–78.

Davies, N.G. 1903. *The Rock Tombs of El-Amarna*, vol. 1, London.

———. 1920. "An Alabastrum Sistrum Dedicated by King Teta," *JEA* 6, pp. 69–72.

Davies, N.G., and N.M.C. Davies. 1933. *The Tombs of Menkheperrasonb, Amenmose and Another*, London.

Derchain, P. 1970. "La réception de Sinouhé à la cour de Sésostris I," *RdE* 22, pp. 79–83.

Duchesne-Guillemin, M. 1981. "Music in Ancient Mesopotamia and Egypt," *WorldArch* 12, pp. 287–297.

Dumbrill, R.J. 1998. *The Musicology and Organology of the Ancient Near East 3500–500 B.C.*, London.

Eaton-Krauss, M., and E. Graefe. 1985. *The Small Golden Shrine from the Tomb of Tutankhamun*, Oxford.

Edgar, C.C. 1915. "Recent Discoveries at Kom El-Hisn," in *Le Musée Egyptien: Recuil de monuments et de notices sur les fouilles d'Egypte*, M.G. Maspero, ed., Cairo, pp. 54–63.

Elphick, G. 1988. *The Craft of the Bellfounder*, Chichester.

Fischer, H.G. 1962. "The Cult and Nome of the Goddess Bat," *JARCE* 1, pp. 7–24.

Gale, N.H. 1991. "Copper Oxhide Ingots: Their Origin, and Their Place in the Bronze Age Metals Trade in the Mediterranean," in *Bronze Age Trade in the Mediterranean. Papers Presented at the Conference Held at Rewley House, Oxford, in December 1989* (*SIMA* 90), N.H. Gale, ed., Jonsered, pp. 202–203.

———. 1999. "Lead Isotope Characterization of the Ore Deposits of Cyprus and Sardinia and Its Application to the Discovery of the Sources of Copper for Late Bronze Age Oxhide Ingots," in *Metals in Antiquity*, S.M. Young, A.M. Pollard, P. Budd, and R.A. Ixer, eds., Oxford, pp. 110–121.

Giumlia-Mair, A. 1988. "The Composition of Copper-based Small Finds from a West-Phoenician Settlement Site and from Nimrud Compared with That of Contemporary Mediterranean Small Finds," *Archaeometry* 34, pp. 1, 107–119.

———. 2001a. "Argento e leghe 'argentee' nell'antichità," in *Atti del 7° convegno "Le Scienze della Terra e l'Archeometria," Taormina, Palermo, Catania 22–26 Febbraio 2000* (*Bollettino dell'Accademia Gioenia di Scienze Naturali* vol. 33, no. 357), pp. 295–314.

———. 2001b. “Technical Studies on the Roman Copper-based Finds from Emona,” *Berliner Beiträge zur Archäometrie* 18, pp. 5–42.

———. 2005, “Tin-rich Layers on Ancient Copper-based Objects,” in *Proceedings of the 18th International Conference on Surface Modification Technologies, Dijon, France, 15–17 November 2004* (*Surface Engineering* 21), A. Giumlia-Mair, ed., pp. 5–6, 359–367.

Giumlia-Mair, A., and S. Quirke. 1997. “Black Copper in Bronze Age Egypt,” *Revue d Égyptologie* 48, pp. 95–108.

Halbherr, F., E. Stefani, and L. Banti. 1977. *Haghia Triada nel periodo tardo-palaziale* (*ASAtene* 55, n.s. 39, 1977 [1980]), Rome.

Hickmann, H. 1949. *Catalogue général des antiquités égyptiennes du Musée au Caire. Instruments de musique*, Cairo.

Klebs, L. 1931. “Die verschiedenen Formen des Sistrums,” *Zeitschrift für ägyptische Sprache und Altertumskunde* 67, pp. 60–63.

Kroeber, A.L. 1940. “Stimulus Diffusion,” *American Anthropologist* 42, pp. 1–20.

Laffineur, R., and E. Greco, eds. 2005. *Emporia: Aegeans in the Central and Eastern Mediterranean. Proceedings of the 10th International Aegean Conference, Italian School of Archaeology in Athens, 14–18 April 2004* (*Aegaeum* 25), R. Laffineur and E. Greco, eds., Liège.

Lambropoulou, A. 1999. “A Bronze Sistrum from the Sanctuary of Syme/Crete,” *AA* 4, pp. 515–521.

Lilyquist, C. 2003. *The Tomb of Three Foreign Wives of Tuthmosis III*, New York.

Manniche, L. 1976. *Musical Instruments from the Tomb of Tutankhamun*, Oxford.

Marinatos, S. 1929. “Πρωτομινωϊκὸς θολωτὸς τάφος παρὰ τὸ χωρίον Κράσι Πεδιάδος,” *ArchDelt* 12, pp. 102–141.

Mathis, F. 2005. *Croissance et propriétès des couches d'oxydation et des patines à la surface d'alliages cuivreux d'intérêt archéologique ou artistique*, Ph.D. diss., University of Paris.

Mikrakis, M. 2000. “Μουσική στην Κρήτη και την Αίγυπτο, Ενα ιδιαιτερό πεδίο ανάπτυξης πολιτισμικόν δεσμών,” in *Κρήτη-Αίγυπτος, Πολιτισμικοί δεσμοί τριών χιλιετιών*, Heraklion, pp. 162–169.

Nagy, E. 1977. “Fragments de sistres au Musée des Beaux-Arts,” *Bulletin du Musée Hongrois des Beaux-Art* 48–49, pp. 49–70.

Platon, N. 1971. *Zakros: The Discovery of a Lost Palace of Ancient Crete*, New York.

Pulak, C. 1988. “The Bronze Age Shipwreck at Ulu Burun, Turkey: 1985 Campaign,” *AJA* 92, pp. 1–37.

———. 2000. “The Cargo of Copper and Tin Ingots from the Late Bronze Age Shipwreck at Uluburun,” in *Anatolian Metal* I (*Der Anschnitt* 13), Ü. Yalçin, ed., Bochum, pp. 137–157.

Quibell, J.E. 1908. *Tomb of Yuaa and Thuiu*, Cairo.

Renfrew, C. 1973. *Before Civilization: The Radiocarbon Revolution and Prehistoric Europe*, London.

Riederer, J. 2002. *Die Berliner Datenbank von Metallanalysen kulturgeschichtlicher Objekte* (*Berliner Beiträge zur Archäometrie* 19), pp. 72–226.

Sakellarakis, Y., and E. Sakellaraki. 1991. “Ανασκαφή Αρχανών 1986–1988,” *ArchEph* 1991, pp. 169–218.

———. 1997. *Archanes, Minoan Crete in a New Light*, vol. 1, Athens.

Schmidt-Colinet, C. 1981. *Die Musikinstrumente in der Kunst des Alten Orients: Archäologisch-philologische Studien*, Bonn.

Seager, R. 1912. *Explorations in the Island of Mochlos*, Boston and New York.

Soles, J.S. 1978. “Mochlos. A New Look at Old Excavations: The University Museum's Work on Crete,” *Expedition* 20, pp. 5–15.

———. 2003. *Mochlos* IA: *Period III. The Neopalatial Settlement on the Coast: The Artisans' Quarter and the Farmhouse at Chalinomouri. The Sites* (*Prehistory Monographs* 7), Philadelphia.

———. 2004. “Conclusions,” in Soles et at. 2004, pp. 139–143.

———. 2005. “From Ugarit to Mochlos, Remnants of an Ancient Voyage,” in Laffineur and Greco, eds., pp. 429–439.

———. 2008. “Metal Hoards from LM IB Mochlos, Crete,” in *Aegean Metallurgy in the Bronze Age. Proceedings of an International Symposium, Held at the University of Crete, Rethymno, Greece, on November 19–21, 2004*, I. Tzachili, ed., Athens, pp. 143–156.

Soles, J.S., and C. Davaras. 1996. “Excavations at Mochlos 1992–1993,” *Hesperia* 65, pp. 175–230.

Soles, J.S., and Z.A. Stos-Gale. 2004. “The Metal Finds and Their Geological Sources,” in Soles et al. 2004, pp. 52–59.

Soles, J.S., C. Davaras, J. Bending, T. Carter, D. Kondopoulou, D. Mylona, M. Ntinou, A.M. Nicgorski, D.S. Reese, A. Sarpaki, W.H. Schoch, M.E. Soles, V. Spatharas, Z.A. Stos-Gale, D.H. Tarling, and C. Witmore. 2004. *Mochlos* IC: *Period III. The Neopalatial Settlement on the Coast: The Artisans' Quarter and the Farmhouse at Chalinomouri. The Small Finds* (*Prehistory Monographs* 9), Philadelphia.

Stoddert, K., ed. 1968. "Anatolia," *Bulletin Metropolitan Museum of Art*, New York.

Theobald, W. 1933. *Technik des Kunsthandwerks im 10. Jht, des Theophilus Presbyter Diversarum Artium Schedula in Auswahl neu herausgegeben, übersetzt und erläutert*, Berlin.

Tiussi, C., and A. Giumlia-Mair. 2002. "Una fossa per il getto di una campana ad Aquileia," in *I Bronzi Antichi: Produzione e tecnologia. Proceedings of the XV International Conference on Ancient Bronzes, Grado and Aquileia, 22–26 May 2001*, A. Giumlia-Mair, ed., pp. 642–651.

Vermeule, E.T. 1979. *Aspects of Death in Early Greek Art and Poetry*, Berkeley, Los Angeles, and London.

Watrous, L.V. 1998. "Egypt and Crete in the Early Middle Bronze Age: A Case of Trade and Cultural Diffusion," in *The Aegean and the Orient in the Second Millennium* (*Aegaeum* 18), E.H. Cline and D. Harris-Cline, eds., Liège and Austin, pp. 19–28.

Ziegler, C. 1979. *Catalogue des instruments de musique égyptiens au musée du Louvre*, Paris.

Zimmermann, T. 2005. "Anatolische 'Isisklappern'—Eine Randnotiz zu einigen bronzezeitlichen Sistren aus Alaca Höyük und 'Horoztepe,' Türkei," *ArchKorrBl* 35, pp. 467–472.

PART III

Metallurgical Technology

CHAPTER

15

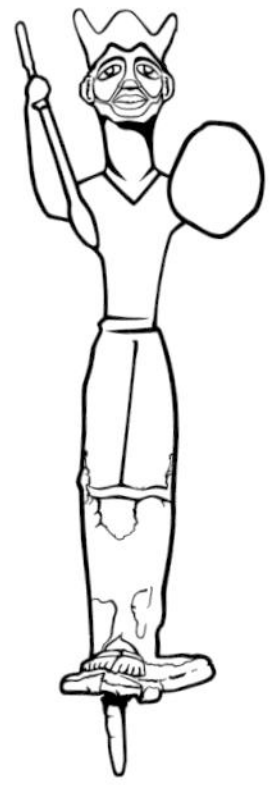

Akrotiraki and Skali: New Evidence for EBA Lead/Silver and Copper Production from Southern Siphnos

Zozi D. Papadopoulou

Siphnos was renown in antiquity for its wealth owing to silver and gold.* It is thanks to their rich mineral resources that the Siphnians were able to build their treasury for Apollo's oracle in Delphi (Hdt. 3.57–58; Paus. 10.11.2). However, neither the travelers of the 17th, 18th, and 19th centuries, who were inspired by the ancient writers to visit the island, nor the miners that reactivated the ancient mines at the end of the 19th century encountered any precious metal. A large scale field program by a team of reseachers from the Max Planck Institute in Heidelberg, Germany worked on Siphnos in the 1970s in order to deal with this controversy. According to the results of this program, there are two mineralization belts on the island (Fig. 15.1): the main one, running from northeast to southwest (Hagios Sostis, Hagios Silivestros, Vorini, Kapsalos, and Xero Xylo), with deposits containing silver and lead; and a second belt, along the southeast coast (Hagios Ioannis, Apokofto, and Aspros Pyrgos) containing gold.

One of the sites examined during the research program was Platy Gialos in southern Siphnos. On the cliff west of Platy Gialos, obviously at the site known as Akrotiraki, Fiedler had first mentioned litharge, heavy slags, and abundant obsidian, but no traces of ancient mining nearby (Fiedler 1841, 134), observations also confirmed by the research team from the Max Planck Institute (Gale 1980, 187; Wagner et al. 1980, 72, 73). Litharge has also

* I would like to thank the editors of this volume for kindly inviting me to participate in Professor J.D. Muhly's festschrift. His inspired approach and multifaceted perspective regarding the metallurgical past of the eastern Mediterranean has contributed to a deeper understanding of this intriguing domain and encouraged junior researchers to undertake work in it.

been found at Hagios Sostis and Kapsalos, proving ancient cupellation activities on the island.

Recently, a rescue excavation undertaken by the 21st Ephorate of Antiquities produced important new evidence regarding metallurgical operations at Akrotiraki. At the same time, a field survey on the southern part of the island resulted in the discovery of an unkown Early Bronze Age (EBA) copper-smelting site, so far unique for Siphnos. The new evidence prompted the undertaking of an archeometallurgical project in this under-explored part of Siphnos. The project involved researchers from various disciplines and was focused on two sites that were representative of the two different contexts within which metallurgical activities appear in the EBA Aegean: (1) Akrotiraki, a domestic site related to lead/silver production, and (2) Skali, a recently localized non-domestic copper-smelting site.

Akrotiraki

Akrotiraki is located on a low hill promontory (ca. 20 m asl) separating the bay of Platy Gialos to the northeast from the small cove of Lazarou to the south. At the end of the 19th century C. Tsountas had explored a number of graves on the site of Akrotiraki (Tsountas 1899, 73–74). They were first mentioned as a prehistoric cemetery a few years earlier (Pollak 1896, 210). The study of the findings from the graves relates the cemetery to the Early Cycladic (EC) I, EC II, and the Kastri periods (Tsountas 1899, 73–74, fig. 28, pls. 8–10; see also Rambach 2000, 60–65, pls. 23–25; Renfrew 2005). The EC settlement linked to the cemetery must be located at the summit of the promontory, as observed by Tsountas and confirmed by later surveys (Renfrew 2005, 45).

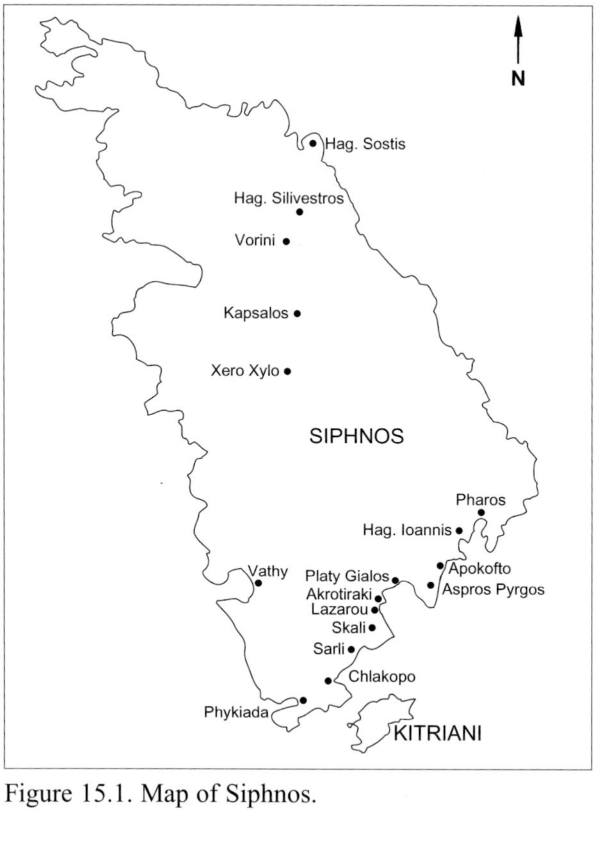

Figure 15.1. Map of Siphnos.

On the northeastern part of the promontory, in a plot close to the seaside (Pittis plot), at the edges of the settlement, a rescue excavation took place that produced important new evidence with regard to EBA lead/silver production. Within extended mixed EC levels with very few structural remains, abundant pottery and lithics (saddle querns, pounders/grinders, obsidian tools, and fragments of stone vases) as well as various metallurgical materials were unearthed (Figs 15.2–15.5). As can be concluded from the preliminary study of the pottery, which is not yet restored, the occupation of the site extends from the Final Neolithic (FN)–EC I to early EC III, but most of the dated material belongs to EC II (Fig. 15.2). It consists mainly of coarse domestic ware, while medium and fine ware is not absent. It is particularly worth mentioning that the site has many pieces of the fabric containing a soft, soapy talc-rich filler, known with the term "talc ware," a fact that reinforces the suggestion that Akrotiraki was the center of production and export of this fabric (Broodbank 2000, 296–297; Renfrew 2005, 46). The distribution of talc-ware pottery has been connected with local sea-trade networks based on the exchange of metals (Vaughan and Wilson 1993, 182–183).

Among the findings, the metallurgical remains are of great interest, particularly the discovery of at least 46 pieces of litharge (some joining) in the shape of a shallow bowl (Fig. 15.3). The largest one has an extant diameter of 0.18 m and a weight of approximately 2,700 g, while the total weight of the litharge is over 18.5 kilos. The denser concentration

Figure 15.2. Sherds from Akrotiraki dating to the EC period.

Figure 15.3. Litharges from Akrotiraki.

Figure 15.4. The semi-circular hearth from Akrotiraki.

Figure 15.5. Lead objects from Akrotiraki.

of litharge (24 fragments) was traced in the same trench (I 4) where a semicircular hearth, possibly related to the smelting operation, was revealed (Fig. 15.4). To date, although we knew that cupellation took place on Siphnos during antiquity, an accurate date could not be established. Prehistoric litharge examples in the shape of a shallow bowl are plentiful in EBA contexts in southeastern Attica (Kakavogianni et al. 2006; Kakavogianni, Douni, and Nezeri 2008 and references within), while one fragment is known from the site of Limenaria on Thasos (Papadopoulos 2002). The intriguing presence of depressions on the inner surface of the examples from southeastern Attica is not observed in the finds from Siphnos. Although it is not possible to determine the beginning of the metallurgical activity at Akrotiraki, it is very important that the majority of the metallurgical remains derives from a clear EC context. Further study of the pottery material in combination with laboratory analyses will hopefully provide additional chronological resolution.

During the examination of the vast quantity of the collected pottery, a considerable number of sherds (130) connected with metallurgical activity was recognized. These sherds, coming mainly from phiales, bear lead oxide on either their inner or both inner and outer surfaces. It is worth mentioning that the larger number of these sherds (88 pieces) comes from Trench I 4, where the greatest concentration of litharge fragments had been observed. Laboratory analyses will contribute to clarify their function.

The evidence for metal working from Pittis plot also includes several lead slags, a few copper slags—among which one is modulated as a tool—and a few furnace fragments, as well as several stone tools, some of which could be connected with metallurgical activities. Finally, parts of three finished products were revealed: a small handle, part of the rim(?) of a vessel, and a triangular object with two diagonal incisions on one side (Fig. 15.5).

The importance of the area for early metallurgy is further proved by the discovery of a considerable number of metallurgical remains in the deeper strata of another rescue excavation, at the southwestern part of Platy Gialos, in the vicinity of the settlement of Akrotiraki. Four long walls were unearthed in this area (Vassilopoulos plot). The function of these walls cannot be determined with certainty based on the existing data. Interestingly, a large number of lead slags and pieces of litharge were discovered. They were mainly unshaped, but some were in the shape of shallow bowls. Several lead and bronze small objects were also uncovered. The metallurgical remains can be dated to the Archaic/Classical period, based on the relatively small quantity of pottery found near the walls. The comparative study of the metallurgical material coming from different plots in the area of Akrotiraki and Platy Gialos, dating both from prehistoric and historic times, will contribute to a better understanding of the development of Siphnian metal technology and production throughout antiquity.

Skali

The significant role of the southern part of Siphnos in prehistoric metallurgy is further proved by the discovery of a copper-smelting site that adds an important element that has been missing from the archeometallurgical research on Siphnos: the existence of at least one copper-smelting site on the island dated with certainty to the EBA. To date, despite the presence of copper ores in varying quantities on several Cycladic islands—particularly in the metalliferous western Cyclades such as Kythnos (Hadjianastasiou and MacGillivray 1988; Bassiakos and Philaniotou-Hadjianastasiou 2007) and Seriphos (Georgakopoulou 2005, 2007; Zacharias et al. 2006)—the proven evidence for EC copper smelting did not include Siphnian sites.

The site is known to the inhabitants with the name of Skali (meaning "step" in modern Greek), and it is located on a ridge between Platy Gialos and Chlakopo, approximately 130 masl. The installation overlooks Akrotiraki as well as part of Platy Gialos Bay and the southwestern coasts of Siphnos and the islet of Kitriani. The nearest known EBA habitation settlement so far is Akrotiraki, which, as mentioned above, has a direct distance from the metallurgical site of about 500 m. It is worth mentioning that a number of ancient towers have been discovered recently in the region around Akrotiraki and Skali; and though they were built at a later date, they probably indicate the boundaries of the territory utilized by the inhabitants (Papadopoulou 2005).

The surface of the metallurgical area is covered with a large quantity of slags, which are highly ferrous and contain small prills of copper. Mixed with the slags are furnace fragments and obsidian objects as well as a few coarse ware sherds. Iron ore (which often accompanies copper minerals) is still visible on the surface, probably a remnant of open-pit mining exploitation, which was common in the EBA (Wagner and Weisgerber 1985).

In the framework of the archeometallurgical project on the south part of Siphnos, two trial trenches were undertaken: one on the cliff-top (Trench I) where a large number of furnace wall fragments and scattered slags were visible on the surface, and one on the south slope of the cliff (Trench II). It must be noted that the area is very rocky with little soil.

In Trench I, the remains of two smelting furnaces were located; the first one consisted of a thin layer of vitrified earth on the bedrock along with a furnace wall fragment found in situ. From the second furnace a piece of wall was preserved in situ, placed almost on the bedrock, which appeared to be carved in order to form a surrounding structure around the furnace. Furthermore, at the northeast part of the trench, after the removal of stones that were put in an upright position to form an improvised recent construction (probably a shelter against the north wind), a layer of scattered pieces of furnace wall fragments and small stones was revealed. This layer was limited by a semicircular construction that was formed by the carved rock on the west and by a series of stones on the north/northeast (Fig. 15.6). During the excavation of this construction, a thick layer of furnace wall

Figure 15.6. Northeastern side of Trench I at Skali.

Figure 15.7. Some copper slags, furnace fragments, and stone tools (one for crushing slag) from Skali.

fragments, some of which had traces of fire, was discovered. Its density decreased at the level of the bedrock. This semicircular construction was probably aimed to control the strong wind coming from the north. To the northeast of this feature, two stones (a circular piece of granite and a trapezoid stone) were on the bedrock and formed a kind of floor (it must be noted that this kind of stone does not exist on the island); their position near the furnaces probably facilitated the transportation of the raw material to the furnaces and the manipulation of the fire.

A very large quantity of slags and furnace wall fragments was collected. In the lower passes the quantity and the size of the slags decreased, while the quantity of the furnace walls increased. A small number of EC sherds and pieces of obsidian were also found. No traces of perforated furnace fragments were found. An important find in the same trench was a stone tool bearing two grooves, probably used for slag crushing (Fig. 15.7).

Trench II was excavated on the south slope of the cliff, not far from the cliff-top, where a large concentration of slags was visible on the surface. The excavation went down to the bedrock which is strongly slopes; no trace of construction was revealed. A very large number of furnace wall fragments and slags was collected (ca. 150 kilos of slag). The volume indicates that the site was possibly used as a deposit for slag. A tool consisting of slag was also found, as well as several sherds characteristic of the EC period.

A field expedition undertaken in 2007 resulted in the location of a second small copper-smelting site on the south edge of the Skali ridge, at the same altitude as the main smelting area.

Conclusions

The ongoing archeometallurgical research in the southern part of Siphnos has already given information of considerable interest concerning the early metallurgy on the island. The findings from Akrotiraki represent some of the earliest evidence for lead/silver production using the practice of cupellation on Siphnos, indisputably dated to the EC period. Equally important is the fact that this evidence is associated with a settlement. On the other hand, the presence of a copper-smelting installation at a distance of 500 m suggests that this area was important for metal production during the EBA. Its geographical position on the straits formed by the southeastern coast of Siphnos and the inhabited islet of Kitriani, on the maritime route connecting the western Cyclades with the central and the north/northeastern part of the Aegean (Pantou and Papadopoulou 2005), obviously played an important role for the development of the settlement and its metallurgical profile.

The site of Skali was a specialized, nondomestic, copper smelting site, the first so far known on Siphnos, indisputably dated to the EBA. Like other Aegean copper-smelting installations, it is located on an exposed cliff and is not easily accessible from the sea, where ancient metallurgists could benefit from the strong winds. The fieldwork in the area surrounding the main smelting installation showed that smelting activity also took place at a second location, within a short distance from the main site

The provenance of the raw material processed in the area of Akrotiraki and Skali cannot be determined with certainty at present. However, there are strong indications for the occurrence of metals (lead/silver and copper) in close proximity to Akrotiraki; further investigation is scheduled to draw out certain conclusions.

The evidence collected from Akrotiraki and Skali suggests that an important industry of metal production existed on the southern part of the island during the EBA, and it shows a high level of specialization. It also makes apparent that the pattern of isolation of the EBA settlements from the metal production sites as attested elsewhere in the southern Aegean (e.g., Georgakopoulou 2005; Betancourt 2006; Bassiakos and Philaniotou-Hadjianastasiou 2007) is not observed in this case (Georgakopoulou 2007). We hope that the ongoing field research and study of the material in connection with the investigation of interactions between metallurgical installations within Siphnos (especially between Akrotiraki and Hagios Sostis) will further improve our knowledge regarding local technology and organization of metal production and will contribute a better understanding of the role of Siphnos in EBA Aegean metallurgy.

Acknowledgments

I would like to thank Prof. P. Betancourt for inviting me to participate in this volume. The main participants of the Siphnos project are, apart from the author, Dr. Y. Bassiakos, Prof. G. Wagner, Dr. M. Georgakopoulou, and M. Chrysoverghi. Equally important at all stages is the contribution of A. Karavis, guard of the Museum of Siphnos, who has a deep knowledge of the countryside of the island, and to whom we owe the localization of Skali. We are thankful to Dr. Marisa Marthari, Director of the 21st Ephorate of Antiquities, for her support, as well as to the Conservation Directorate of the Hellenic Ministry of Culture for the necessary permits. Major funding for the project is provided by the Institute for Aegean Prehistory.

References

Bassiakos, Y., and O. Philaniotou-Hadjianastasiou. 2007. "Early Copper Production on Kythnos: Archaeological Evidence and Analytical Approaches to the Reconstruction of Metallurgical Progress," in Day and Doonan, eds., 2007, pp. 19–56.

Betancourt, P.P. 2006. *The Chrysokamino Metallurgy Workshop and Its Territory* (*Hesperia Suppl.* 36), Princeton.

Broodbank, C. 2000. *A Island Archaeology of the Early Cyclades*, Cambridge.

Day, P., and R. Doonan. eds. 2007. *Metallurgy in the Early Bronze Age Aegean* (*Sheffield Studies in Aegean Archaeology* 7), Oxford.

Fiedler, K.G. 1841. *Reise durch alle Theile des Königreiches Griechenland in den Jahren 1834 bis 1837*, Vol. 2, Leipzig.

Gale, N.H. 1980. "Some Aspects of Lead and Silver Mining in the Aegean," in *Thera and the Aegean World II*, C. Doumas, ed., London, pp. 161–195.

Georgakopoulou, M. 2005. *Technology and Organisation of Early Cycladic Metallurgy: Copper on Seriphos and Keros, Greece*, Ph.D. diss., University College of London.

———. 2007. "Metallurgical Activities within Early Cycladic Settlements: The Case of Daskaleio-Kavos," in Day and Doonan, eds., 2007, pp. 123–134.

Hadjianastasiou, O., and S. MacGillivray. 1988. "An Early Bronze Age Copper Smelting Site on the Aegean Island of Kythnos. Part Two: The Archaeological Evidence," in *Aspects of Ancient Mining and Metallurgy*, J. Ellis-Jones, ed., Bangor, pp. 31–39.

Kakavogianni, O., K. Douni, and F. Nezeri. 2008. "Silver Metallurgical Finds Dating from the End of the Final Neolithic Period until the Middle Bronze Age in the Area of Mesogeia," in *Aegean Metallurgy in the Bronze Age*, I. Tzachili, ed., Athens, pp. 45–57.

Kakavogianni, O., K. Douni, F. Nezeri, M. Georgakopoulou, and Y. Bassiakos. 2006. "Απόπειρα τεχνολογικής προσέγγισης της παραγωγής αργύρου και μολύβδου κατά την Τελική Νεολιθική και Πρωτοελλαδική Ι Περίοδο στα Μεσόγεια," in *Αρχαία Ελληνική Τεχνολογία. Πρακτικά 2ου Διεθνούς Συνεδρίου, οκτώβριος 2005*, Athens, pp. 77–83.

Pantou P., and Z. Papadopoulou. 2005. "Επιφανειακή έρευνα στις νησίδες Σεριφοπούλα και Κιτριανή," in *Πρακτικά Β΄ Διεθνούς Σιφναϊκού Συμποσίου* A', Athens, pp. 87–102.

Papadopoulos, S. 2002. "Early Silver and Copper Metallurgy on the Island of Thassos in the North Aegean," in *Abstracts from the 2002 Annual Meeting of the European Association of Archaeologists, Thessaloniki, September 2002*, Thessaloniki, pp. 194–195 (abstract).

Papadopoulou, Z.D. 2005. "Δεκατρείς άγνωστοι αρχαίοι πύργοι στη Σίφνο," in *Πρακτικά Β΄ Διεθνούς Σιφναϊκού Συμποσίου* A', Athens, pp. 103–122.

Pollak, L. 1896. "Von griechischen Inseln," *AM* 21, pp. 188–228.

Rambach, J. 2000. *Kykladen* I: *Die Frühe Bronzezeit, Grab und Siedlungs-funde*, Bonn.

Renfrew, C. 2005. "The Duration of the Cycladic Cemeteries: The Case of Akrotiraki on Siphnos," in *Πρακτικά Β΄ Διεθνούς Σιφναϊκού Συμποσίου* A', Athens, pp. 43–60.

Tsountas, Ch. 1899. "Κυκλαδικά ΙΙ," *ArchEph* 1899, pp. 73–134.

Vaughan, S., and D. Wilson. 1993. "Interregional Contacts in the Aegean in Early Bronze II: The Talc Ware Connection," in *Wace and Blegen: Pottery as Evidence for Trade in the Aegean Bronze Age 1939–1989*, C.W. Zerner, P. Zerner, and J. Winder, eds., Amsterdam, pp. 169–186.

Wagner, G.A., W. Gentner, H. Gropengiesser, and N.H. Gale. 1980. "Early Bronze Age Lead-Silver Mining and Metallurgy in the Aegean. The Ancient Workings on Sifnos," in *Scientific Studies in the Early Mining and Extractive Metallurgy* (*BMOP* 20), London, pp. 63–86.

Wagner, G.A., and G. Weisgerber. 1985. *Silber, Blei und Gold auf Siphnos: Prähistorische und antike Metallproduktion* (*Der Anschnitt* 3), Bochum.

Zacharias, N., C.T. Michael, M. Georgakopoulou, V. Kilikoglou, and Y. Bassiakos. 2006. "Quartz TL Dating on Selected Layers from Archaeometallurgical Kiln Fragments: A Proposed Procedure to Overcome Age Dispersion," *Geochronometria* 25, pp. 29–35.

CHAPTER

16

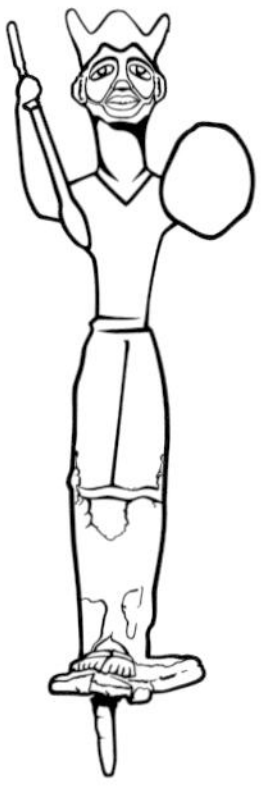

Early Bronze Age Copper Smelting on Seriphos (Cyclades, Greece)

Olga Philaniotou, Yannis Bassiakos, and Myrto Georgakopoulou

Archeometallurgical studies on the islands of Kythnos (Bassiakos and Philaniotou 2007) and Siphnos (Wagner and Weisgerber, eds., 1985) have brought to light conclusive evidence for Early Bronze Age (EBA) copper and lead-silver production, respectively. This evidence emphasizes the important, although not exclusive (Georgakopoulou 2007), role of the western Cycladic islands as early Aegean metal suppliers. Seriphos, which is situated between Kythnos and Siphnos, has long been postulated as another potential metal-producing island during the EBA (e.g., Barber 1987, 14; Broodbank 2000, 292). The two previously briefly reported slag heaps of Avessalos and Kephala, however, remained undated and largely unstudied until recently (for preliminary analyses of a small number of slags see Dimou and Oikonomou 1997; Gale et al. 1985).

This paper presents the results of a collaborative archeometallurgical project initiated in 2000 between the 21st Ephorate of Prehistoric and Classical Antiquities and the Archaeometry Laboratory of the National Centre for Scientific Research (NCSR) "Demokritos." The aims of this project were to examine the known metallurgical sites and to investigate potential ore sources on the island. Plans for excavation were not made at that stage, but a substantial suite of surface samples of the different types of material was collected and studied as part of a Ph.D. dissertation (Georgakopoulou 2005). In addition to the two previously known sites of Avessalos and Kephala, a third smaller slag heap was identified and studied at Phournoi. The association of these sites with copper production is clear even from a macroscopic examination of the slags and has been confirmed by the results of the analytical study (Georgakopoulou et al. 2011). This paper presents a detailed description of the three sites and the main types of finds encountered and provides important evidence that proves the role of Seriphos in EBA Cycladic copper production.

Archaeological and Geological Background

Seriphos remained until recently archaeologically one of the least studied islands, particularly regarding the prehistoric periods. With the exception of a few artifacts attributed to a possible Seriphian origin (Renfrew 1972, 511) and the suspected early date of the reported slag heaps, no further evidence for Bronze Age habitation had been presented (Broodbank 2000, 52; Davis, Tzonou-Herbst, and Wolpert 2001, 88). During our field investigations we identified an extensive EBA site in the area of Plakalona (see below), where a few trial trenches were subsequently opened by the 21st Ephorate of Prehistoric and Classical Antiquities (Pantou, forthcoming).

The main geological units of Seriphos are depicted in the simplified geological map given in Figure 16.1. Seriphos is relatively rich in metal resources, particularly iron, which has been extensively mined in modern times. Three types of ore deposits are reported on the island (Marinos 1951): magnetite deposits are mainly located in the contact zone between the granodiorite and the metamorphic rocks; extensive hematite and limonite deposits are found predominantly within the marble units of the western and southwestern parts; and a rich mixed sulphidic deposit (pyrite, galena, sphalerite, and to a lesser extent chalcopyrite) is known in the area of Moutoula in the northeastern part of Seriphos. The existence of primary and secondary copper minerals that usually accompany the iron deposits is mentioned in some of the modern geological reports (Marinos 1951; Salemink 1980) and has been confirmed in the field during archeometallurgical explorations by our team as well as others (Gale et al. 1985). These are small, local mineralizations of no importance to modern economic geology, but they may have been suitable for prehistoric exploitation, although conclusive evidence for prehistoric mining has not been identified to date.

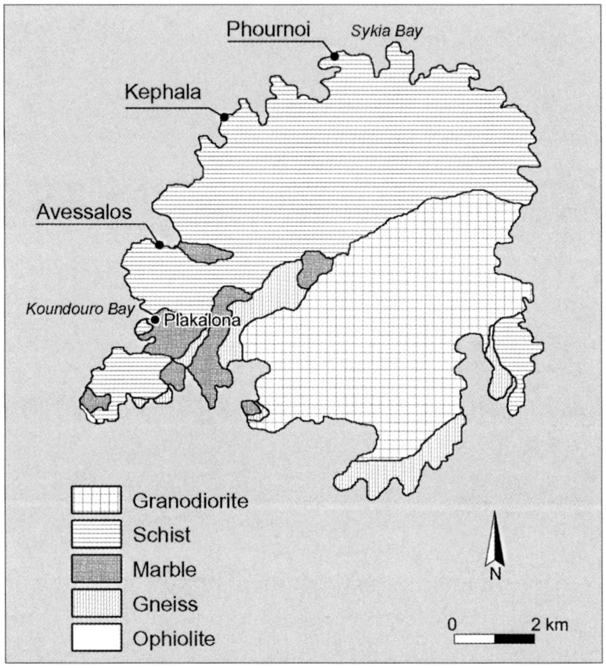

Figure 16.1. Simplified geological map of Seriphos showing the sites mentioned in the text (modified from Salemink 1980, fig. 2).

Smelting Sites

Phournoi

This newly discovered metallurgical site lies on a coastal promontory at approximately 50 m above sea level in northern Seriphos to the west of the bay of Sykia (Fig. 16.1). The location has three toponyms (P. Chrysoloras, pers. comm.): Phournoi (Greek for "furnaces"), Skala, and Gaidourospilia (we use the first for the purposes of this study). Metallurgical remains, primarily slags and furnace wall fragments, extend from just below the ridge down the north-facing slope. They are divided into two main deposits where the distribution is more dense. These deposits measure approximately 150 m^2 (Phournoi 1: eastern concentration) and 200 m^2 (Phournoi 2: western concentration) and are separated by roughly 30 m of barren schist bedrock. The depth of the deposits varies depending on location, but it does not generally seem to exceed 40 cm. Hundreds more slag fragments appear to have rolled down the steep north-facing slope and lie dispersed on horizontal planes of the schist bedrock, while additional material may have been lost to the sea. Very few pieces are seen on the south-facing side, while some stray pieces of slag also were observed on the path that leads to the site (ca. 1,500 m southeast). Using the methodology proposed by Bachmann (1982, 5) and assuming an average heap thickness of 10 cm, slag specific gravity of 3.5 g/cm^3, and a packing factor of 0.4, the total mass of material present in the two main deposits is estimated to be about 50 tons. This is a very rough approximation of the total weight, because values for average thickness and packing factor cannot be estimated accurately in the absence of excavation. Nevertheless, it provides a preliminary surface assessment and allows a broad comparison with the site of Kephala. Similar to the other two slag heaps discussed below, the slag is found mixed with abundant ceramic hearth or furnace wall fragments, while occasionally isolated metal prills or fragments of discarded copper ore were recovered.

Kephala

Our examination of the previously known slag heap of Kephala brought forward several new interesting observations. The first relates to the size of the heap, which appears to be substantially larger than previously estimated. The main concentration of material is found on the north-facing slope of the promontory. It is again divided into two deposits (Fig. 16.2), measuring roughly 2,500 m^2 (Kephala 1: eastern concentration) and 4,500 m^2 (Kephala 2: western concentration), with a distance of about 100 m between them. In general, the two deposits do not appear to exceed 50 cm in depth, although this approximation presently is based only on surface observations. Using the same methodology as that for Phournoi, with an average thickness of 30 cm and a packing factor of 0.7, the total mass of slag in these two main deposits is estimated to be ca. 5,000 tons. The amount is nearly double that previously reported (Gale et al. 1985, 83), probably because the farther deposit of Kephala 2 had not been identified. Scatters of metallurgical remains, however, are observed also on the south-facing slope in substantial quantities but in a far more dispersed distribution. In some parts on this side, concentrations of centimeter-size slag fragments are observed, possibly representing areas where slag was deliberately crushed for the extraction of prills.

Just below the crest of the promontory, south of Kephala 1 and toward the south-facing slope, we identified the remains of two furnaces carved into the schist bedrock (Fig. 16.3). The diameter of their bottom surface is ca. 30 cm each. An obvious layer of slag is seen on the side of one of these furnaces, while smaller droplets of slag are also preserved on other parts of their sides, together with traces of green stain. Other than that, the furnaces are almost clear of slag or other metallurgical remains, although dispersed scatters of material are seen in the surrounding area. It is interesting to note that although the largest concentration of material on Kephala is found on the north-facing slope, these two installations face toward the southeast.

A small number of furnace fragments (approximately 10 examples) with evidence for perforations were found within a relatively confined area of the Kephala 2 deposit (Fig. 16.4). The majority of pieces only bear the imprint of a single hole, but one specimen had two. The fragments only had part of the hole intact, with the exception of one example where the entire hole is preserved. The diameter of this hole is 3 cm on the external side of the fragment and approximately 2.5 cm on the internal side. The perforations are inclined inward and slightly downward, as if formed by pushing the wet clay from outside toward the inside. Perforated furnace fragments are known from a number of EBA southern Aegean copper-smelting sites, the most prominent being the ones from Kythnos (Bassiakos and Philaniotou 2007) and Chrysokamino (Betancourt 2006).

Figure 16.2. The slag heap of Kephala (two arrows point at Kephala 1 and Kephala 2 deposits).

Figure 16.3. Furnaces carved into bedrock at Kephala.

Figure 16.4. Perforated furnace fragments from Kephala. Arrows indicate edges of perforations.

Figure 16.5. The slag heap of Avessalos.

Contrary to these sites, however, the perforated examples at Kephala only represent a minority of the total furnace fragments visible on the surface, the majority of fragments do not show indications for perforations.

A single igneous hammer stone has been found at Kephala so far. In addition, two round slag fragments were recovered, whose intriguing shape raises the possibility that they also were used as tools.

Avessalos

Avessalos is by far the largest slag heap presently known on the island and more broadly in the Cyclades. The main deposit extends from the crest down the north-facing slope of a hill on the southern side of the Avessalos Bay (Fig. 16.5). Smaller quantities of material also were found on the south-facing side, separated from the main deposit by flat schist bedrock and a modern dirt track. The heap is cut twice by a modern road on the north-facing side, creating sections through the slag layer. The deposit exposed in the upper section sits on the schist bedrock about 1.5 m from the level of the road. It extends approximately 170 m in length along the road and varies considerably in width from roughly 15 cm at the edges to 80 cm at its thickest part. Overall the depth of the Avessalos heap appears quite variable. In some parts, particularly toward the top, the deposit appears to exceed 2 m in depth. Gale et al. (1985, 83) estimate that the Avessalos heap is at least 100,000 tons. At this stage it is not possible to offer a more precise assessment because further and more systematic measurements have not been carried out.

The road section presented an opportunity to partly assess the vertical characteristics of this heap. In some of the thicker parts of the section, two to three superimposed layers can be discerned depending on the locality, and these are distinguished by differences in the size of the slag in each one. Three types of layers were observed: small slag (1–3 cm), medium slag (6–10 cm), and larger slag, which usually sit on the top. The sequence between the first two is not consistent, and although the lower layer usually consisted of small slags, occasionally a layer of small slags sat on top of a layer of medium ones. The fragmented visibility of this stratification and the absence of a clear soil separation layer makes it impossible to decide at

Figure 16.6. Rock-carved pits at the top of the Avessalos slag heap.

Figure 16.7. Cluster of small rock-carved pits and grooves on a schist outcrop at the top of the Avessalos slag heap.

this stage whether these layers correspond to distinct operations possibly carried out in different chronological periods, or whether the phenomenon is simply the result of natural distribution of the material as it eroded from the hilltop.

On the plateau on top of the hill, just above the north-facing slag deposit, we found a number of small, rounded, relatively shallow pits, usually about 10 cm in diameter and up to 8 cm deep, that have been carved into the schist bedrock. These are usually seen in small clusters where the bedrock is exposed. Larger pits were also noted (ca. 40 cm in diameter), and their perimeter bears a number of smaller pits. One of these larger pits was found alone (Fig. 16.6), while in a different part of the exposed bedrock three large pits were found close to each other. We believe these features represent installations for ore or slag crashing. The association is further supported by the identification of numerous lithic tools, either intact or fragmented, in this part of the site (see below). In the same area peculiar grooves were carved on a steeply sloping schist outcrop, and they were located close to a cluster of small pits (Fig. 16.7). It is not clear at present whether these cuts were made purposefully, and what purpose, if any, they may have served.

Similar to Kephala, a minority of the furnace fragments (approximately 10 have been recovered so far) observed on the surface of the heap bear part of a single perforation, with only one specimen showing two. Also particularly noteworthy at Avessalos is the relative abundance of stone tools compared to the other slag heaps on Seriphos, as well as those on the neighboring island of Kythnos (Bassiakos and Philaniotou 2007). Seven almost-intact stone tools were recovered from this site, while at least 10 more possible fragments were noted in the field. They were all found at the top of the slag heap on the schist bedrock, in close proximity to the rock-carved pits. Besides these tools, a roughly orthogonal schist slab with six shallow holes carved in two parallel lines was identified above the top road section. It measures roughly 25 x 35 cm, and the holes have a diameter of ca. 4 cm and a depth ca. 1 cm. A similar find has been identified at the site of Pounta on Kythnos (Bassiakos and Philaniotou 2007, 28).

Dating of the Smelting Sites

One of our primary aims during this project was to date the activities associated with these sites. A few EBA sherds have been identified at Phournoi and Avessalos (Fig. 16.8), but diagnostic pottery fragments were not found at Kephala. The Avessalos slag heap is frequently thought to be of post–Bronze Age date based on its large size in comparison to other known EBA Aegean smelting sites (e.g., Gale et al. 1985), as well as on previous reports of fourth-century A.D. coins found among the slags (Davies 1935, 260). The identification of several EBA sherds during our field study of the heap confirms

Figure 16.8. Pottery sherds from Avessalos.

an earlier account for possible prehistoric pottery on the site (Weisgerber 1985, 112 n. 28). These finds, together with other features presently associated with EBA copper production in the Cyclades (e.g., the setting of the site on the slope of a windy promontory and the presence of perforated furnace fragments), strongly suggest that at least part of the heap may actually be prehistoric. At the same time, the size of the heap is indeed impressive, and the possibility of multi-period activities cannot be disregarded. The dating of the Avessalos slag heap presents a challenging problem that must be addressed appropriately in the future.

A program of thermoluminescence dating of furnace fragments has, however, been undertaken for the sites of Kephala and Phournoi. Charcoal remains in the slags are notoriously rare at Cycladic smelting sites (Bassiakos and Philaniotou 2007), and the same dearth was noted on Seriphos thereby precluding the application of radiocarbon dating. Six samples were submitted for analysis: two from Phournoi 1, two from Kephala 1, and two from Kephala 2. A larger number of samples was considered necessary from Kephala in view of the much larger size of the heap, the identification of two clearly spatially separated deposits, and possibly also two different types of furnace wall fragments (perforated and non-perforated). Our sampling therefore also aimed to investigate possible chronological differences within Kephala and between the perforated and non-perforated furnace fragments (a single perforated sample [KEF2B] was included among the four Kephala samples).

A detailed account of the methodology followed and a full table of results has been presented elsewhere (Zacharias et al. 2006), and only a summary is given here. A mean value of 4640±240 years B.P. is calculated for Kephala, and 4940±360 years B.P. for Phournoi. Both sites therefore fall roughly within the first half of the third millennium B.C., corresponding to the Aegean EBA, with the site of Phournoi possibly a little earlier than Kephala, although the associated errors do not allow a secure distinction. Within the analyses of Kephala, the analyses from Kephala 2 consistently gave a somewhat earlier date than the two analyses from Kephala 1, but the small number of samples analyzed and the large errors attached to individual analyses also preclude at this stage anything more than a qualitative observation. Significant differences were not noted between the perforated and non-perforated furnace fragments from Kephala.

EBA Habitation Site at Plakalona

In the southwestern part of the island, above the bay of Koundouro and approximately 2 km south of Avessalos in the area known as Plakalona, we located a site with abundant EBA pottery and obsidian as well as visible remains of walls (Fig. 16.9). The identification of Plakalona is of particular importance, as it is the first definite EBA habitation site to be recognized so far on Seriphos. The region is among the most iron-rich on the island, and remains of intense modern iron-mining activities are clearly visible at Koundouro and surrounding areas. Copper minerals are frequently found within these iron ores.

Gale and Stos-Gale (2002) report a scatter of slags close to Koundouro, but as this has not been located during our field investigations, its connection with the Plakalona settlement is presently unclear. Nevertheless, the setting of this site within a metal-rich region and possible presence of slags in its vicinity raise the possibility of a connection with local metallurgical activities. Further clarification will hopefully emerge following the study of the finds of the recent excavations by the 21st Ephorate of Prehistoric and Classical Antiquities (Pantou, forthcoming).

Figure 16.9. Remains of EBA walls at Plakalona.

Conclusion

The present work on Seriphos offers a significant contribution to the study of EBA southern Aegean metal production. Kephala, Phournoi, and possibly part of Avessalos, can now be securely dated to this period, while the field investigation of the three sites has brought forward interesting features and objects that in many cases find parallels at other copper-smelting sites in this region. Seriphos's recent history is intimately connected with mining, mainly for iron, while several references to possible exploitation of the island's resources in the early historic periods can be found in the literature (e.g., Davies 1935, 260–261; Galanos 1962, 146; Healy 1978, 118). Continuing archeometallurgical investigations on the island promise a diachronic glimpse into a rich and varied mining and metallurgical past.

Acknowledgments

We are grateful to the late Mr. Panayiotis Chrysoloras who informed us of the site of Phournoi and guided us through the strenuous path. Expenses for fieldwork on Seriphos were partially covered by awards from the Institute of Archaeology and the Graduate School, University College, London (UCL). Thermoluinescence dating was undertaken by Dr. N. Zacharias at the Laboratory of Archaeometry (NCSR "Demokritos").

References

Bachmann, H.-G. 1982. *The Identification of Slags from Archaeological Sites*, London.

Barber, R.L.N. 1987. *The Cyclades in the Bronze Age*, London.

Bassiakos, Y., and O. Philaniotou. 2007. "Early Copper Production on Kythnos: Archaeological Evidence and Analytical Approaches to the Reconstruction of the Metallurgical Process," in *Metallurgy in the Early Bronze Age Aegean* (*Sheffield Studies in Aegean Archaeology* 7), P.M. Day and R.C.P. Doonan, eds., Oxford, pp. 19–56.

Betancourt, P.P. 2006. *The Chrysokamino Metallurgy Workshop and its Territory* (*Hesperia Suppl.* 36), Athens.

Broodbank, C. 2000. *An Island Archaeology of the Early Cyclades*, Cambridge.

Davies, O. 1935. *Roman Mines in Europe*, Oxford.

Davis, J.L., I. Tzonou-Herbst, and A.D. Wolpert. 2001. "Addendum: 1992–1999," in *A Review of Aegean Prehistory* (*AJA Suppl.* 1), T. Cullen, ed., Boston, pp. 77–94.

Dimou, E., and G. Oikonomou. 1997. "Ορυκτολογική-Ορυκτοχημική Σύσταση των Αρχαίων Σκωριών από διάφορες θέσεις του Ελλαδικού Χώρου," in *Αρχαία Ελληνική Τεχνολογία. Πρακτικά Πρώτου Διεθνούς Συνεδρίου*, Thessaloniki, pp. 313–323.

Galanos, F. 1962. *Seriphos,* Part 1, Athens.

Gale, N.H., A. Papastamataki, Z.A. Stos-Gale, and K. Leonis. 1985. "Copper Sources and Copper Metallurgy in the Aegean Bronze Age," in *Furnaces and Smelting Technology in Antiquity* (*BMOP* 48), P.T. Craddock and M.J. Hughes, eds., London, pp. 81–101.

Gale, N.H., and Z.A. Stos-Gale. 2002. "Archaeometallurgical Research in the Aegean," in *The Beginnings of Metallurgy in the Old World*, M. Bartelheim, E. Pernicka, and R. Krause, eds., Rahden/Westf., pp. 277–302.

Georgakopoulou, M. 2005. *Technology and Organisation of Early Cycladic Metallurgy: Copper on Seriphos and Keros, Greece*, Ph.D. Diss., University College, London.

———. 2007. "The Metallurgical Remains," in *Keros, Dhaskalio Kavos: The Investigations of 1987–88* (*McDonald Institute Monographs*), A.C. Renfrew, C. Doumas, L. Marangou, and G. Gavalas, eds., Cambridge, pp. 380–401.

Georgakopoulou, M., Y. Bassiakos, and O. Philaniotou. 2011. "Seriphos Surfaces: A Study of Copper Slag Heaps and Copper Sources in the Context of Early Bronze Age Aegean Metal Production," *Archaeometry* 53, pp. 123–145.

Healy, J.F. 1978. *Mining and Metallurgy in the Greek and Roman World*, London.

Marinos, G. 1951. "Γεωλογία και μεταλλογένεια της νήσου Σερίφου," *The Mineral Wealth of Greece* 1, pp. 95–127.

Pantou, P. Forthcoming. "Evidence of Prehistoric Habitation and Metallurgical Activities on Seriphos: The Site of Plakalona," in *Prehistoric Metal Production in the Aegean: Material Evidence and Analysis*, Y. Bassiakos, ed., Heidelberg.

Renfrew, C. 1972. *The Emergence of Civilisation: The Cyclades and the Aegean in the Third Millennium B.C.*, London.

Salemink, J. 1980. "On the Geology and Petrology of Seriphos Island (Cyclades, Greece)," *Annales Géologiques des Pays Helléniques* 30, pp. 342–365.

Wagner, G.A., and G. Weisgerber, eds. 1985. *Silber, Blei und Gold auf Sifnos* (*Der Anschnitt Beiheft* 3), Bochum.

Weisgerber, G. 1985. "Bemerkungen zur prähistorischen und antiken Bergbautechnik," in Wagner and Weisgerber, eds., 1985, pp. 86–112.

Zacharias, N., C.T. Michael, M. Georgakopoulou, V. Kilikoglou, and Y. Bassiakos. 2006. "Quartz TL Dating on Selected Layers from Archaeometallurgical Kiln Fragments: A Proposed Procedure to Overcome Age Dispersion," *Geochronometria* 25, pp. 29–35.

CHAPTER

17

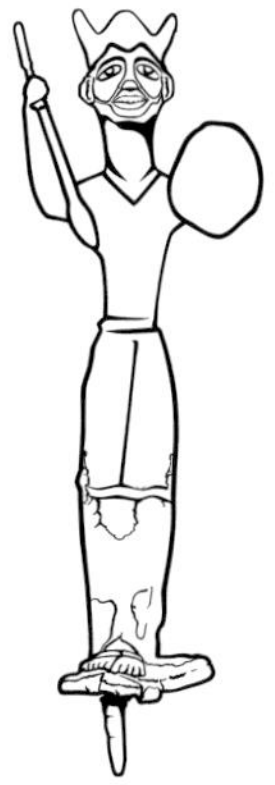

Searching for the Early Bronze Age Aegean Metallurgist's Toolkit

Christos G. Doumas

As Keith Branigan had pointed out in his pioneering study of metallurgy in the Early Bronze Age (EBA) Aegean, facilities and equipment for smelting metal ores should be sought "in the vicinity of the ore sources themselves" (Branigan 1974, 69), from where the extracted metal was brought in the form of small ingots to the settlements (Branigan 1974, 70; Renfrew and Slater 2003, 307). Recent surveys and excavations have demonstrated not only that smelting was, indeed, conducted both near and far from the mining areas (Betancourt 1997; 2006, 179–184; 2007, 64; Bassiakos and Philaniotou 2007; Catapotis 2007; Catapotis and Bassiakos 2007; Tsipopoulou 2007), but also that melting activities were absent from smelting sites (Day and Doonan, eds., 2007). In this paper, by reviewing the evidence for the practice of metallurgy at settlement sites in the EBA Aegean, an attempt is made to reconstruct the facility needed for melting and/or alloying metals. It is a great pleasure for me to offer this tentative reconstruction in honor of Jim Muhly, a scholar who has contributed enormously to the advancement of Aegean and Near Eastern studies, and who specifically with his numerous publications on the history and technology of early metallurgy in these regions has promoted archaeology as a humanistic discipline going far beyond typology.

Since ready metal was evidently brought to the settlements in the form of ingots or ingot fragments, the procedure for its transformation into artifacts or tools would include the following stages: (1) breaking, pounding, and grinding; (2) melting and alloying; (3) casting; and (4) refining and metal working. It is obvious that in each of these stages specific tools were required, and their identification is necessary in order to document the practice of metallurgy at a specific site.

1. Breaking, Pounding, and Grinding Metal

This first step in the whole operation was important because ground metal was easier to heat to the melting point. Although stone tools—such as heavy choppers (Fig. 17.1), hammers (Fig. 17.2), and anvils—in EBA settlement sites do not constitute direct evidence for the practice of metallurgy there, the presence of these in sites where metallurgical activity is well documented can hardly be ignored for this first step. For example, a stone hammer-axe from Poliochni on Lemnos, although its function has been debated (Bernabò-Brea 1964, 672, pl. 184:9), would be an ideal tool for breaking metal into small pieces. Stone pounders with percussion marks from the same site (Bernabò-Brea 1964, 672, pl. 189:1–5) could also be used for grinding metal fragments in combination with anvils/mortars (Bernabò-Brea 1964, 672, pl. 188:1–7). A type of ellipsoidal grinder, encountered in large numbers at Poliochni and bearing percussion marks on both sides, probably had a similar function (Bernabò-Brea 1964, 672, pl. 188:9, 11, 13, 14). Such artifacts—stone hammers, pounders, anvils/mortars, grinders—have also been found at Thermi on Lesbos (Lamb 1936, 188–190, figs. 56, 59) as well as at Emporio on Chios (Hood 1982, 648–649, pls. 134–135) and Akrotiri on Thera (Fig. 17.2), where, too, there is evidence of metallurgy. Some wasted stone pieces from Markiani on Amorgos were identified primarily as weights, while the percussion marks preserved on them are attributed to their secondary use as "hammers or pounders" (Scarre 2006, 179–189). Still, in the Cyclades the site of Kastri on Syros, with abundant evidence for metallurgical activity, has also yielded heavy stone hammers (Bossert 1967, 60–63; Marthari 1998, 28; Suto 1989, 216; Tsountas 1899:123). A similar picture emerges from mainland sites, such as Lithares in Boeotia (Tzavella-Evjen 1984, 171–172, 184, pls. 86:α–γ, 87:α, β, 96:β–ι, π–ρ, 97:δ, ε, η–λ, ν–ξ). It is highly unlikely that these stone artifacts would be considered tools for metallurgical activity if contextual evidence of the other steps of the operation did not exist.

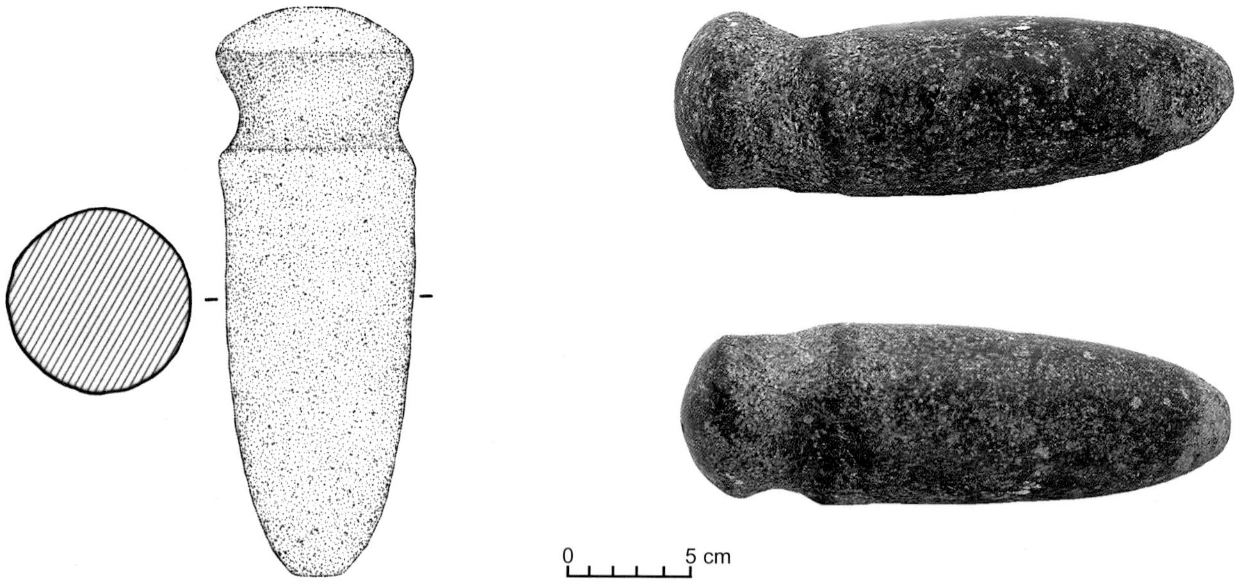

Figure 17.1. Drawing and two views of a metal "chopper" (NM 8987) from Hagios Kosmas in Attica.

Figure 17.2. Two views of a heavy stone hammer (AKR 485) from Akrotiri, Thera, weighing ca. 14 kg and 0.303 m in length. The hard volcanic material (andesite) was difficult to perforate for a haft. It was probably bound to the hammer with rope wound around the grooves. The four depressions possibly accommodated knots to prevent the rope from slipping.

2. Melting and Alloying Metal

The stage of melting was more complex, requiring special equipment including a hearth for heating, a crucible to hold the metal, and a furnace. A specific mechanism for achieving the melting temperature would also be necessary.

The fact that no special fire installations have been reported from EBA settlement sites in the Aegean suggests that ordinary hearths, either permanent or portable, might be used. Permanent hearths, consisting of either stone slabs or clay braziers embedded in the floor, are known from various sites, including Kastri on Syros (Bossert 1967, 60; Marthari 1998, 27) and Markiani on Amorgos (Birtacha 2006, 133–134). Portable hearths are shallow circular clay vessels, usually called baking pans, which are known from sites such as Korakou (Blegen 1921, 13, fig. 15), Hagios Kosmas (Mylonas 1959, 24, fig. 119), Markiani (Birtacha 2006, 133), Akrotiri on Thera (Fig. 17.3), and elsewhere. Their height is 4–5 cm, while their rim slopes downward at one end until it coincides with the bottom of the vessel. Hearths, whether permanent or portable (Renar 1995, 260), could have served dual domestic and industrial functions: cooking/baking and melting/alloying. Portable ones had the advantage of use outdoors, if required.

Crucibles have been classed as two types: the "saucer type" without a handle and the type with a handle (Branigan 1974, 77). Different varieties have been recognized according to the form and position. The saucer type, known from Troy IIg, is rather rare (Branigan 1974, 70, fig. 3). Two other examples from Hagia Photia in Crete are the only crucibles known to date from grave contexts (Betancourt and Muhly 2007). Their thick base has a hole into which a long wooden haft could be inserted, to assist in handling the crucible while hot (Davaras and Betancourt 2004, 18, 50–51, figs. 26:10B.10, 111:45.17; Betancourt and Muhly 2007, 147, 150, fig. 9.1). Of the crucible with handle, an early variety from the islet of Giali near Nisyros in the Dodecanese is claimed to be of Late Neolithic date (Fig. 17.4; Sampson 1988, 30, 108). Another variety, known from Thermi Town I, has a handle pierced with a hole at the side, perhaps to insert a stick to help manipulate the vessel (Lamb 1936, 157–158, fig. 44:31–72). A more elaborate variety is the crucible with the handle set at a right angle in relation to the spout. Also present in Thermi Towns I, III, and IV, its handle is provided with a socket, obviously for a wooden haft to manipulate the vessel while hot (Lamb 1936, 157, pls. 24:30–43, 31:71; Branigan 1974, 70, fig. 3). The crucible from the Heraion on Samos is perhaps unique: it has a concave handle of U-shaped section, inside which a haft could be affixed (Milojčič 1961, 51, pl. 33:1). The crucibles from Kastri on Syros (Fig. 17.5) seem to be an even more sophisticated version: their handle has a downsloping socket in which a wooden haft

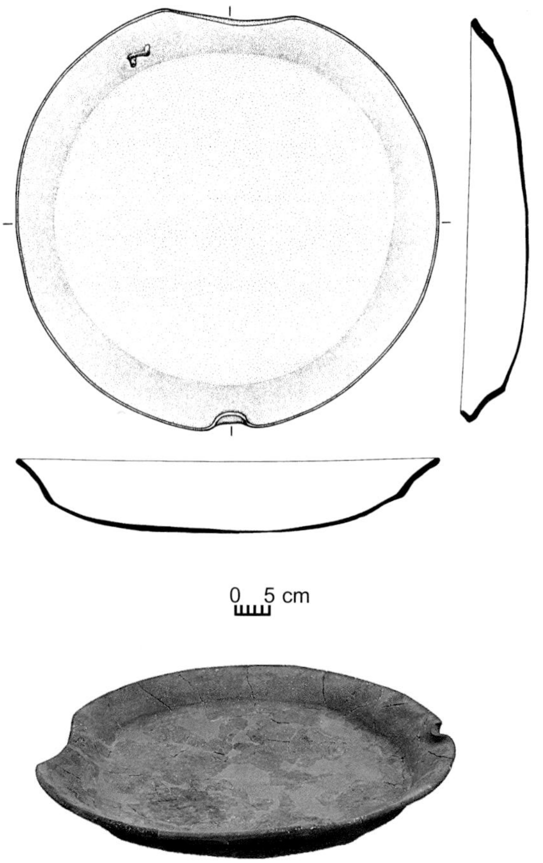

Figure 17.3. Portable hearth (AKR 10,157) from the Early Cycladic horizon at Akrotiri, Thera.

Figure 17.4. Crucible from the islet of Giali near Nisyros, Dodecanese, with slag remains. Height 3.8 cm, rim diameter 8 cm (photo courtesy of T. Marketou).

could be inserted (Bossert 1967, 60–61, fig. 3:5, 6; Tsountas 1899, 124–126). All these varieties of crucibles may give an insight into the inquisitive mind of the early metalworkers in their search for the optimum solution.

Crucibles were not of standard size. As a rule, they were small with an estimated capacity of 100–250 cc (Branigan 1974, 70). The largest examples known so far come from Poliochni, and their estimated capacity of 10–15 kg (Branigan 1974, 69) is perhaps indicative of the scale of metallurgical activity there. Although the site of Poros-Katsambas near Herakleion on Crete appears to have been a center for almost exclusively melting or alloying copper and casting bronze artifacts (Doonan and Day 2007; Doonan, Day, and Dimopoulou-Rethemiotaki 2007), the largest crucibles there do not seem to have held more than 500 g of molten copper alloy, a quantity sufficient "to cast virtually any artifact encountered in the EM I–IIA period in a single pour" (Doonan, Day, and Dimopoulou-Rethemiotaki 2007, 104).

Whatever the size and type of crucibles, special supports would be needed to accommodate them on the hearth. A category of clay objects found at various Early Helladic sites has been described as "spit holders," "supports," and "stands" among other terms. The best-preserved and most-elaborate version of this kind of object appears to be the "zoomorphic stand" from Hagios Kosmas (Fig. 17.6), which comprises four thick legs supporting a thick or rounded body, over which extends a flat handle (Mylonas 1959, 146). The excavator, acknowledging that the function of the object is "not entirely clear" and confessing that he was not happy in associating it with "funeral rites," also argued against its use as a firedog or spit support (Mylonas 1959, 147). Numerous fragments found at Lithares in Boeotia seem to belong to similar objects (Tzavella-Evjen 1984, 172–173, pls. 88–89), and Hetty Goldman attempted the reconstruction of what she called a "tripod" from fragments of this kind of utensil (Goldman 1931, 195, figs. 267, 268). It is interesting that she remarked that this type was common in mainland Greece, particularly in Boeotia and Attica. Though of a slightly different variety, it was not unknown in the Peloponnese either, as attested by examples from Tiryns (Müller 1938, 65–66, fig. 51, pl. 25:11, 12) and Tsoungiza in Nemea (Harland 1951). Described as "made up

Figure 17.5. Fragment of a crucible (NM 5238) from Kastri on Syros. A: rear view showing the hole for the haft. B: side view showing remains of copper.

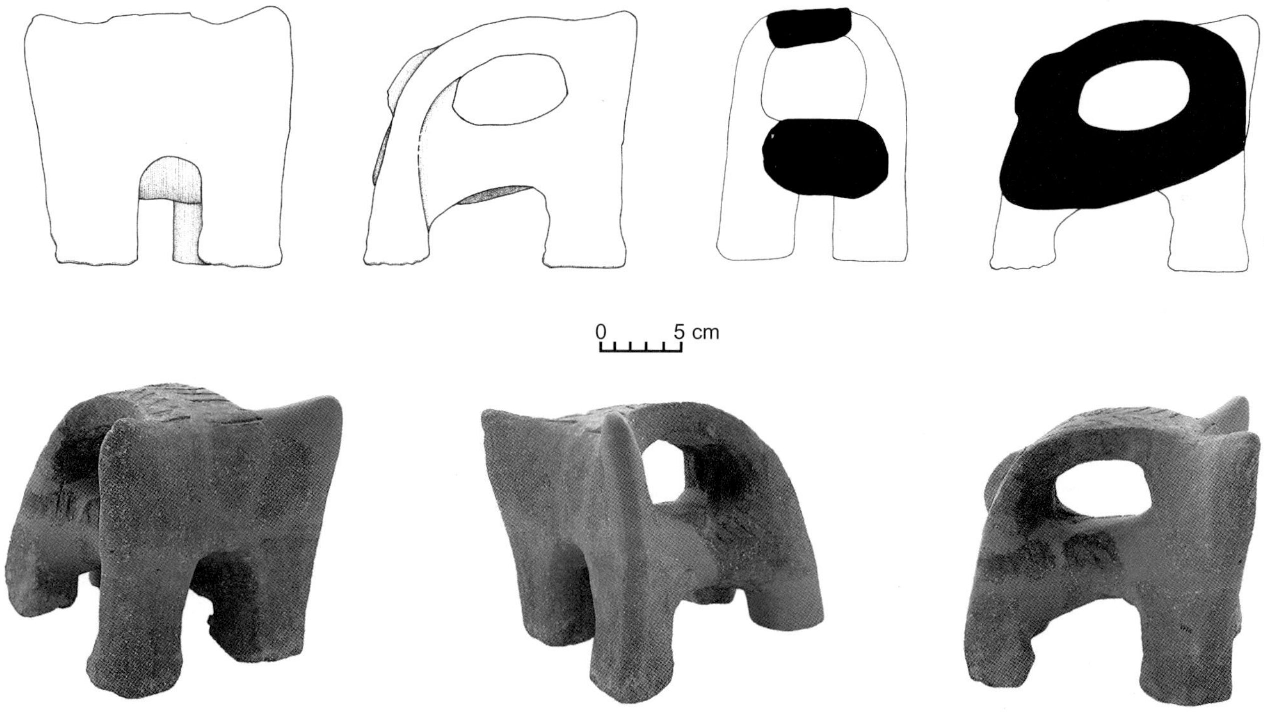

Figure 17.6. Drawing and three views of a crucible support (NM 8977) from Hagios Kosmas in Attica.

of a curved, downward sloping and tapering flat member, the end of which forms a rear base, and of two cylindrical legs set at the two upper and front corners," this type has a vertical handle at its highest point (Harland 1951, 106, pl. 3:c, d). Arguing, quite convincingly, against its interpretation as a spit holder, Harland provides very interesting information by saying that ". . . the front of the object, all of the right leg, and the top of the handle, are brown in color, while the rest of the support is brick-red, all of which indicates that the front was turned toward the heat . . ." (Harland 1951, 106).

Although some scholars persist in designating these objects as "jar supports" (Renar 1995, 259), Harland's observation that their "front was turned toward the heat" seems to provide the answer to the riddle of their use: they were designed on the one hand to support a vessel and on the other to keep the burning charcoal in place. As a more elaborate version, the Hagios Kosmas variety would allow an air draft between the legs of the object to feed the charcoal (Fig. 17.6). If the vessel was a crucible supported above burning charcoal, then Harland's perplexity at why "a people of so early a period and in a back-country, mountain valley, should have made and used so convenient, but so unnecessary, a utensil . . ." (Harland 1951, 106) is rebutted: the utensil was not only convenient but also necessary.

The draft of air between the legs of these supports was obviously not sufficient to heat the crucible to the temperature required for melting metal, and extra air input would be needed. This need was probably served by another category of utensils, for which, again, no satisfactory explanation has been offered as yet. Such objects found at Askitario in Attica (Fig. 17.7) were described as "four-legged vessels, probably imitating saddles or pack animals" (Theocharis 1955, 114). Very similar "saddle-like" clay objects came to light recently at Akrotiri, Thera, from Early Cycladic horizons (Figs. 17.8, 17.9). The funneled ends of these objects would be quite convenient for accommodating a blowpipe nozzle, or tuyère. Two or at the most three of the aforementioned supports would be enough to keep the crucible in position over burning charcoal, while leaving sufficient room for placing "saddle-like" nozzle holders between them (Fig. 17.10). An extra supply of air thus could be generated by blowpipes, thereby increasing the temperature under the crucible.

Analysis of copper slag remains on crucibles from Poros-Katsambas revealed the absence of iron, thus confirming that the copper used in settlements was smelted elsewhere (Betancourt and Muhly 2007, 150). Moreover, from the oxidizing scorch marks observed in certain fragments, it has been deduced that the respective crucibles rested on hot charcoal, while their heavily vitrified inner surfaces indicate that the "crucibles were heated from above, the fuel and metal being in intimate association" (Doonan, Day, and Dimopoulou-Rethemiotaki 2007, 104). This information implies that a specific kind of furnace was used—applied as a lid—on the rim of the crucible. A hemispherical "curious pot-stand" from Thermon in Aetolia, with a central hole in its domed top and a row of another four holes below it (Romaios 1916, 183–184, figs. 4, 5), appears to satisfy the requirement of the furnace used on top of a crucible. That this vessel was directly associated with fire is suggested by the information that fragments of such vessels were found in three different buildings ($\alpha1$, $\alpha4$, and $\alpha6$, respectively) along with vestiges of hearths and considerable quantities of black ash (Romaios 1915, 234–235; 1916, 183).

A similar function could be ascribed to the "mask-like object" (Figs. 17.11, 17.12) recorded for

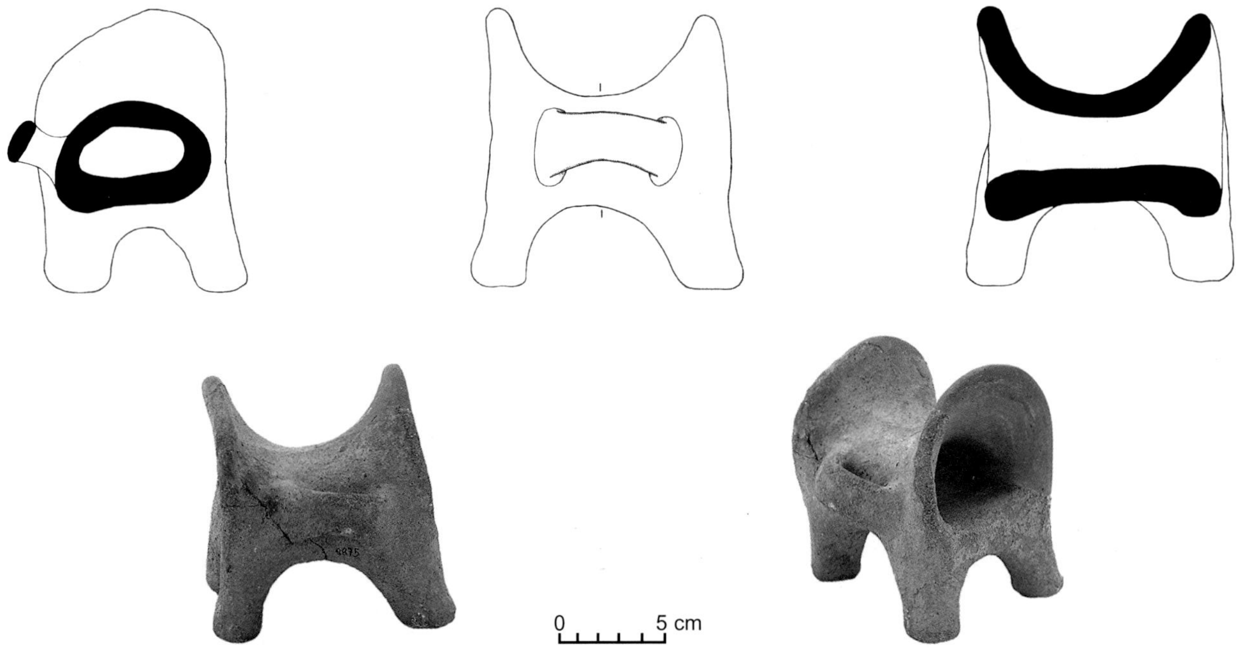

Figure 17.7. Drawing and two views of a nozzle holder (NM 8875) from Askitario in Attica.

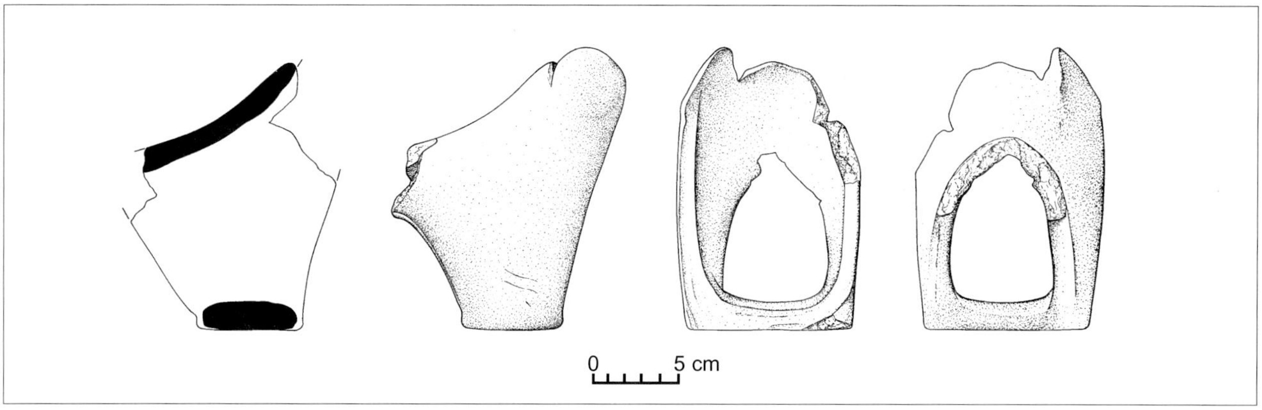

Figure 17.8. Nozzle holder (AKR 10,734) from Akrotiri, Thera.

Figure 17.9. Three views of a nozzle holder (AKR 10,155) from Akrotiri, Thera.

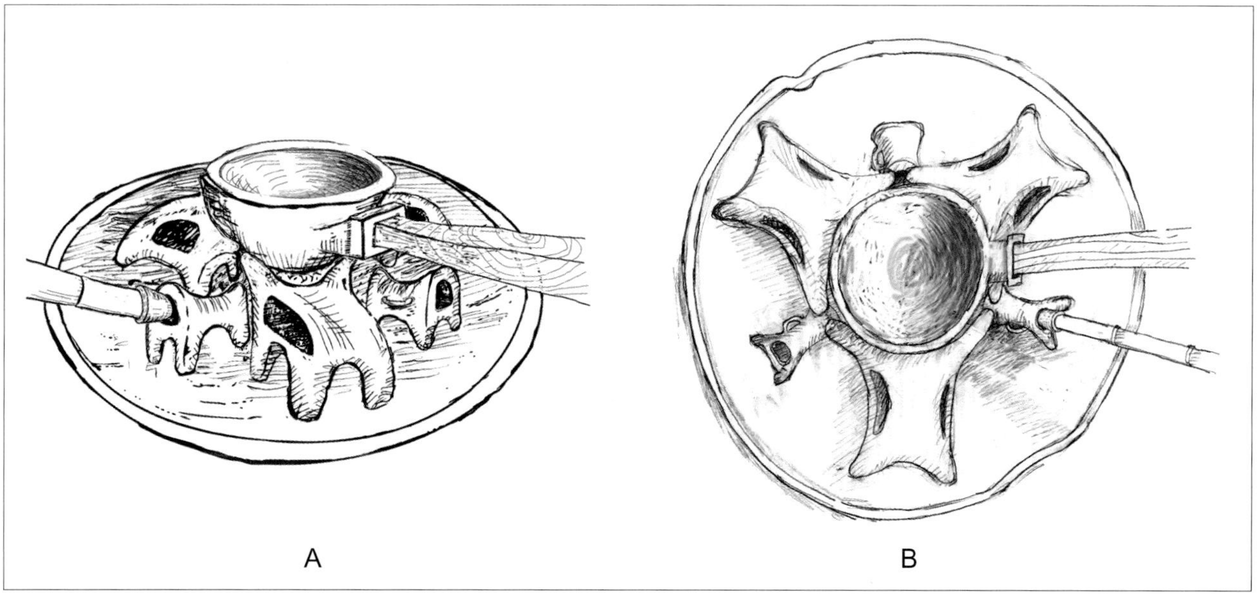

Figure 17.10. Crucible resting on clay supports over a portable hearth and nozzle holders. A: side view. B: view from above. Sketch by Manolis Zacharioudakis.

the first time among the rich contents of an Early Cycladic grave excavated in the nineteenth century at Dokathismata on Amorgos (Dümmler 1886, 19, fig. 2:C1). Dümmler's description of the objects as a "sort of mask-like lid crowning a large vessel" has always been ignored, and reference to a "strange mask-like" object has prevailed ever since (Zervos 1957, pls. 99, 100). The context of another two examples found later in Naxos (Sakellariou and Papathanassopoulos 1964, 65–66) was too vague to support a satisfactory interpretation. Likewise, the fragment from Phylakopi, recovered accidentally in 1958 from the dump of the early excavations, is bereft of context (Zapheiropoulou 1965, 513). Failure to understand the function of this object, probably due to the erroneous position from which it was viewed (Fig. 17.12:D), led to a variety of interpretations, such as "mask" (Doumas 1965, 60), "brazier," "support vessel," and "firedog" among others (Marangou, ed., 1990, 56). The recent discovery of a good number of such objects at Markiani on Amorgos demonstrated their direct association with fire and grounded an interpretation of them as "hearth equipment, probably used to support a spit employed in cooking" (Renfrew and Evans 2007, 155, pl. 19:a, b). Occurring at Markiani as early as Phase II, the vessel evolved during Phases III and IV displaying different varieties (Karantzali 2006, 122; Birtacha 2006, 137; Eskitzioglou 2006, 153–155).

Dümmler's suggestion that his find was a sort of lid seems to have been based on the presence of mat impressions on the rim of its base (Fig. 17.12:B–C), obviously created when the object was set on a mat to dry before firing (Eskitzioglou 2006, 155). This standard position (Fig. 17.11) may provide a clue to its use and perhaps solve the problem of the missing furnace for heating the crucible from above. For this "mask-like object," filled with fuel and metal grits and adjusted on the rim of the crucible (Fig. 17.13), could act as the agent that brought fuel and metal into "intimate association." The side holes could easily accommodate the nozzles of blowpipes, while the top ones could serve as exhausts. Although later in date, the utensils recovered from the horizon of LBA I Miletus (period IV) seem to be similar to the Cycladic "mask-like" furnace. Found together with fire-boxes in an area "reserved for industrial purposes," these "horned stands" have been interpreted as possible "holders for hot pots or trays in a domestic, cultic, or industrial process" (Kaiser 2005, 195–196, pl. XLVII:d).

Concerning the nozzles or tuyères, which could be inserted into both the nozzle supports under the crucible and the lateral holes of the "mask-like" furnace, the evidence from EBA settlement sites is quite satisfactory, despite the various interpretations proposed. For example, those found in Thermi Towns II, III, IV, and V were described as "hollow cones with hole at top" and were considered to have a rather "obscure" function (Lamb 1936, 161, pl. 23). Similar utensils recovered from a pit containing abundant fragments of copper slag at Askitario near Rafina in Attica (Figs. 17.14:B, 17.15) were interpreted as "little funnels" used for pouring fluid metal into a closed mold, despite the fact that no traces of metal were observed on them (Theocharis 1951, 80, fig. 3; 1952, 131). Similar finds from Troy were associated by Schliemann with metallurgy, a view rejected by Bernabò-Brea on the grounds that they present no traces of slag. Such "cones," however, were in use at Poliochni as early as period Blue (Bernabò-Brea 1964, 589, 657, pls. 83:r–t, 167:1, 2, 170:7, 8). Finally, another example of such a nozzle was found at Daskalio Kavos on Keros (Fig. 17.14:A) where metallurgical activity has been documented (Georgakopoulou 2007). The frequent presence of such objects at sites where metallurgical activity is well documented constitutes a strong argument for their interpretation as nozzles for blowpipes rather than as funnels.

The above discussion on each individual object as an accessory of a domestic melting/alloying device led me to propose the reconstruction of this device as illustrated in Figure 17.16.

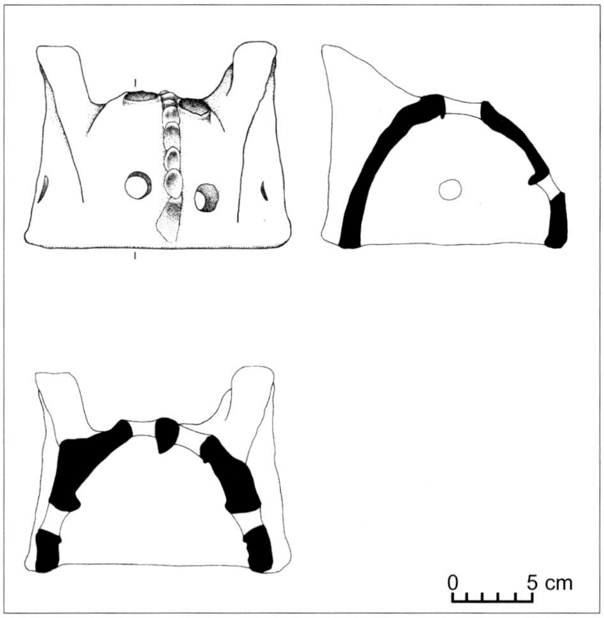

Figure 17.11. Drawing of a mask-like furnace (NM 6113.1) from Dokathismata, Amorgos.

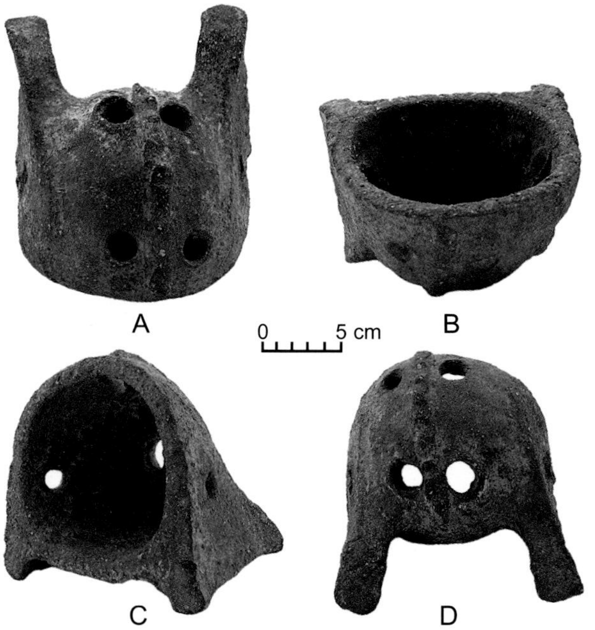

Figure 17.12. Mask-like furnace (NM 6113.1) from Dokathismata, Amorgos. A: standing. B: upside down. C: view of the base with mat impressions on the rim. D: the misleading position.

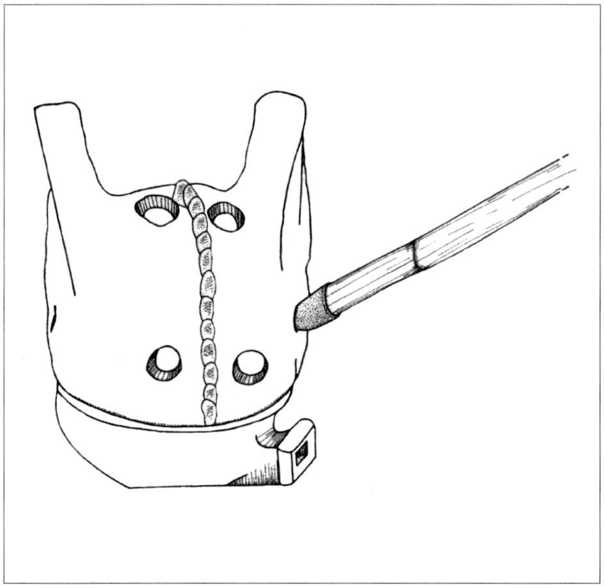

Figure 17.13. Mask-like furnace placed on the rim of a crucible as a furnace. Sketch by Manolis Zacharioudakis.

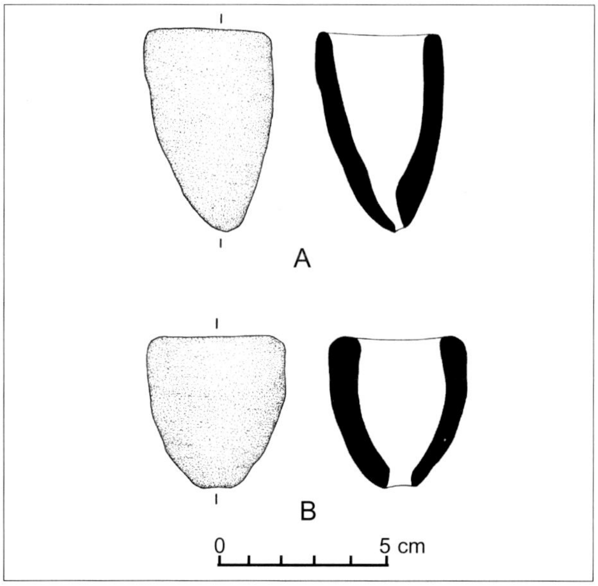

Figure 17.14. Clay nozzles. A: Daskalio Kavos, Keros, MN 2278. B: Askitario in Attica, NM 5242.

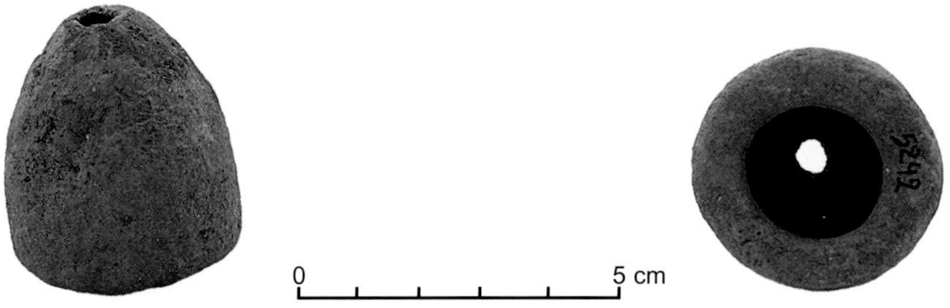

Figure 17.15. Two views of a clay nozzle (NM 5242) from Askitario in Attica.

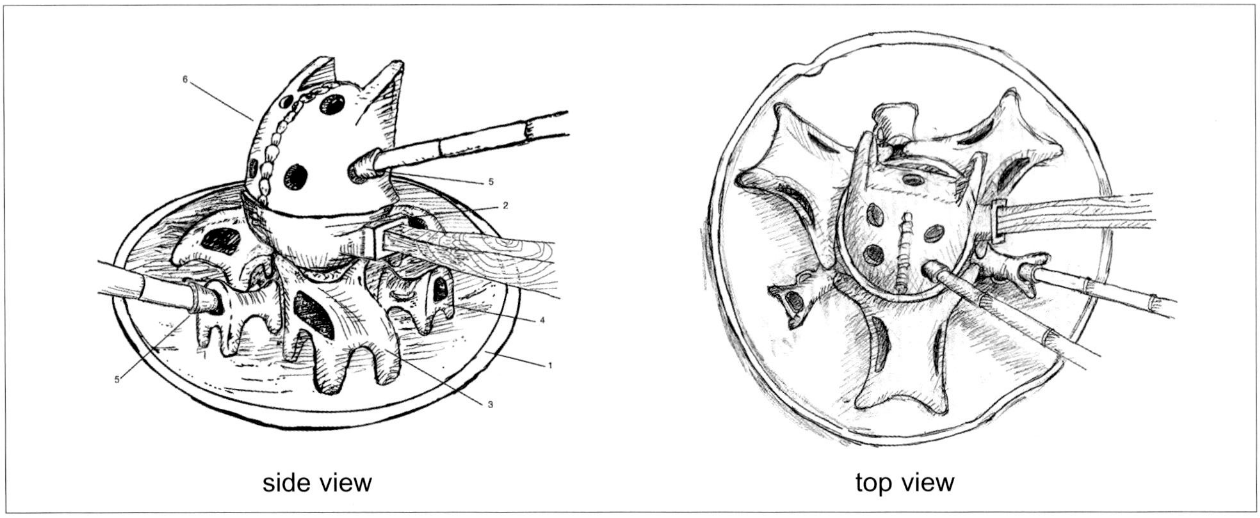

Figure 17.16. Reconstruction of the metal melting/alloying device. Sketch by Manolis Zacharioudakis.

3. Casting Metal

Melting/alloying would not be necessary inside the settlements unless it was followed by casting. Indeed, it is the presence of molds that proves without any doubt the in situ practice of metallurgy. Most common are the open molds for casting mainly chisels, flat axes, daggers, and knives among other objects. They occur either with a single casting surface, in which one or more matrices were cut, or as "block moulds, with individual matrices cut into between two and five sides of a stone block" (Figs. 17.17, 17.18; Branigan 1974, 78). A closed mold from Poliochni (Fig. 17.19) appears to be the earliest evidence of the practice of the lost-wax technique of casting in the Aegean (Branigan 1974, 82). The type of axe illustrated in Figure 17.20, also from Poliochni, seems to have been cast in a similar mold.

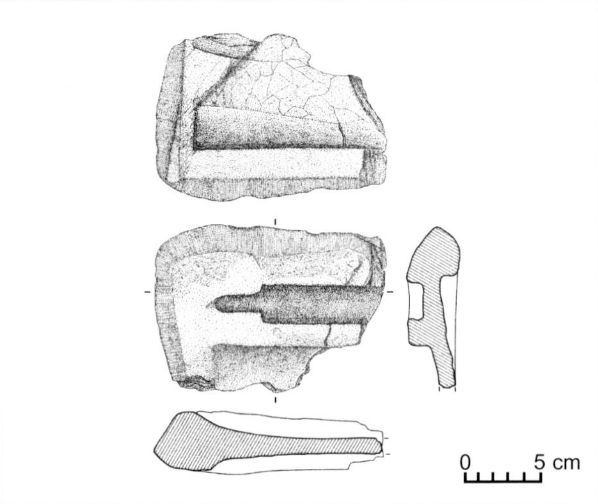

Figure 17.17. Drawing of an open mold (NM 5236) from Kastri, Syros.

Figure 17.18. Fragments of an open mold (NM 5236) from Kastri, Syros.

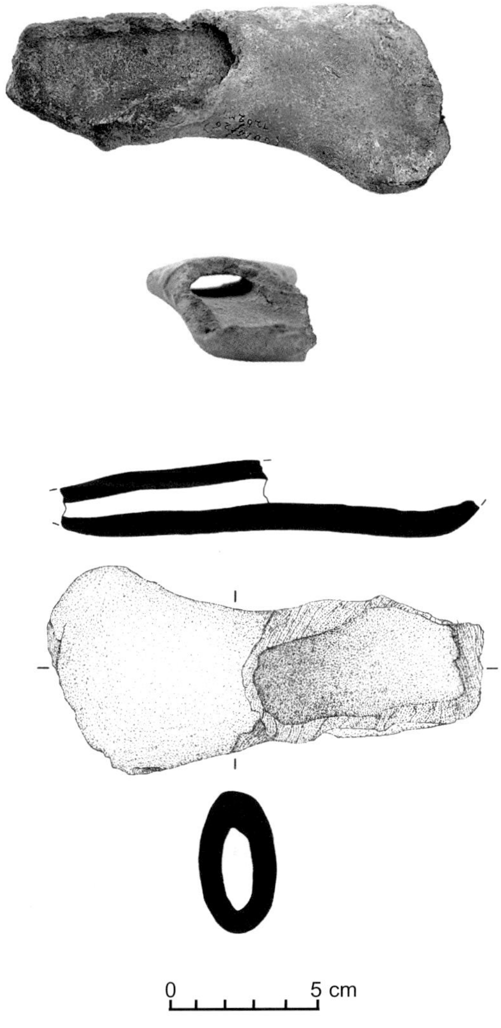

Figure 17.19. Two views and drawing of a closed mold (NM 7202) for casting a shaft-hole axe in the lost-wax technique from Poliochni, Lemnos.

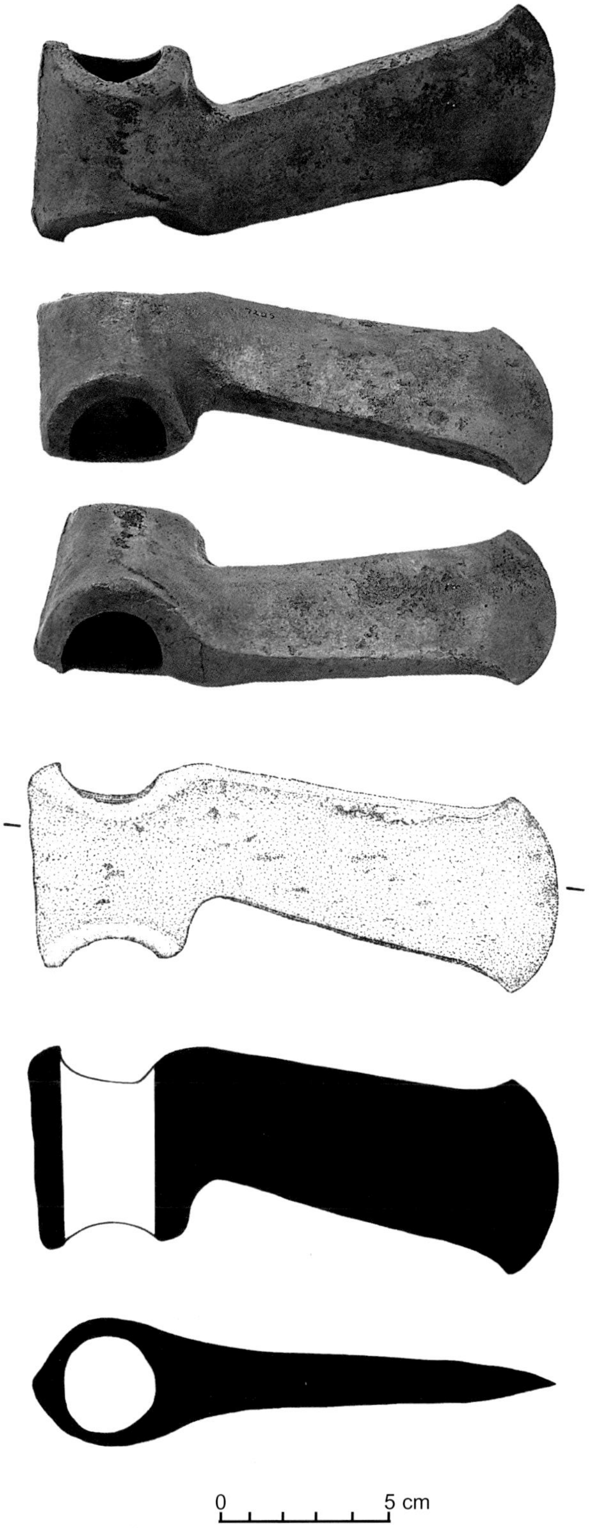

Figure 17.20. Three views and drawing of a copper shaft-hole axe (NM 7205) from Poliochni, Lemnos.

4. Refining and Working Metal

Because "forging was followed by annealing and that by a final cold-hammering on an anvil" (Branigan 1974, 83), some of the lighter stone hammers and anvils mentioned above could have been used for this purpose. More refined hammers would be required for working the cutting edges of axes and chisels. The stone hammer-axes with a shaft-hole (Fig. 17.21) could satisfy this need. Stone hammer-axes of this type have been found at Poliochni from at least the advanced phase of period Blue onward (Bernabò-Brea 1964, 591), and in unprecedented quantities both as finished pieces (Fig. 17.21) and unfinished ones (Figs. 17.22, 17.23). Fragmented finished examples indicate that they were used locally, while the fact that the unfinished ones represent almost all stages of their production indicates that they were made locally (Bernabò-Brea 1964, 671). This type of artifact is very widely distributed throughout Bronze Age Europe from Scandinavia to the Mediterranean and from the Urals to the Atlantic Ocean. It occurs mainly in funerary contexts where it is known as a "battle-axe," but it may well have been a metal-working tool for hammering metal sheet and making metal vessels (Doumas 1998).

Another activity for which there is some evidence from EBA settlement sites seems to have been jewelry-making to form rings, bracelets, and pins among other objects made of gold or silver wire (Branigan 1974, 91). The presence of such objects in settlement sites does not necessarily imply their local production, but there are cases for which this possibility cannot be excluded. From the Heraion on Samos come two clay funnel-like perforated vessels interpreted as fumigators or braziers (Fig. 17.24; Milojčič 1961, 11, 46, pls. 21:2, 24:6, 36:17). An experiment conducted by the British archaeologist, Jacqui Wood, however, showed that this type of vessel—with apparently wide distribution throughout Bronze Age Europe—might have been a kind of Bunsen burner in the service of metalworkers. The flame issues through the central hole at the top (Fig. 17.25), the burner was quite safe for the craftsman to handle, and it could create the temperature needed for softening gold and silver wire to make pieces of jewelry (Denison 1999).

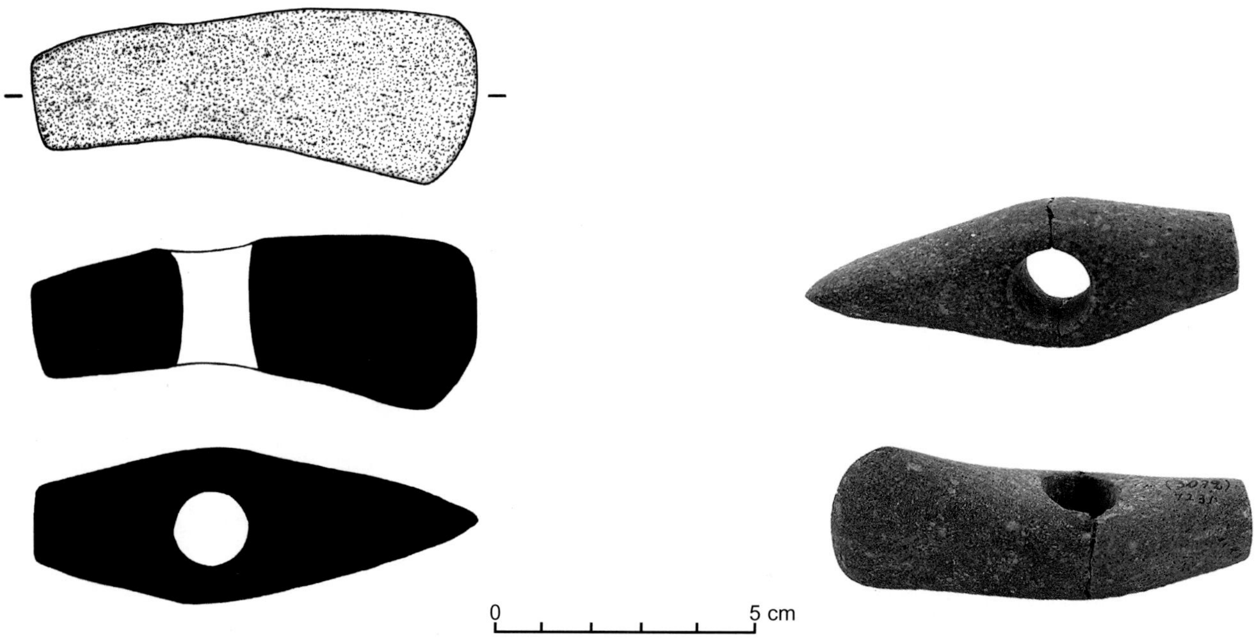

Figure 17.21. Drawing and two views of a stone hammer-axe ("battle axe"; NM 7231) from Poliochni, Lemnos.

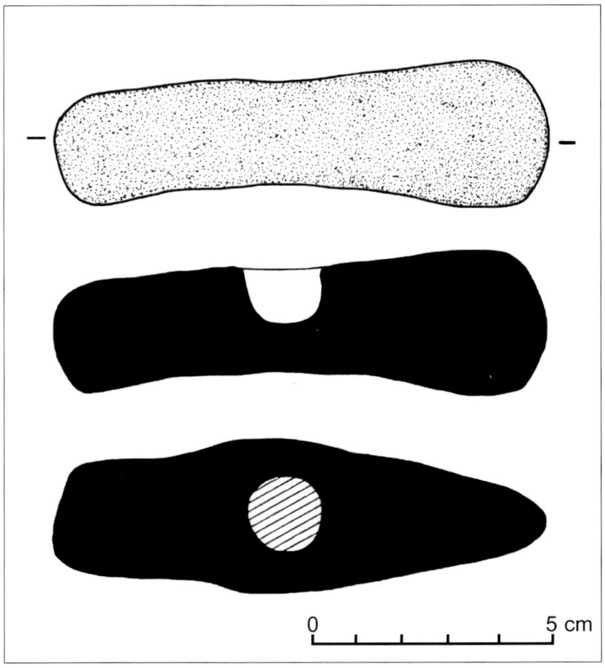

Figure 17.22. Unfinished stone hammer-axe ("battle axe"; NM 4476) from Poliochni, Lemnos.

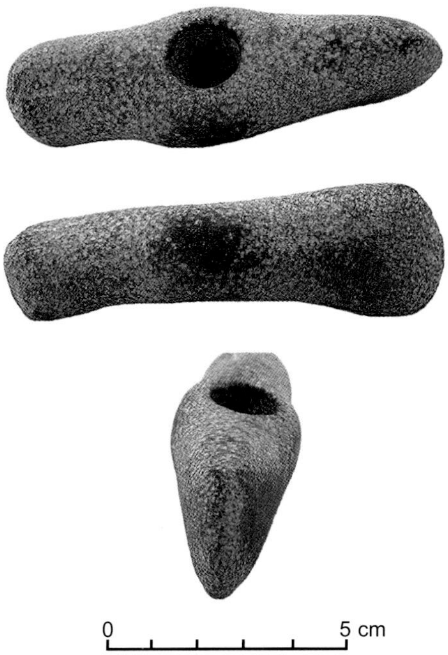

Figure 17.23. Three views of an unfinished stone hammer-axe ("battle axe"; NM 4476) from Poliochni, Lemnos.

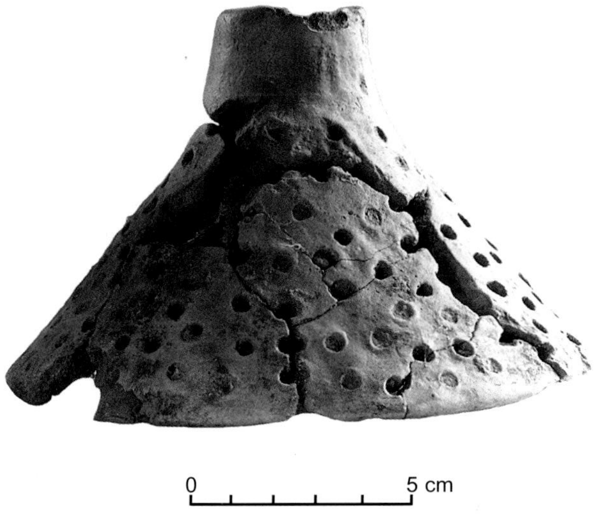

Figure 17.24. Funnel-shaped perforated clay utensil (inv. no. 2831) from the Heraion, Samos (photograph courtesy of the German Archaeological Institute at Athens).

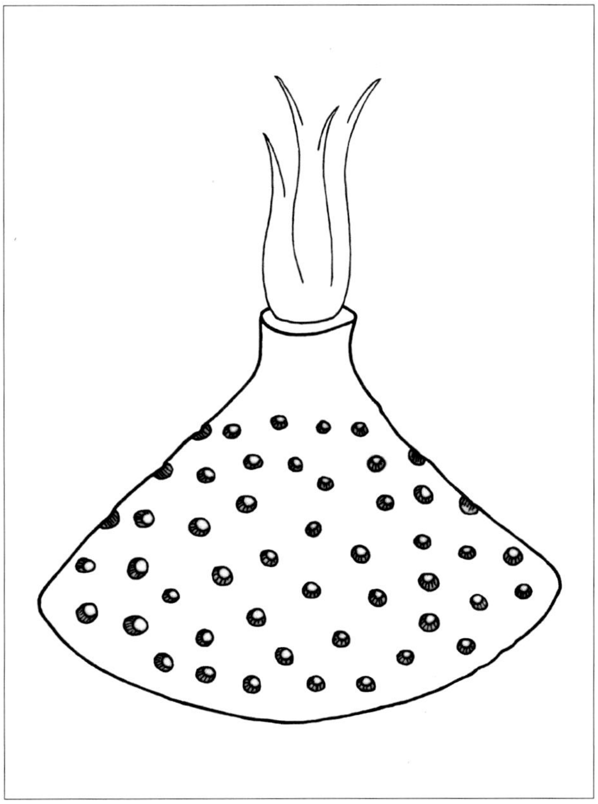

Figure 17.25. Reconstruction drawing of a funnel-shaped perforated clay utensil (inv. no. 2831) from the Heraion, Samos. Possibly a "Bunsen burner."

Conclusion

From the preceding discussion, it becomes clear that melting/alloying metals, casting and finishing artifacts, and making metal sheet and vessels as well as jewelry was feasible in the EBA Aegean settlements. The objects studied, mostly enigmatic in their function and meaning, have been puzzling archaeologists for generations. I do not claim to propose a definite answer for each one of them, but their classification as parts of the metallurgist's toolkit seems plausible. Time and further research will tell.

References

Bassiakos, Y., and O. Philaniotou. 2007. "Early Copper Production on Kythnos: Archaeological Evidence and Analytical Approaches to the Reproduction of Metallurgical Process," in Day and Doonan, eds., 2007, pp. 19–56.

Bernabò-Brea, L. 1964. *Poliochni: Città preistorica nell' isola di Lemnos* II, Rome.

Betancourt, P.P. 1997. "The Copper Smelting Site at Chrysokamino, Crete," in *Αρχαία Ελληνική Τεχνολογία, 1ο Διεθνές Συνέδριο: Πρακτικά. Θεσσαλονίκη, Εταιρεία Μελέτης Αρχαίας Ελληνκής Τεχνολογίας–Τεχνικό Μουσείο Θεσσαλονίκης*, Thessaloniki, pp. 51–54.

———. 2006. *The Chysokamino Metallurgy Workshop and Its Territory* (*Hesperia Supple*. 36), Princeton.

———. 2007. "The Final Neolithic to Early Minoan III Metallurgy Site at Chrysokamino, Crete," in Day and Doonan, eds., 2007, pp. 56–67.

Betancourt, P.P., and J.D. Muhly. 2007. "The Crucibles from the Aghia Photia Cemetery," in Day and Doonan, eds., 2007, pp. 146–153.

Birtacha, K. 2006. "The Pottery from Markiani Phase III," in Marangou et al., eds., 2006, pp. 130–138.

Blegen, C. 1921. *Korakou: A Prehistoric Settlement near Corinth*, Boston and New York.

Bossert, E.-M. 1967. "Kastri auf Syros," *ArchDelt* 22, pp. 53–76.

Branigan, K. 1974. *Aegean Metalwork of the Early and Middle Bronze Age*, Oxford.

Catapotis, M. 2007. "On the Spatial Organisation of Copper Smelting Activities in the Southern Aegean during the Early Bronze Age," in Day and Doonan, eds., 2007, pp. 207–223.

Catapotis, M., and Y. Bassiakos. 2007. "Copper Smelting at the Early Minoan Site of Chrysokamino on Crete," in Day and Doonan, eds., 2007, pp. 68–83.

Davaras, C., and P.P. Betancourt. 2004. *The Hagia Photia Cemetery*. I: *The Tomb Groups and Architecture* (*Prehistory Monographs* 14), Philadelphia.

Day, P.M., and R.C.P. Doonan, eds. 2007. *Metallurgy in the Early Bronze Age Aegean* (*Sheffield Studies in Aegean Archaeology* 7), Oxford.

Denison, S. 1999. "Des Becs Bunsen de l'Âge du Bronze," *Archéologia* 354 (March), p. 7.

Doonan, R.C.P., and P.M. Day. 2007. "Mixed Origins and the Origins of Mixing: Alloys and Provenance in the Early Bronze Age Aegean," in Day and Doonan, eds., 2007, pp. 1–18.

Doonan, R.C.P., P.M. Day, and N. Dimopoulou-Rethemiotaki. 2007. "Lame Excuses for Emerging Complexity in EBA Crete," in Day and Doonan, eds., 2007, pp. 98–122.

Doumas, G.C. 1965. "Κορφή τ' Αρωνιού," *ArchDelt* 20 (Meletai), pp. 41–64.

———. 1998. "Metallurgy and the So-Called Battle-Axe: The Tool of a Trade?" in *Mensch und Umwelt in der Bronzezeit Europas*, B. Hänsel, ed., pp. 157–162.

Dümmler, F. 1886. "Mitteilungen von den griechischen Inseln," *AM* 11, pp. 15–46.

Eskitzioglou, P. 2006. "The Pottery of Markiani Phase IV," in Marangou et al., eds., 2006, pp. 139–157.

Georgakopoulou, M. 2007. "Metallurgical Activities within Early Cycladic Settlements: The Case of Daskaleio Kavos," in Day and Doonan, eds., 2007, pp. 123–134.

Goldman, H. 1931. *Excavations at Eutresis in Euboia*, Cambridge, MA.

Harland, J.P. 1951. "An Early Helladic Kitchen Utensil," in *Studies Presented to David Moore Robinson*, vol. I, G.E. Mylonas, ed., St. Louis, pp. 106–107.

Hood, S. 1982. *Prehistoric Emporio and Ayio Gala* II, London.

Kaiser, I. 2005. "Minoan Miletus: A View from the Kitchen," in *Emporia: Aegeans in the Central and Eastern Mediterranean* (*Aegaeum* 25), R. Laffineur and E. Greco, eds., Liège, pp. 193–197.

Karantzali, E. 2006. "The Pottery of Markiani Phases I and II," in Marangou et al., eds., 2006, pp. 101–130.

Lamb, W. 1936. *Excavations at Thermi in Lesbos*, Cambridge, UK.

Marangou, L., ed. 1990. *Cycladic Culture: Naxos in the 3rd Millennium B.C.*, Athens.

Marangou, L., C. Renfrew, C. Doumas, and G. Gavalas, eds. 2006. *Markiani, Amorgos* (*BSA Suppl.* 40), London.

Marthari, M. 1998. *Syros, Chalandriani, Kastri*, Athens.

Milojčič, V. 1961. *Die prähistorische Siedlung unter dem Heraion: Grabung 1853 und 1955* (*Samos* I), Bonn.

Müller, K. 1938. *Die Urfirniskeramik* (*Tiryns* IV), Munich.

Mylonas, G.E. 1959. *Aghios Kosmas: An Early Bronze Age Settlement and Cemetery in Attica*, Princeton

Renar, J. 1995. *Le Péloponnèse au bronze ancien* (*Aegaeum* 13), Liège.

Renfrew, C., and R. Evans. 2007. "The EBA Pottery," in *Excavations at Phylakopi in Melos 1974–1977* (*BSA Suppl.* 42), C. Renfrew, N. Brodie, C. Morris, and C. Scarre, eds., pp. 129–180.

Renfrew, C., and E. Slater. 2003. "Metal Artifacts and Metallurgy," in *Prehistoric Sitagroi* 2 (*Monumenta Archaeologica* 20), E. Elster and C. Renfrew, eds., Los Angeles, pp. 301–322.

Romaios, A.K. 1915. "Ἐκ τοῦ προϊστορικοῦ Θέρμου," *ArchDelt* 1, pp. 225–279.

———. 1916. "Ἔρευναι ἐν Θέρμῳ," *ArchDelt* 2, pp. 179–189.

Sakellariou, A., and G. Papathanassopoulos. 1964. *Εθνικό Αρχαιολογικό Μουσείο: Α΄ Προϊστορικές Συλλογές*, Athens.

Sampson, A. 1988. *The Neolithic Settlement at Yali, Nisyros*, Athens.

Scarre, C. 2006. "Stone Vessels and Implements," in Marangou et al., eds., 2006, pp. 174–191.

Suto, Y. 1989. "Kastri Group and the Later Early Bronze Age in the Aegean," *Bulletin of the Archaeological Institute, Tokyo* 8, Dec. 1989, pp. 145–219.

Theocharis, D.R. 1951. "Ἀνασκαφὴ ἐν Ἀραφῆνι," *Prakt* 1951, pp. 77–92.

———. 1952. "Ἀνασκαφὴ ἐν Ἀραφῆνι," *Prakt* 1952, pp. 120–151.

———. 1955. "Ἀνασκαφὴ ἐν Ἀραφῆνι," *Prakt* 1955, pp. 109–117.

Tsipopoulou, M. 2007. "Aghia Photia–Kouphota: A Centre for Metallurgy in the Early Minoan Period," in Day and Doonan, eds., 2007, pp. 135–145.

Tsountas, C. 1899. "Κυκλαδικά," *ArchEph* 1899, pp. 136–219.

Tzavella-Evjen, C. 1984. *Λιθαρές*, Athens.

Zapheiropoulou, F. 1965. "Μῆλος," *ArchDelt* 20 (Chronika B3), pp. 508–513.

Zervos, C. 1957. *L'Art des Cyclades*, Paris.

CHAPTER

18

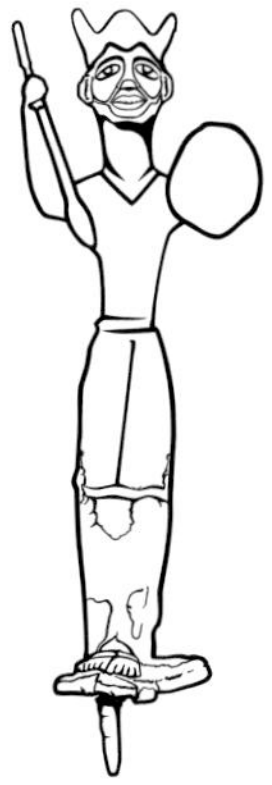

Technological Aspects of Bronze Age Metallurgical Ceramics in the Eastern Mediterranean

Anno Hein and Vassilis Kilikoglou

Introduction

The development of metallurgy has been associated with the selection of adequate materials for the production of pyrotechnical tools and their adaptation for use at high temperatures. The flawless function of heat resistant tools, such as furnaces, tuyères, molds, and crucibles, was necessary for effective processing of the metal ores and metals. Failure would have resulted possibly in the loss of valuable resources. Ceramics were the most common materials used for the production of pyrotechnical tools due to their high thermal capacity and their heat resistance at considerably high temperatures. The common pottery clays that were used in antiquity, however, were still far from perfect in terms of refractoriness, with regard to temperatures reached during metallurgical processes (Freestone and Tite 1986). These clays were fired usually at temperatures up to 1,100°C, as they would normally get distorted at higher ones. The reason for this behavior was the nature of the raw materials, comprising considerable amounts of basic components such as CaO, MgO, K_2O, or Na_2O, which were reacting at these temperatures (Noll 1991, 23–29). For the production of pyrotechnical or metallurgical ceramics, therefore, either more refractory clays had to be selected, or the clay paste preparation and ceramic construction had to be modified toward increased resistance at high temperatures.

A variety of approaches to improve the thermal properties of archaeological ceramics used at high temperatures have been reported in the scholarship. A common approach was the fabrication of rather thick walls to stabilize the ceramic body. Furthermore, the heat transfer was reduced, and in this way thermal stresses were suppressed due to smaller temperature gradients. Another common technique for the reduction of heat transfer was the generation of porosity. This was done either by adding organic material to the unfired clay body, which was burned out during firing, or by appropriate processing and wedging of the clay. Apart from reducing heat transfer, pores and voids at the same time improved the toughness of the ceramic body by interrupting the propagation of cracks in the matrix (Kilikoglou et al. 1998). Finally, high volume fractions of non-plastic inclusions were added in order to improve the refractory properties of the ceramic body. Alternatively, coarse clays with high concentrations of non-plastic inclusions were selected for the fabrication of the pyrotechnical ceramics. Thermally inert inclusions, such as quartz, stabilized the ceramic matrix at temperatures at which clay minerals started to decompose. Furthermore, due to heat expansion coefficients that were different from the ceramic matrix, inclusions generated microcracks and therefore additional porosity.

The required material properties of the ceramics, however, depended also on the type of heat source and on the direction of the heat flux. For example, internal heat, like in a smelting furnace required low thermal conductivity in order to maintain the heat. On the contrary, in the case of an external heat source, like in a crucible, high thermal conductivity is required to accelerate heating of the contents. At the same time, molds or tuyères should exhibit high resistance to thermal shock. Furthermore, mechanical properties, such as strength and toughness, had to be regarded as they are both related to thermal performance.

Heat Transfer in Ceramics

Heat Transfer Processes

In order to assess the suitability of traditional approaches for the fabrication of pyrotechnical ceramics, the heat transfer processes in ceramics have to be considered. There are basically two types of heat transfer that are relevant for archaeological metallurgical ceramics: heat conduction and thermal convection.

Heat conduction describes the heat transfer, mainly in solid materials like in a furnace wall. The thermal conductivity k of a specific material describes the heat flux dQ/dt normal to an area A under a particular temperature gradient T. In the case of ceramics, k is usually in the range of 0.5 to 1.5 $W \cdot (m \cdot K)^{-1}$ normally with decreasing values at higher temperatures. If the conditions are not constant, however, the temperature development is described with the "heat conduction equation" (Kingery, Bowen, and Uhlmann 1976):

Equation 1 $$\frac{dT}{dt} = \alpha \cdot \frac{d^2T}{dx^2}$$

The so-called thermal diffusivity α is the ratio of the thermal conductivity k and the heat capacity per unit volume $\rho \cdot c_p$ where ρ is the density of the material and c_p is the heat capacity. For ceramics, its values are typically in the range of 0.005 to 0.01 m^2s^{-1}.

The second heat transfer process to be considered is thermal convection, which describes the heat exchange between the surface of a solid and a moving fluid. It depends on the temperature of the surface T_s, the bulk temperature of the fluid T_∞, and the so-called film coefficient h, which is a material property of the fluid:

Equation 2 $$\frac{dQ}{dt} = h \cdot A \cdot (T_s - T_\infty)$$

In the case of a smelting furnace, the heat exchange between ceramics and the ambient air is subject to thermal convection rather than heat conduction. Air presents a film coefficient of 10–100 $W \cdot (m^2 \cdot K)^{-1}$ depending on humidity, wind speed, and shape of the furnace. For example, in the case of a cylindrical furnace, 4 Beaufort wind force corresponds to a film coefficient of ca. 30 $W \cdot (m^2 \cdot K)^{-1}$.

Finally, there is a third heat transfer process: heat radiation. In comparison to the two described processes, however, heat radiation is still relatively small at the temperatures used in ancient metallurgy, and it therefore can be neglected (Litovsky and Shapiro 1992).

Influence of Ceramics' Microstructure

In comparison to modern ceramics, ancient ceramics were more or less complex composite materials, i.e., natural mixtures of various components, possibly modified by adding temper materials. The actual ceramic matrix consisted, in most cases, of a mixture of clay minerals and, when fired adequately, of respective high temperature phases. The thermal properties of the ceramics, as well as other material properties, were affected by the texture and by the properties of the particular components. The heat transfer was restricted by pores and voids, which acted as heat barriers. On the other hand, different kinds of mineral and rock inclusions, either natural components of the raw material or added as temper, could increase the thermal conductivity. Non-plastic inclusions, such as quartz, calcite, feldspars, and micas usually shift thermal properties of the ceramic matrix due to their higher density and more compact texture. The thermal conductivity, however, also was affected by the firing of the ceramics, either intentionally before use or during the operation of the originally unfired clay furnaces. An increase of thermal conductivity with firing temperature could be observed, probably subject to changes in texture and to higher thermal conductivity of high temperature phases (Hein et al. 2008).

While properties like density or heat capacity depended simply on volume percentages of the particular components and on the absolute porosity, thermal conductivity was affected also by shape and orientation of inclusions and voids. The effective thermal conductivity, however, had a strong influence on the temperature development during use of the metallurgical ceramics. Apart from the estimation of its effect on the actual performance, an accurate assessment of this parameter can contribute to the reconstruction of operation conditions based on archaeological evidence (Kingery and Gourdin 1976; Tite et al. 1990; Hein, Kilikoglou, and Kassianidou 2007). The effective thermal conductivity of a particular ceramic microstructure can be assessed either with laboratory measurements (Hein et al. 2008) or by computer simulations (Hein and Kilikoglou 2007).

Archaeological Evidence

Former studies of Bronze Age metallurgical ceramics primarily concerned furnace linings, but crucibles and molds also were examined. Kingery and Gourdin (1976) presented a study on stone furnaces coated internally with a clay-based sandstone lining from the Early Bronze Age copper-smelting site of Wadi Zaghra in southeastern Sinai. Based on refiring experiments and observations of the degree of vitrification at different depths by SEM, the operation time of a furnace was estimated. This estimation was based on reference values for thermal conductivity of typical ceramics and a numerical solution of Equation 1. Another important study was focused on Bronze Age and Iron Age metallurgical ceramics, furnace linings, tuyères, and crucibles from the copper-smelting site of Timna in the Levant (Tite et al. 1990). The analytical results indicated that the high refractoriness of the materials was achieved with non-calcareous clays containing a high proportion of kaolinite. Furthermore, the clays contained quartz and shale inclusions as natural components, and some ceramics were tempered with crushed slag in order to further enhance refractoriness. In this case study, the operation conditions of the furnaces, following the approach of Kingery and Gourdin (1976), were estimated as well.

Bronze Age crucibles were fired from above with fuel and metal in contact, as indicated in a study of Early Bronze Age crucibles from Poros Katsambas, Crete (Doonan, Day, and Dimopoulou-Rethemiotaki 2007). These crucibles presented wall thicknesses between 7 and 28 mm and were fabricated from a buff, calcareous clay with frequent small quartz inclusions. They were tempered with organic material, which was burned out either during firing or use. An example from the Classical period is represented in the study of a bronze workshop in Olympia, Greece (Schneider and Zimmer 1984). For the production of crucibles, which were also heated from inside, local non-calcareous to low-calcareous clays were used, tempered with angular quartz particles and organic fibers, leaving large voids. Also in this case the ceramics were not particularly refractory, and heavy alterations due to the process temperature were observed.

Several other studies of archaeological metallurgical ceramics revealed a variety of non-plastic inclusions (Tylecote 1982; Tite et al. 1985; Freestone 1989; Bayley et al. 1991), in most cases quartz, but also slag remains, grog, or carbonaceous materials, such as charcoal or graphite. Carbonaceous materials additionally were assumed to inhibit the oxidation of the molten metal (Freestone 1989). In many cases, tempering with organic materials was observed, though the related high porosity was not generally assumed to be an advantage, especially in view of the vessel strength. The addition of organic temper, particularly fibrous materials, was usually assessed as advantageous for strengthening the unfired ceramic body and, at the same time, acting as a carbon source (Freestone 1989).

The widespread and systematic use of kaolonitic clays, i.e., fireclays, became common only from the Roman period (Freestone 1989). These effectively refractory clays enabled the development of thin-walled pyrotechnical ceramics with high thermal conductivity, which could be externally heated. Recent studies on post-medieval crucibles revealed that apart from the chemical composition, with a low content of fluxes, the firing at extremely high temperatures and the corresponding formation of high temperature phases, such as mullite or cristoballite, was crucial for the quality of the ceramics (Martinón-Torres, Rehren, and Freestone 2006; Martinón-Torres and Rehren 2009).

Finally, there are also historical sources with descriptions and instructions for preparation of furnaces or crucibles. For example, Agricola in his *De re metallica* provided a detailed description for the preparation of smelting furnaces in preindustrial Renaissance Europe (Agricola 1928, 328–330). It is suggested that they should have been constructed either from bricks or preferably from stones and coated inside with a mixture of clay and powdered coal or charcoal (two parts powdered coal and one part clay), obviously to increase porosity. In contrast, crucibles should be fabricated from a plastic and dense clay mixed with powdered fragments of old broken crucibles or fired bricks, i.e., grog or ground chamotte (Agricola 1928, 198).

Copper Smelting Furnaces from Seriphos

In the framework of a recent study of Early Bronze Age copper metallurgy on Seriphos (Georgakopoulou 2005), an assemblage of 54 furnace fragments from three different smelting sites was collected and examined: Kephala, Phournoi, and Avessalos. Most of the fragments present clear indications for their use in metallurgical processes in terms of slag traces on the surface, color variations due to varying redox conditions, and visibly bloating pores in areas that were exposed to extreme temperatures. In general, the ceramic matrix gave a coarse impression with frequent and large inclusions and evidence of incomplete clay mixing. The majority of furnace wall fragments presented a thickness of 20–30 mm, with some clearly thicker fragments (>40 mm), possibly presenting furnace bases. The inner diameter of the furnaces was estimated to be in the range of 320–340 mm.

The base material according to SEM-EDS analysis was low-calcareous, but it exhibited comparably high concentrations of Na, Mg, and K, indicating the use of raw materials deriving from the local schists. The studied furnace fragments provide the basis for a reconstruction of the typical configuration of the smelting furnaces on Seriphos, which were at least in terms of the collected material, generally the same at all three sites.

*Copper Smelting Furnaces from Politiko-*Phorades*, Cyprus*

The Late Bronze Age copper-smelting site of Politiko-*Phorades*, which is the earliest primary copper-smelting site discovered in Cyprus to date, is located in the northeastern foothills of the Troodos Mountains. It was probably one of a chain of similar sites close to the mines where the copper sulphide ores were exploited. Politiko-*Phorades* represents the start of a period of large-scale copper production in Cyprus, acting as a supplier to the eastern Aegean and other regions of the Mediterranean (Knapp, Donnelly, and Kassianidou 1998; Knapp, Kassianidou, and Donnelly 1999). Out of a large assemblage of metallurgical ceramics discovered on the site, 15 tuyère fragments and 25 furnace fragments were selected for more detailed examination in order to obtain basic information about the construction of furnaces and tuyères and their use (Hein, Kilikoglou, and Kassianidou 2007).

According to the analytical results, local non-calcareous clays were used for the fabrication of the metallurgical ceramics, which were similar to the illitic clays used in contemporary pottery production. The metallurgical ceramics, however, were considerably coarser than comparable pottery, and their microstructure was dominated by inclusions of volcanic rock fragments. These inclusions apparently were related to the local geological environment and probably were a natural component of the selected raw materials. Another prominent feature observed in the microstructure was the high porosity that was intentionally generated in the ceramic body. Part of it was related to tempering with organic material as indicated by clear traces of burned out organic fibers. Other frequent and elongated voids were apparently related to processing of the coarse clay paste without extensive mixing or compressing (Bayley et al. 1991). The open porosity was estimated to be approximately 25% for a typical tuyère and up to almost 40% for some furnace wall fragments. Furthermore, in all samples a preferred orientation of pores and voids parallel to the surface was observed, apparently generated during construction. Furnaces and tuyères presented no significant difference in terms of raw materials and fabrication technique, although the furnace fragments generally gave a coarser impression.

Based on the observed microstructure, a two-dimensional multiphase model of the ceramics was developed and evaluated with the finite element method (FEM; Hein and Kilikoglou 2007). Shape and orientation of the pores proved to be a decisive parameter for the reduction of heat transfer in the furnace walls. Furthermore, the estimations of the thermal conductivity could be applied in a computer model of the entire furnace in order to simulate its operation conditions with FEM. In contrast to former studies on furnace linings (Kingery and Gourdin 1976; Tite et al. 1990), the heat convection on the outer surface of the freestanding furnace walls also could be considered.

Discussion

In all the studies summarized above, the porosity, which can be observed in ancient metallurgical ceramics, appears to have been intentionally generated in order to reduce heat transfer during operation of metallurgical tools. In combination with thick walls and large amounts of non-plastic inclusions, it thus was possible to use the ceramics at considerably high temperatures, which obviously exceeded the ones that the actual refractoriness of the matrix would permit. Due to the insulating properties, however, distortion was restricted to the ceramic surface, which was in contact with the hot load. Furthermore, thermal stresses in the ceramics were reduced. In this way, the tools remained intact during use, and also their handling was possibly improved.

Another aspect that might have been influenced by porosity and wall thickness was the economization of fuel. In Figure 18.1 the temperature development on the outer surface of a 30-mm furnace wall is presented. In this estimation a temperature of

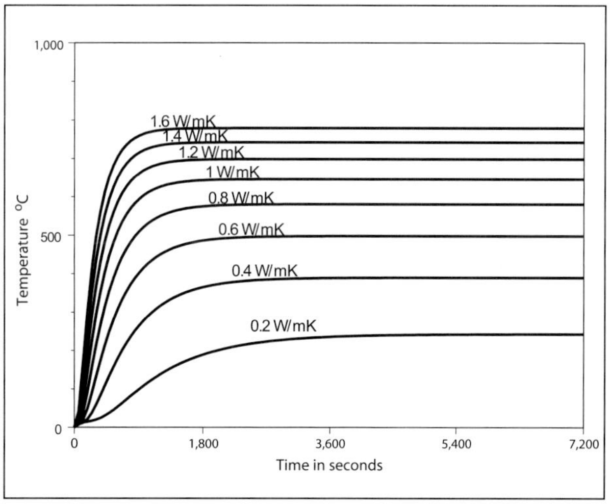

Figure. 18.1. Effect of thermal conductivity: temperature development on the outer surface of a 30-mm furnace wall with varying thermal conductivity; heat of 1,200°C is simulated on the inner surface. For the ceramics, a density of 1,500 kg/m^3 and a heat capacity of 900 J/(kg·K) are assumed. For the ambient air, a bulk temperature of 30°C and a film coefficient of 30 W·(m^2·K)$^{-1}$ are assumed.

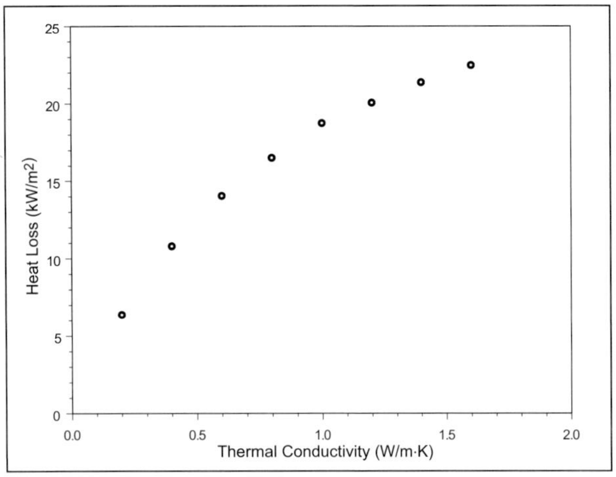

Figure 18.2. Effect of thermal conductivity: corresponding heat loss through thermal convection on the outer surface.

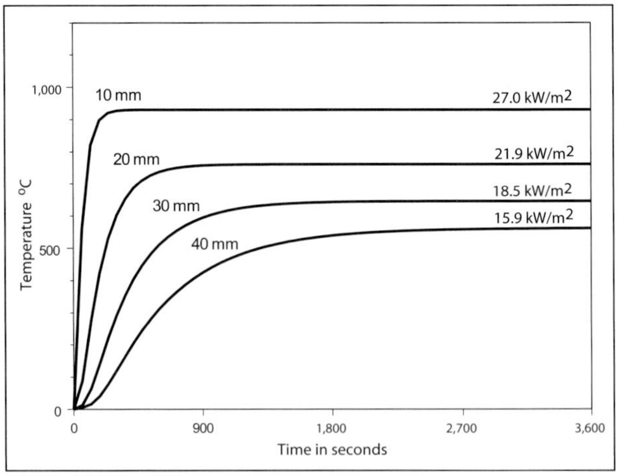

Figure 18.3. Effect of the thickness of a furnace wall. For the ceramics, thermal conductivity of 1 W/(m·K) is assumed. All other constraints correspond to the previous simulation.

1,200°C was assumed for the inner surface of the ceramic wall and a bulk temperature of 30°C for the ambient air with a film coefficient of 30 $W\cdot(m^2\cdot K)^{-1}$. The curves correspond to different thermal conductivities of the furnace wall from 0.2 to 1.6 $W\cdot(m^2\cdot K)^{-1}$. Ceramics with an effective thermal conductivity of 0.2 $W\cdot(m^2\cdot K)^{-1}$ could be produced from ceramics with a thermal conductivity of 1.0 $W\cdot(m^2\cdot K)^{-1}$ of the actual matrix by generating 35%–40% porosity with a suitable orientation perpendicular to the expected heat flux (Hein and Kilikoglou 2007). This appears quite realistic with regard to the archaeological evidence. In each case the surface temperature reaches equilibrium after approximately 30 minutes to one hour depending on the thermal conductivity. Furthermore, the thermal conductivity obviously affects the surface temperature and thus also the heat loss through heat convection. In Figure 18.2 the corresponding heat loss in kW/m^2 after reaching the equilibrium is presented against the thermal conductivity. Because the heat flux is correlated to the temperature difference between surface and air (Equation 2), the heat loss would have been reduced considerably by generating porosity and thus reducing thermal conductivity. In Figure 18.3 the corresponding temperature development for varying wall thickness is presented. Even though the increment of wall thickness obviously was also a way to reduce the heat loss, the potential improvement decreased toward larger wall thicknesses. Moreover, the handling of the furnace probably became more difficult due to the larger mass. The examination of the archaeological ceramics in combination with the simulation and modeling, therefore, reveals that the use of porous ceramics was apparently the best strategy, with reasonably thick furnace walls of 20 to 30 mm. In this way a compromise between porosity and weight toward the optimum operation of the furnace was achieved.

References

Agricola, G. 1928. *De re metallica libri XII – Zwölf Bücher vom Berg- und Hüttenwesen*, new edition (with comments), Berlin.

Baylay, J., I.C. Freestone, A. Jenner, and A. Vince. 1991. "Metallurgy," *London and Middlesex Archaeological Society Special Paper* 12, pp. 398–405.

Doonan, R.C.P., P.M. Day, and N. Dimopoulou-Rethemiotaki. 2007. "Lame Excuses for Emerging Complexity in Early Bronze Age Crete: The Metallurgical Finds from Poros Katsambas and Their Context," in *Metallurgy in the Early Bronze Age Aegean*, P.M. Day and R.C.P. Doonan, eds., Oxford, pp. 98–122.

Freestone, I.C. 1989. "Refractory Materials and Their Procurement," in *Archaeometallurgie der Alten Welt—Old World Archaeometallurgy*, A. Hauptmann, E. Pernicka, and G.A. Wagner, eds., Bochum, pp. 155–162.

Freestone, I.C., and M.S. Tite. 1986. "Refractories in the Ancient and Preindustrial World," in *Ceramics and Civilization, High-Technology Ceramics—Past, Present and Future*, Volume III, W.D. Kingery, ed., pp. 35–63.

Georgakopoulou, M. 2005. *Technology and Organization of Early Cycladic Metallurgy: Copper on Seriphos and Keros, Greece*, Ph.D. diss., University College, London.

Hein, A., and V. Kilikoglou. 2007. "Modeling of Thermal Behavior of Ancient Metallurgical Ceramics," *Journal of the American Ceramic Society* 90 (3), pp. 878–884.

Hein, A., V. Kilikoglou, and V. Kassianidou. 2007. "Chemical and Mineralogical Examination of Metallurgical Ceramics from a Late Bronze Age Copper Smelting Site in Cyprus," *JAS* 34, pp. 141–154.

Hein, A., N.S. Müller, P.M. Day, and V. Kilikoglou. 2008. "Thermal Conductivity of Archaeological Ceramics: The Effect of Inclusions, Porosity and Firing Temperature," *Thermochimica Acta* 480 (1–2), pp. 35–42.

Kilikoglou, V., G. Vekinis, Y. Maniatis, and P.M. Day. 1998. "Mechanical Performance of Quartz-Tempered Ceramics: Part I, Strength and Toughness," *Archaeometry* 40 (2), pp. 261–279.

Kingery, W.D., and W.H. Gourdin. 1976. "Examination of Furnace Linings from Rothenberg Site #590 in Wadi Zaghra," *JFA* 3, pp. 351–353.

Kingery, W.D., H.K. Bowen, and D.R. Uhlmann. 1976. *Introduction to Ceramics*, New York.

Knapp, A.B., M. Donnelly, and V. Kassianidou. 1998. "Excavations at Politiko-*Phorades* 1997," *RDAC* 1998, pp. 247–268.

Knapp, A.B., V. Kassianidou, and M. Donnelly. 1999. "Excavations at Politiko-*Phorades* 1998," *RDAC* 1999, pp. 125–146.

Litovsky, E.Y., and M. Shapiro. 1992. "Gas Pressure and Temperature Dependences of Thermal Conductivity of Porous Ceramic Materials: Part I. Refractories and Ceramics with Porosity Below 30%," *Journal of the American Ceramic Society* 75 (12), pp. 3425–3439.

Martinón-Torres, M., and T. Rehren. 2009. "Post-Medieval Crucible Production and Distribution: A Study of Materials and Materialities," *Archaeometry* 51 (1), pp. 49–74.

Martinón-Torres, M., T. Rehren, and I.C. Freestone. 2006. "Mullite and the Mystery of Hessian Wares," *Nature* 444, pp. 437–438.

Noll, W. 1991. *Alte Keramiken und ihre Pigmente*, Stuttgart.

Schneider, G., and G. Zimmer. 1984. "Technische Keramik aus antiken Bronzegußwerkstätten in Olympia und Athen," *Berliner Beiträge zur Archäometrie* 9, pp. 17–60.

Tite, M.S., I.C. Freestone, N.D. Meeks, and P.T. Craddock. 1985. "The Examination of Refractory Ceramics from Metal-Production and Metalworking Sites," in *The Archaeologist and the Laboratory*, P. Phillips, ed., London, pp. 50–55.

Tite, M.S., M.J. Hughes, I.C. Freestone, N.D. Meeks, and M. Bimson. 1990. "Technological Characterisation of Refractory Ceramics from Timna," in *The Ancient Metallurgy of Copper*, B. Rothenberg, ed., London, pp. 158–175.

Tylecote, R.F. 1982. "Metallurgical Crucibles and Crucible Slags," in *Archaeological Ceramics*, J.S. Olin and A.D. Franklin, eds., Washington, D.C., pp. 231–243.

CHAPTER

19

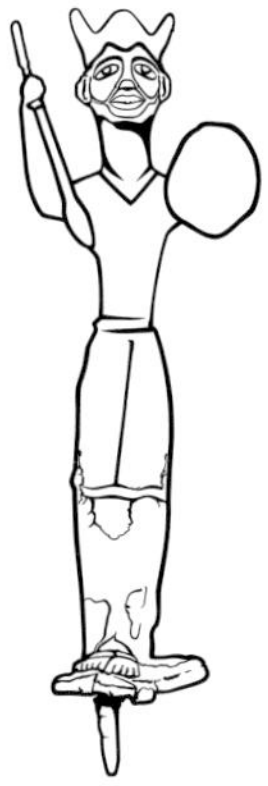

Slags from the Late Bronze Age Metal Workshops at Kition and Enkomi, Cyprus

Andreas Hauptmann

Introduction

There are two major issues in archaeometallurgy.* The first one is the question regarding the provenance of metal objects, or, in addition, of the raw sources that were utilized to produce metal. In a larger context, provenance studies are needed to reconstruct trade routes, exchange of goods, shipment centers, and the spatial distribution of metal. The aim of the second issue is to decipher technologies applied in ancient mining and metallurgy. What was the composition of the ores used, from which part of an ore deposit were they mined, and which metallurgical operations were applied to smelt them to finally produce an artifact? All of these questions are basic for a proper understanding of analytical data and also for provenance studies.

Both provenance studies and technological questions need an interdisciplinary approach of archaeologists and natural scientists. Even if many of those studies led to controversial debates on the understanding of archaeometallurgical issues—e.g., the long-term discussion about the lead-silver-copper-deposit of Laurion (Lohmann 2005)—many mining and metallurgical districts were successfully investigated during the last few decades. The copper districts of Faynan and Timna were intensively investigated by Hauptmann (2007) and Rothenberg (1990). They are today two paradigms for ancient copper production from *oxidic* ores. As a paradigma for the early production of copper from *sulfidic* ores, the exploitation of the numerous Volcanogenic Massive Sulfide deposits (VMS) of Cyprus should be specified (Stos-Gale, Maliotis, and Gale 1998; Given and Knapp 2003). Very often,

* I would like to dedicate this article sincerely to the archaeologically sided colleague of the archaeometallurgical twins James D. Muhly and Robert Maddin.

for an understanding and interpretation of archaeometallurgical finds, technological descriptions and observations were used from the 18th/19th centuries A.D., for example during the first investigations of Bronze Age copper production at Mitterberg, near Salzburg, Austria, by Zschocke and Preuschen (1932).

Faced with the gigantic masses of slag at Cyprus, it was rather difficult to compile a classification of slags and to assign them to specific metallurgical processes (Zwicker, Rollig, and Schwartz 1972; Bachmann 1982a; Koucky and Steinberg 1982). Also, it was, and it is a problem to identify roasting processes and to pinpoint debris of such operations. It is hard to distinguish such debris from material found in the oxidation zone of the many sulfidic ore deposits at Cyprus, neither in its chemical, nor in its mineralogical composition. To this category of finds, in part, belong those that were excavated within the fortifications and temple constructions of the Late Bronze Age urban settlements at Kition and Enkomi.

A large variety of metallurgical operations was suggested for these settlements (Tylecote 1982; Zwicker 1985): primary smelting of ores, roasting, further processing of matte and/or copper, refining dross, re-melting scrap metal, and the production of final (alloyed) objects. Worth mentioning is a publication of Stech (1982), who suggested a treatment of copper-rich slag within the Late Bronze Age urban settlements at Cyprus. But due to their macroscopic appearance, and due to their phase content (fayalite and Fe-oxide), the slags found at Kition, Level II, were also compared with those from 7th/6th century B.C. Murlo, Etruria, and even iron production at Kition was considered (Stech 1982).

A term was coined during the discussions about the early Late Bronze Age metallurgy to address the slag found at Kition and Enkomi and some other sites. Frequently, these pieces of slag were found in fist-sized lumps crusted by a whitish-yellow layer, and because a clear identification was not possible, they were named "furnace conglomerate." In the course of time, this term was used very often, and it was used contradictorily.

It is the aim of this study to find out what is meant by "furnace conglomerate." Is it possible to identify the metallurgical process where this material was produced? Does it derive from a roasting operation or from a smelting or a melting process? What was the final product?

"Furnace Conglomerate"

The term furnace conglomerate (abbrevated in the following text as "fc") was coined by Bachmann (1980) to distinguish this material from molten slags in the true sense. He suggested that fc represents the very first step of metallurgical treatment of ores and that they are a result of a partial reduction of ores to metal. Furnace conglomerate is cinder-like material. As exemplified by a sample from Wadi Amram in the Timna region, it consists of decomposed ore, fragments of gangue, pieces of charcoal, and globules of copper, all held together by a "slaggy" matrix.

At the settlement of Kition, large chunks of fc were found from 13th/12th century B.C. contexts (Stech 1982). Stech, Maddin, and Muhly (1985, 395, 397) used the term fc to describe heterogeneous chunks of slag from Rooms 14 and 39 at Kition. These chunks proved "that mixed ores were used." They contained copper prills entrapped in a mass of "non metallic copper," iron, and sulfur, which actually means matte (Cu-Fe-sulfides). Such chunks would have been discarded at the smelting sites, and the impure copper and copper-bearing fc would be sent to the cities for further processing.

Muhly wrote on fc (1991, 183–184): "What we find at all these urban centers [Enkomi, Kition, Athienou, Kalavassos–*Ayios Dhimitrios*, Maroni-*Vournes*, Myrtou-*Pigades*] are large chunks of a material that H.G. Bachmann (1982[b]) has designated furnace conglomerate. This material was first recognized by Tylecote (1971, 53) at Kition. I have now identified this distinct material at Enkomi, Kalavasos–Ayios Dhimitrios, Maroni-Vournes, and Myrtou-Pigadhes. The theory is that this furnace conglomerate represents what was produced at mining-smelting sites such as Apliki-Karamallos, the conglomerate then being shipped to the coastal urban centres and there further refined to produce metallic copper."

Knapp, Kassianidou, and Donnely (1999) stated that Muhly wrongly associated fc from Kition with the large slag cakes that Tylecote (1982) described as plano-convex blocks of slag of fayalitic composition with yellow matte. The authors argued that fc would be the product of roasting, i.e., an intermediate product of the metallurgical treatment of ore.

Karageorghis and Kassianidou (1999, 179) follow Knapp, Kassianidou, and Donnely (1999) and described fc from Kition, Room 14 as ". . . a kind of find well known from the Late Bronze Age Cyprus. These unfortunately have not been analyzed, although it is generally assumed that they were a product of roasting operation." For Given and Knapp (2003, 67), fc "consists of pieces of ore, charcoal, slag, and related material fused by fire. . . ." They described material from spoil heaps with the term "furnace or roasting conglomerate."

The Slag Samples from Kition and Enkomi

The slag samples were taken from the Archaeological District Museum at Larnaca during a visit in 2002. They derived from excavations at Kition and, subordinated, at Enkomi. Altogether, the slags observed in the archives did not exceed a few hundred kilograms or so. In comparison to many other localities where ancient slags were found, all the slag is considerably affected by corrosion.

A small number of samples were selected, which, from the macroscopic aspect, represented the different types of slag identified in the collections (Figs. 19.1–19.13). They range from nut- to fist-sized up to a few that were more than skull-sized chunks weighing more than 8 kg. Most of the slag samples consist of sharped-edged chunks cut by deep fissures that eventually lead to breaking and crumbling into small pieces. The slag samples were almost totally encrusted by layers of different minerals (up to one cm thick), and the slag fragments were baked together to form some sort of a breccia or conglomerate (breccia is a geological term as it is a conglomerate: the first means a solid clastic rock consisting of *angular* components; the second term is a consolidated sedimentary gravel rock consisting of *rounded* components). These encrustations can also be observed inside gas bubbles, along vein cuttings, and in the inter-porous volume of the slag samples. They form white, and in part, brown, and brilliant yellow efflorescent layers, dotted with green spots. X-ray diffraction analyses showed the existence of the following phases:

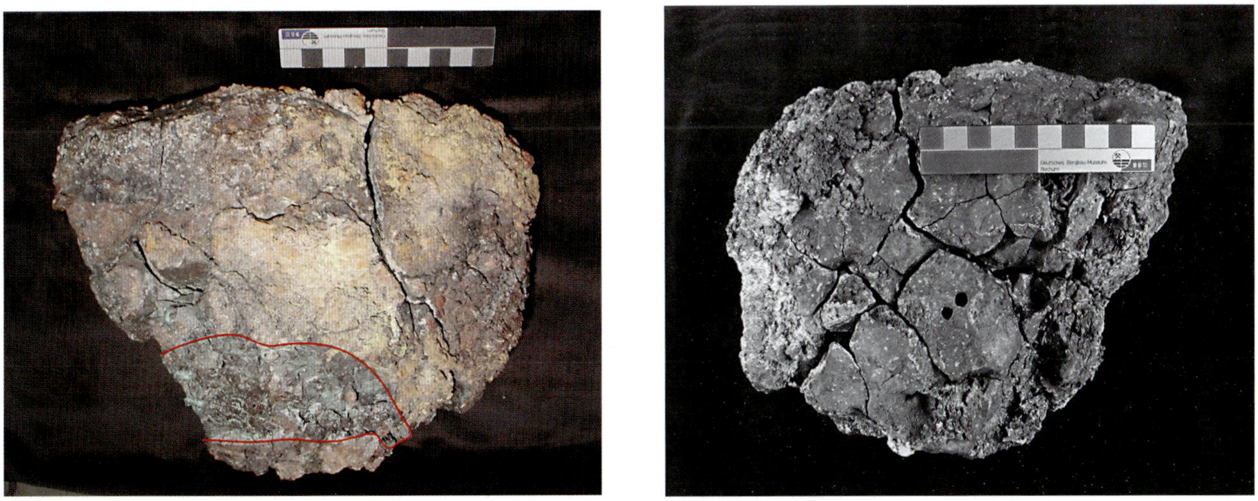

Bottom Top

Figure 19.1. Sample CY-1/1, Kition, showing a negative impression on the bottom (see red line) and gas bubbles on top. Four samples were taken from different spots of the slag: bottom of the slag, negative impression; lower sidewall of the slag; top of slag; top slag.

white: calcite, $CaCO_3$; gypsum, $CaSO_4{\cdot}2H_2O$; and quartz, SiO_2

brown: jarosite, $KFe_3(SO_4)_2(OH)_6$; goethite, limonite, $FeO(OH){\cdot}nH_2O$, and akaganeite, $FeO(OH_1Cl)$

yellow: melanterite, $FeSO_4{\cdot}7H_2O$

green: atacamite, $Cu_2Cl(OH)_3$; malachite, $Cu_2(CO_3)(OH)_2$

The white encrustation is due to the natural soil in the Kition area. This coating is built up by decomposition of marl and chalk, and it is rich in $CaCO_3$ and in gypsum. The iron-rich sulfates point to acidic solutions rich in SO_4^{2-}, formed either by the existence of sulfates in the soil, by the decomposition of sulfides, or, alternatively, one cannot exclude that the original source for the precipitation of Fe-sulfates could be iron-rich (cuprous) ores stored inside the settlement, or roasting products. In the geological context, jarosite and melanterite are characteristically and frequently found in the gossans of sulfidic ore deposits on Cyprus. Here, they are formed by the decomposition of Fe-rich sulfides. Goethite, limonite, and akaganeite are formed from Fe-oxides and from Fe-silicates in the slag. Finally, malachite

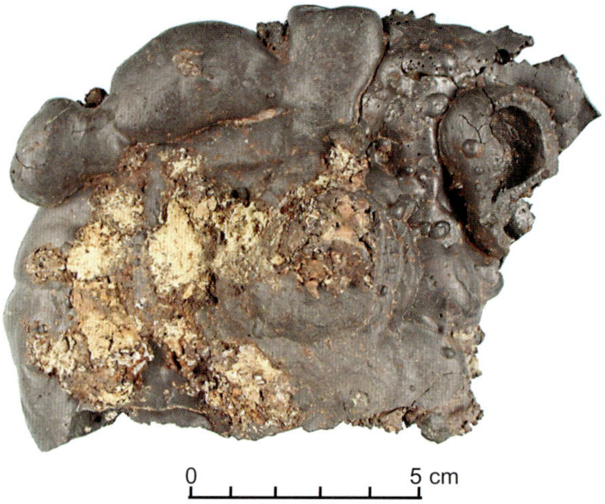

Figure 19.2. Sample CY-1/2, Kition. Tap slag with flow structures, dense silicate slag, and thick efflorescence of Fe-sulfates that penetrate deep into the slag. The weight of the slag is 0.5 kg (note charcoal inclusions).

Figure 19.3. Sample CY-1/3, Kition. Semi-globular chunk, irregular surface. Section shows brecciated texture with inclusions of Fe-sulfates, hostrock, and charcoal in an iron-rich silicate slag, weight ca. 1 kg.

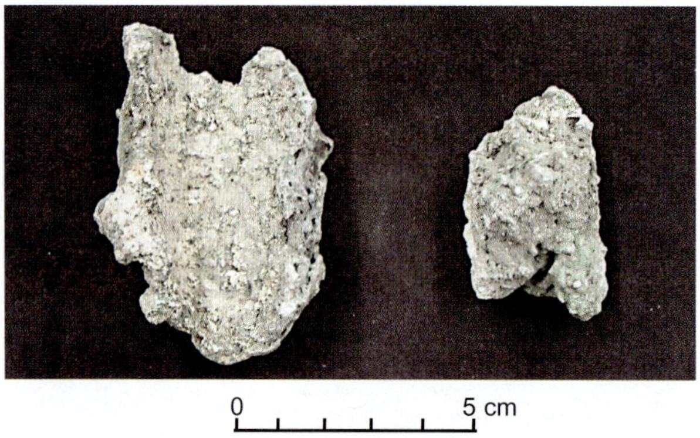

Figure 19.4. Sample CY-1/6, Kition. Irregular chunks, rough surface, attached to tuyères. Section shows high porosity with mm-sized gas bubbles. Remarkably heavy, weight 0.1–0.2 kg.

Figure 19.5. Sample CY-2/2c, Kition. Fragment of a slag showing yellow secondary fillings mainly of iron-sulfate in gas bubbles, which successively created radial cracks and fissures in the silicate slag body. Width of the surface polished thin section: 27 mm.

and atacamite are typical secondary copper minerals in a limy environment, partially enriched by chlorides from the sea close by. In summary, the encrustations result from an intense corrosive interaction between the slag material itself and the surrounding soil where it was embedded for millennia.

After carefully cleaning the surfaces of the slag samples using a sand blasting unit, an enrichment of green secondary copper minerals at the lower part of sample CY-1/1 became visible along with a dark reddish tint (Fig. 19.1). This phase was identified by X-ray diffraction to be cuprite (Cu_2O). Finally, we could distinguish macroscopically between four types of slags that occur at Kition.

Sampling and Measurement

To reconstruct metallurgical operations carried out at Kition and Enkomi, the chemical and mineralogical compositions of slag samples were studied at the Research Laboratory for Materials Science and Archaeology at the German Mining Museum, Bochum. In addition, a study of the petrographic texture of the entire slag chunk was carried out to gain information on both the macro- and microstructure. We prepared several series of surface polished thin sections, covering the whole section of a few small slag cakes (CY-1/3, CY-1/6, CY-2/2). Four thin sections were prepared from the large slag no. CY-1/1 to show its composition at the top, the middle part, the side wall, and the bottom. X-ray diffraction was performed using PANalytical X'Pert PRO MPD, and the chemical composition was analyzed by inductively coupled plasma optical emission spectrometry (ICP-OES Thermo Jarrell Ash; Table 19.1).

In addition, three samples from Kition and Enkomi (CY-1/1b, CY-1/6, CY-2/2a) were selected for lead isotope analyses. These investigations were determined by thermal ionization mass spectrometry (TIMS) at the Laboratory of Geochronology, University of Muenster. From these samples, only two contained sufficient lead to provide clear signals of isotope compositions. Sample no. CY-1/6, which consisted totally of Fe-sulfates, did not contain sufficient lead for analysis.

Catalog of Samples

CY-1/1 (Fig. 19.1). Kition, Area II, Room 14, Floor II. Bowl-shaped, dense silicate slag with a negative impression at the bottom (see red outline in Fig. 19.1), gas

Sample Numbers	Percent (%)																	
	SiO_2	TiO_2	Al_2O_3	FeO	MnO	MgO	CaO	K_2O	Na_2O	P_2O_5	S	Cu	Zn	Pb	As	Sn	Co	Total
CY-1/1a	12.8	0.20	3.49	51.1	0.04	0.82	2.45	0.42	0.82	0.12	0.79	13.2	0.020	0.013	0.064	<0.001	0.027	**86.5**
CY-1/1b	9.76	0.16	2.38	44.4	0.02	0.54	1.35	0.24	0.55	0.09	1.39	25.9	0.022	0.016	0.14	<0.001	0.085	**87.1**
CY-1/1c	28.9	0.45	7.03	54.2	0.09	1.71	3.03	0.78	1.90	0.15	0.58	0.51	0.023	<0.001	0.002	<0.001	0.003	**99.4**
CY-1/1d	28.6	0.42	6.55	54.0	0.08	1.74	2.82	0.75	1.67	0.14	0.52	0.73	0.022	<0.001	0.002	<0.001	0.004	**98.1**
CY-1/3	15.4	0.24	3.86	60.4	0.03	1.61	1.45	0.26	0.28	0.08	1.67	4.00	0.007	0.001	0.036	0.13	0.026	**89.5**
CY-1/6	11.8	0.09	3.41	50.4	0.07	1.69	4.47	0.47	0.29	0.11	0.14	20.1	1.18	0.050	0.15	0.086	0.16	**94.8**
CY-2/2a	16.5	0.26	3.68	63.5	0.02	0.44	1.12	0.31	0.13	0.19	1.52	2.54	0.004	<0.001	0.001	<0.001	0.015	**90.2**

Table 19.1. Chemical composition of slag samples from Kition (CY-1) and from Enkomi (CY-2). Sample nos. CY-1/1a–d are taken from different spots at the top (CY-1/1c), middle part (CY-1/1d), and bottom (CY-1/1a, b) of the large slag block (see catalog). All values are given in weight %. The total iron content was calculated as FeO. Note that copper is calculated as Cu_2O (see text). Not shown in the table are Ba, Sr, Cr, Ni, Sb, Bi, and Ag, which also were analyzed, but due to constantly low concentrations (<0.01 wt.%) were decided to be of no significiance.

bubbles at top (Fig. 19.1). The slag is scarred and deeply ruptured. The weight is ca. 10 kg. Similar large slag chunks were also found at Atheniou, Kition, and Enkomi. The slag is comparable with those from Medieval matte smelting from Oman (Hauptmann 1985). It was cast from the furnace in a fore hearth and solidified after an insufficient separation of a cake composed of matte and copper. Four samples were taken from different spots of the slag: 1/1a) bottom of the slag, negative impression; 1/1b) lower sidewall of the slag; 1/1c) top of slag; 1/1d) top slag.

CY-1/2 (Fig. 19.2). Kition, 1963, Area I, Square γ4, Room 41. Fragment of a plate-like tap slag with flow structures, dense silicate slag, thick efflorescence of Fe-sulfates that penetrate deep into the slag. Weight 0.5 kg. Note charcoal inclusions.

CY-1/3 (Fig. 19.3). Kition, 1979, Area II, Square Δ5, Room 12, Floor III. Semi-globular chunk, fist-sized, with irregular surface. Section shows texture like breccia with inclusions of Fe-sulfates, hostrock, slag fragments, and charcoal. Weight ca. 1 kg.

CY-1/4. Kition, Area II, Room 15, Floor II. Slag layer around tuyère with inner diameter of 2.5 cm. Dense heavy layer of ca. 0.5 cm thickness.

CY-1/6 (Fig. 19.4). Kition, Area II, Room 15, Floor II. Irregular small chunks, rough surface, attached to tuyères. Similar to **CY-1/4**. Section shows high porosity with mm-sized gas bubbles. Remarkably heavy, weight 0.1–0.2 kg.

CY-2/2. Enkomi, 1953, Area III, T-Δ, 58–60, between Floors I and II. Plate-like slag samples (Ø 10 cm), with humps at the bottom. Inclusions of quartz and goethite (see Fig. 19.5).

Results and Discussion

Chemical Composition of the Slag Samples

Main components of all six slag samples analyzed are iron oxide and SiO_2 that are close to the stoichiometric composition of the mineralogical phase of fayalite ($Fe_2^{2+}SiO_4$). As this phase is overwhelmingly the main constituent in the slags, Fe_{tot} was calculated as FeO (Table 19.1). Admittedly, this is not fully correct, as many slag samples also contain magnetite ($Fe^{2+}Fe^{3+}_2O_4$) as another phase.

All of the samples are virtually free of MnO. The concentrations of this oxide are below 0.1 wt.%, and, herewith, they are in line with compositions described otherwise from the workshops of Kition, Hala Sultan Tekke, Idalion, Kalavassos, Skouriotissa, and Toumba tou Skourou (Bachmann 1982a; Koucky and Steinberg 1982). In comparison with other copper smelting slags, e.g., those from Oman (Hauptmann 1985), Al_2O_3, CaO, and MgO are low. This is especially true for MgO (<2 wt.%), and for CaO (mostly <3 wt.%), which can be seen as a very limited input of host rock of the copper deposits. Similar to Oman, the host rocks of the copper deposits at Cyprus consist of basic volcanics, but in contrast to Oman, slag from the Bronze Age to the Medieval period is consistently higher in these two oxides.

The slag samples are also very low in P_2O_5, which does not exceed 0.2 wt.%. I herewith exclude that material such as bone ash (as found at the workshops at Kition) was added to improve slag formation and properties (Stech 1982). If such a technique would have been applied, the slag from Kition and Enkomi would be comparable in composition to that from Faynan, Jordan. Hauptmann showed (2007) that P_2O_5-containing copper ores as they occur at Faynan would result not only in an enrichment of this oxide in smelting slag up to several wt.%, but also in the metal. Twenty-five analyses of copper ingots from Uluburun, whose origin from Cyprus was proved by lead isotope analyses (Gale and Stos-Gale 2005), did not show evidence for a deliberate input of phosphorus (Hauptmann, Maddin, and Prange 2002).

All the slag samples contain small amounts of sulfur (up to 1.5 wt.%) and herewith point to the presence of sulfidic materials during the metallurgical operation. The concentrations of copper are worth discussing in detail. In sample CY-1/1 from Kition, the bulk composition of the upper part of the bowl slag shows Cu <1 wt.%, while the metal increases up to 26 wt.% at the bottom of the sample. This indicates an enrichment of copper in the lower part of the slag, but it also shows that the metal—in whatever state—was far away from being separated from the slag body.

Texture and Phase Content

Major constituents in all slag samples investigated are fayalite ($Fe_2^{2+}SiO_4$), wüstite ("FeO"), and Fe-rich spinel that shows compositions between magnetite ($Fe^{2+}Fe^{3+}{}_2O_4$) and hercynite ($Fe^{2+}Al_2O_4$). In none of the slag we discovered remains of possibly added fluxing agents such as sand or quartz.

The shape of the fayalite crystals varies from spinifex to porphyritic (so-called hopper olivine), lattice and platy texture. All these textures reflect the history of a changing, basically rapid cooling of an Fe-rich silicate slag from a hot, super-cooled liquid. Such crystallographic features are frequently described in detail in archaeometallurgy, i.e., to set up a classification of cooling rates or so, but all these discussions show that it is hardly possible to use them for more than a general description. It is just in sample CY-1/1 where spinifex texture was observed at the top of the slag (Fig. 19.6) and hopper-fayalite at the side wall (Fig. 19.7), which points to the trivial fact that the slag was cooled outside of a furnace in a tapping hole. The top was cooled rapidly, and the side wall cooled a little bit slower. Also, slag samples CY-1/2 and CY-2/2 cooled rapidly outside a furnace.

Wüstite, in varying amounts, forms dendritic skeletons in many of the slag bodies. It is difficult to differentiate between this Fe-oxide and magnetite, but the latter preferably forms euhedral crystals and agglomerations. Magnetite sometimes seems to be associated with finely dispersed Fe-sulfide, and it marks internal pillow-like cooling rims of different slag layers or at the surface of slag (Fig. 19.8). In slag sample CY-1/1, magnetite is an excellent marker to make the inhomogeneous internal structure visible where the slag body is built up. While the upper part of the slag contains only small amounts of magnetite and is solidified from a rather homogeneous fayalitic melt, it is mainly the bottom of the slag where we note only partially dissolved mm-sized agglomerations and sharply edged rims of magnetite. Here also, abrupt changes in the crystal shape of fayalite occur. Magnetite agglomerations can contain inclusions of chalcocite and copper (Fig. 19.9). These magnetite agglomerations are in contradiction to wuestite in the upper part of the slag.

Figure 19.6. Sample CY-1/1d, Kition. Section from the surface of slag no. CY-1/1. Fayalite as the main constituent crystallizes in long and thin needles in a characteristic so-called spinifex texture. This term was coined in petrology for volcanic rocks that usually cooled rapidly. The same is true, of course, for such parts of a slag that were in contact with air after tapping. Surface-polished thin section, transmitted light, crossed nicols. Scale: 500 µm.

Figure 19.7. The same slag as shown in Figure 19.6 shows fayalite in a different shape and habitus: due to a lower cooling rate, the phase forms thicker crystals in the form of hopper olivines. The section was taken from the side wall of the slag chunk, which, after tapping, was protected from quenching by its contact in a tapping hole. Surface-polished thin section, transmitted light, crossed nicols. Scale: 500 µm.

Figure 19.8. Sample CY-2/2a, Enkomi. Fayalitic slag high in magnetite (gray crystals). Note the close association of Fe-sulfides (white, randomly distributed) with magnetite. At the upper part of the image, magnetite and hematite (white) form a massive cooling rim typical for solidification of an iron-rich silicate liquid under a surplus of oxygen. Mounted sample, reflected light, oil immersion. Scale: 100 μm.

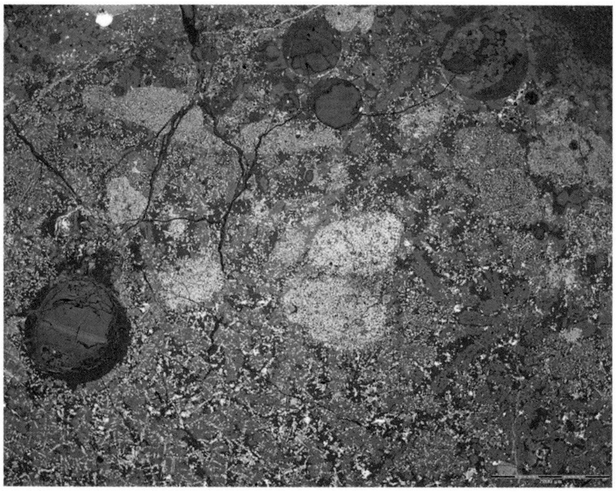

Figure 19.9. Sample CY-1/1a, Kition. The image shows partially dissolved magnetite agglomerations in a fayalitic matrix at the bottom of the bowl slag. Finely dispersed are globules of chalcocite and remaining particles of copper. Most of the copper is transformed to cuprite. Gas bubbles are filled with Fe-hydroxides. Mounted sample, reflected light. Scale: 2,000 μm.

Figure 19.10. Sample CY-1/6, Kition. The slag is built up almost completely by thick dendrites of magnetite. The crystals are rimmed by fine needles of delafossite among a glassy silicate matrix (black). Copper prills (light) show inclusions of cuprite. The phase association points to an only slightly reducing condition during its formation and is compatible to slags near a tuyère. Mounted sample, reflected light, oil immersion. Scale: 100 μm.

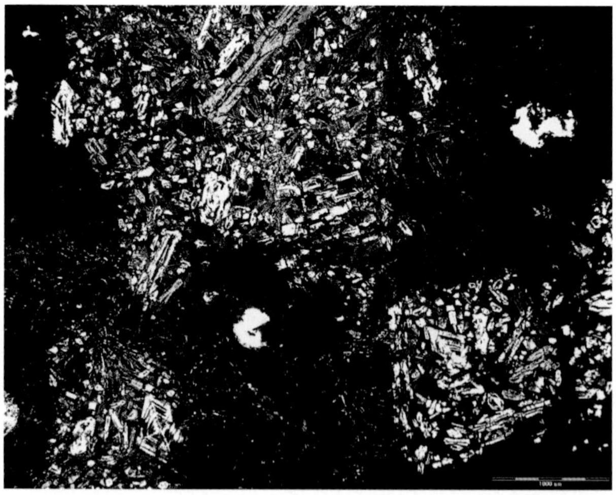

Figure 19.11. Sample CY-1/3, Kition. Breccia-like slag with angular inclusions of fayalitic liquid embedded in a limonitic matrix (black). The angular fayalitic inclusions are rich in Cu-sulfides. They result from crushing slag in order to extract copper and/or matte for further treatment. Surface-polished thin section, transmitted light, crossed nicols. Scale: 1,000 μm.

Slag samples CY-1/4 and CY-1/6 are built up almost completely by magnetite. The crystals are rimmed by fine needles of delafossite ($Cu^{1+}Fe^{3+}O_2$) (Fig. 19.10), and frequently we observed copper inclusions. Due to the comparably high oxygen pressure in the gas atmosphere in a smelting furnace near the tuyères (Hauptmann 2007), where these slags were formed, fayalite is not stable and was not determined.

The semiglobular fist-sized sample CY-1/3 is heterogeneously composed. It shows a breccia-like texture interspersed with inclusions of charcoal and iron hydroxides visible with the naked eye (see Fig. 19.3). Microscopic observations revealed angular fragments of a fayalite slag with crystals precipitated from a liquid, and it is full of Cu-rich sulfides (Fig. 19.11). The texture of the sample indicates a chunk of crushed slag that, together with charcoal, was baked in the soil. The texture reminds one of slag-breccias found from the processing of iron in La Tène–period Black Forest sites and in Siegerland/Germany (Gassmann, Rösch, and Wieland 2006; unpublished data, Deütsches Bergbaü-Müseüm Bochüm).

Cuprite, Copper, and Copper-Sulfide

Due to the heavy corrosion to which the slag was exposed, fissures and cracks, which are a characteristic macroscopic feature and which created a decomposition of mineral phases, continue into the micro level. This is true especially for Fe-oxides, and it is true to a lesser extent for Fe-silicates, which are partly decomposed to Fe-hydroxides. In the bowl slag (CY-1/1) copper is mostly decomposed to cuprite (Cu_2O), atacamite, and malachite. The bottom part of the bowl slag (CY-1/1) is especially infiltrated by such red-colored veins of secondary cuprite, i.e., cuprite that was *not* precipitated from a liquid (Fig. 19.12). The oxide also occurs inside of large, crescent-shaped Cu-sulfides like chalcocite, but it does not form eutectic intergrowth with sulfide and/or copper, as it occurs during roast-reaction smelting processes. I suggest this phase association to be a "ghost structure" where copper, in its metallic, state was transformed to oxide by corrosion.

The first reason for this proposition is that many of the mm-sized inclusions of Cu-sulfides

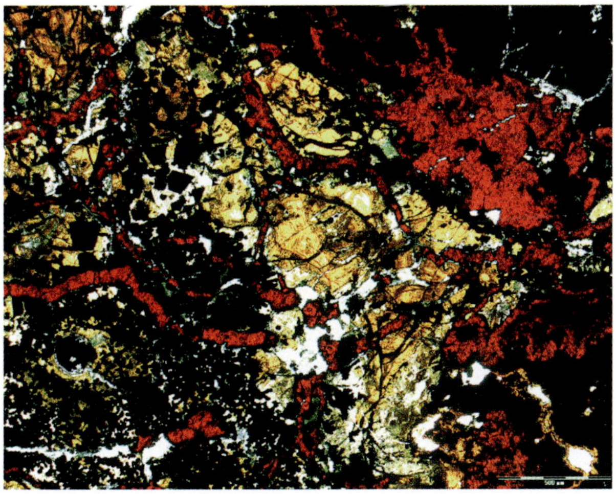

Figure 19.12. Sample CY-1/1b, Kition. The bottom of the slag chunk is infiltrated by secondary cuprite (red), subordinated malachite (green dots), and limonite. These minerals are formed by corrosion. Black dots are magnetite agglomerations (see Fig. 19.9). Surface-polished thin section, transmitted light. Scale: 500 μm.

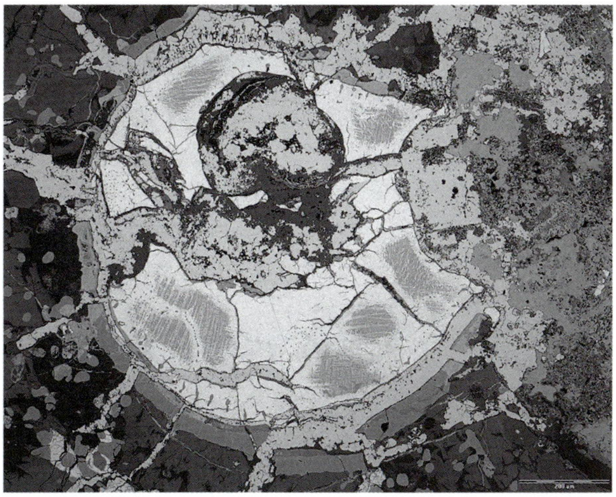

Figure 19.13. Sample CY-1/1b, Kition. Large droplet of matte from the bottom of the slag with a composition near chalcocite (~ Cu_2S). The exsolution texture is due to thin lamellae of bornite (Cu-Fe-sulfide). The core of the matte droplet consists of (secondary) cuprite, formed by corrosion from metallic copper. Mounted sample, reflected light. Scale: 200 μm.

are associated with metallic copper, and they once formed a more or less complete envelope around tiny metal prills. The second reason is that the association of cuprite + chalcocite in the slag contradicts the stability of an environment where the slag was deposited. In the Eh-pH-diagrams, such a milieu is given in the system Cu-Fe-S-O-H (Garrels and Christ 1965). After Sato (1992), however, such associations arise from the metastable persistency of sulfide minerals in oxidizing conditions beyond stability boundaries. At temperatures prevailing on the surface, metastable co-existence of chemical substances is common because chemical reaction rates are often so low as to allow phases to coexist metastably. In other words, once a Cu-sulfide phase is formed, it can persist without actively undergoing redox reactions. This is in contrast to metallic copper, which is easily soluble in any solution.

Consequently, the cuprite veins and fillings point to an originally much higher concentration of copper in its metallic state, i.e., a precipitation of a regulus composed of Cu-sulfide *and* copper.

Sulfide Phases

Sulfide phases of the system Cu-Fe-S are omnipresent in slag from Kition and Enkomi. They range in composition from the copper-rich member (chalcocite, Cu_2S, and related modifications), and covellite (CuS) over bornite (Cu_5FeS_4), and chalcopyrite ($CuFeS_2$), to iron rich phases such as pyrrhotite ($Fe_{1-x}S$). In slag sample CY-1/1, an indistinct imprint at the bottom of the slag is visible (see Fig. 19.1), which results from an incomplete separation of a matte-metal regulus from the slag. Such imprints are known in an almost perfect shape from Middle Bronze Age matte smelting at Politiko-*Phorades* (Kassianidou 2003), and they are also known from other periods of copper production: slags from EBA Shahr-i Sokhta, Iran (Hauptmann, Rehren, and Schmitt-Strecker 2003) reveal such imprints, and they were observed in Bronze Age and Medieval copper smelting in Oman as well (Hauptmann 1985).

All the sulfides show characteristic exsolution textures of high-temperature origin, which were described in detail elsewhere (Schlegel and Schüller 1952; Hauptmann 1985; Schreiner 2002; Metten 2003).

While all of the slags investigated are characterized by inclusions of Cu-rich sulfides such as chalcocite, covellite, and partially with exsolutions of bornite, two most interesting textural features occur in the slag chunk CY-1/1:

1. The slag is interspersed all over by <1-mm-sized yellowish Fe-rich sulfides (see Fig. 19.9). They are xenomorphous and are enriched in the residual parts of the melts. Mostly, they are associated with wüstite/magnetite, and they solidified among the last phases during the cooling of the slag. These sulfides are of chalcopyrrhotite composition, i.e., pyrrhotite with exsolutions of chalcopyrite. As minor constituents we also identified chalcocite (~ Cu_2S) and covellite (CuS), but these could be due to weathering processes.
2. The largest inclusions are droplets of Cu_2S and CuS at the bottom of the slag (Fig. 19.13). They are 10–100 times larger than the Fe-sulfide inclusions. Occasionally, they contain exsolution lamellae of bornite and corroded remains of metallic copper. Obviously, sulfides with a bulk composition like that were the main products of the smelting process.

These inclusions are evidence for a smelting process producing matte with a fraction of metal. The principles of matte smelting are based upon quick liquefaction of sulfides. Cu-rich sulfides melt at temperatures around or below 900°C (chalcopyrite: ca. 880°C, Dutrizac 1976; covellite: 813°C, Kullerud and Yund 1960), while the liquidus temperature of $Fe_{1-x}S$ is at 1170°C (Dutrizac 1976). Oxidic copper ores are quickly reduced to metal at ca. 800°C, and this is liquefied at 1086°C. Silicate slag formation usually needs a longer time period of heat access so that liquefied sulfides and metal may be separated soon from the slowly, successively liquefied slag. To the contrary of modern flotation ore slurry, we know from the archaeological record that coarse-grained ores were smelted in antiquity. This enabled a formation of larger sulfidic droplets that quickly sank to the bottom of the slag, supported by the gravity effect of Cu-sulfides, which are of a higher density than Fe-sulfides.

During the rapid cooling of a slag, equilibria conditions are hardly reached, and slag phases are often quenched in metastable conditions. Nevertheless, their development during cooling usually follows equilibrium conditions (Keesmann 1989). According to the system Cu-Fe-S (Schlegel and Schüller 1952), chalcocite/bornite$_{ss}$ coexist (and form phase associations under non-equilibria) on the one hand with metallic copper, and on the other hand with $Fe_{1-x}S$. Metallic copper ("hair copper") segregates from chalcocite/bornite$_{ss}$ by dissociation and slight desulfurization; but due to the existence of wuestite in the slag, we exclude the formation by roast-reaction. Copper associated with Cu-sulfides in sample CY-1/1, also present in "ghost structures," clearly exceeds the amount of "hair copper" and suggests a smelting whenever of a mixed sulfide-oxide charge.

Also, Fe-rich sulfide may coexist with Cu-rich sulfide by exsolution in the solid state (Fig. 19.14). Nevertheless, it came as a surprise that Cu- and Fe-rich sulfides occur continuously side by side in the same samples. The sulfides are clearly separated by their spatial distribution, and the texture excludes an origin from a homogeneous liquid. This situation indicates that they were incorporated into the slag at different moments or time periods, and perhaps for different purposes.

Lead Isotope Analyses

The results of two lead isotope analyses from two slag samples are shown in Table 19.2.

The isotope ratios of slag chunk CY-1/1b from Kition matches the isotopic composition of the ore district at Apliki-Skouriotissa-Mavrovouni (Gale and Stos-Gale 2005). This is at a distance of some 80 km from Kition, in the northeastern foothills of the Troodos mountain range. This ore mining district was one of the major sources for copper. Based on their databank of lead isotope analyses, Stos-Gale et al. (1997) and Gale (1999) concluded that all oxhide ingots from various localities in the Mediterranean, which were produced from 1250 B.C. onward perhaps for 100 years, were produced from Cypriot ores of this mining district, most probably from Apliki itself. They suggested, in addition, that ores from this location were also smelted to a small extent only, and perhaps for assaying purposes in the Late Bronze Age urban settlements of Kition, Enkomi, and others.

Slag sample CY-2/2a from Enkomi is consistent with the isotopic composition of ores from the Kalavassos mine and almost consistent with a slag from the smelting site at Petromoutti in the southwestern part of the Troodos Mountains, some 30 km from Kition, but some 80 km from Enkomi (Gale and Stos-Gale 2005).

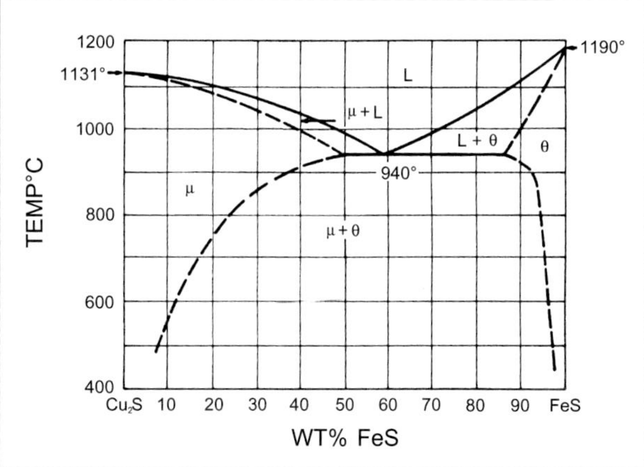

Figure 19.14. Binary phase diagram of $FeS\text{-}Cu_2S$ (after Chang, Lee, and Neumann 1976, fig. 3). This system is a section of the system Cu-Fe-S that describes in detail most of the compositions of matte produced from sulfidic copper-(iron-)ores. The binary section shows that—under equilibrium—a Cu_2S-rich matte may coexist with Fe-sulfide. In sample CY-1/1, however, the distribution and texture of Cu-rich and of Fe-rich sulfides excludes an origin from one liquid. L = liquid; μ = digenite (~ chalcocite; $Cu_{2-x}S$); θ = pyrrhotite ($Fe_{1-x}S$).

Sample Number	Ratio		
	$^{208}Pb/^{206}Pb$	$^{207}Pb/^{206}Pb$	$^{204}Pb/^{206}Pb$
CY-1/1b	2.077	0.8436	0.0542
CY-2/2a	2.051	0.8230	0.0527

Table 19.2. Lead isotope ratios of two slag samples from Kition (CY-1/1) and Enkomi (CY-2/2a).

Conclusion

It was shown that the metallurgical debris found in the Late Bronze Age urban settlements of Kition and Enkomi, which was named with the (preliminary) term "furnace conglomerate," consists of iron silicate slag. The consistency of this material was considerably affected, both on the surfaces and in its inner part as well, by severe alterations after its deposition in the ground.

The analytical data, despite the fact that these are in line with well known fayalitic copper smelting slags, do not provide clear evidence that the slag samples necessarily derived from such an *ore smelting* operation. This seeming contradiction was observed previously, and it led to repeated confusion among numerous scholars. Stech, Maddin, and Muhly (1985), and Zwicker (1985), who investigated metallurgical material from Kition, in fact, discussed different possibilities and models of copper smelting operations at Kition, but copper ores were not found in the metal workshops, and the presence of roasting products was just assumed, but not proven.

What happened at Kition? If the slags do not result from a smelting operation, from which process do they come? It was suggested, e.g., by Stech (1982) and Bachmann (1982a), that while smelting ore would have been performed at the smelting sites near the mines, copper-rich ores, "impure copper and copper bearing conglomerate" would have been sent to the cities for further metallurgical treatment, i.e., for smelting or melting.

This model can be explained more precisely by this study. It could be shown that not only matte, but matte with changing amounts of metallic copper was produced in the workshops of Kition and Enkomi as well. Furthermore, it can be proved that slags high in Cu-(Fe-)sulfides were crushed and re-melted again. In addition, it was observed that there are slag samples (CY-1/1) of heterogeneous composition that point to the use of a variety of components in the charged material. The texture shows that these components—Fe-sulfides, Cu-sulfides, magnetite, and wuestite—are parts of original slag grains that were re-melted.

We know from other Late Bronze Age smelting sites and copper workshops such as Acqua Fredda (Metten 2003; Cierny 2008), and Mitterberg (Eibner 1982), but also from many other sites from other periods, e.g., the Early Bronze Age metal workshop at Khirbet Hamra Ifdan in the Faynan district (Levy et al. 2002; Hauptmann 2007), that crushing slag was a firmly established step in metal production throughout the ages. The purpose of such an operation was to mechanically extract inclusions of copper or copper-sulfide from slag, because an ideal separation from slag was not possible from the very beginning of extractive metal production in the Chalcolithic period up until the Medieval period (Hauptmann 1985; Bartels et al. 2007) and later, as shown by traditional copper smelting in Japan at the beginning of the 19th century A.D. (Tsuna [1801] 1983, [1801] 1984).

To go back to the Bronze Age, it is especially elucidated by Khirbet Hamra Ifdan that slags with entrapped copper prills and full of cuprite (Cu_2O), produced at various smelting sites at a distance of several kilometers, were transported to this central metal factory and were crushed and re-melted *inside* the settlement again to extract the metal. The composition of the slag produced by such a re-melting process chemically does not show any differences to overall smelting slags from "primary" smelting sites, except a lower concentration in copper, of course.

In summary, it is suggested that the slags found in the workshops of Kition and Enkomi may be attributed at least in part to a re-melting of Cu-rich slag that was collected together with pure matte and/or copper at the "primary" smelting sites and was transported to the urban settlements for further treatment. The lead isotope analyses confirm a statement of Stos-Gale, Maliotis, and Gale (1998) that it would be too monolinear in thinking if the nearest mineralization or smelting site would be considered as a source of copper (or copper-rich slag) for a given site.

Acknowledgments

This study was made possible by a generous grant from the Institute for Aegean Prehistory (INSTAP, Philadelphia, USA) to Prof. Dr. Robert Maddin and the author of this contribution. My thanks go to Prof. Maddin who continuously followed this study by many valuable discussions and suggestions. Lead isotope analyses were performed by Dr. Michael Bode, Laboratory for Geochronology, University of Muenster. We gratefully acknowledge the support of the General Director of the Department of Antiquities of Cyprus, Dr. S. Hadjisavvas who gave permission to export samples to Germany for laboratory analysis. Many thanks to Dr. G. Constaninou, former General Director of the Cyprus Geological Survey, and Prof. Dr. Vassili Kassianidou who contributed much to the paper by critical discussions.

References

Bachmann, H.-G. 1980. "Early Copper Smelting Techniques in Sinai and in the Negev as Deduced from Slag Investigations," in *Scientific Studies in Early Mining and Extractive Metallurgy* (*BMOP* 20) P.T. Craddock, ed., London, pp. 103–134.

———. 1982a. "Copper Smelting Slags from Cyprus: Review and Classification of Analytical Data," in Muhly, Maddin, and Karageorghis, eds., 1982, pp. 81–103.

———. 1982b. *The Identification of Slags from Archaeological Sites*, London.

Bartels, C., M. Fessner, L. Klappauf, and F.A. Linke. 2007. *Kupfer, Blei und Silber aus dem Goslarer Rammelsberg von den Anfängen bis 1620: Die Entwicklung des Hüttenwesens von den frühmittelalterlichen Schmelzplätzen im Wald bis zur Metallerzeugung in grossem Massstab am Beginn des 17. Jahrhunderts nach den archäologischen und schriftlichen Quellen* (*Montanregion Hertz* 8), Bochum.

Chang, Y.A., Y.E. Lee, and J.P. Neumann. 1976. "Phase Relationships and Thermodynamics of the Ternary Copper-Iron-Sulfur System," in *Extractive Metallurgy of Copper An International Symposium*, J.C. Yannopoulos and J.C. Agarwal, eds., New York, pp. 21–48.

Cierny, J. 2008. *Prähistorische Kupferproduktion in den südlichen Alpen: Region Trentino Orientale* (*Der Anschnitt Beiheft* 23), Bochum.

Dutrizac, J.E. 1976. "Reactions in Cubanite and Chalcopyrite," *Canadian Mineralogist* 14, pp. 172–181.

Eibner, C. 1982. "Kupfererzbergbau in Österreichs Alpen," in *Südosteuropa zwischen 1600 und 1000 v. Chr.* (*Prähistorische Archäologie in Südosteuropa* 1), B. Hänsel, ed., Berlin, pp. 399–407.

Gale, N. 1999. "Lead Isotope Characterization of the Ore Deposits of Cyprus and Sardinia and Its Application to the Discovery of the Sources of Copper for Late Bronze Age Oxhide Ingots," in *Metals in Antiquity* (*BAR-IS* 792), S.M.M. Young, M. Pollard, P. Budd, and R. Ixer, eds., London, pp. 110–121.

Gale, N.H., and Z.A. Stos-Gale. 2005. "Zur Herkunft der Kupferbarren aus dem Schiffswrack von Uluburun und der spätbronzezeitliche Metallhandel im Mittelmeerraum," in *Das Schiff von Uluburun: Welthandel vor 3000 Jahren. Ausstellungskatalog deutsches Bergbau-Museum*, Ü. Yalçin, C. Pulak, and R. Slotta, eds., Bochum, pp. 117–131.

Garrels, R.M., and C.L. Christ. 1965. *Solutions, Minerals, and Equilibria* (*Harper International Student Reprint*), London and Tokyo.

Gassmann, G., M. Rösch, and G. Wieland. 2006. "Das Neuenbürger Erzrevier im Nordschwarzwald als Wirtschaftsraum während der Späthallstatt- und Frühlatènezeit," *Germania* 84 (2), pp. 273–306.

Given, M., and B. Knapp. 2003. *The Sydney Cyprus Survey Project: Social Appoaches to Regional Archaeological Survey* (*Monumenta Archaeologica* 21), Los Angeles.

Hauptmann, A. 1985. *5000 Jahre Kupfer in Oman.* 1: *Die Entwicklung der Kupfermetallurgie vom 3. Jahrtausend bis zur Neuzeit* (*Der Anschnitt Beiheft* 4), pp. 1–137.

———. 2007. *The Archaeometallurgy of Copper: Evidence from Faynan, Jordan*, Heidelberg.

Hauptmann, A., R. Maddin, and M. Prange. 2002. "On the Texture and Composition of Copper and Tin Ingots Excavated from the Shipwreck of Uluburun," *BASOR* 328, pp. 1–30.

Hauptmann, A., T. Rehren, and S. Schmitt-Strecker. 2003. "Early Bronze Age Copper Metallurgy at Shahr-i Sokhta (Iran), Reconsidered," in *Man and Mining. Studies in Honor of Gerd Weisgerber on Occasion of His 65th Birthday* (*Der Anschnitt Beiheft* 16), Th.

Stöllner, G. Körlin, G. Steffens, and J. Cierny, eds., Bochum, pp. 197–213.

Karageorghis, V., and V. Kassianidou. 1999. "Metalworking and Recycling in Late Bronze Age Cyprus—The Evidence from Kition," *OJA* 18 (2), pp. 171–188.

Kassianidou, V. 2003. "Archaeometallurgy: Data, Analysis, and Discussion," in Given and Knapp, eds., 2003, pp. 214–221.

Keesmann, I. 1989. "Chemische und mineralogische Detailuntersuchungen zur Interpretation eisenreicher Schlacken," in *Archaeometallurgy of Iron. Proceedings of the Symposium of the Comité pour la siderurgie ancienne de l'UISPP. Liblice 1987*, R. Pleiner, ed., Prague, pp. 17–34.

Knapp, B., V. Kassianidou, and M. Donnelly. 1999. "Excavation at Politiko-*Phorades*," *RDAC* 1999, pp. 125–146.

Koucky, F., and A. Steinberg. 1982. "The Ancient Slags of Cyprus," in Muhly, Maddin, and Karageorghis, eds., 1982, pp. 117–141.

Kullerud, G., and R.A. Yund. 1960. "Cu-S System," *Geological Society of America Bulletin* 71, pp. 1911–1912.

Levy, T.E., R.B. Adams, A. Hauptmann, M. Prange, S. Schmitt-Strecker, and M. Najjar. 2002. "Early Bronze Age Metallurgy: A Newly Discovered Copper Manufactory in Southern Jordan," *Antiquity* 76, pp. 425–437.

Lohmann, H. 2005. "Zum Forschungsstand des prähistorischen und antiken Blei-Silberbergbaus im Laurion," in *Anatolian Metal* III (*Der Anschnitt Beiheft* 18), Ü. Yalçın, ed., Bochum, pp. 105–136.

Metten, B. 2003. "Beitrag zur spätbronzezeitlichen Kupfermetallurgie im Trentino (Südalpen) im Vergleich mit anderen prähistorischen Kupferschlacken aus dem Alpenraum," *Metalla* 10 Heft 1 (2), pp. 1–122.

Muhly, J.D. 1991. "The Development of Copper Metallurgy in Late Bronze Age Cyprus," in *Bronze Age Trade in the Mediterranean. Papers Presented at the Conference Held at Rewley House, Oxford, in December 1989* (*SIMA* 90), N.H. Gale, ed., Göteborg, pp. 180–196.

Muhly, J.D., R. Maddin, and V. Karageorghis, eds. 1982. *Early Metallurgy in Cyprus, 4000–500 B.C. Acta of the International Archaeological Symposium, Larnaca, Cyprus, 1–6 June 1981*, Nicosia.

Rothenberg, B., ed. 1990. *Researches in the Arabah 1959–1984* II: *The Ancient Metallurgy of Copper*, London.

Sato, M. 1992. "Persistency-Field Eh-pH Diagrams for Sulfides and Their Application to Supergene Oxidation and Enrichment of Sulfide Ore Bodies," *Geochimica and Cosmochimica Acta* 56 (8), pp. 3133–3156.

Schlegel, H., and A. Schüller. 1952. *Die Schmelz- und Kristallisationsgleichgewichte im System Kupfer-Eisen-Schwefel und ihre Bedeutung für die Kupfergewinnung* (*Freiberger Forschungshefte Reihe* B), Berlin, pp. 3–32.

Schreiner, M. 2002. *Mineralogical and Geochemical Investigations into Prehistoric Smelting Slags from Tepe Sialk/Central Iran*, Freiberg.

Stech, T. 1982. "Urban Metallurgy in Late Bronze Age Cyprus," in Muhly, Maddin, and Karageorghis, eds., 1982, pp. 105–116.

Stech, T., R. Maddin, and J.D. Muhly. 1985. "Copper Production at Kition in the Late Bronze Age," in *Excavations at Kition* V: *The Pre-Phoenician Levels*, Part I, by V. Karageorghis and M. Demas, pp. 388–402.

Stos-Gale, S., G. Maliotis, and N.H. Gale. 1998. "A Preliminary Survey of the Cypriot Slag Heaps and Their Contribution to the Reconstruction of Copper Production on Cyprus," in *Metallurgica Antiqua. In honor of Hans-Gert Bachmann and Robert Maddin* (*Der Anschnitt Beiheft* 8), T. Rehren, A. Hauptmann, and J.D. Muhly, eds., pp. 235–262.

Stos-Gale, Z.A., G. Maliotis, N.H. Gale, and M. Annetts. 1997. "Lead Isotope Characterization of the Cypriot Copper Ore Deposits Applied to Provenance Studies of Copper Oxhide Ingots," *Archaeometry* 31 (1), pp. 83–123.

Tsuna, M. (1801) 1983. *Kodō Zuroku: Illustrated Book on the Smelting of Copper*, Z. Shirakawa, trans., Norwalk, CT.

———. (1801) 1984. *Kodō-zuroku: Illustrierte Abhandlung über die Verhüttung des Kupfers,* B. Lewin and A. Hauptmann, trans., Bochum.

Tylecote, R.F. 1971. "Observations on Cypriot Copper Smelting," *RDAC* 1971, pp. 53–58.

———. 1982. "The Late Bronze Age: Copper and Bronze Metallurgy at Enkomi," in Muhly, Maddin, and Karageorghis, eds., 1982, pp. 81–103.

Zschocke, K., and E. Preuschen. 1932. *Das urzeitliche Bergbaugebiet von Mühlbach-Bischofshofen* (*Materialien zur Urgeschichte Österreichs* 6), Vienna.

Zwicker, U. 1985. "Investigations of Samples from the Metallurgical Workshops at Kition," in *Excavations at Kition* V: *The Pre-Phoenician Levels*, Part I, by V. Karageorghis and M. Demas, Nicosia, pp. 403–429.

Zwicker, U., H. Rollig, and U. Schwartz. 1972. "Investigations on Prehistoric and Early Historic Copper Slag from Cyprus (Preliminary Report)," *RDAC* 1972, pp. 34–45.

CHAPTER

20

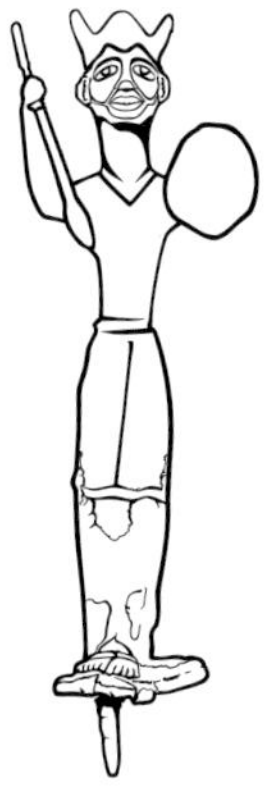

The Metallurgy of Iron during the Early Years of the Iron Age

Robert Maddin

I first became aware of Prof. J.D. Muhly in 1972 on reading an article he had written for *The American Scientist*, a publication of the scientific research society, Sigma Xi. He wrote of the trade in metals (copper and tin) during the Bronze Age in the eastern Mediterranean. I found the article fascinating and was even more interested to learn that he was at the same institution, the University of Pennsylvania; his office was only about 350 meters from mine. At that time, my interests in early metallurgy were nascent; my primary commitment was to studies in materials science. His publication, however, aroused my interest in the early use of metals. Attempts to make contact by telephone were unsuccessful, so I wrote a letter and received an answer some weeks later; he was on sabbatical leave in Athens. He suggested that we meet for lunch when he returned in the fall. We had that lunch in September of 1972 and have collaborated ever since. Understanding trade in the oddly shaped (oxhide) copper ingots was among the first studies we undertook; soon we branched into many different aspects of metals and alloys. It immediately became apparent that his phenomenal memory of the ancient literature, even to the details of the specific areas from the excavations where finds were made, is astounding but is just one of the many aspects of his stature as a great scholar. We have worked together for more than 35 years and still continue today. In addition to being a great scholar, he is an entertaining traveling companion, a bon vivant and an excellent raconteur. I welcome the privilege of contributing to a festschrift for him.

As the title indicates, the topic is about the metallurgy of the early years of the Iron Age. To fully understand why iron began to replace bronze during the early years of the Iron Age, it is necessary to also discuss the economic, social, and cultural as well as political contributions made from all of these influences in addition to knowledge of the technological developments. This brief discourse simply draws attention to one aspect of the technological developments, that of how increasing the strength may have helped lead to the adoption of iron. The enhancement of the strength is best shown through metallographic examination of excavated iron artifacts. Still other important technological contributions would be examinations of the how iron was actually smelted. Reports such as those from recent excavations are now being published (Veldhuijzen and Rehren 2007). There are, however, few reports of the metallographic examination of excavated iron artifacts. This relative absence simply may be because most excavated iron artifacts, especially from the Middle East in the time period 1500–1000 B.C.E., have corroded beyond possible metallographic examination. On occasion, and for reasons not fully understood, iron artifacts that are not completely corroded, with much viable iron, appear. At times, corroded artifacts are found in which remnant iron grains are observed under high magnification merging into their corroded relics (ghosts). On still other (rare) occasions, iron artifacts with only a veneer of corrosion have been uncovered. Useful information can be extracted under all of these conditions (Knox 1963). The discussion here is concerned primarily with metallographic results that are or should be uncontested; doubtful interpretations are noted.

Iron is a fascinating element. Its atom arrangement changes with a change in temperature. At temperatures just below its melting point (1537°C), the crystal structure is body-centered cubic and stays so until 1380°C where it changes to a face-centered cubic and then back again to the body-centered structure at 910°C, remaining in that arrangement below that temperature. This behavior, along with introducing carbon atoms as an alloying element, permits manipulation of the iron/carbon alloy to become unusually hard. Although a few other metal elements behave similarly, iron is unique because of the ready availability and abundance of its smeltable ores, as was carbon in the form of charcoal. Copper ores relative to iron ores are scarce, and even scarcer are tin ores from which the tin used to make bronze is extracted.

Copper corrodes slowly, usually into a pleasant adherent green patina, while iron corrodes quickly into a reddish non-adherent powder. In early times, iron could not be cast; even today, special equipment is needed to reach its melting point. Copper and its alloy, bronze, could readily be cast from 1200°C–1050°C, temperatures readily attainable in early times. During the beginning of the Iron Age, the smelted iron was relatively pure with less than about 2% slag remaining after squeezing the slag from the mixture of iron and slag. The resulting bloom was initially harder than pure copper and could be made even harder by hammering. Adding tin to copper made the copper even harder because of the presence of tin, but also because the rate at which the bronze could be hardened by hammering is greater. Hammered bronze (10% tin) is a stronger alloy than hammered bloomery iron. My conclusion is that a most important factor leading to the adoption of iron once the technology was known was not only the ready availability of iron ores but also the production of harder artifacts.

There is general consensus that iron first appears in ca. 1500 B.C.E. in or near the present city of Boğazköy (the ancient capital of the Hittite empire, Hattusas). There are reports of iron well before 1500 B.C.E. (for compilations, see Waldbaum 1978, 1980, 1999). There are, of course, some artifacts made from meteoritic iron experimentally well documented. What sort of iron was the iron of the Hittites? Since the melting point of iron (1537°C) was much beyond what could be attained in 1500 B.C.E., smelting iron ores below that temperature yields a sponge-like mass, the interstices of which are filled with a viscous substance that solidifies into a glass at room temperature (see Fig. 20.1). If attempts are made to hammer this mass at room temperature, the glass shatters and the mass falls apart. The discovery of hammering to squeeze out the slag from the mixture at a temperature where the slag is again viscous was the answer to producing a malleable iron. This discovery we now attribute to the Hittites, although we cannot really be sure of the role played by this culture. Iron from which most of the slag has been eliminated is called bloomery iron. This readily available and easily malleable metal, though soft, was nevertheless far cheaper than other metal artifacts in use before then.

During the period from the introduction of this technology until about the 11th century B.C.E., another development—i.e., how to harden iron by producing the thin layer of "steel"—must have been made. The metallurgy of the process involves allowing carbon atoms to diffuse into the surface layers of the bloomery iron. This most likely happened when the blacksmith continually had to re-heat the tool in order to shape the object. The continuing exposure of the tool in the charcoal fire allowed carbon atoms to diffuse into the surface layers of the iron, thus producing a layer of "steel." While steel is a uniform iron/carbon alloy containing less than 1.9% carbon, we may think of this layer as a layer of steel, and the artifact being forged can be heat treated as one would heat treat steel. The thickness of this "steel" layer is a function of the temperature, the time, and the contact between the iron and the carbonaceous material. The resulting hardness is greatly dependent upon the amount of carbon dissolved. This process is time/temperature dependent, such that the penetration of the carbon atoms into the iron is slow and often unsuccessful. In modern times, accelerators are added to hasten the process. Bloomery iron at room temperature can dissolve very small amounts (~0.001%) of carbon. If, however, the iron is heated to temperatures between 710°C and 910°C, it can dissolve up to 1.9% carbon. Slowly lowering the temperature to below 710°C allows the excess carbon to precipitate out to the equilibrium value of ~0.001%. Lowering the temperature quickly, however, such as by quenching the hot iron with carbon in solution into a cold medium (e.g., water), does not permit time for the carbon atoms to precipitate. This results in a structure far harder than the slowly cooled iron. The amount of hardening thus can be controlled by varying the rate of cooling from the higher temperature. Heat treatment consists of holding a steel (or carburized iron) at temperatures between 723°C and 910°C, quenching it into cold water, followed by reheating the artifact at temperatures of about 500°C. This last step is called "tempering." It renders the artifact more malleable but reduces its hardness. To the best of my knowledge, this "tempering" step has never been metallographically confirmed as having been consciously used in early times.

An associated technology would decrease the time for the diffusion process by preparing a carburized thin sheet of bloomery iron that could be hammer welded to the bloomery tool, thus doubling the surface area in contact with the charcoal, and thus decreasing the time for the diffusion of the carbon atoms. Such a process is thought to have been used to smith the ax/adze (Fig. 20.2) from Protogeometric Sardis, ca. 1000 B.C.E. (Maddin, Muhly, and Waldbaum 1983).

Aside from finding a chance iron ore body containing impurity minerals such as phosphorus and other metal minerals (i.e., nickel, manganese, and chromium) that could be co-smelted with the iron mineral, such elements could not be added to alloy with the iron simply because of the inability to attain the high temperatures necessary for alloying. Ores containing such elements have been reported yielding a harder iron (Piaskowski 1982). There is evidence, however, that the use of bog iron ores high in phosphorus resulted in higher strength iron (Pigott 2003). Though phosphorus is an effective hardener of iron, phosphorus sufficient to harden the iron often results in a more brittle iron.

Limitation in allotted space permits me to cite only those studies I consider to be seminal in understanding the metallurgy of the hardening of iron. In order to revive metallurgical studies of

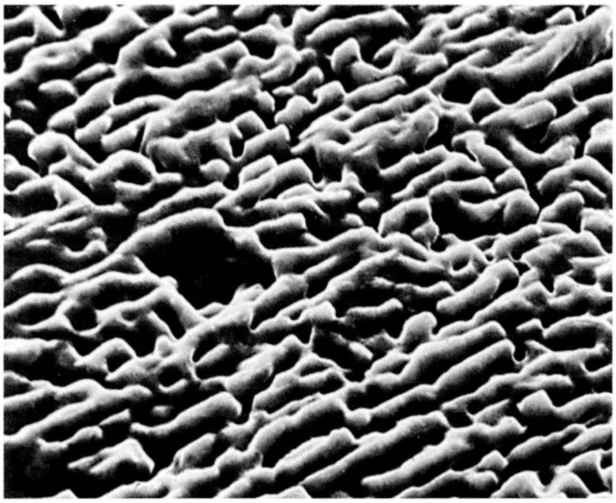

Figure 20.1. Iron reduced in the solid state as observed in the scanning electron microscope (SEM). Original magnification x2400 (pers. comm., Y.K. Rao, University of Washington).

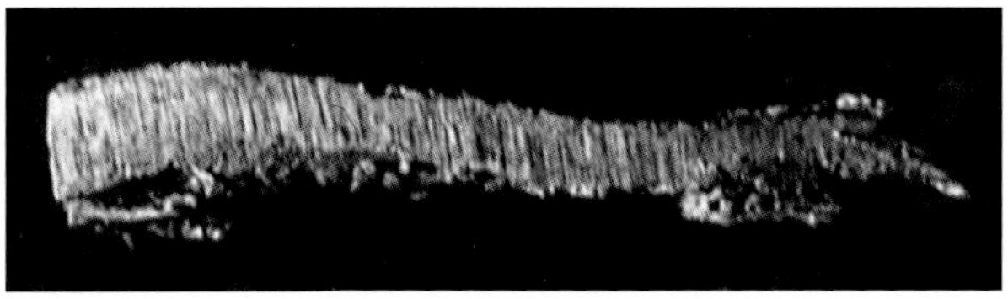

Figure 20.2. Layered structure in ax/adze from Sardis.

iron artifacts, my colleagues J.D. Muhly and T. Stech and I initiated a program in 1975 to examine iron artifacts from excavations of contexts within the range 1500–1000 B.C.E. Unfortunately, iron artifacts from Boğazköy as well from Alaca Höyük (considered to come from this time frame) showed neither viable iron remnants nor observable relic (ghost) structures (Muhly et al. 1985).

We were aware of the report by the Swedish metallurgist, E. Tholander (1971), who showed the presence of martensite in two knives from the site of Idalion (Cyprus) said to be 11th century B.C.E. Martensite is a product of the quenching of a carburized iron or "steel," and, as such, is *prima facie* evidence of pre-existing carburized iron. We examined those knives along with other objects stored in the National Museum of Antiquities in Stockholm. In addition to confirming Tholander's results, we reported the results (Åström et al. 1986) of our examination of 14 other objects including knives, daggers, rods, and hafts of knives. The blacksmith, because of the consistency of the carburized sections, appeared to be aware of the use of the artifact he was constructing. Perhaps even earlier there is the pick excavated by D. Davis from Mt. Adir in Israel (Fig. 20.3; Davis et al. 1985), tentatively assigned a date as the early 12th century B.C.E., that shows a clearly martensitic structure. No archaeological context of this pick has ever been published; hence the date cited must remain uncertain. Hardness measurements were all consistent with a eutectoid, quenched, hardened "steel." Composition at the eutectoid is 0.8% carbon. The microstructure components of many iron artifacts in the process of corrosion showed sections of the complete phases with continuity into the associated parts of the phase already in relic (ghost) form. It therefore was possible to reconstruct the original microstructure. In every knife (nine, all 11th–10th century B.C.E.) the end or near the end, the point, and the middle part of the knife showed moderate to extensive carburization. In many cases there was clear evidence of hammering. We found it striking that in two cases the hafts of the same knives showed no or little carburization. I am inclined to suggest that the blacksmith(s) understood what properties he wanted certain parts of the knife to have. Yet, at other sites on the island, considered contemporaneous, at Paleopaphos-*Skales* and Kition from which we examined metallographically 21 tools as well as weaponry, we observed few cases clearly showing carburization either in metallic iron remnants or in relic structures (Stech, Muhly, and Maddin 1985). Other than the ax from Sardis, there appears to be no case of tools or other objects so far reported to document hardening efforts.

Not until the excavation at Kinneret by Volkmar Fritz (Fritz et al. 1991) near the Sea of Galilee in Israel does a clear case of carburization and quenching appear. We reported a metallographic study of a knife from Stratum V, dated to the 10th century B.C.E., clearly showing the microstructure associated with hardening by carburization followed by quenching (see Fig. 20.4).

J. Waldbaum (1999) updated her earlier compilation of excavated iron to include an impressive summary of more recent finds of iron (her appendices A and B). She cites some intriguing studies that may add to our accounts of deliberate hardening by carburization and quenching, but we must await the publication of these studies.

In his article "How Iron Technology Changed the Ancient World—And Gave the Philistines a Military Edge" (Muhly 1982), Muhly describes the episode from Homer's *Odyssey* where Odysseus and his crew, after inducing the giant Polyphemus (the Cyclops) to drink enough wine to make him drunk, plunge a heated olive branch into his single eye to blind him, enabling Odysseus and his crew to escape. Muhly's description of the event says that, "[s]o graphic a simile could only have been created by someone who had himself watched the blacksmith at work." Scholars agree that *The Odyssey*, although incorporating older material, was first written in the late 8th century B.C.E. This scene clearly depicts a hardening process because little increase in strength comes from quenching a bloomery iron (quenching a bloomery iron does have a minimal affect on the strength by refining the grain). It presents the most graphic description of quenching to be found in the early Greek literature.

From at least the late 8th century but most likely from the time when the blacksmith fashioned the Kinneret knife in the 10th century, carburized tools, occasionally if not routinely, were being produced in other parts of the eastern Mediterranean. The question immediately arises as to why carburization occurred there only and not in nearby regions, such as in other parts of Cyprus or even in other parts of the eastern Mediterranean. One possible

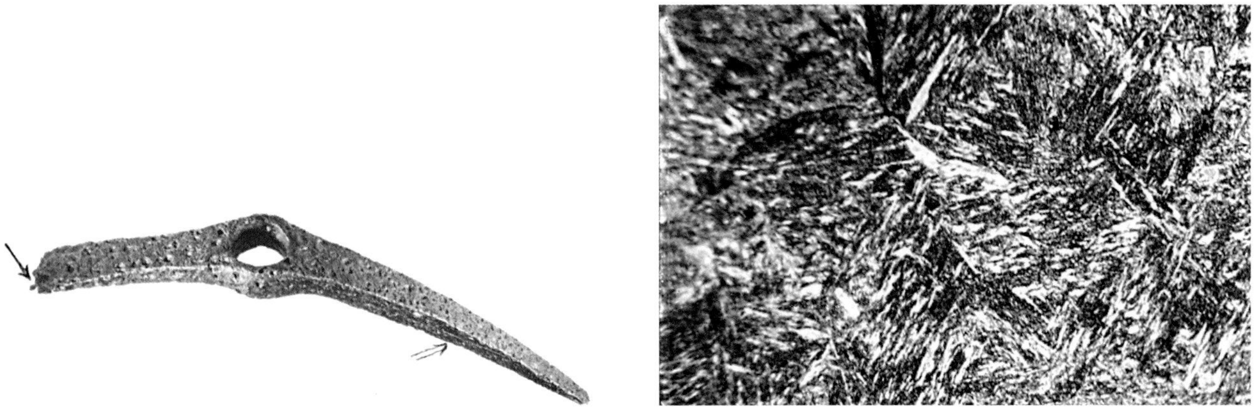

A B

Figure 20.3. Pick (A) from Mt. Adir and its martensitic structure (B).

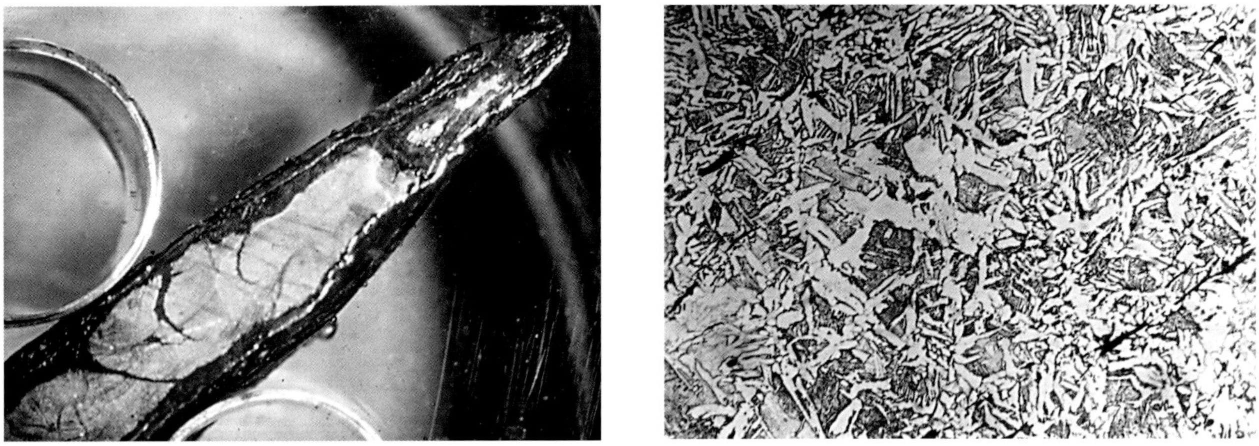

A B

Figure 20.4. Knife (A) from Kinneret and its microstructure (B).

answer may be found in the success, or lack of success, in the carburization process that is so necessary a step in the hardening. Carburization most likely occurred because the blacksmith continually had to immerse the tool into the hot coals to reheat, in order to continue to hammer; he could not effectively hammer a "cold" iron. Each time he plunged the tool into the hearth and left it there to heat, further carburization occurred. Recipes for successful carburization may at first have been a protected technology. Recipes of how to carburize tools, such as that for hardening files, continue to appear, for example, in the Theophilus treatise, "On Divers Arts," by Theophilus Presbyter (ca. 1100 C.E.; Hawthorne and Smith 1963). We must await metallographic studies on many excavated iron artifacts from many other contemporary or nearly contemporary areas to draw firm conclusions.

Why the objects from the tombs in Idalion, Lapithos, and Amathus were in a better state of preservation than other iron objects from other sites, especially in these areas of the eastern Mediterranean, is not understood. Generally, excavated iron artifacts from these areas are in an advanced state of corrosion, while iron from slightly newer contexts such as, for example, 5th century B.C.E. Al Mina, contains much viable iron. An Al Mina chisel (Fig. 20.5; Wheeler, Muhly, and Maddin 1976) clearly has been quenched. Here is an excellent example showing cracks from what must have been an attempt to use the chisel; this is characteristic of the use of freshly quenched martensitic "steel." The adze from Al Mina (Fig. 20.6) shows a carburized layer (Muhly, Wheeler, and Maddin 1977). We do not understand why some iron objects from the same areas show low corrosion while others have been almost totally consumed. By Roman times, however, production of tools fashioned to function became routine (Maddin, Hauptmann, and Baatz 1991; Maddin, Hauptmann, and Weisgerber 1996).

There are many excavated iron objects waiting to be studied metallographically, but unfortunately there appear to be few scientists/scholars willing to undertake such studies. For example, there is the hoard of more than 1,100 iron artifacts in the museum in Augsburg, exceptionally well dated to 9–17 C.E. and published by W. Hübener (Hübener 1973). Among the more than 2,000 items excavated between 1911 and 1972 were many iron shields, javelins, knives, helmets, daggers, nails, chains, and architectural parts and tools for masons, carpenters, and agriculturalists. There are also the current studies yet to be published and some recently published. For example, M. McConchie (McConchie 2004) has published a volume of work on northeastern Anatolian iron from the first millennium B.C.E., which has been recently reviewed by Muhly (2007). The McConchie contribution appears to be one of the few published in recent years containing metallographic observations on iron.

In what has been outlined above, there are large gaps in the data. For example, there is a 500-year period from which we have little or no information, especially from regions to the north and east of the Hittite capital. Undoubtedly, there are and may well be excavations in progress or planned from which iron artifacts will be excavated and undergo laboratory studies; these results may well change our view on the history of the metallurgy of iron during the early years of the Iron Age.

References

Åström, P., R. Maddin, J.D. Muhly, and T. Stech. 1986. "Iron Artifacts from Swedish Excavations in Cyprus," *OpAth* 16 (3), pp. 27–41.

Davis, D., R. Maddin, J.D. Muhly, and T. Stech. 1985. "A Steel Pick from Mt. Adir in Palestine," *JNES* 44, pp. 41–51.

Fritz, V., R. Maddin, J.D. Muhly, and T. Stech. 1991. "The Iron from Kinneret," *Materialy Archeologiczne* 26, pp. 97–102.

Hawthorne, J.G., and C.S. Smith, trans. 1963. "The Hardening of Files," in *Theophilus: On Divers Arts*, Chicago, pp. 93–94.

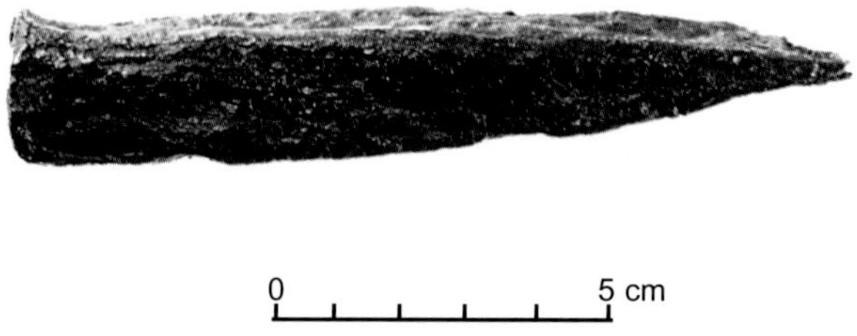

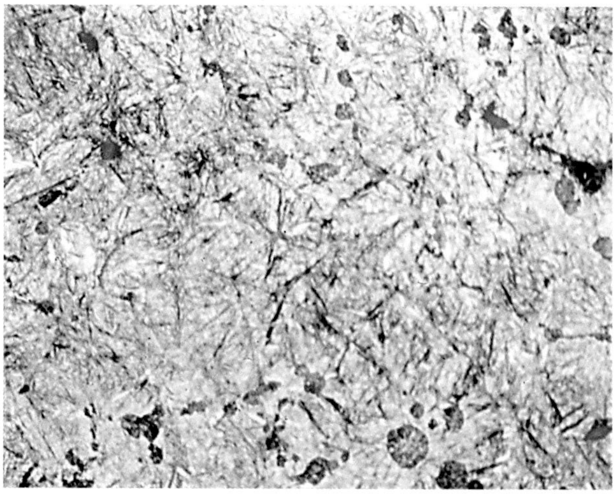

Figure 20.5. Chisel (A) from Al Mina showing martensite structure (B) along with cracks at tip. Ashmolean Museum.

A

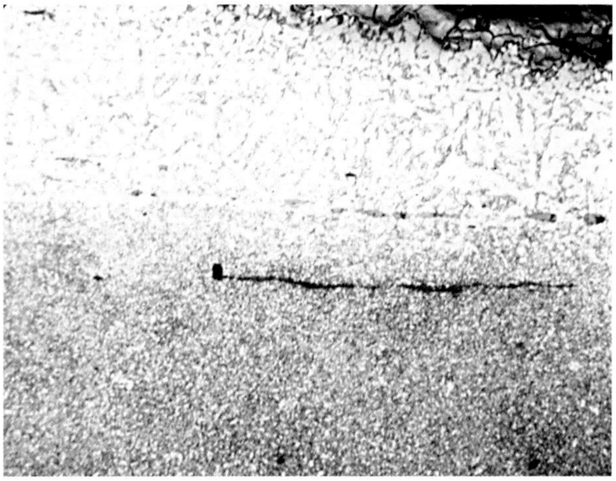

B

Figure 20.6. Adze (A) from Al Mina showing a layer of carburization (B). Ashmolean Museum.

Hübener, W. 1973. *Die römischen Metallfunde von Augsburg-Oberhausen* (*Materialhefte zur Bayerischen Vorgeschichte* 28), Kallmünz.

Knox, R. 1963. "Detection of Iron Carbide Structure in the Oxide Remains of Ancient Steel," *Archaeomertry* 6, pp. 43–45.

Maddin, R., A. Hauptmann, and D. Baatz. 1991. "A Metallographic Examination of Some Iron Tools from the Saalburgmuseum," *SaalbJb* 46, pp. 5–24.

Maddin, R., A. Hauptmann, and G. Weisgerber. 1996. "Metallographische Untersuchungen an römischen Gezähe aus Rio Tinto, Spanien," *Metalla* 3 (1), pp. 27–44.

Maddin, R., J.D. Muhly, and J. Waldbaum. 1983. "Metallurgical Analysis of Early Iron Age Laminated Iron Tool," in *Metalwork from Sardis: The Finds through 1974*, J. Waldbaum, Cambridge, MA, pp. 178–180.

McConchie, M. 2004. *Archaeology at the North-East Frontier* V: *Iron Technology and Iron-Making Communities of the First Millennium* B.C. (*Ancient Near East Studies Suppl.* 13), Louvain.

Muhly, J.D. 1982. "How Iron Technology Changed the Ancient World—And Gave the Philistines a Military Edge," *Biblical Archaeology Review* 8 (6), pp. 40–54.

———. 2007. Review of M. McConchie, *Archaeology at the North-East Anatolian Frontier* V: *Iron Technology and Iron-making Communities of the First Millenium* B.C., Louvain, 2004, in *BASOR* 347, pp. 112–114.

Muhly, J.D., R. Maddin, T. Stech, and E. Özgen. 1985. "Iron in Anatolia and the Nature of the Hittite Iron Industry," *AnatSt* 35, pp. 67–84.

Muhly, J.D., T.S. Wheeler, and R. Maddin. 1977. "An Iron Adze of the Fifth Century B.C. from Al Mina," *Levant* 9, pp. 156–161.

Piaskowski, J. 1982. "A Study of the Origin of the Ancient High-Nickel Iron Generally Regarded as Meteoritic," in *Early Technology*, T.A. Wertime and S.F. Wertime, eds., Washington, D.C., pp. 237–243.

Pigott, V.C. 2003. "Iron and Pyrotechnology at 13th Century Late Bronze Age Tel Yin'am (Israel): A Reinterpretation," in *Man and Mining—Mensch und Bergbau. Studies in Honour of Gerd Weisgerber on Occasion of His 65th Birthday* (*Der Anschnitt Beiheft* 16), T. Stöllner, G. Korlin, G. Steffens, and J. Cierny, eds., Bochum, pp. 365–375.

Stech, T., J.D. Muhly, and R. Maddin. 1985. "The Analysis of Iron Artifacts from Paleopaphos-*Skales*," *RDAC* 1985, pp. 192–202.

Tholander, E. 1971. "Evidence for the Use of Carburized Steel and Quench Hardening in Late Bronze Age Cyprus," *OpAth* 10, pp. 15–22.

Veldhuijzen, H.A., and T. Rehren. 2007. "Slags and the City: Early Iron Production at Tell Hammeh, Jordon, and Tel Beth-Shemish, Israel," in *Metals and Mines: Studies in Archaeometallurgy*, S. La Niece, D. Hook, and P. Craddock, eds., London, pp. 189–120.

Waldbaum, J.C. 1978. *From Bronze to Iron* (*SIMA* 54), Göteborg.

———. 1980. "The First Archaeological Appearance of Iron and the Transition to the Iron Age," in *The Coming of the Age of Iron*, T.A. Wertime and J.D. Muhly, eds., New Haven, pp. 69–98.

———. 1999. "The Coming of Iron in the Eastern Mediterranean: Thirty Years of Archaeological and Technological Research," in *The Archaeometallurgy of the Asian Old World* (*University Museum Monograph* 89), V.C. Pigott, ed., Philadelphia, pp. 27–57.

Wheeler, T.S., J.D. Muhly, and R. Maddin. 1976. "A Steel Tool of the Fourth Century B.C. from Al Mina, Syria," *Levant* 8, pp. 107–112.

PART IV

Trade and Interactions in the History of Metallurgy

CHAPTER

21

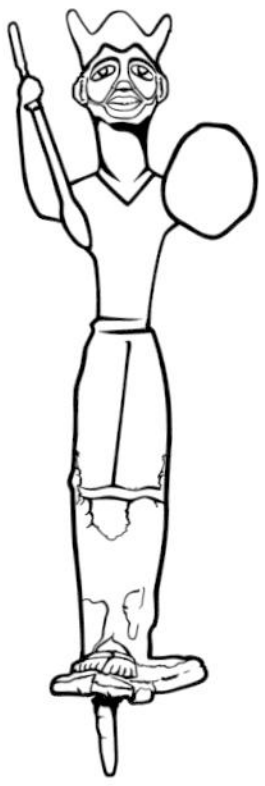

Copper Oxhide Ingots and Lead Isotope Provenancing

Noël H. Gale

The distinguished and multifarious contributions of James Muhly to ancient history, archaeology, and archaeometallurgy have embraced both traditional and very new approaches to research. Muhly long ago perspicaciously advocated that the study of the Late Bronze Age Mediterranean metals trade had to begin with learning all that we could about the copper oxhide ingots, especially *"how they were made, why they were made and where they were made"* (Muhly 1977, 81). Muhly thereby inspired, and made major contributions to, over 30 years of research in this field. He was also quick to appreciate and engage with (Muhly 1983) the new field of applying lead isotope analysis (LIA) to the study of the provenance of lead (Stos-Gale and Gale 1982), silver (Gale, Gentner, and Wagner 1980), and especially copper (Gale and Stos-Gale 1982). Indeed, Muhly (2005, 508) was kind enough to describe the application of LIA to provenancing the copper in oxhide ingots as the second major development in the field of copper oxhide ingot studies during the past 20 years, the first being a major increase in the corpus of known exemplars (especially from the Bronze Age Uluburun and Gelidonya shipwrecks and from Sardinia [Lo Schiavo 2005]).

Another major advance has been the recent intensive study of the metallography and chemical composition of copper oxhide ingots from the Uluburun shipwreck (Hauptmann, Maddin, and Prange 2002). From these studies they argued against the hypothesis of the casting of oxhide ingots directly from molten smelted copper metal from the furnace, because they discovered angular fragments of slag within the body of Uluburun ingots, while the direct use of smelted copper from the furnace should carry with it globular slag fragments, not angular. They found that there was but weak evidence that oxhide ingots were made by melting down bun ingots. Moreover, Hauptmann, Maddin, and Prange (2002, 5) observed microscopic features (internal cooling rims) in some Uluburun ingots, which they write "could be caused by pouring several batches of the metal into a mold to complete an ingot." This supports earlier macroscopic observations by Pulak (1998, 196; 2000, 142) of layered textures, which indicate that at least some of the oxhide ingots were finished by multiple pourings.

Implications for Lead Isotope Provenancing

Problems might ensue for lead isotope provenancing if oxhide ingots were made by several pourings of molten copper into the same mold. In particular, if different batches of copper metal smelted from different ore sources were brought to a central site where the batches were melted down and poured sequentially into the oxhide ingot mold, where they formed a sort of layer cake structure, the lead isotope compositions of single samples that were sawed off the exterior would be representative only of the layer of which they were a part. This situation might be misleading if batches of copper metal from different ore sources with significantly different lead isotopic compositions were utilized.

The existing lead isotope data suggests, however, that such isotopically different batches of copper metal were not used to make oxhide ingots. If they had been, then the existing data for samples sawed off hundreds of analyzed copper ingots would randomly pick up the different batches, and quite different isotopic compositions would be seen. In reality, for the post–1400 B.C. ingots, the limited range of isotopic compositions found matches the range found for Cypriot copper ores only from the Apliki mine region (e.g., Gale 1999, 2006; Gale and Stos-Gale 2005), and it does not reflect the isotopic composition of ores from, say, Sha or Kalavassos, or Limassol Forest, or Mathiati, or any of the other ore deposits around the Troodos (Stos-Gale et al. 1997).

Lead Isotope Analyses of Internal Core Samples

Nevertheless, it seemed important to measure directly the homogeneity or otherwise of lead isotope compositions through the body of some copper ingots. For this Hauptmann and Pulak provided samples taken at various points from cores drilled though ingots from the Uluburun shipwreck. Lead isotope analyses were made on these samples in the Department of Earth Sciences, Oxford, using a Nu Instrument Multicollector Inductively Coupled Plasma mass spectrometer (MC-ICPMS; Belshaw et al. 1998).

Analytical procedures and the numerical data were reported by Gale (2005). In sum, the MC-ICPMS lead isotope data for cores taken through the three Uluburun copper oxhide and one bun ingots support that at least some ingots were made by the successive pouring of different batches of molten copper into the mold. All the MC-ICPMS data further prove that the different batches of copper in a given Uluburun ingot come from the same ore deposit, which has an isotopic composition identical with the earlier thermal ionization mass spectrometry (TIMS) measurements of the lead isotopic composition of surface samples from some 300 Uluburun copper oxhide and bun ingots. Consequently, the evidence derived from these MC-ICPMS determinations of the lead isotope compositions validates the lead isotope provenancing method for determining the ore sources of copper used in making these ingots.

Lead Isotope Provenancing of Post–1400 B.C. Copper Oxhide Ingots

Provenancing of the copper used to make copper oxhide ingots was hampered before 1995 by the relatively small number of lead isotope analyses of ores from Cypriot mines. Before 1995 there existed 33 lead isotope analyses of Cypriot ores made in Oxford together with a further 10 analyses made by Hamelin et al. (1988). The Isotrace Laboratory in Oxford had, before 1997, constructed an overall "Cypriot field" by drawing an ellipse to encompass these 43 analyses, though even then arguments given by Hamelin et al. (1988) and Stos-Gale et al. (1997, 85–86) suggested that this "field" had structure, possibly folding together a number of smaller fields characteristic of individual ore deposits in Cyprus. Moreover, we (Stos-Gale and Gale 1994, 105–108, fig. 3.38; Gale and Stos-Gale 1995, fig. 2) had also pointed out the tendency for oxhide ingots found on Cyprus to plot toward and slightly below the bottom edge of the overall "Cypriot field" in one lead isotope diagram. We had also reported that this position on the edge of the "Cypriot field" was matched by the lead isotope composition of the only two then available ores from the Cypriot Apliki mine given to us by R.F. Tylecote.

We conducted field trips to Cyprus in 1995 and 1996 to increase both the range of Cypriot ore deposits sampled and the number of ore samples from each deposit. Nearly 200 new lead isotope analyses of sulphidic and oxidized ores from 26 copper mines on Cyprus were subsequently published (Stos-Gale et al. 1997; Gale et al. 1997). Lead isotope analyses (LIA) for ores from Cyprus and Lavrion are shown in Figure 21.1, plotted in the usual two diagrams, which represent orthogonal plane sections through the three-dimensional space occupied by LIA data (for details and data for the ores analyzed from Cyprus, see Stos-Gale et al. 1997 and Gale et al. 1997). The LIA for ores from Lavrion are clearly separated from those for any ores from Cyprus. Figure 21.1 shows that the data for ores from the Cypriot Apliki mine fall into a group that is not overlapped by the LIA for other Cypriot ores; nor is the Apliki data overlapped by the LIA data for any other mines in the Mediterranean region or Anatolia or the Near East, generally.

These analyses showed that the mines from different geological regions group in five distinctive isotopic groups, each with a substructure, related to the geological history of the ore formation. It was clear and was explicitly stated (Stos-Gale et al. 1997, 118), that the previous idea of an overall "Cypriot field" had to be given up in favor of point by point comparison of LIA data for artifacts with the individual data for particular Cypriot ore deposits. Stos-Gale et al. (1997) examined the particular case of the provenance of copper for 78 Late Bronze Age copper oxhide ingots found in Cyprus, Crete, Greece, Sardinia, Turkey, and Bulgaria. The data shows that all oxhide ingots then analyzed, dating to 1400 B.C. and later, were made of copper consistent isotopically with only the Apliki mining region in the geographical north of Cyprus.

Figure 21.2 presents the lead isotope analyses of 30 copper oxhide ingots excavated from, or found in, sites in Cyprus, plotted in relation to LIA for Cypriot copper ores in an expanded diagram showing the ores plotting with or near the ingots. The oxhide ingots came from Enkomi (7), Kalavasos-*Ayios Dhimitrios* (4), Maa (1), Maroni (5), Pyla (2), Mathiati (4), and Skouriotissa (7); all date from the Late Cypriot II period. All of these copper oxhide ingots have lead isotope compositions consistent with their production from copper coming from the Apliki deposits alone. Their lead isotope compositions clearly match no other ore deposits in Cyprus. Not shown, for lack of space, is LIA data for LBA copper oxhide ingots from Sardinia, mainland Greece, Crete, Bulgaria, the Cape Gelidonya shipwreck, Anatolia, Syria, Israel, and Egypt—all of which also match the LIA data for ores solely from the Apliki deposits (Gale 1999, 2006).

The new data that we have accumulated both for Cypriot ores and for oxhide ingots (Gale 1999, 2006) shows not an offset from, but a perfect match between the lead isotope compositions of the ingots and particular Cypriot ore bodies. Those who, basing themselves largely on the view that there is no such match, have written about a postulated crisis in lead isotope and metal provenancing studies (Budd et al. 1995, 1996) but no longer have any evidence on which to base their conjectures; there was and is no crisis.

We now refer to the majority of oxhide ingots with lead isotope analyses that overlap those for Apliki ores as post–1400 B.C. Our previous reference to these oxhide ingots as being post–1250 B.C. came from the fact that the great majority of such ingots (from Cyprus, Greece, Turkey, Bulgaria, Sardinia, and Cape Gelidonya), which we analyzed in the earlier stages of our analytical program, did date after approximately 1250 B.C.

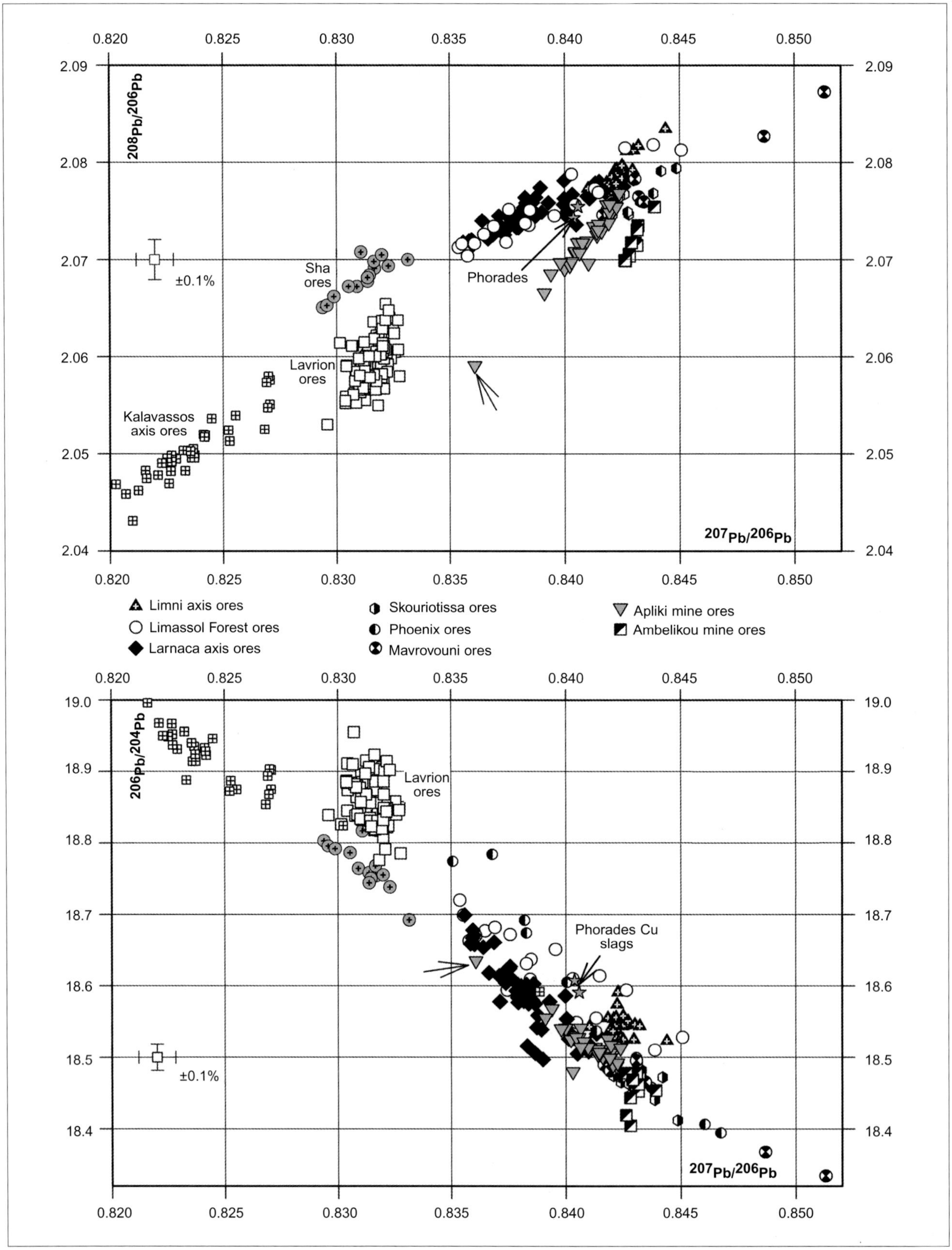

Figure 21.1. Plot of lead isotope analyses for ores from Cyprus and Lavrion (Attica). Axis numbers denote ratios of the number of atoms of lead (Pb) with atomic masses of 204, 206, 207, and 208.

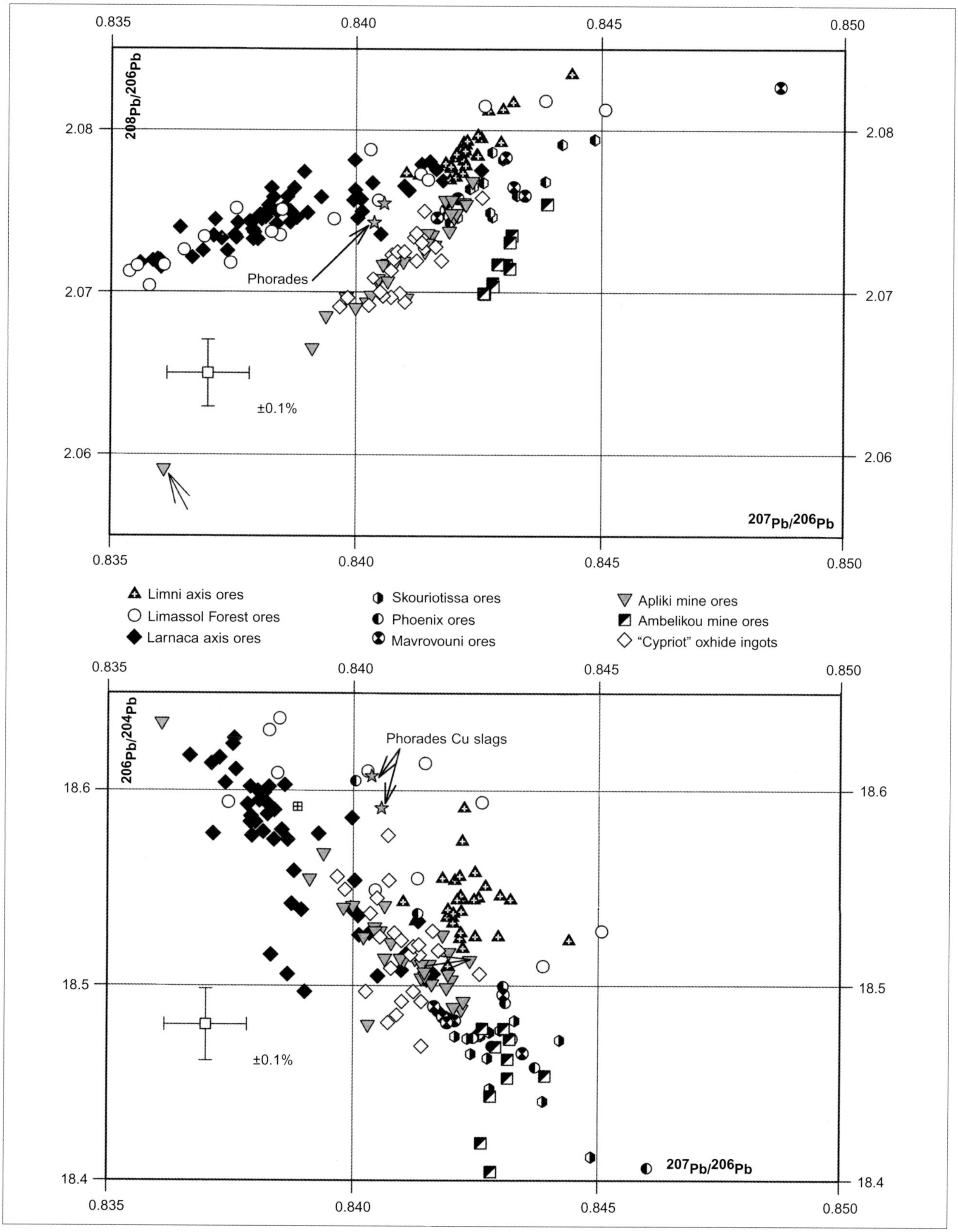

Figure 21.2. Plot of lead isotope analyses of 30 copper oxhide ingots found on Cyprus in relation to lead isotope analyses of ores from Cypriot ore deposits. Axis numbers denote ratios of the number of atoms of lead (Pb) with atomic masses of 204, 206, 207, and 208.

and had lead isotope compositions overlapping those of ores from Apliki, with the exception chiefly of the 19 Late Minoan (LM) IB examples from the villa of Hagia Triada in Crete (Pugliese Caratelli 1945; Gale and Stos-Gale 1986; Cucuzza and Gale 2004). Later work has now shown that LM IB copper oxhide ingots from both Gournia and Mochlos were, however, made of copper from the Apliki ore deposit, so it seems that this deposit was in fact used as a source of copper for oxhide ingots before 1250 B.C., perhaps as early as 1550–1400 B.C. (Absolute dates for LM IB remain controversial [Betancourt 1998; Bietak 2003; Manning et al. 2006].) The mine of Ambelikou, which is the only Cypriot mine with direct evidence of having been worked in the Bronze Age (Merrillees 1984), was exploited in the Early to Middle Cypriot periods, too early to have been a source of copper for oxhide ingots, as is also confirmed by lead isotope analyses of its ores, which do not match those of oxhide ingots (see Figs. 21.1, 21.2). The Cypriot Bronze Age copper smelting site of Politiko-*Phorades*, dating to about 1600 B.C. (Knapp 2003), is also too early to have produced copper for oxhide ingots, and this, too, is confirmed by lead isotope analyses of copper slags from Politiko-*Phorades* shown in Figures 21.1 and 21.2.

Apliki Mine Deposit

The discovery from lead isotope studies (Stos-Gale et al. 1997; Gale 1999, 2006) that the copper from which were made the majority of copper oxhide ingots came from copper ore deposits in or around the Apliki mine in Cyprus, though widely accepted, prompted some reservations from Knapp (1999, 2000), Kassianidou (2001, 2006), and Muhly (2005, 508–509), which we have examined elsewhere (Gale and Stos-Gale, in prep.). We reiterate that our finding—the majority of copper oxhide ingots were made of copper from the Apliki deposit—was an empirical conclusion, not remotely anticipated, which flowed inexorably from our comparison of the lead isotope analyses of a large number of oxhide ingots with those for the copper deposits in Cyprus (e.g., Stos-Gale et al. 1997; Gale 1999, 2006). Among the analyzed oxhide ingots were those excavated in Sardinia, for which two independent laboratories (Gale 1999, 2006; Begemann et al. 2001) have shown that their lead isotope compositions match only the ores from Apliki. The same laboratories (Begemann et al. 2001; Gale and Stos-Gale 1992; Gale 2006) both showed that the Nuragic copper-based artifacts in Sardinia had lead isotope compositions that did not match any Cypriot ores, but rather matched ores from Sardinia. The hypothesis advanced by Kassianidou (2001, 2006) to explain this situation was shown by Gale (2006) to be incorrect because it was inconsistent with the combined lead isotope and chemical analytical data. An explanation apparently consistent with the lead isotope and chemical compositional data for Sardinian artifacts and oxhide ingots was proffered in a recent paper read by R.G. Valera (Marras, Valera, and Valera 2007) at the Second International Conference on Archaeometallurgy held in Aquileia.

Those archaeologists who incline to reject the Apliki hypothesis all affirm their belief that the great majority of copper oxhide ingots were made of copper smelted from Cypriot ores, largely on the basis of lead isotope analyses. They thereby involve themselves in a logical inconsistency. In fact, throughout the long history of the serious study of oxhide ingots (beginning with Buchholz 1959), there was no firm direct evidence for this belief prior to the application of comparative analyses of lead isotope compositions of oxhide ingots and ores from Cyprus (Gale and Stos-Gale 1986; Gale 1991, 1999, 2006; Stos-Gale et al. 1997). But the lead isotope compositions for the post–1400 B.C. copper oxhide ingots do not in fact match the lead isotope compositions of most Cypriot ores (Stos-Gale et al. 1997; Gale 1999, 2006; Gale and Stos-Gale 2005). Most Cypriot ores are in fact excluded as the source of copper for oxhide ingots; the match is solely with the ores from the Cypriot ore deposit of Apliki (see Figs. 21.1, 21.2). It follows that acceptance regarding comparative lead isotope analyses showing the great majority of copper oxhide ingots were made of copper smelted from Cypriot ores necessarily carries with it the corollary that they were made of copper smelted from ores from the Apliki ore deposit. The lead isotope analyses do not allow acceptance of a Cypriot origin for the copper used to make oxhide ingots without accepting that the source in Cyprus is the Apliki ore deposit. It seems possible that this northeastern Troodos mining area had a monopoly on providing copper metal for making the large copper oxhide ingots, though LIA shows that many copper-based artifacts from Bronze Age Cypriot sites were made from copper that came from other copper ore deposits around the Troodos. This is a new factor that needs to be incorporated into socioeconomic discussions of the organization of copper production on Bronze Age Cyprus.

Acknowledgments

I am grateful to the Science and Engineering Research Council (SERC), the Institute for Aegean Prehistory (INSTAP), the British Academy, and the University of Oxford for funding laboratory and fieldwork, and to Dr. G. Maliotis for his invaluable and extensive help. The MC-ICPMS analyses of samples from Uluburun ingots were made in the Department of Earth Sciences, Oxford, with the help of Dr. N. Belshaw and the kind permission of Prof. Sir R.K. O'Nions and Dr. D. Porcelli.

References

Begemann, F., S. Schmitt-Strecker, E. Pernicka, and F. Lo Schiavo. 2001. "Chemical Composition and Lead Isotopy of Copper and Bronze from Nuragic Sardinia," *EJA* 4 (1), pp. 43–85.

Belshaw, N.S., P.A. Freedman, R.K. O'Nions, M. Frank, and Y. Guo. 1998. "A New Variable Dispersion Double-focusing Plasma Mass Spectrometer with Performance Illustrated for Pb Isotopes," *International Journal of Mass Spectrometry* 181, pp. 51–58.

Betancourt, P.P. 1998. "The Chronology of the Aegean Late Bronze Age: Unanswered Questions," in *Sardinian and Aegean Chronology* (*Studies in Sardinian Archaeology* 5), M.S. Balmuth and R.H. Tykot, eds., Oxford, pp. 291–296.

Bietak, M. 2003. "Science versus Archaeology: Problems and Consequences of High Aegean Chronology," in *The Synchronization of Civilizations in the Eastern Mediterranean in the Second Millennium* B.C., M. Bietak, ed., Vienna, pp. 23–33.

Buchholz, H.-G. 1959. *Keftiubarren und Erzhandel im zweiten vorchristlichen Jahrtausend* (*PZ* 37), pp. 1–40.

Budd, P., R. Haggerty, A.M. Pollard, B. Scaife, and R.G. Thomas. 1996. "Rethinking the Quest for Provenance," *Antiquity* 70, pp. 168–174.

Budd, P., A.M. Pollard, B. Scaife, and R.G. Thomas. 1995. "Oxhide Ingots, Recycling and the Mediterranean Metals Trade," *JMA* 8 (1), pp. 1–32.

Cucuzza, N., and N.H. Gale. 2004, "Il mezzo lingotto oxhide da Haghia Triada," *Creta Antica* 5, pp. 137–153.

Gale, N.H. 1991. "Copper Oxhide Ingots: Their Origins and Their Place in the Bronze Age Metals Trade in the Mediterranean," in *Bronze Age Trade in the Mediterranean. Papers Presented at the Conference Held at Rewley House, Oxford, in December 1989* (*SIMA* 90), N.H. Gale, ed., Jonsered, pp. 197–239.

———. 1999. "Lead Isotope Characterisation of the Ore Deposits of Cyprus and Sardinia and Its Application to the Discovery of the Sources of Copper for Late Bronze Age Oxhide Ingots," in *Metals in Antiquity* (*BAR-IS* 792), S. Young, M. Pollard, P. Budd, and R.A. Ixer, eds., Oxford, pp. 110–121.

———. 2005. "Die Kupferbarren von Uluburun. Teil 2: Bleiisotopenanalysen von Bohrkernen aus den Barren," in Yalçin, Pulak, and Slotta, eds., 2005, pp. 141–148.

———. 2006. "Lead Isotope Studies—Sardinia and the Mediterranean. Provenance Studies of Artefacts Found in Sardinia," *Instrumentum* 23, pp. 29–34.

Gale, N.H., W. Gentner, and G.A. Wagner. 1980. "Mineralogical and Geographical Silver Sources of Archaic Greek Coinage," in *Metallurgy in Numismatics* 11 (*Royal Numismatic Society Special Publication* 13), D.M. Metcalf, ed., pp. 3–50.

Gale, N.H., and Z.A. Stos-Gale. 1982. "Bronze Age Copper Sources in the Mediterranean: A New Approach," *Science* 216, pp. 11–19.

———. 1986. "Oxhide Ingots in Crete and Cyprus and the Bronze Age Metals Trade," *BSA* 81, pp. 81–100.

———. 1992. "New Light on the Provenance of the Copper Oxhide Ingots Found on Sardinia," in *Sardinia in the Mediterranean: A Footprint in the Sea*, R.H. Tykot and T.K. Andrews, eds., Sheffield, pp. 317–346.

———. 1995. "Comments on Oxhide Ingots Recycling and the Mediterranean Metals Trade," *JMA* 8 (1), pp. 33–41.

———. 2005. "Zur Herkunft der Kupferbarren aus dem Schiffswrack von Uluburun und der spätbronzezeitliche Metallhandel im Mittelmeeraum," in Yalçin, Pulak, and Slotta, eds., 2005, pp. 117–132.

———. In prep. "Oxhide Ingots from Gournia, East Crete," in *Gournia Copper and Bronze: Studies of Minoan Metal Artifacts in the Penn Museum*, P.P. Betancourt and S.C. Ferrence, eds.

Gale, N.H., Z.A. Stos-Gale, G. Maliotis, and N. Annetts. 1997. "Lead Isotope Data from the Isotrace Laboratory, Oxford: Archaeometry Database 4, Ores from Cyprus," *Archaeometry* 39 (1), pp. 237–246.

Hamelin B., B. Dupre, O. Brevart, and C.J. Allegre. 1988. "Metallogenesis at Paleo-Spreading Centers: Lead Isotopes in Sulfides, Rocks and Sediments from the Troodos Ophiolite (Cyprus)," *Chemical Geology* 68 (3–4), pp. 229–238.

Hauptmann, A., R. Maddin, and M. Prange. 2002. "On the Structure and Composition of Copper and Tin Ingots Excavated from the Shipwreck of Uluburun," *BASOR* 328, pp. 1–30.

Kassianidou, V. 2001. "Cypriot Copper in Sardinia: Yet Another Case of Bringing Coals to Newcastle?" in *Italy and Cyprus in Antiquity: 1500–450 B.C.*, L. Bonfanti and V. Karageorghis, eds., Nicosia, pp. 97–120.

———. 2006. "The Production, Use and Trade of Metals in Cyprus and Sardinia: So Similar and So Different," *Instrumentum* 23, pp. 34–37.

Knapp, A.B. 1999. "The Archaeology of Mining: Fieldwork Perspectives from the Sydney Cyprus Survey Project (SCSP)," in *Metals in Antiquity* (*BAR-IS* 792), S. Young, M. Pollard, P. Budd, and R.A. Ixer, eds., Oxford, pp. 98–109.

———. 2000. "Archaeology, Science-Based Archaeology and the Mediterranean Bronze Age Metals Trade," *EJA* 3 (1), pp. 31–56.

———. 2003. "The Archaeology of Community on Cyprus: Politiko-*Phorades* in Context," *AJA* 107 (4), pp. 559–580.

Lo Schiavo, F. 2005. "Metallhandel im zentralen Mittelmeer," in Yalçin, Pulak, and Slotta, eds., 2005, pp. 399–414.

Manning, S.W., C.B. Ramsey, W. Kutschera, T. Higham, B. Kromer, P. Steier, and E.M. Wild. 2006. "Chronology for the Aegean Late Bronze Age 1700–1400 B.C.," *Science* 312, no. 5773, pp. 565–569.

Marras, G., P.G. Valera, and R.G. Valera. 2007. "Metal Supply and Domestic Trade in Nuragic Sardinia." Paper read at the 2nd International Conference of Archaeometallurgy in Europe, 17–21 June 2007, Aquileia. Pre-conference publication on CD-ROM, Associazione Italiana di Metallurgia, Milan.

Merrillees, R.S. 1984. "Ambelikou-*Aletri*: A Preliminary Report," *RDAC* 1984, pp. 1–13.

Muhly, J.D. 1977. "The Copper Ox-Hide Ingots and the Bronze Age Metals Trade," *Iraq* 39, pp. 73–82.

———. 1983. "Lead Isotope Analysis and the Kingdom of Alashiya," *RDAC* 1983, pp. 210–212.

———. 2005. "Kupfer und Bronze in der spätbronzezeitliche Ägäis," in Yalçin, Pulak, and Slotta, eds., 2005, pp. 503–513.

Pugliese Caratelli, G. 1945. "Le iscrizione preelleniche di Haghia Triada in Creta e della Grecia peninsulare," *MonAnt* 40, pp. 421–610.

Pulak, C. 1998. "The Uluburun Shipwreck: An Overview," *IJNA* 27(3), pp. 188–224.

———. 2000. "The Copper and Tin Ingots from the Late Bronze Age Shipwreck at Uluburun," in *Anatolian Metal* I (*Der Anschnitt Beiheft* 13), Ü. Yalçin, ed., Bochum, pp. 137–157.

Stos-Gale, Z.A., and N.H. Gale. 1982. "The Sources of Mycenaean Silver and Lead," *JFA* 9, pp. 467–485.

———. 1994. "The Origin of Metals Excavated on Cyprus," in *Provenience Studies and Bronze Age Cyprus: Production, Exchange and Politico-Economic Change*, A.B. Knapp and J. Cherry, eds., Madison, pp. 92–122, 210–216.

Stos-Gale, Z.A., G. Maliotis, N.H. Gale, and N. Annetts. 1997. "Lead Isotope Characteristics of the Cyprus Copper Ore Deposits Applied to Provenance Studies of Copper Oxhide Ingots," *Archaeometry* 39 (1), pp. 83–123.

Yalçin, Ü., C. Pulak, and R. Slotta, eds. 2005. *Das Schiff von Uluburun: Welthandel vor 300 Jahren*, Bochum.

CHAPTER

22

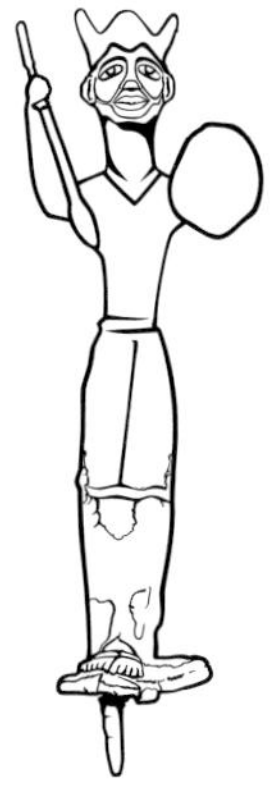

"Biscuits with Ears:" A Search for the Origin of the Earliest Oxhide Ingots

Zofia Anna Stos-Gale

The Uniqueness of Oxhide Ingots

Copper oxhide ingots are generally flat, rectangular ingots measuring 4–8 cm thick, 30–60 cm long, and 20–45 cm wide. The weight of these ingots varies greatly, from about 10 kg to 37 kg. The most prominent features of the "oxhide" ingots are elongated, pulled-out corners that would facilitate carrying them by two people or tying them together, or to the back of an animal, with a rope. The shape of these "ears" and the ingot sides vary a great deal, from very long hornlike "ears" and a small waist as on the ingot found at Mycenae, to gently concaved sides and small protrusions at the corners as seen on the "pillow" ingots from Kyme, Kato Zakros, Tylissos, and Hagia Triada (Fig. 22.1). The other prominent feature of these ingots is a fleecelike bubbly top surface and smooth pitted sides and underside resulting from pouring freshly smelted copper into the molds made of (most likely) sand or stone. This process left the ingots full of gas holes. Ingots of this shape never have the smooth surface appearance of refined, alloyed, or re-melted metal; the purity of copper in all tested cases is over 99%. Some of them have signs, either stamped or imprinted during casting, or as in many cases, incised after the ingot had cooled. So far there is no firm conclusion as to the meaning of these signs. Similarly, the shapes seem to be without a particular meaning. The best study case for these ingots is the collection of several hundred discovered on the Uluburun shipwreck, which also carried oxhide-shaped tin ingots (Pulak 2000).

To 19th-century archaeologists these ingots resembled an "oxhide," but George Bass and Cemal Pulak often tell the story of discovery of the Uluburun shipwreck, after the sponge divers mentioned to one of their colleagues a site where they have spotted copper "biscuits with ears" at the bottom of the sea. This description seems very evocative. The shape of these ingots made them a symbol

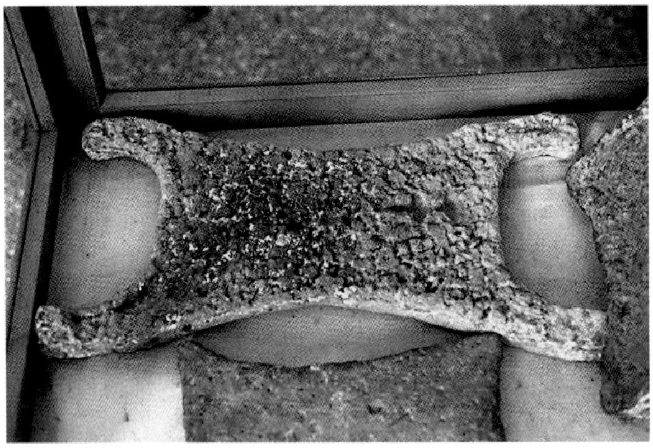

Figure 22.1. Oxhide ingot from Mycenae and ingots from Kyme in the Numismatic Museum, Athens.

of the Late Bronze Age trade in the Mediterranean. Many papers in the last 50 years have been devoted to the discussion of the origin and significance of these ingots. Noël Gale has summarized in detail the early publications concerning these objects, adding the information that was becoming available from the lead isotope (LI) analyses (Gale 1991). Since then, little has been discovered about the ingots from the archaeological point of view, but much more of the analytical data has become available.

The program of LI analyses of metals, ores, and slags related to oxhide ingots was mainly developed in Oxford and Heidelberg-Mainz in the 1980s and 1990s. The Oxford Isotrace Laboratory produced during that time a relatively comprehensive LI database of oxhide ingots that was published simultaneously with a database of ores from Cyprus (Gale et al. 1997; Stos-Gale et al. 1997). More recently, the LI analyses of the ingots from the Uluburun (Gale and Stos-Gale 2005) and Gelidonya (Stos-Gale et al. 1998; Stos 2009) shipwrecks were published. Further survey and collection of ores and slags added more LI and ^{14}C data for Cypriot slag heaps (Stos-Gale, Maliotis, and Gale 1998), while the much more daunting task of surveying the ore deposits in Anatolia was carried out by the Heidelberg-Mainz team (Wagner and Öztunali 2000, with references), with a significant contribution from Yener (Yener et al. 1991).

From the present evident, it seems that the earliest ingots of the oxhide shape are still those found on Crete in the Hagia Triada palace and at Tylissos, Gournia, Zakros, and coastal Mochlos. All these finds are dated to 1500–1450 B.C. (Late Minoan [LM] IB). Most of these ingots and fragments were analyzed for their LI compositions. Perhaps the time is right to look again at the evidence as to the origin of these ingots. The Festschrift to Jim Muhly, who devoted much of his time to these ingots and the Bronze Age metal trade, is an appropriate place for this summary.

The Origin of the Oxhide Ingot Metal

The LI and trace elemental compositions of all analyzed oxhide ingots dated between the 14th and 11th century B.C. are fully consistent with their origin from Cypriot ores. There can be no doubt that Cyprus in this period was a prolific producer of copper ingots of this shape, and a good example of this is the cargo of the Uluburun shipwreck, dated to ca. 1314–1307 B.C. (Gale and Stos-Gale 2005). These ingots of Cypriot copper travelled as far west as Sardinia, north to the coast of the Black Sea, and into central Anatolia (Stos-Gale et al. 1997, 110–111, table 6). The only fragment of an oxhide ingot so far excavated in Egypt, in Qantir, is also fully consistent with the origin from the Cypriot ores (see Table 22.1). It seems that the oxhide ingot shape is what we could call a Cypriot "brand." However, some of the earlier (LM IB) ingots found on Crete are not isotopically consistent with their origin from the Cypriot ores (Table 22.2).

The discussion of the origin of copper of the Hagia Triada ingots was published some 20 years ago (Gale and Stos-Gale 1986). (The ingots discussed here are the "pillow"-shaped whole ingots. Half of an oxhide ingot found in Hagia Triada, discussed by Gale [1991, 202], is of a possibly later date and consistent with Cypriot ores). Because the LI ratios of ores are strongly related to the geological age of the mineral deposits, it was quite clear that these LI ratios do not reflect ore deposits that were created about 100 million years (Ma) ago, as it is known for Cyprus, but from a mineralization of a much older age, around 400–700 Ma (Silurian to Precambrian). On the same

Object Number	Site	Description	Date	$^{208}Pb/^{206}Pb$	$^{207}Pb/^{206}Pb$	$^{205}Pb/^{204}Pb$
CA 20/1	Mochlos, Crete	Corner of a "pillow" ingot 790 g	LM IB	2.0697	0.83956	18.525
CA 20/10	Mochlos, Crete	Oxhide fragment	LM IB	2.06735	0.83897	18.543
CA 20/11	Mochlos, Crete	Oxhide fragment	LM IB	2.0672	0.83969	18.571
CA 20/12	Mochlos, Crete	Oxhide fragment	LM IB	2.08168	0.84364	18.573
CA 20/13	Mochlos, Crete	Tip of a horn	LM IB	2.06821	0.83909	18.569
CA 44/9	Mochlos, Crete	Oxhide fragment, side	LM IB	2.07046	0.84114	18.505
CA 45/4	Mochlos, Crete	Corner of a "pillow" ingot	LM IB	2.06547	0.83851	18.548
CA 60/1	Mochlos, Crete	Fused metals: horn of oxhide	LM IB	2.07499	0.84120	18.494
CA 70/1	Mochlos, Crete	1/4 of oxhide HT type	LM IB	2.06541	0.83724	18.647
CA 70/2	Mochlos, Crete	Horn of an oxhide ingot	LM IB	2.07166	0.841	18.520
CA 71/3	Mochlos, Crete	Corner of a pillow	LM IB	2.06652	0.83909	18.566
HM 2602	Kato Zakros, Crete	Oxhide ingot	LM IB	2.07622	0.84595	18.408
HM 2603	Kato Zakros, Crete	Oxhide ingot	LM IB	2.07413	0.84154	18.5394
HM 2604	Kato Zakros, Crete	Oxhide ingot	LM IB	2.06974	0.84056	18.533
HM 2606	Kato Zakros, Crete	Oxhide ingot	LM IB	2.08042	0.84836	18.359
KS	Kato Syme, Crete	Oxhide ingot	LM IB	2.0709	0.83536	18.791
MeP 87/45	Qantir, Egypt	Oxhide ingot fragment	Rameses II	2.07394	0.84124	18.527

Table 22.1. Lead isotope compositions of LM I oxhide ingots not published in Stos-Gale et al. 1997, or in other papers, consistent with their origin from Cypriot ores. Samples courtesy: J. Soles (Mochlos), Herakleion Museum (Kato Syme and Zakros), and T. Rehren (Qantir).

Object Number	Site	Date	$^{208}Pb/^{206}Pb$	$^{207}Pb/^{206}Pb$	$^{206}Pb/^{204}Pb$
AT A	Hagia Triada	LM IB	2.11368	0.87139	17.821
AT 721 C	Hagia Triada	LM IB	2.12248	0.87788	17.698
AT 722 I	Hagia Triada	LM IB	2.12962	0.88594	17.488
AT 723 F	Hagia Triada	LM IB	2.11838	0.87479	17.767
AT L	Hagia Triada	LM IB	2.12243	0.87928	17.647
AT M	Hagia Triada	LM IB	2.12411	0.87871	17.685
AT 724 P	Hagia Triada	LM IB	2.12582	0.87842	17.718
AT 726 B	Hagia Triada	LM IB	2.10081	0.85911	18.267
AT 726 B patch	Hagia Triada	LM IB	2.09979	0.86032	18.110
AT 726 E	Hagia Triada	LM IB	2.12782	0.88180	17.641
AT 726 H	Hagia Triada	LM IB	2.13086	0.88109	17.742
AT 726 J	Hagia Triada	LM IB	2.13187	0.88631	17.495
AT 726 O	Hagia Triada	LM IB	2.11965	0.87783	17.674
HM 1763/87	Tylissos	LM IB	2.09794	0.85841	18.138
HM 2601	Kato Zakros	LM IB	2.10254	0.86324	18.093
HM 2605	Kato Zakros	LM IB	2.12275	0.88116	17.593

Table 22.2. Lead isotope compositions of oxhide ingots from three sites in Crete: Hagia Triada, Kato Zakros, and Tylissos. They are not consistent with their origin from Cypriot ores (samples courtesy of the Herakleion Museum).

grounds, the ore deposits from mainland Greece, Crete, and other Aegean islands were rejected as a possible ore source for these ingots.

It has been suggested that the copper ore used for the Hagia Triada ingots might be found in Anatolia. This suggestion seems somewhat strengthened by the fact that out of many hundreds of Bronze Age (BA) copper-based artifacts from the eastern Mediterranean analyzed for their LI compositions, the only other two sites where metals of similar LI compositions have been located were Kastri on Syros (Early Cycladic [EC] III) and Troy (Early Bronze [EB] II). A speculative map of the Anatolian deposits was presented in the paper reporting these results, indicating copper ore deposits of a relevant age (Stos-Gale, Gale, and Gilmore 1984, 27–28). A few years later a speculative possibility of the origin of these ores in the Precambrian copper deposits in Afghanistan, Iran, or southern Russia (Transcaucasus) was suggested by Gale (1991, 226).

Further Search for the Ores Consistent with the Hagia Triada Ingots

Before its closure in 2001, the Isotrace Laboratory in Oxford produced over 2,600 LI analyses of ore and slag samples collected during numerous archaeometallurgical and geological surveys in Greece, Cyprus, Italy, Spain, and Bulgaria. Together with LI data published by other researchers, the Oxford LI database includes over 3,000 LI data for ores relevant to the Mediterranean Bronze Age metals and over 2,600 for the BA metal artifacts (for a summary, see Stos-Gale 2001, 197, table 10.1). The LI database for Cypriot ores and slags includes now over 500 entries. The other very important development is the availability of more analytical data for the ores from Sardinia (over 300) and Wadi Arabah (about 100 from Timna and Feinan), where there are ore deposits of approximately comparable age to that indicated by the LI compositions of the unusual Hagia Triada ingots. Additionally, some 70 ore data from the coast of the Red Sea and about 40 from central Iran have been published. Finally, over 320 LI data for ores from various deposits in Turkey also have been published in the last 15 years. Unfortunately, there are still very important gaps including Egyptian and Middle Eastern ore sources.

The interpretation of the LI data obtained for archaeological artifacts has evolved over the years from simple plotting on LI ratios diagrams. It is currently done by using a three-stage procedure:

1. First, the Euclidean distances in the three-dimensional space with axes defined by the three LI ratios are calculated between each of the artifact's LI ratios and all currently available LI data points for ore and slag samples. Software called TestEuclid sorts the data in the order of increasing Euclidean distances. The LI ratios of the artifact and an ore sample are regarded as identical if all three ratios for both are within the analytical error for each of the three LI ratios. At this stage it is possible to find that several ore samples from two or three geographically different regions have all three LI ratios identical with an artifact.
2. Second, the geochemical, geographical, and historical (archaeological) information is considered.
3. Third and finally, the data points are compared on two-dimensional plots of LI ratios of the artifacts and ores selected in the previous two steps.

Usually, at the end of these procedures, all but one or two ore sources can be eliminated.

These procedures were applied to check if any group of the currently available LI data for ores and slags from the mines in the Mediterranean and the Near and Middle East is consistent with the LI ratios for the oxhide ingots from Hagia Triada. It is not possible to discuss in detail the results of these comparisons in this short paper, but the conclusion is still the same—none of the available groups of LI data for ore deposits match the main group of these ingots. LI ratios in the range for $^{208}Pb/^{206}Pb$ of 2.10–2.13 are quite rare among the published LI data for copper ore deposits surrounding the Mediterranean. The closest LI compositions are for the copper ores from Sardinia and Wadi Arabah.

Current LI evidence for ores from Sardinia (Boni and Koeppel 1985; Stos-Gale et al. 1995; Begemann et al. 2001) definitely excludes the Sardinian ores as sources of the copper for these ingots. The same, so far, is true for the copper ores from Timna and Feinan Bronze Age copper mines in the Arabah Valley (Hauptmann et al. 1992; unpublished data from the Isotrace Laboratory). Among the published LI data for ores from Turkey, there are some minerals that have $^{208}Pb/^{206}Pb$ ratios in this range, but they are from lead-zinc, not copper, deposits. Graphic presentation of this conclusion is presented on Figure 22.2. The LI range of this diagram is nearly the same as in the original publication (Gale and Stos-Gale 1986, figs. 5, 94).

These unusual LM IB ingots fall into two groups: one ingot from Hagia Triada (inv. no. AT 726B) and the ingots from Tylissos (HM 1763) and Kato Zakros HM 2601) have lower LI ratios (from deposits formed around 375 Ma), while the other non-Cypriot ingot from Zakros and the remaining ingots from Hagia Triada form a rather dispersed group related to the earlier ore formation (about 640 Ma). These groups do not necessarily mean that the metal for both of them was derived from geographically different areas. It is not uncommon for certain regions in Europe and the Near East to have mining regions where ore deposits were formed in different geological periods (for example Sardinia).

Who Introduced the Oxhide Ingot Shape?

The ingots from Hagia Triada are not necessarily the ancestors of all other oxhide ingots. As is demonstrated in Table 22.1, a number of LM IB ingots and ingot fragments found on Crete are isotopically fully consistent with their origin from Cypriot ores. The analyzed group of six ingots from Kato Zakros includes four that are fully consistent with Cypriot ores from the Solea region, which includes the big copper mines of Skouriotissa, Apliki, Phoenix, and Mavrovouni.

Oxhide ingots were cast from freshly smelted, unrefined copper, so it is interesting to see in Kato Zakros a group of similarly looking oxhide ingots made from copper mined in geographically different regions. That might suggest that the ore had to be transported to the location where the smelting and then casting of ingots was taking place. This would seem somewhat inefficient, because the ore is much more bulky than metal obtained from it, so it means that the transport of the ore over long distances is not the most economical approach.

However, there are examples of the smelting of copper during the BA in places remote from the mines. An early example of copper extraction far away from the source of the ore is the Early Minoan (EM) site of Chrysokamino on Crete (Stos and Gale 2006). On Cyprus there are copper slag heaps scattered in the mountains in the vicinity of the mines (most of them date to the Roman or Medieval times, see Stos-Gale, Maliotis, and Gale 1998), but there are also quite substantial finds of copper slag in the Bronze Age settlements of Enkomi and Kalavassos–*Ayios Dhimitrios*. LI ratios of slags from both of these sites proved them to be consistent with copper ores from various outcrops in the Troodos Mountains. Some of the slags from Enkomi have LI compositions identical to the ingots carried by the Uluburun ship. An unpublished study of the Uluburun ingot molds carried out by Pulak combined with the LI data for these ingots suggests that some of the ingots were cast in the vicinity of the mines, but some might have been made, for example, in Enkomi or Ayios Dhimitrios (Pulak and Stos-Gale, unpublished data).

The presence of ingots at Hagia Triada in a storeroom of the palace suggests that they were put there as an asset or a status symbol. Perhaps these ingots were a diplomatic gift to the Minoan king? The depictions of oxhide ingots in the procession of gift bearers in the tombs of the Egyptian pharaohs (e.g., Useramon in Thebes; see Wachsmann 1987, pl. 33A) implies that ingots of this shape were well-regarded gifts. It also might suggest that the Egyptians were not casting such ingots themselves, and therefore they were valuable.

From about 1550 B.C. the Anatolian coast facing Cyprus on the northeast was coming under the influence of the emerging Hurrian state of Mitanni, which in the next two centuries became a power equal to Egypt and the Hittite Empire (Kuhrt 2000, 286–287).

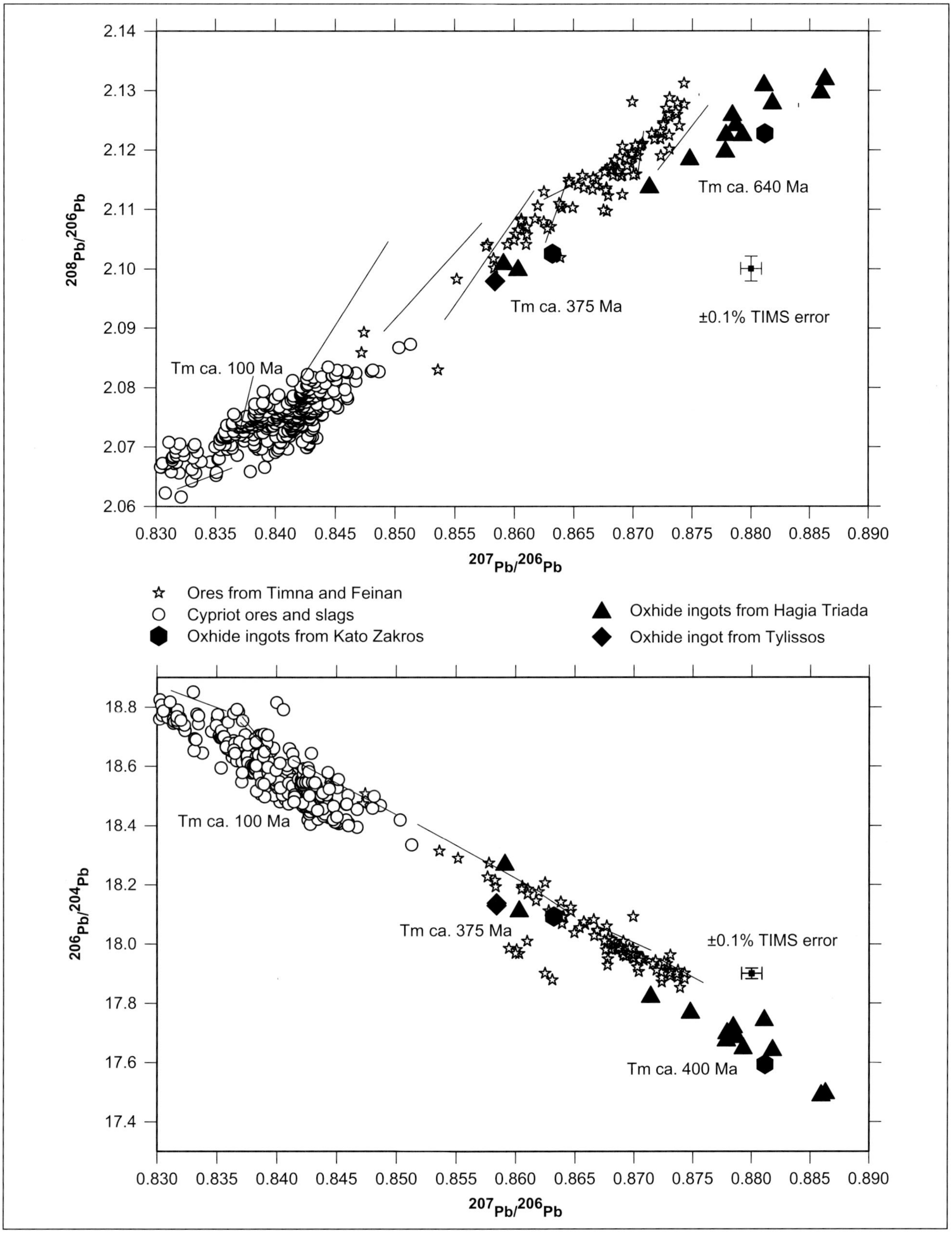

Figure 22.2. LI compositions of the LM I copper oxhide ingots of non-Cypriot origin compared with Cypriot and Near Eastern copper ores. The position of Sardinian ore data is indicated by lines. TIMS = Thermal Ionization Mass Spectometry. Tm = geological age of ore deposits.

At its greatest extent the kingdom of Mitanni controlled an area stretching from Lake Van to the Middle Euphrates and from the Zagros Mountains to the Syrian coast—a region rich in mineral resources and with a long tradition of copper metallurgy. In his paper on the Late Bronze Age technology transfer between Western Asia and Egypt, P.R.S. Moorey (2001, 3–4) summarizes the evidence for the important role that the Hurrians played in this region in propagating innovative technologies. It is very likely that they were also involved in facilitating the tin trade from central Asia to the eastern Mediterranean. Did the Hurrians invent the oxhide ingot shape?

The fact that the ingots from Hagia Triada remained intact supports the conclusion obtained from the LI work on Minoan metals that shows extensive use of copper from Aegean sources (Stos-Gale 2001, 202); there was no need to use these ingots, because the supply of copper was secure. Tin had to be imported, however, and in Crete the tin bronzes were made earlier than in the other parts of the Aegean (Stos-Gale 2001, 204–206). Perhaps these ingots came to Crete together with tin ingots that were immediately used? It seems likely that tin was coming to Crete together with copper from Cyprus (Muhly 1977, 44). About 18% of LM copper-based metals analyzed are consistent with Cypriot ores, and the presence of Cypriot oxhide ingot fragments in the workshop at Mochlos proves that they were used (Soles and Stos-Gale 2004, 45–59).

Thus, we still have no answer based on their LI compositions as to the origin of these unusual LM IB ingots listed in Table 22.2. Their LI ratios do not resemble any of the ores analyzed so far, and they are similar only to a few other Bronze Age metal artifacts from the eastern Mediterranean. However, there is no comparative material from the second millennium B.C. from the areas east of the Syrian coast, from the Kingdom of Mitanni, and from Babylonia. Also, there are no LI analyses of the Transcaucasian copper ores where the metallurgy on the eastern coast of the Black Sea (in present day Georgia and Armenia) was well developed from the Early Bronze Age (Chernykh 1992). Until more LI analytical work is done, we can only speculate about who invented the oxhide ingot shape.

Acknowledgments

The work in the Isotrace Laboratory, Oxford, was made possible thanks to the support of the UK Research Councils, Leverhulme Trust, British Academy, and grants from the Institute for Aegean Prehistory (INSTAP). Generous help came from the geologists of the Greek Institute of Geological and Mineral Exploration (IGME), Hellenic Mining, Ltd. (Cyprus), Agip, Ltd. (Italy), and numerous archaeologists.

References

Begemann, F., S. Schmitt-Strecker, E. Pernicka, and F. Lo Schiavo. 2001. "Chemical Composition and Lead Isotopy of Copper and Bronze from Nuragic Sardinia," *EJA* 4, pp. 43–85.

Boni, M., and V. Koeppel. 1985. "Ore-lead Isotope Pattern from the Iglesiente-Sulcis Area (SW Sardinia) and the Problem of Remobilization of Metals," *Mineralium Deposita* 20, pp. 185–193.

Chernykh, E.N. 1992. *Ancient Metallurgy in the USSR: The Early Metal Age*, Cambridge.

Gale, N.H. 1991. "Copper Oxhide Ingots: Their Origin and Their Place in the Bronze Age Metals Trade in the Mediterranean," in *Bronze Age Trade in the Mediterranean* (*SIMA* 90), N.H. Gale, ed., Jonsered, pp. 197–239.

Gale, N.H., and Z.A. Stos-Gale. 1986. "Oxhide Ingots in Crete and Cyprus and the Bronze Age Metals Trade," *BSA* 81, pp. 81–100.

———. 2005. "Zur Herkunft der Kupferbarren aus dem Schiffswrack von Uluburun und der spätbronzezeitliche Metallhandel im Mittelmeerraum," in *Das Schiff von Uluburun—Welthandel vor 3000 Jahren*, Ü. Yalçın, C. Pulak, and R. Slota, eds., Bochum, pp. 117–132

Gale, N.H, Z.A. Stos-Gale, G. Maliotis, and N. Annetts. 1997. "Lead Isotope Data from the Isotrace Laboratory, Oxford: Archaeometry Data Base 4, Ores from Cyprus," *Archaeometry* 39 (1), pp. 237–246.

Hauptmann, A., F. Begemann, R. Heitkemper, E. Pernicka, and S. Schmitt-Strecker. 1992. "Early Copper Produced in Feinan, Wadi Araba, Jordan: The Composition of Ores and Copper," *Archeomaterials* 6 (1), pp. 1–33.

Kuhrt, A. 2000. *The Ancient Near East—c. 3000–330 B.C.* (*Routledge History of the Ancient World*, 2 vols.), London and New York.

Moorey, P.R.S. 2001. "The Mobility of Artisans and Opportunities for Technology Transfer between Western Asia and Egypt in the Late Bronze Age," in *The Social Context of Technological Change*, A. Shortland, ed., Oxford. pp. 1–14.

Muhly, J.D. 1977. "New Evidence for Sources and Trade in Bronze Age Tin," in *The Search for Ancient Tin*, A.D. Franklin, J.S. Olin, and T.A. Wertime, eds., Washington D.C., pp. 43–48.

Pulak, C. 2000. "The Copper and Tin Ingots from the Late Bronze Age Shipwreck at Uluburun," in *Anatolien Metal* I (*Der Anshnitt Beiheft* 13), Ü. Yalçın, ed., Bochum, pp. 137–157.

Soles, J.S., and Z.A. Stos-Gale. 2004. "The Metal Finds and Their Geological Sources," in *Mochlos* IC: *Period III. Neopalatial Settlement on the Coast: The Artisans' Quarter and the Farmhouse at Chalinomouri. The Small Finds* (*Prehistory Monographs* 9), Philadelphia, pp. 45–59.

Stos, Z.A. 2009. "Across the Wine-Dark Seas . . . Sailor Tinkers and Royal Cargoes in the Late Bronze Age Eastern Mediterranean," in *From Mine to Microscope: Advances in the Study of Ancient Technology*, A.J. Shortland, I.C. Freestone, and T. Rehren, eds., Oxford, pp. 163–180.

Stos, Z., and N.H. Gale. 2006. "Lead Isotope and Chemical Analyses of Slags from Chrysokamino," in *The Chrysokamino Metallurgy Workshop and Its Territory* (*Hesperia Suppl.* 36), by P.P. Betancourt, Princeton, pp. 298–319.

Stos-Gale, Z.A. 2001. "Minoan Foreign Relations and Copper Metallurgy in MM III–LM III Crete," in *The Social Context of Technological Change*, A. Shortland, ed., Oxford. pp. 195–210.

Stos-Gale, Z.A., N.H. Gale, G. Bass, C. Pulak, E. Galili, and J. Sharvit. 1998. "The Copper and Tin Ingots of the Late Bronze Age Mediterranean: New Scientific Evidence," in *Proceedings of the Fourth International Conference on the Beginning of the Use of Metals and Alloys (BUMA-IV): May 25–27, 1998*, H. Kimura, ed., Matsue, Shimane, Japan, pp. 115–126.

Stos-Gale, Z.A., N.H. Gale, and G.R. Gilmore. 1984. "Early Bronze Age Trojan Metal Sources and Anatolians in the Cyclades," *OJA* 3 (3), pp. 23–44.

Stos-Gale, Z.A., N.H. Gale, J. Houghton, and R. Speakman. 1995. "LI Analyses of Ores from the Western Mediterranean," *Archaeometry* 37 (2), pp. 407–415.

Stos-Gale, Z.A., G. Maliotis, and N.H. Gale. 1998. "A Preliminary Survey of the Cypriot Slag Heaps and Their Contribution to the Reconstruction of Copper Production on Cyprus," in *Metallurgica Antiqua, in Honour of Hans-Gert Bachmann and Robert Maddin*, T. Rehren, A. Hauptmann, and J. Muhly, eds., Bochum, pp. 235–262.

Stos-Gale, Z.A., G. Maliotis, N.H. Gale, and N. Annetts. 1997. "LI Characteristics of the Cyprus Copper Ore Deposits Applied to Provenance Studies of Copper Oxhide Ingots," *Archaeometry* 39 (1), pp. 83–124.

Wachsmann, S. 1987. *Aegeans in the Theban Tombs* (*Orientalia Lovaniensa Analecta* 20), Leuven.

Wagner, G.A., and Ö. Öztunali. 2000. "Prehistoric Copper Sources in Turkey," in *Anatolian Metal* I (*Der Anschnitt Beiheft* 13), Ü. Yalçın, ed., Bochum, pp. 31–68.

Yener, K.A, E.V. Sayre, H. Ozbal, E.C. Joel, I.L. Barnes, and R.H. Brill. 1991. "Stable LI Studies of Central Taurus Ore Sources and Related Artefacts from Eastern Mediterranean Chalcolithic and Bronze Age Sites," *JAS* 18 (5), pp. 541–577.

CHAPTER

23

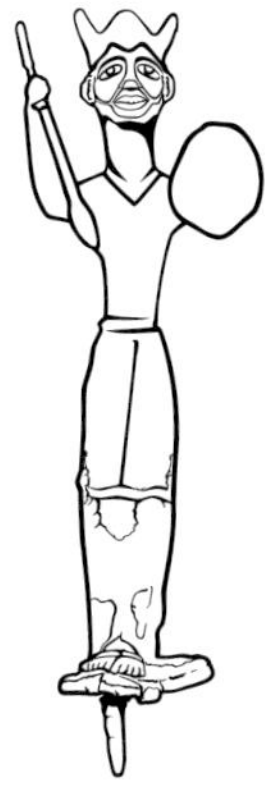

Metal Exchange in Italy from the Middle to the Final Bronze Age (14th–11th Century B.C.E.)

Reinhard Jung, Mathias Mehofer, and Ernst Pernicka

Introduction

The supply and interregional exchange of metal in the Italian Bronze Age is much discussed. Distribution maps of bronze objects, the classical tool of such studies, show their geographical distribution as it is known from archaeological excavations or chance finds. However, many studies have shown that concentrations of specific types in certain regions and their absence in others are not necessarily representative for the areas in which they were made and/or used. Apart from chance finds (e.g., in industrialized and densely populated regions), ancient customs regulating the deposition of bronze objects may have acted as filters for the appearance of certain types in the archaeological record (cf. Hansen 1994; Maraszek 2006). Distribution maps thus can provide only *indications* for possible production regions of those types.

While the find location of an object is the place of its deposition (intentional or not) and may also be the place of its last use, metal analysis offers a different kind of information. It allows conclusions regarding the provenance of the raw material used for its production, if comparative analytical data from ore deposits are available. In combination with the types of the objects and their distribution areas, analytical results can be further used to infer the production regions of different types. If the object is an ingot, its find place in combination with its analytical results allows conclusions regarding the distribution of metal (either as raw metal or already in the form of an alloy) of a specific provenance to other areas.

In this study we present some preliminary results of a research project entitled "Dissemination of War Technology in the Mediterranean at the End of the 2nd Millennium B.C.—or Why the Bronzesmith Had to Have Some Knowledge about Fighting." This project is financially supported by the Institute for Aegean Prehistory (INSTAP), which is gratefully acknowledged. It is carried out by Reinhard Jung and Mathias Mehofer; chemical and isotope analyses are performed at the Curt-Engelhorn-Zentrum Archäometrie (Mannheim, Germany). The program is related to another research project entitled "Cultural, Economic, and Political Contacts between the Aegean and Italy from 1600 to 1000 B.C.," which was carried out by Reinhard Jung at the Mykenische Kommission of the Austrian Academy of Sciences and financially supported from 2005 to 2007 by the Austrian Science Fund (FWF). For sampling permissions, we wish to thank Alberto Cazzella (Rome), Renato Peroni (Rome), Giulia Recchia (Foggia), Luciano Salzani (Verona), Maria Tunzi-Sisto (Taranto), and Alessandro Vanzetti (Rome). We thank them all for facilitating our work and for sharing with us their intimate knowledge of the sites and material they are studying.

Inferences from distribution maps were dominating in previous discussions of Middle to Final Bronze Age metal exchange in Italy, because only very few analyses were available for this time span. For our project, small samples of less than 20 mg from the interior of the objects were taken with stainless steel drills. Major and trace element concentrations were determined by energy dispersive X-ray fluorescence spectrometry (XRF), and lead isotope ratios were analyzed by using Multiple-Collector Inductively-Coupled Plasma Mass Spectrometry (MC-ICP-MS). For the procedures of sample preparation and measurement see Lutz and Pernicka (1996) and Niederschlag et al. (2003).

At this stage of the project we have analyzed 22 artifacts (mostly weapons, but also some implements) from northern Italy (Veneto and Lombardy), six ingots (pick-ingots and plano-convex ingots) from northern Italy, and seven artifacts from southern Italy (Apulia and Calabria). The find locations of these 35 analyzed objects are shown in Figure 23.1. They derive from 11 sites in northern and southern continental Italy (see Table 23.1; note that several objects were sampled twice). The objects cover an overall time span from Middle Bronze Age (MBA) 3 to Final Bronze Age (FBA) 2, or, in absolute dates, from the 14th to the beginning of the 11th century B.C.E. Results of analyses are shown in Figures 23.2 to 23.9, which are discussed in detail below.

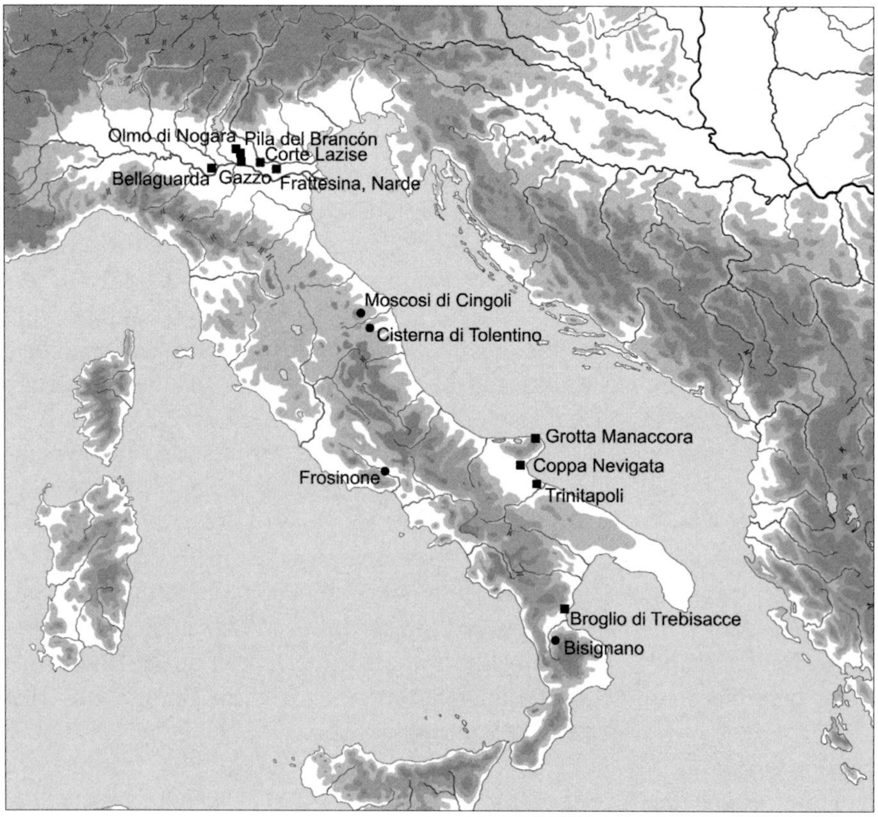

Figure 23.1. Sites of the analyzed objects (black squares); sites of objects with published analytical results, which are discussed in comparison (black circles).

Sample Number	Inventory or Object Number	Object Description	Site	Publication
MA-071041	inv. no. IG.VR 26503	spearhead with cast socket	Pila del Brancón, hoard	Salzani 1994, 88, fig. 4:32
MA-071042	inv. no. IG.VR 26511	spearhead with cast socket	Pila del Brancón, hoard	Salzani 1994, 87, fig. 3:16
MA-071043	inv. no. IG.VR 26506	spearhead	Pila del Brancón, hoard	Salzani 1994, 89, fig. 5:34
MA-071044	inv. no. IG.VR 26491	sword blade	Pila del Brancón, hoard	Salzani 1994, 84, fig. 1:7
MA-071045	inv. no. VR 26489	Naue II sword, type C/Allerona	Pila del Brancón, hoard	Salzani 1994, 84, fig. 1:5
MA-071046	inv. no. VR 40288	Naue II sword, type A/Cetona	Pila del Brancón, hoard	Salzani 1998, 70, no. 149
MA-071047	inv. no. IG 65193	Naue II sword, type C/Allerona	Gazzo Veronese	Salzani, ed., 1987, 63, figs. 43, 44
MA-071526 (blade); MA-071529 (hand-grip)	inv. no. 227	Naue II sword, type C/Allerona	Narde, tomb 227	Salzani 1989b, 38, fig. 16:3
MA-071527 (blade); MA-071528 (rivet)	—	Naue II sword, type C/Allerona	Narde, tomb 168	Salzani 1989b, 34, fig. 12:5
MA-071530	inv. no. IG 306912	knife, type Matrei	Frattesina	—
MA-071531	inv. no. IG 297915	knife, type Matrei	Frattesina	—
MA-071532	inv. no. IG 254735	knife, type Matrei	Frattesina	Salzani 1989a, 67, fig. 1:3
MA-071533 (blade); MA-071534 (rivet)	—	Naue II sword, type A/Cetona	Olmo di Nogara, tomb 41	Salzani 2005, 336, pl. VI:Tb41.A
MA-071535	object no. 13	dagger, type Pertosa	Corte Lazise, Villa Bartolomea	Salzani 2006, 28, no. 13
MA-071536	inv. no. IG.VR 26523	Naue II sword, type A/Cetona	Pila del Brancón	Salzani 1994, 84, fig. 1:6
MA-071537	inv. no. IG.VR 26492	dagger, type Pertosa	Pila del Brancón	Salzani 1994, 85, fig. 2:8
MA-072833	inv. no. Manfredonia 1872	spearhead with cast socket	Coppa Nevigata	Cinquepalmi and Radina, eds., 1998, 269, fig. 809
MA-072834	inv. no. 13040	knife, type Matrei	Broglio di Trebisacce	Luppino, Peroni, and Vanzetti 2005, 731, pl. 97
MA-072835	inv. no. 33138	Griffplattenschwert	Trinitapoli	Tunzi Sisto 1999, 254, 255, no. 1496, pl. 1:1496

Table 23.1. Summary of the investigated objects.

Sample Number	Inventory or Object Number	Object Description	Site	Publication
MA-072836	—	sword, type Pertosa	Grotta Manaccora	Bianco Peroni 1970, pl. 6:39
MA-072837	—	sword blade	Grotta Manaccora	—
MA-072838	—	sword, type Montegiorgio	Grotta Manaccora	Bianco Peroni 1970, pl. 17:127
MA-072839a; MA-072839b	—	sword blade	Grotta Manaccora	Bianco Peroni 1970, pl. 24:171
MA-073179	inv. no. IG 272483	pick-ingot	Frattesina, Hoard 2	Salzani 2000, 42, no. 28
MA-073180	inv. no. IG 272496	plano-convex ingot	Frattesina, Hoard 2	Salzani 2000, 43, no. 41
MA-073181	inv. no. IG 272498	pick-ingot	Frattesina, Hoard 2	Salzani 2000, 42, no. 29
MA-073182; MA-073183	inv. no. B 6	spearhead with cast socket	Bellaguarda, urn burial	Bruno 2007, pl. A54:C, 657
MA-073184	inv. no. IG 272492	socketed shovel	Frattesina, Hoard 2	Salzani 2000, 39, no. 6
MA-073185	inv. no. IG 272469	javeline (inside socket of socketed shovel)	Frattesina, Hoard 2	Salzani 2000, 41, no. 16
MA-073186	inv. no. IG 272469	socketed shovel	Frattesina, Hoard 2	Salzani 2000, 41, no. 16
MA-073187	inv. no. IG 272482	anvil	Frattesina, Hoard 2	Salzani 2000, 43, no. 45
MA-073188	inv. no. IG 272495	pick-ingot	Frattesina, Hoard 2	Salzani 2000, 42, no. 26
MA-073189	inv. no. IG 272491	pick-ingot	Frattesina, Hoard 2	Salzani 2000, 42, no. 27
MA-073190	inv. no. IG 80909	winged axe, type Ponte San Giovanni	Frattesina, Hoard 2	Salzani 2000, 39, no. 1
MA-073191	inv. no. IG 272521	plano-convex ingot	Frattesina, Hoard 2	Salzani 2000, 43, no. 42

Table 23.1, cont. Summary of the investigated objects.

Lead isotope ratios together with chemical analyses are now widely used for material classification and discussion of the provenance of the raw materials. Jim Muhly was at the forefront of the application of these tools in archaeological research, and he acknowledged, from very early on, their potential, but he also has pointed out some pitfalls and possible misconceptions (e.g., Muhly 1983). Here we present an example from Italy where typological and material classification point in the same direction and support each other. More often the material classification and suggestions for provenance are at variance, and then a joint effort by scientists and archaeologists is necessary to explain the differences. In this case this is not necessary, and we hope that Jim will appreciate this. Of course, not all questions can be answered, especially at this preliminary stage of investigation, but the prospects look very good.

Chemical Composition of the Investigated Objects

For the interpretation of the minor and trace element concentrations, we use diagrams of As, Ag, Bi, Ni, and Sb concentrations, because these elements follow the copper during all stages of the metal production processes (e.g., Pernicka 1987, 638; 1990; Christoforidis and Pernicka 1988, 252) and, therefore, the ratios of these elements to copper do not change very much, especially if one can assume that oxide ores were used. These element ratios thus are best suited to detect differences between objects containing copper of different origin. In comparing the trace element concentrations in artifacts from southern Italy with those found in northern Italy, we note that both groups (Fig. 23.2) have overall similar compositions, although in most cases the southern Italian objects have higher Ni contents than those from the north. A clear separation is not observed, however, and similar variations are also found within the copper ingots from Frattesina in northern Italy.

Generally, in trace element studies of ancient metal objects three major complications have to be considered. The first complication is that ores are generally inhomogeneous on all scales; they show zoning. The objection has been raised that this precludes any correlation between artifacts and ores, but this need not be so, as the prehistoric mining region of the Mitterberg in Salzburg, Austria, shows. Here, trace element concentrations in chalcopyrite-rich ores vary over more than two orders of magnitude, yet this variation is not a random one. It is known, for example, that nickel occurs mainly in the form of Ni-As minerals such as gersdorffite (NiAsS), so that copper produced from Mitterberg ores is characterized by a combination of nickel and arsenic as major impurities with variable concentrations combined with relatively low amounts of antimony, silver, and bismuth. In a recent study of some 800 Late Bronze Age copper artifacts from the adjacent area, it was found that about 80% of the copper alloys conformed to this general pattern (Sperber 2004). Knowing that the peak production period at the Mitterberg was during the Late Bronze Age Urnfield Culture, it is only reasonable to assume that a correlation between ores and artifacts does exist.

A second complication arises from the smelting process, as this will differ depending on the type of ore being used. Reduction of oxide ores is quite straightforward compared with the processing of sulfide ores. Although it cannot be expected that reduction happened under chemical equilibrium conditions, it is possible to use thermodynamic data for equilibrium reactions to predict the general behavior of certain trace elements during smelting (Pernicka 1987). Sulfide ores cannot be reduced directly but must be oxidized (roasted) before reduction. This is usually accomplished at around 700°C, well below the melting point of copper and most ore minerals. Nevertheless, it has been suggested that volatile elements like zinc, arsenic, antimony, and, presumably, selenium and tellurium, are partly lost during roasting (Tylecote, Ghaznavi, and Boydell 1977). This is certainly true for high concentrations of these elements, but it has been demonstrated that the losses are much less at low concentrations (Pernicka 1999).

The third complication arises from the treatment of the smelted copper to produce the finished object. Nowadays, raw copper is generally refined to remove sulfur, iron, and other impurities, and this may also have been true in prehistoric times. This is an easy process because all that is needed is to remelt the copper. Iron and other easily oxidized impurities are then removed as dross. It has been suggested that the refining of copper erases most of the characteristics that survived from the ore (Merkel 1990). However, using Merkel's experimental data it can be shown that this is not the case for elements like silver, nickel, and antimony.

Returning to the pattern of minor elements in the samples studied (Fig. 23.2), it can be noted that the variation of all elements considered is less than two orders of magnitude, and silver varies even less. Therefore, it is entirely possible that the copper of all artifacts studied derive from the same deposit or at least from the same region.

The lead (Pb) content of the vast majority of sampled objects is well below 1%. This means that no lead has been added to the alloy. Even in those cases with a somewhat higher lead content (between 1.19% and 2.14%, mainly in some of the artifacts and ingots from Frattesina), it is not likely that we are dealing with intentionally added lead because such a small percentage of added lead would reduce the melting temperature by just a few degrees centigrade (cf. Pernicka 1987, 688, fig. 35). The lead isotope ratios of the sampled objects, therefore, can be related to the lead present in the copper ore deposits. They thus may be used for discussing the provenance of the copper.

Lead Isotope Ratios of the Investigated Objects

The lead isotope results will be published elsewhere, but Figure 23.3 shows that nearly all analyzed artifacts are aligned along a trend line that suggests variable U/Pb ratios in the geological source region. This means that the objects derive either from one deposit that is chemically rather homogeneous but variable as to lead isotope ratios or from several different deposits that happen to be chemically similar. Although the second possibility cannot be completely ruled out, the first seems much more likely, because there are two regions with copper mineralizations known that would match such a description, namely southwest Sardinia and the eastern Alps. Thus one could regard the majority of the samples as a single group because of their chemical similarity. An artifact from southern Italy, to which we will return later, is an exception. The lead isotope ratios of the different ingot types show a larger spread than those of the weapons and implements, but in the isotope diagram they cover the same region as the artifacts (with one exception, see below). This might indicate that the copper of the majority of the weapons and implements as well as the ingots derives from the same mining region.

Now, one would of course like to know where this hypothetical mining region might be located. Few Italian copper mines are sufficiently sampled to offer comparative data. Basically, we have at hand lead isotope data of polymetallic ore deposits from Tuscany and Sardinia, which were analyzed by Stos-Gale and Gale (Stos-Gale et al. 1995, 411–413, table 1), and further lead isotope data of Sardinian ores were analyzed by Begemann, Schmitt-Strecker, Pernicka, and Lo Schiavo (Begemann et al. 2001, 72, 75–76, table 5, fig. 14) and Boni and Köppel (1985). When comparing the central Italian ores from mines in Tuscany with our analyses, we can easily exclude those as sources for the northern and southern Italian copper-based artifacts of this study (Fig. 23.4).

Comparison with Sardinian Ores and Artifacts

The case of Sardinia is a little more complicated. First, one has to note that the ores from the island exhibit a much larger isotopic variability than the ores from Tuscany. Of course it is necessary to verify that the sampled ore deposits were actually exploited during the prehistoric periods, which are relevant for the specific research issue. For this purpose Begemann et al. (2001) collected samples from Sardinian bronze hoards with weapons and implements of Sardinian types (Begemann et al. 2001, 45–50, 60–61, 70–71, tables 3, 4; cf. Lo Schiavo 1998, 196–197). Those bronze hoards can at least partly be dated to the FBA (for the dating to FBA 1 of the pottery vessels containing the hoard of Ittireddu at Nuraghe Funtana, cf. Begemann et al. 2001, 47, fig. 3 with Campus and Leonelli 2000, 207, 248, 435, 444, pls. 143:9, 252:1; 2006, 386, 388, 389, fig. 2:8.28) and are therefore roughly contemporary with the latest of the northern Italian artifacts and with all of the northern Italian ingots analyzed in our project. The lead isotope data of the Sardinian artifacts confirm the wide range of the lead isotope ratios detected in the ore data. It is, therefore, difficult to prove the use of Sardinian copper, if one is dealing with a single object.

If, however, one regards the artifacts from northern and southern Italy as a single group and compares it with Sardinian ores and artifacts, then a difference is observed, especially in the $^{208}Pb/^{206}Pb$ ratios, which are systematically lower in the

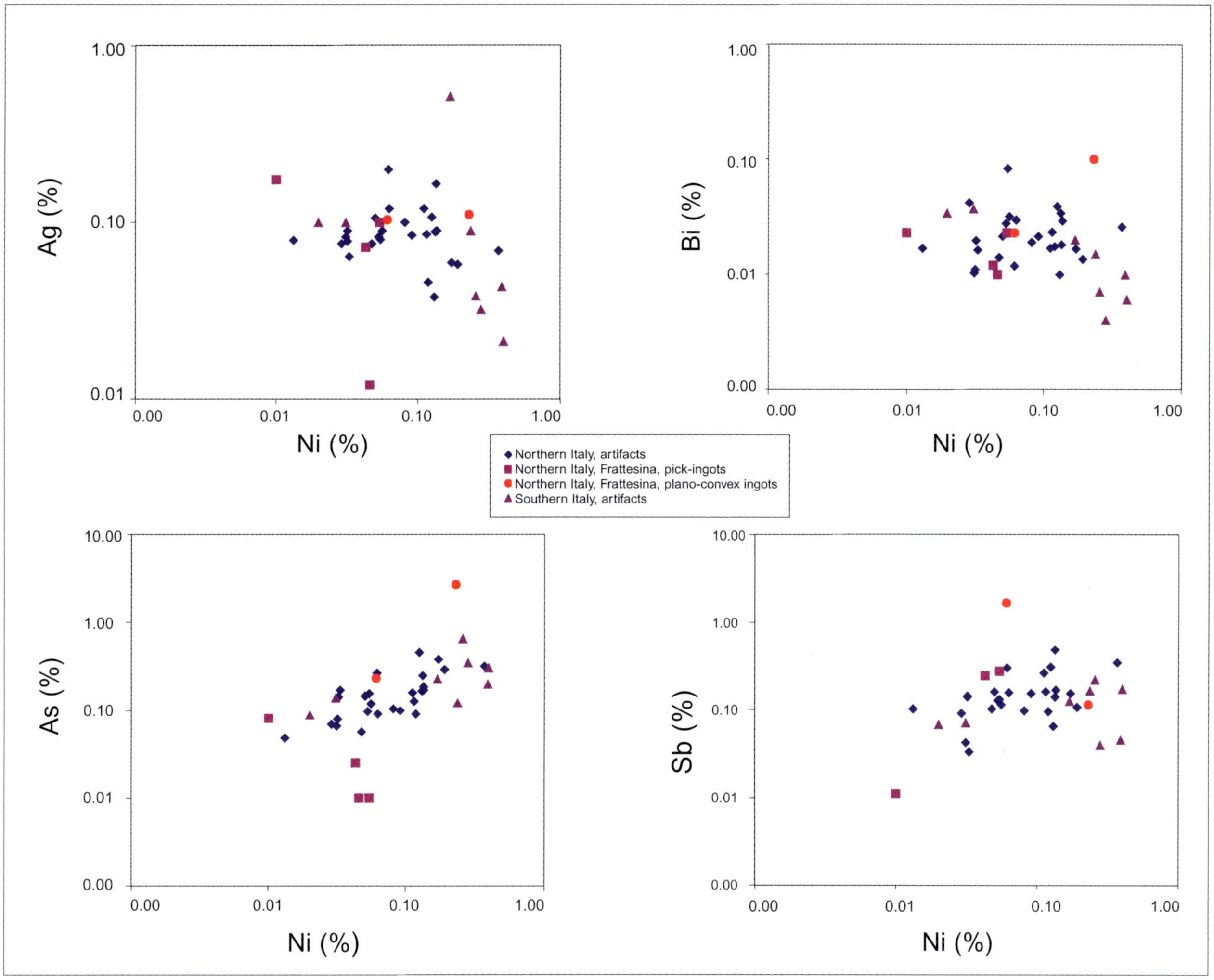

Figure 23.2. Logarithmic plots of element concentrations.

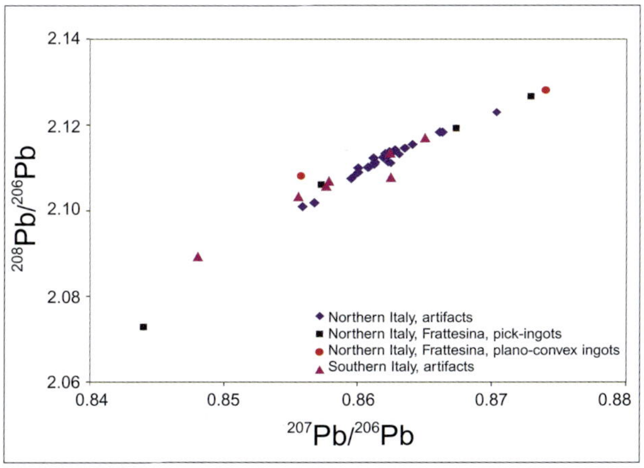

Figure 23.3. Lead isotope ratios of the objects studied.

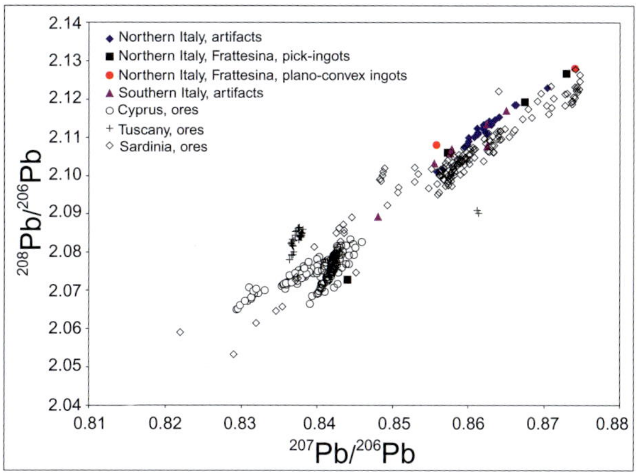

Figure 23.4. Lead isotope ratios of the objects studied compared with copper and lead ores from Tuscany, Sardinia, and Cyprus.

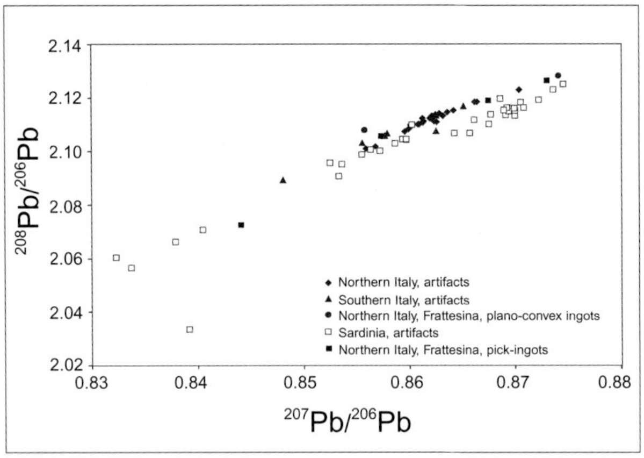

Figure 23.5. Lead isotope ratios of the objects studied compared with Bronze Age (mostly FBA) artifacts from Sardinia (Begemann et al. 2001).

Sardinian ores and artifacts, suggesting that the lead in Sardinian deposits derives from an environment with lower Th/U ratios.

In comparing the artifacts from southern Italy with those found in northern Italy, we have already noted that both groups have overall more or less similar compositions (Fig. 23.2). Generally, these results are consistent with the lead isotope data and allow the conclusion that copper ores from the same mining region were used to produce the majority of the analyzed MBA–FBA weapons and implements from continental Italy (Fig. 23.5).

On the other hand, the Sardinian artifacts appear separated quite well from both the northern and the southern Italian artifacts regarding their silver, nickel, arsenic, and antimony concentrations (Fig. 23.6). This confirms the isotopic separation of the copper used to produce the Sardinian artifacts from the one used in the production of the continental (northern and southern) Italian artifacts. One may now conclude that Sardinian copper mines did not supply the workshops in which the weapons and implements from the Veneto and Lombardy and from Apulia and Calabria were made. This is in accordance with the fact that also the typology of these sampled objects from northern and southern Italy does not reveal any tangible connection to the contemporary Sardinian bronze industry.

Comparison with Ores and Artifacts from the Alpine Regions

On the other hand, the typology of some artifacts from the Veneto region would suggest an Alpine connection. Objects such as the Matrei knives have a clear northern Italian and Alpine distribution. Indeed, the Adige valley is one of the main traffic routes leading from the Veneto plain into the Alps, to the regions of Trentino and Alto Adige/Südtirol, which are rich in copper. These deposits were exploited during the Recent Bronze Age (RBA) and FBA periods, as several smelting sites close to the mining areas demonstrate (Cierny et al. 2004a, 2004b). In addition, the pottery distribution of the Alpine Laugen/Luco Culture illustrates contacts of the mountainous regions with the plains (Cierny et al. 2004a, 136, fig. 18). Unfortunately, chemical and isotopic analyses from the northern Italian ore deposits in Trentino and Alto Adige/Südtirol are not yet available. However, there are a few chemical analyses of copper droplets from a smelting site, Acqua Fredda in Trentino, that can be dated approximately from advanced RBA to the FBA by its pottery and metal artifacts (Cierny et al. 2004a, 134–138; 2004b, 160–163). Various finds (e.g., slag, remains of furnaces) prove that it was a workshop area where copper smelting was carried out. One metal fragment (1 x 2 x 0.5 cm) and three slag fragments with small copper inclusions were analyzed by wavelength dispersive X-ray (WDX) analysis (Metten 2003, 61, 116, table 27). All four pieces consist of copper (95.4%–98.5%). Judging from the trace element patterns, they are compositionally similar to the contemporary northern Italian artifacts analyzed in our project (Fig. 23.7). This suggests that the ores processed at Aqua Fredda and those used to produce the artifacts from the Veneto and Lombardy originate from the same ore deposits in the wider Trentino and Alto Adige/Südtirol regions.

Farther north in the eastern Alpine region, there are also many copper deposits that were worked in the Bronze Age. They produced fahlore copper (Sperber 2004), which is chemically distinct and different from the northern Italian artifacts, and

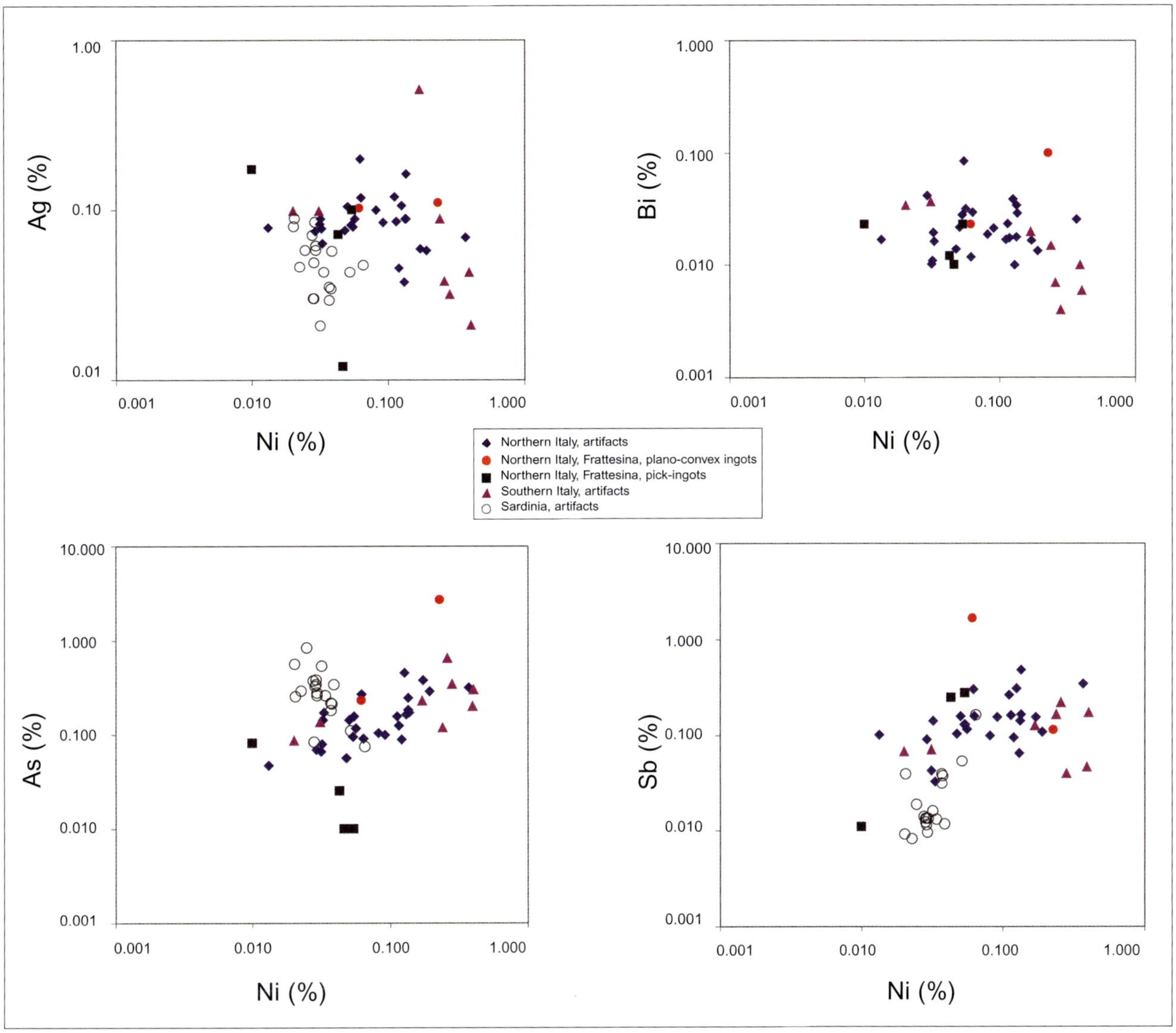

Figure 23.6. Comparison of silver and nickel in objects from this study with Sardinian artifacts (Begemann et al. 2001). The signatures are the same as in Figure 23.2. Sardinian artifacts are shown as open circles in black.

copper with arsenic and nickel as major impurities at concentrations between 0.1% and 1%. However, those deposits also contain uranium, so that the lead isotope composition is highly radiogenic, resulting mostly in negative model ages. In contrast, from the lead isotope ratios in the northern Italian artifacts, one calculates model ages in the range between 350 and 500 million years ago (Fig. 23.8), so that eastern Alpine copper deposits should be excluded as sources for the northern Italian bronze industry.

As mentioned above, there are also very few metal analyses of artifacts available, which could be used for comparison with the new data from the Veneto, Lombardy, Apulia, and Calabria. However, two sites in the Marche region of eastern central Italy provide a fair number of new chemical analyses conducted with energy dispersive X-ray fluorescence (EDXRF), atomic absorbtion spectroscopy (AAS), scanning electron microscopy (SEM), and an unspecified technique (De Marinis et al. 2005, 680–682. However, in this publication it is not specified which single measurements are the results of which analytical methods). The two sites are the settlements of Moscosi di Cingoli and Cisterna di Tolentino, both located several dozens of kilometers inland from the Adriatic coast. They date to a time range from MBA 3 to RBA 2 (Percossi, Pignocchi, and Sabbatini 2005; Sabbatini and Silvestrini 2005).

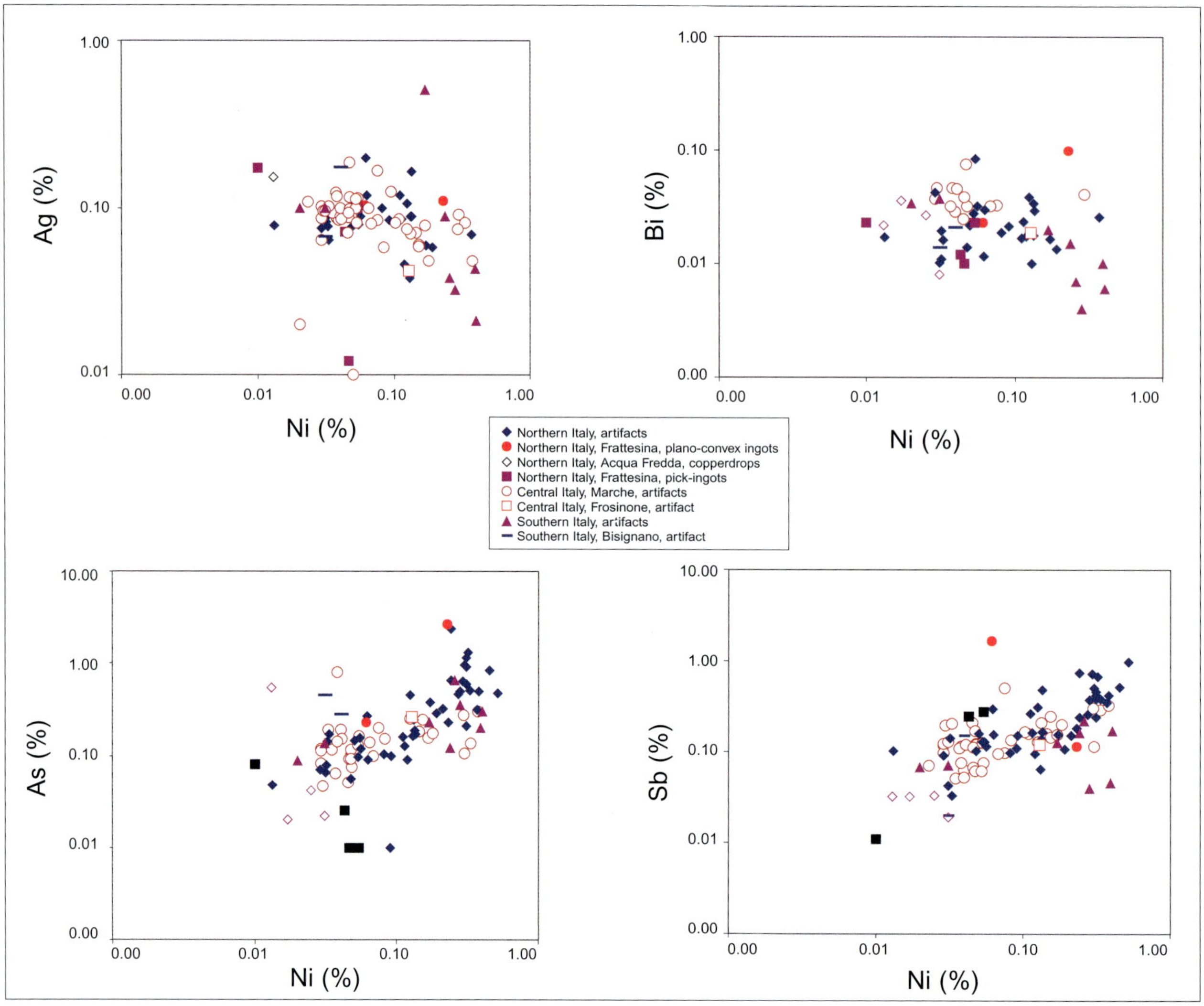

Figure 23.7. Comparison of the minor element composition of the objects from this study with literature data of artifacts from the continental Italian regions of Trentino, Marche, Latium, and Calabria (for references see text).

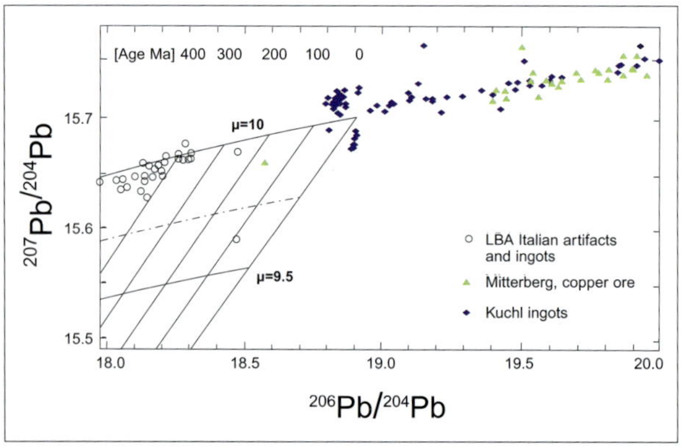

Figure 23.8. Lead isotope ratios (normalized to ^{204}Pb of copper ores from the Mitterberg area in Salzburg, Austria) of LBA copper ingots from the Salzach valley near Mitterberg (Pernicka, unpublished) and artifacts and ingots from Italy (this study). Ma = million years ago.

Thus, they cover the same periods as the artifacts from the Veneto, Lombardy, Apulia, and Calabria. Lead isotope ratios have not been measured, but the similarity between the northern Italian and the Marche group is noteworthy (Fig. 23.7). This good agreement may indicate that the metal workshops in the Marche region were using copper of the same origin as the workshops farther north in the Adige river plain. An interesting detail is the fact that among the published analyses from Moscosi di Cingoli two measurements are from bronze droplets. These should hint at the local production of bronze objects. Both droplets cluster with the artifacts from the Veneto region and thus provide further evidence for our hypothesis of copper export from northern to central Italy.

Copper Circulation in Italy from the Middle to the Final Bronze Age

From the discussion of the analytical results, there is a strong probability that bronze weapons and implements in Apulia and Calabria in southern Italy consist of copper with the same provenance as the bronze weapons and implements in the Veneto and in Lombardy. Theoretically, one can imagine two kinds of explanation for this phenomenon. First, copper was regularly transported from the estuary region of the rivers Po and Adige along the Adriatic coast to Marche, Apulia, and even to the coasts of the Ionian Sea. Second, only certain classes of objects were traded on this route. In this case, our analytical results would not be representative for the local bronze production of southern Italy, but, by coincidence, they only detect those exported objects of northern Italian manufacture.

Hence, it is necessary to take a closer look at the typology of the analyzed objects and their exact chronology. In northern Italy we sampled objects from two large hoard finds, Pila del Brancón, dated to RBA 2/FBA 1 (Salzani 1994, 1998; Jung 2006, 148–150), and Frattesina, Hoard 2, dated to FBA 2 (Salzani 2000; Carancini and Peroni 1999, pls. 29, 32). Swords of types Allerona (type Stätzling in central Europe) and Cetona (types Nenzingen or Reutlingen in central Europe), spear heads with plastic ribs, and a dagger of type Pertosa are among the objects analyzed from Pila del Brancón. These types have a wide geographical distribution inside and outside Italy. Sampled objects of Hoard 2 from Frattesina include implements rather typical for northern and central Italy, like socketed shovels and a winged axe of type Ponte San Giovanni. Additional objects included in this study are mainly swords and daggers of types that also appear at Pila del Brancón. The analyzed spectrum is a good sample of the northern Italian bronze industry from the 13th to 11th century B.C.E.

In southern Italy the possibilities of sampling were more restricted for technical reasons. Nevertheless a certain range of weapons and implements is covered by our analyses. The burial cave Grotta Manaccora on the coast of the Mount Gargano peninsula is one of the most important find places for MBA (and partly RBA) bronze objects in southern Italy. Four swords from that burial cave could be sampled. One of them belongs to type Montegiorgio (Bianco Peroni 1970, 59, no. 127, pl. 17:127), equivalent to the central European types Sprockhoff Ia and Annenheim, with variants. In Italy this type is predominantly found in the northeastern and central regions of the peninsula. The find from Grotta Manaccora is the southernmost occurrence and indeed the only one in southern Italy (Bianco Peroni 1970, 60–61, pl. 69:D). In fact, it plots with the objects found in northern Italy in our analytical diagrams. One might conclude that this weapon was imported from a region farther to the north. However, in the light of the other analytical results of weapons from the same burial cave, which all plot with the artifacts from northern Italy, the case of this sword does not seem to be straightforward.

Two more sword blades from Grotta Manaccora cannot be assigned to any specific type because of their bad state of preservation (one is cataloged by Bianco Peroni [1970, 74, no. 171, pl. 24:171]; the other one cannot be identified with certainty, because the hand grip is entirely lost). The last one of the four sampled pieces belongs to a sword of type Pertosa (Bianco Peroni 1970, 24, no. 39, pl. 6:39). We have sampled a second sword of this type. It was also found in Apulia, but to the south of the Gargano peninsula, inside the artificial burial cave (*hypogaeum*) of Trinitapoli (Tunzi Sisto 1999, 254, 255, no. 1496, pl. 1:1496 [other swords from Tirintapoli could not be sampled, because they were totally corroded]). These two swords are dated to MBA 3 by their stratigraphical contexts (Recchia 1993, 394, fig. 27:L.13; Peroni 1999). The Pertosa type is characteristic of southern and central Italy and does not occur farther north than Latium (Bianco Peroni 1970, 26–27, pl. 68:D; cf. Bettelli 2006). Thus, the specimens from Grotta Manaccora and Trinitapoli cannot be classified as imported objects produced in northern Italy. We are well informed about contemporary sword types of northern Italy (see Bianco Peroni 1970; Cupitò 2006, 62–88). The absence of type Pertosa swords in the north therefore should reflect regional use patterns of Bronze Age sword types in continental Italy. One sword assigned to the type Pertosa (or its Sicilian variety called Thapsos) was found in the eastern Mediterranean, in the Uluburun shipwreck

(Pulak 1988, 21–23, fig. 22; Vagnetti and Lo Schiavo 1989, 222–224, fig. 28.2). A lead isotope analysis by Gale and Stos-Gale (2005, 128) has shown that it was probably produced from Cypriot copper. Therefore, it may be interpreted as a product made in the eastern Mediterranean. In combination with the inferior military quality (when compared to the grip-tongue swords of Mycenaean and Levantine types from the same wreck), this result opens up the possibility that the sword was produced in the east for some warrior with a migration background from the central Mediterranean travelling aboard the ship (Jung 2009, 130–133). This interpretation now gains support by our analytical results showing that Pertosa swords were produced in southern mainland Italy without using Cypriot copper (cf. Fig. 23.4). However, to exclude an Italian production region for the Uluburun sword, it is furthermore necessary to analyze the swords of the Thapsos variety found in Sicily. Meanwhile, these swords have been sampled, and the analyses are under way.

The last two analyzed objects from southern Italy are dated to the Recent Bronze Age. A knife of type Matrei was found in a RBA 2 context at the Calabrian settlement of Broglio di Trebisacce (Luppino, Peroni, and Vanzetti 2005, 731, pl. 97). It represents a shape, which seems to be characteristic of the Italian north and the Alpine regions in general (Jung 2006, 54–55, with bibliography). Our sample of finds from northern Italy includes three knives of the type, which were found at Frattesina. In the lead isotope diagrams, the Broglio knife shows a slightly lower $^{208}Pb/^{206}Pb$ ratio than the rest of the artifacts from continental Italy. It overlaps with measurements of Sardinian artifacts. However, in the various trace element diagrams, it plots with the continental artifacts and not with the Sardinian ones. In conclusion, the knife from Broglio might be an object imported from northern continental Italy. However, one has to admit that we are not well informed about the shapes of RBA knifes in southern Italy (cf. Bianco Peroni 1976, pl. 61:A, B). Moreover, for several reasons there appears a general research gap concerning the southern Italian bronze industry of that period. In contrast to northern Italy, the south is more sparsely settled nowadays, which certainly has a negative effect on the frequency with which chance finds are discovered. (All over Europe most of the known bronze hoards are chance finds; only very few were found during controlled archaeological excavations. The rarity of hoard finds in southern Italy severely affects our knowledge of the bronzes in use in those regions.) Few RBA cemeteries are known in southern Italy and even those excavated are not rich in bronzes. Finally, contemporary settlements are excavated only to a small extent. Therefore, we do not know if the absence of other examples of knives of type Matrei reflects an ancient reality or not. In consequence, one cannot easily declare the knife from Broglio as a northern Italian import. It remains a possibility that it was produced locally in Calabria rather than in the Veneto.

The last southern Italian object is a socketed javeline head from the Apulian settlement of Coppa Nevigata. It has been found in a level dated to RBA 2 (Cinquepalmi and Radina, eds., 1998, 269, fig. 809; for the context, see Cazzella and Moscoloni 1998, 239, 245, 246, figs. 1, 2:b). This is the only real outlier in our data set, which cannot be convincingly compared with any other ore or artifact data that either for archaeological or geographic reasons seems to be relevant in our case. This analytical result suggests that southern Italian workshops had direct or indirect access to more than one copper producing region.

Finally, two measurements of FBA objects from central and southern Italy can be compared with our data. They are part of the British Museum collections and were analyzed by Inductively Coupled Plasma–Auger Electron Microscopy (ICP-AES) and/or AAS for trace elements (Hook 2007, 308, 314, table 1, nos. 127, 219; unfortunately, it is not stated which method was used for each analyzed object). Both are swords of type Allerona: a long one from Frosinone in Latium and a short one from Bisignano in Calabria (Bietti Sestieri and Macnamara 2007, 63, 72, 81–82, 101, pls. 27:127 [drawing incorrect, cf. Schauer 1974, pl. 9:2], 45:219; Giardino 1994, pls. 167:1, 168:1. In the case of the Bisignano sword two samples were measured, one from the blade and one from a rivet [Hook 2007, 314, table 1, no. 219]). Both plot with the other artifacts from continental Italy (north, central, and south, see Fig. 23.7).

The ingots analyzed in our project belong to two different classes: plano-convex ingots and pick ingots. All were found at Frattesina. Two of the former class were analyzed and consist of 92% and 95% copper, respectively, with small quantities of other elements, but without added tin. The four analyzed pick-ingots are made of bronze with a tin content between 10.2% and 14.7%. They were cast in an open mold.

While plano-convex ingots were used all over prehistoric Europe and in the Mediterranean during various periods of the Bronze Age, pick-ingots are much more restricted in space and time. They are found in southern central Europe and not farther south than central Italy (Bietti Sestieri 1997, 387–389, fig. 5). In Italy pick-ingots do not turn up in contexts earlier than FBA 2 (Carancini and Peroni 1999, 19, pls. 29:63, 30:63). For these reasons they were frequently discussed in terms of FBA metal trade (Borgna 1992, 77–87; Borgna and Turk 1996; Bietti Sestieri 1997, 387–390). It was supposed that pick-ingots represent a second step of ingot production, in which copper of plano-convex ingots was cast into pick-ingots (Borgna and Turk 1996, 352).

We have already referred to the lead isotope ratios of the analyzed ingots (Fig. 23.3). Concerning their chemical composition, the diagrams show that the two plano-convex ingots have higher contents of Ni, As, and Bi than the four pick-ingots, while the contents of Sb and Ag of just one plano-convex ingot are higher than in the case of the pick-ingots (Fig. 23.2). These differences are within the range of the variation within ore mineralizations. It should be noted, however, that the plano-convex ingots have higher Fe concentrations than the pick-ingots (not shown in these diagrams). A pair consisting of one pick ingot and one plano-convex ingot is indicative, because both ingots have very similar lead isotope ratios as well as Ag, Ni, and Bi contents. Just As and Sb show a somewhat larger difference. In conclusion, it is very likely that the pick-ingots from Frattesina consist of copper with the same provenance as the plano-convex ingots.

The aforementioned hypotheses of metal exchange by means of exporting pick- and plano-convex ingots are difficult to verify because of the lack of analytical data. Apart from the chemical compositions of two plano-convex ingots from the Marche region, only chemical analyses (carried out with ICP-AES) of pick-ingots and plano-convex ingots from Slovenia (Trampuž-Orel 1996, 213–234) are published.

These chemical analyses show first that the pick-ingots from Slovenia have a very broad concentration range for several elements (e.g., ranging from 0.03% to 15.69% for Ni or from 0.01% to 15.25% for As; Trampuž-Orel 1996, 229–234 [analyses nos. 725, 730, 906]). Second and even more important, these Slovenian pick-ingots (as well as the Slovenian plano-convex ingots) do not contain intentionally added tin. The Sn content is less than 1% for most of them, while the Pb content reaches concentrations of about 30% (Trampuž-Orel 1996, 227 [e.g., analysis no. 582], 234 [analyses nos. 909, 911, 914, 917]). It is a well known phenomenon that Pb was added to bronze or copper in the bronze workshops of central and southeastern Europe from Ha A2/B onward—i.e., in the 11th century B.C.E. and later (Trampuž-Orel 1999, 417–418; Hansen 2005, 97). In contrast, all the pick-ingots from Frattesina analyzed in our project are made of tin bronze (see above). (In this respect they can be compared to two pick-ingots from Madriolo in Friuli, analyzed in the 19th century, which contain 12.27% and 14.77% of Sn [Pigorini 1895, 18]. Four other pick-ingots from Redipuglia [also in Friuli] were analyzed by AES and have Sn contents between 3.2% and 4.8% [Borgna 1992, 29–30]). In addition, the modes of the trace elements are much higher for the Slovenian pick-ingots than for the Slovenian plano-convex ingots or the Italian pick- and plano-convex ingots. One may conclude that the pick-ingots from Frattesina and those from Slovenia (hoards from Veliki Otok I, Kanalski Vrh I and II) do not represent

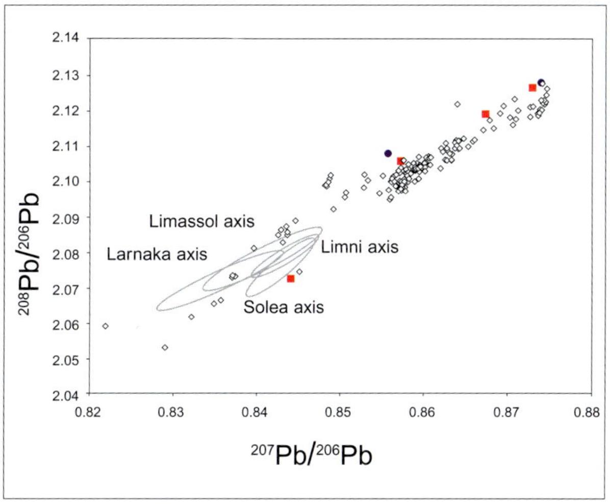

Figure 23.9. Lead isotope ratios of the northern Italian pick-ingots (red squares) and plano-convex ingots (blue dots) compared with copper and lead ores from Sardinia (open squares) and copper ores from Cyprus (outlined by the ellipses).

metal imported from the other side of the Caput Adriae to the regions in which they were found. Thus, they cannot be taken as direct proof of interregional copper trade in this wider geographical area. The few analyses available today show a considerable variation of metal composition of pick-ingots and seem to support Borgna's hypothesis of manifold production and distribution processes reflected by this ingot type (cf. Borgna 1992, 83).

The analytical results of the Slovenian plano-convex ingots are more difficult to interpret. They show an irregular trace element distribution pattern, which is overlapping with the Italian pick- and plano-convex ingots and partially also with the Slovenian pick-ingots. Further conclusions concerning their ore sources as well as their relations to other regions can only be made once lead isotope analysis on these objects become available (of course lead isotope analyses can only be used for provenancing copper of artifacts without intentionally added lead; this excludes many of the pick-ingots from Slovenia).

Returning to the Italian data set, as already stated above, the different Frattesina ingots show a lead isotope pattern that is generally quite similar to the one of the northern and southern Italian artifacts. The pick-ingot and the plano-convex ingot, which closely resemble each other in their chemical and isotopic characteristics, have higher $^{208}Pb/^{206}Pb$ and $^{207}Pb/^{206}Pb$ ratios than the main group (Fig. 23.3), but it is possible that we do not yet grasp the whole extension of the isotopic field of the relevant ore deposits due to an insufficient number of objects analyzed so far. (One socketed shovel from Frattesina plots between the main field of the northern Italian objects and the two outlying ingots. This could be a hint that the isotopic field extends farther and in fact bridges that gap between the main northern Italian group and the two ingots.)

However, there is one true outlier among the pick-ingots (sample no. MA-073188, Fig. 23.4). Morphologically it does not differ from the other ones (Salzani 2000, 40, 42, no. 26), but it plots near Cypriot ores of the Solea axis (Fig. 23.9; for the Cypriot ores, see Gale et al. 1997, 241–244, 246, tables 2–4, 6; Gale, Stos-Gale, and Maliotis 1998, 247, table 6). In the same region of the lead isotope diagram there is also one outlier of the Sardinian copper ores. However, with the large spread of lead isotope ratios in Sardinian deposits, the coincidence may well be fortuitous. The general purity of the copper in this ingot and especially the low silver concentration might suggest favoring a Cypriot origin rather than a Sardinian one. This would be surprising, as the pick-ingot is not a Cypriot type (Lo Schiavo, Macnamara, and Vagnetti 1985, 14), nor are there any other indications for copper importation from Cyprus to north-eastern Italy. (Only one very indirect hint at Cypriot copper transport into the Adriatic exists: a fragment of an oxhide ingot found in Romania. But even for this ingot, an alternative route from the Aegean through the river valleys of the southern Balkans or through the Black Sea is possible [see Jung 2005, 57]). However, the chemical composition and the trace element pattern of this pick-ingot does not really match the signature of artifacts that derive most likely from Cyprus (e.g., Begemann et al. 2001), so that the provenance of this metal must remain unsolved at present.

Conclusions

With all due caution we try to outline some historical-archaeological conclusions that may be drawn from the evaluation of our analytical and archaeological data. We stress that these conclusions are preliminary at present and will most probably need some modification, once more analyses are available. Such future analyses should combine both chemical and lead isotope data, and the samples should be taken from objects from clear stratigraphic contexts.

Our analyses of stratified bronze finds from different sites in the Veneto and in Lombardy provide strong evidence for the interpretation that the bronze workshops in this region received their raw material from the same copper ore deposits throughout the whole RBA and during the first two phases of the FBA. This suggests stable supply conditions.

In addition, it seems that during MBA 3 and probably also in the following periods down to the phase RBA 2 or even into the FBA, southern Italian bronze workshops had access to copper/bronze of the same provenance and that this metal was shipped from northern Adriatic harbor sites. One may guess that the provenance of the raw metal is to be sought in the southern Alpine copper ore deposits, but here we are clearly in need of more analytical data. It seems reasonable to exclude the ore deposits of Tuscany as potential copper sources for that material, while it is more difficult to prove, but still highly probable that Sardinian copper can be excluded as well.

Northern Italian metal may have been transported in the form of ingots to the southern Italian workshops. These ingots were most probably of plano-convex type, as this type is attested even in the southernmost regions of Apulia, on Lipari, and on Sicily (Bernabò-Brea and Cavalier 1980, 756–757, 785–787, pl. 319:e, j, k; Maggiulli 2009, 212–213, fig. 2:241).

References

Begemann, F., S. Schmitt-Strecker, E. Pernicka, and F. Lo Schiavo. 2001. "Chemical Composition and Lead Isotopy of Copper and Bronze from Nuragic Sardinia," *Journal of European Archaeology* 4, pp. 43–85.

Bernabò-Brea, L., and M. Cavalier. 1980. *L'acropoli di Lipari nella preistoria* (*Meligunìs Lipára* IV), Palermo.

Bettelli, M. 2006. "Fogge simili ma non identiche: Alcune considerazioni sulle spade tipo Thapsos-Pertosa," in *Studi di protostoria in onore di Renato Peroni*, Florence, pp. 240–245.

Bianco Peroni, V. 1970. *Die Schwerter in Italien/Le spade nell'Italia continentale* (*Prähistorische Bronzefunde* IV [1]), Munich.

———. 1976. *Die Messer in Italian/I coltelli nell'Italia continentale* (*Prähistorische Bronzefunde* VII [2]), Munich.

Bietti Sestieri, A.-M. 1997. "Italy in Europe in the Early Iron Age," *PPS* 63, pp. 371–402.

Bietti Sestieri, A.-M., and E. Macnamara. 2007. *Prehistoric Metal Artefacts from Italy* (*3500–720* B.C.) *in the British Museum*, London.

Boni, M., and V. Köppel. 1985. "Ore-Lead Isotope Pattern from the Iglesiente-Sulcis Area (SW Sardinia) and the Problem of Remobilization of Metals," *Mineralium Deposita* 20, pp. 185–193.

Borgna, E. 1992. *Il ripostiglio di Madriolo presso Cividale e i pani a piccone del Friuli-Venezia Giulia* (*Studi e ricerche di protostoria mediterranea* 1), Rome.

Borgna, E., and P. Turk. 1996. "Metal Exchange and the Circulation of Bronze Objects between Central Italy and the Caput Adriae (XI–VIIIth B.C.): Implications for the Community Organization," in *Unione Internazionale delle Scienze Preistoriche e Protostoriche* (*Atti del XIII Congresso*), Forli, pp. 351–364.

Bruno, A. 2007. *Punte di lancia nell'età del bronzo nella terraferma italiana. Per una loro classificazione tipologica* (*Fonti Archeologiche per la Protostoria Italiana* 2), Lucca.

Campus, F., and V. Leonelli. 2000. *La tipologia della ceramica nuragica. Il materiale edito*, Viterbo.

———. 2006. "La Sardegna nel Mediterraneo fra l'età del Bronzo e l'età del Ferro. Proposta per una distinzione in fasi," in *Studi di protostoria in onore di Renato Peroni*, Florence, pp. 372–392.

Carancini, G.L., and R. Peroni. 1999. *L'età del bronzo in Italia: Per una cronologia della produzione metallurgica* (*Quaderni di Protostoria* 2), Perugia.

Cazzella, A., and M. Moscoloni. 1998. "Il passaggio Bronzo Recente–Bronzo Finale a Coppa Nevigata e nella Puglia nord-orientale," in *Protovillanoviani e/o Protoetruschi. Ricerche e scavi. Atti del Terzo Incontro di Studi "Preistoria e Protostoria in Etruria," Manciano-Farnese 12/14 maggio 1995*, N. Negroni Catacchio, ed., Florence, pp. 239–247.

Christoforidis, A., and E. Pernicka. 1988. "II. Gruppierung von Metallanalysen mit Hilfe der Clusteranalyse," in *Die endneolithischen und frühbronzezeitlichen Grabfunde auf der Nordstadtterrasse von Singen am Hohentwiel* (*Forschungen und Berichte zur Vor- und Frühgeschichte in Baden-Württemberg* 32), R. Krause, ed., Stuttgart, pp. 252–262.

Cierny, J., F. Marzatico, R. Perini, and G. Weisgerber. 2004a. "La riduzione del rame in località Acqua Fredda al Passo del Redebus (Trentino) nell'età del Bronzo Recente e Finale," in *Alpenkupfer—Rame delle Alpi,* (*Der Anschnitt Beiheft* 17), G. Weisgerber and G. Goldenberg, eds., Bochum, pp. 125–154.

———. 2004b. "Der spätbronzezeitliche Verhüttungsplatz Acqua Fredda am Passo Redebus (Trentino)," in

Alpenkupfer—Rame delle Alpi (*Der Anschnitt Beiheft* 17), G. Weisgerber and G. Goldenberg, eds., Bochum, pp. 155–164.

Cinquepalmi, A., and F. Radina, eds. 1998. *Documenti dell'età del Bronzo: Ricerche lungo il versante adriatico pugliese*, Fasano di Brindisi.

Cupitò, M. 2006. *Tipocronologia del Bronzo medio e recente tra l'Adige e il Mincio sulla base delle evidenze funerarie* (*Saltuarie dal laboratorio del Piovego* 7), Padua.

De Marinis, G., A. Giumlia-Mair, M. Miccio, and P. Pallecchi. 2005. "Tecnologie produttive nei siti dell'Età del Bronzo di Moscosi di Cingoli e Cisterna di Tolentino," in *Atti della XXXVIII Riunione scientifica "Preistoria e Protostoria delle Marche," Portonovo, Abbadia di Fiastra, 1–5 ottobre 2003*, Florence, pp. 679–694.

Gale, N., and Z. Stos-Gale. 2005. "Zur Herkunft der Kupferbarren aus dem Schiffswrack von Uluburun und der spätbronzezeitliche Metallhandel im Mittelmeerraum," in *Das Schiff von Uluburun. Welthandel vor 3000 Jahren. Katalog der Ausstellung des Deutsches Bergbau-Museums Bochum*, Ü. Yalçın, C. Pulak, and R. Slotta, eds., Bochum, pp. 117–131.

Gale, N.H., Z.A. Stos-Gale, and G. Maliotis. 1998. "A Preliminary Survey of the Cypriot Slag Heaps and Their Contribution to the Reconstruction of Copper Production on Cyprus," in *Metallurgica antiqua: In Honour of Hans Gert Bachmann and Robert Maddin* (*Der Anschnitt Beiheft* 8), Bochum, pp. 235–262.

Gale, N.H., Z.A. Stos-Gale, G. Maliotis, and N. Annetts. 1997. "Lead Isotope Data from the Isotrace Laboratory, Oxford: Archaeometry Data Base 4, Ores from Cyprus," *Archaeometry* 39, pp. 237–246.

Giardino, C. 1994. "Bronzi protostorici dalla Sibaritide conservati presso il British Museum," in *Enotri e Micenei nella Sibaritide*, R. Peroni and F. Trucco, eds., Taranto, pp. 779–782.

Hansen, S. 1994. *Studien zu den Metalldeponierungen während der älteren Urnenfelderzeit zwischen Rhônetal und Karpatenbecken* (*Universitätsforschungen zur prähistorischen Archäologie* 21), Bonn.

———. 2005. "Neue Forschungen zur Metallurgie der Bronzezeit in Südosteuropa," in *Anatolian Metal III* (*Der Anschnitt Beiheft* 18), Ü. Yalçın, ed., Bochum.

Hook, D. 2007. "The Composition and Technology of Selected Bronze Age and Early Iron Age Copper Alloy Artefacts from Italy," in *Prehistoric Metal Artefacts from Italy (3500–720 B.C.) in the British Museum*, A.-M Bietti Sestieri and E. Macnamara, eds., London, pp. 308–323.

Jung, R. 2005. "Aspekte des mykenischen Handels und Produktenaustauschs," in *Interpretationsraum Bronzezeit. Festschrift Bernhard Hänsel* (*Universitätsforschungen zur Prähistorischen Archäologiue* 121), B. Horejs, R. Jung, E. Kaiser, and B. Teržan, eds., Bonn, pp. 45–70.

———. 2006. *Χρονολογία comparata. Vergleichende Chronologie von Südgriechenland und Süditalien von ca. 1700/1600 bis 1000 v.u.Z.* (*Veröffentlichungen der mykenischen Kommission* 26), Vienna.

———. 2009. "I 'bronzi internazionali' ed il loro contesto sociale fra Adriatico, Penisola balcanica e coste levantine," in *Dall'Egeo all'Adriatico: organizzazioni sociali, modi di scambio e interazione in età postpalaziale (XII–XI sec. a. C.). From the Aegean to the Adriatic: Social Organisations, Modes of Exchange and Interaction in Postpalatial Times (12th–11th c. B.C.). Atti del Seminario internazionale 1–2 Dicembre 2006 (Udine)* (*Studi e ricerche di protostoria mediterranea* 8), E. Borgna and P. Cassola Guida, eds., Rome, pp. 129–157.

Lo Schiavo, F. 1998. "Zur Herstellung und Distribution bronzezeitlicher Metallgegenstände im nuraghischen Sardinien," in *Mensch und Umwelt in der Bronzezeit Europas. Abschlußtagung der Kampagne des Europarates: Die Bronzezeit. Das erste goldene Zeitalter Europas an der Freien Universität Berlin, 17–19 März 1997*, B. Hänsel, ed., Kiel, pp. 193–216.

Lo Schiavo, F., E. Macnamara, and L. Vagnetti. 1985. "Late Cypriot Imports to Italy and their Influence on Local Bronzework," *PBSR* 53, pp. 1–71.

Luppino, S., R. Peroni, and A. Vanzetti. 2005. "Broglio di Trebisacce, campagna 2004," in *Tramonto della Magna Grecia. Atti del XLIV Convegno di studi sulla Magna Grecia, Taranto 24–28 settembre 2004*, Taranto, pp. 729–731.

Lutz, J., and E. Pernicka. 1996. "Energy Dispersive X-ray Fluorescence Analysis of Ancient Copper Alloys: Empirical Values for Precision and Accuracy," *Archaeometry* 38 (2), pp. 313–323.

Maggiulli, G. 2009. "I ripostigli di Roca Vecchia (Lecce): analisi dei materiali e problematiche archeologiche," in *Dall'Egeo all'Adriatico: organizzazioni sociali, modi di scambio e interazione in età postpalaziale (XII–XI sec. a. C.). From the Aegean to the Adriatic: Social Organisations, Modes of Exchange and Interaction in Postpalatial Times* (*12th–11th c. B.C.*). *Atti del Seminario internazionale 1–2 Dicembre 2006 (Udine)* (*Studi e ricerche di protostoria mediterranea* 8), E. Borgna and P. Cassola Guida, eds., Rome, pp. 205–218.

Maraszek, R. 2006. *Spätbronzezeitliche Hortfundlandschaften in atlantischer und nordischer Metalltradition* (*Veröffentlichungen des Landesamtes für Denkmalpflege und Archäologie Sachsen-Anhalt, Landesmuseum für Vorgeschichte* 60), Halle.

Merkel, J. 1990. "Experimental Reconstruction of Bronze Age Copper Smelting Based on Archaeological Evidence from Timna," in *The Ancient Metallurgy of Copper*, B. Rothenberg, ed., London.

Metten, B. 2003. "Beitrag zur spätbronzezeitlichen Kupfermetallurgie im Trentino (Südalpen) im Vergleich mit anderen prähistorischen Kupferschlacken aus dem Alpenraum," *Metalla* 10, pp. 1–122.

Muhly, J.D. 1983. "Lead Isotope Analysis and the Kingdom of Alashiya," *RDAC* 1983, pp. 210–218.

Niederschlag, E., E. Pernicka, T. Seifert, and M. Bartelheim. 2003. "Determination of Lead Isotope Ratios by Multiple Collector ICP-MS: A Case Study of Early Bronze Age Artefacts and Their Possible Relation with Ore Deposits of the Erzgebirge," *Archaeometry* 45, pp. 61–100.

Percossi, E., G. Pignocchi, and T. Sabbatini. 2005. "Un sito dell'età del bronzo a Cisterna di Tolentino," in *Atti della XXXVIII Riunione Scientifica "Preistoria e Protostoria delle Marche," Portonovo, Abbadia di Fiastra, 1–5 ottobre 2003*, Florence, pp. 659–678.

Pernicka, E. 1987. "Erzlagerstätten in der Ägäis und ihre Ausbeutung im Altertum: Geochemische Untersuchungen zur Herkunftsbestimmung archäologischer Metallobjekte," *JRGZM* 34 [1989], pp. 607–714.

———. 1990. "Gewinnung und Verbreitung der Metalle in prähistorischer Zeit," *JRGZM* 37 [1995], pp. 21–129.

———. 1999. "Trace Element Fingerprinting of Ancient Copper: A Guide to Technology or Provenance?" in *Metals in Antiquity* (*BAR-IS* 792), S.M.M. Young, A.M. Pollard, P. Budd, and R.A. Ixer, eds., Oxford, pp. 163–171.

Peroni, R. 1999. "La cronologia e il contesto storico-culturale," in *Ipogei della Daunia: Preistoria di un territorio*, A.M. Tunzi Sisto, Foggia, pp. 217–219.

Pigorini, L. 1895. "Antichi pani di rame e di bronzo da fondere rinvenuti in Italia," *BPI* 21, pp. 5–38.

Pulak, C. 1988. "The Bronze Age Shipwreck at Ulu Burun, Turkey: 1985 Campaign," *AJA* 92, pp. 1–37.

Recchia, G. 1993. "Grotta Manaccora (Peschici). Considerazioni sulla Grotticella funeraria e sull'area antistante (scavi Rellini-Baumgärtel)," *Origini* 17, pp. 317–401.

Sabbatini, T., and M. Silvestrini. 2005. "Piano di Fonte Marcosa, Moscosi di Cingoli: un sito pluristratificato dell'Appennino marchegiano. Le fasi del Bronzo Recente," in *Istituto Italiano di Preistoria e Protostoria: Atti della XXXVIII Riunione Scientifica "Preistoria e Protostoria delle Marche," Portonovo, Abbadia di Fiastra, 1–5 ottobre 2003*, Florence, pp. 639–657.

Salzani, L. 1989a. "Fratta Polesine, Frattesina," *Quaderni di Archeologia del Veneto* 5, pp. 66–68.

———. 1989b. "Necropoli dell'Età del Bronzo Finale alle Narde di Fratta Polesine, prima nota," *Padusa* 25, pp. 5–42.

———. 1994. "Nogara. Rinvenimento di un ripostiglio di bronzi in località 'Pila del Brancón'," *Quaderni di Archeologia del Veneto* 10, pp. 83–94.

———. 1998. "Segnalazioni di rinvenimenti archeologici nel Veronese. Nogara. Nuovi dati sul ripostiglio della Pila del Brancón," *Quaderni di Archeologia del Veneto* 14, pp. 66–71.

———. 2000. "Fratta Polesine. Il ripostiglio di bronzi n. 2 da Frattesina," *Quaderni di Archeologia del Veneto* 16, pp. 38–46.

———. 2005. *La necropoli dell'età del Bronzo all'Olmo di Nogara* (*Memorie del Museo Civico di Storia Naturale di Verona 2. Ser. Sezione dell'Uomo* 8), Verona.

———. 2006. "L'area votiva di Corte Lazise a Villa Bartolomea (Verona). Nuovi rinvenimenti e considerazioni generali," in *...ut...rosae...ponerentur. Scritti di archeologia in ricordo di Giovanna Luisa Ravagnan* (*Quaderni di Archeologia del Veneto Serie Speciale 2*), E. Bianchin Citton, and M. Tirelli, eds., Venice, pp. 25–34.

Salzani, L., ed. 1987. *La preistoria lungo la valle del Tartaro*, Isola della Scala.

Schauer, P. 1974. "Zu einigen italischen Schwertern in westeuropäischen Museen," in *Beiträge zu italienischen und griechischen Bronzefunden* (*Prähistorische Bronzefunde* 20 [1]), H. Müller-Karpe, ed., Munich, pp. 27–32.

Sperber, L. 2004. "Zur Bedeutung des nördlichen Alpenraumes für die spätbronzezeitliche Kupferversorgung in Mitteleuropa," in *Alpenkupfer—Rame delle Alpi* (*Der Anschnitt Beiheft* 17), G. Weisgerber and G. Goldenberg, eds., Bochum, pp. 303–345.

Stos-Gale, Z., N.H. Gale, J. Houghton, and R. Speakman. 1995. "Lead Isotope Data from the Isotrace Laboratory, Oxford: Archaeometry Data Base 1, Ores from the Western Mediterranean," *Archaeometry* 37, pp. 407–415.

Trampuž-Orel, N. 1996. "Spektrometriène raziskave depojskih najdbe pozne bronaste dobe/Spectrometric Reasearch of the Late Bronze Age Hoard Finds," in *Depojske in posamezne kovinske najdbe bakrene in bronaste dobe na Slovenskem 2/Hoards and Individual Metal Finds from the Eneolithic and Bronze Ages in Slovenia* 2 (*Katalogi in monografije* 30), B. Teržan, ed., Ljubljana, pp. 165–243.

———. 1999. "Archaeometallurgic Investigations in Slovenia. A History of Research on Non-Ferrous Metals," *ArhVest* 50, pp. 407–429.

Tunzi Sisto, A.M. 1999. *Ipogei della Daunia: Preistoria di un territorio*, Foggia.

Tylecote, R.F., H.A. Ghaznavi, and P.J. Boydell. 1977. "Partitioning of Trace Elements between the Ores, Fluxes, Slags and Metal during the Smelting of Copper," *JAS* 4, pp. 305–333.

Vagnetti, L., and F. Lo Schiavo. 1989. "Late Bronze Age Long Distance Trade in the Mediterranean: The Role of the Cypriots," in *Early Society in Cyprus*, E. Peltenburg, ed., Edinburgh, pp. 217–243.

CHAPTER

24

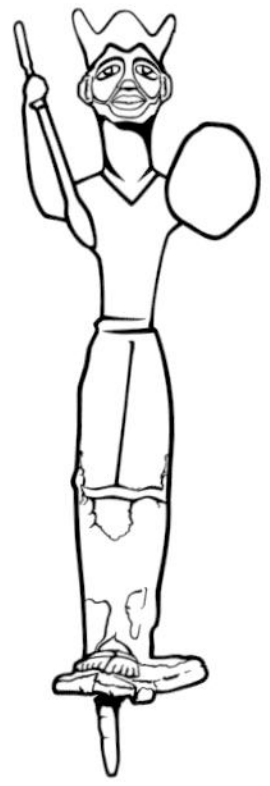

Cyprus, Copper, and Alashiya

A. Bernard Knapp

The Alashiya Issue

Cyprus, copper, and Alashiya have formed the subject of academic inquiry for well over a century, but it was not before James Muhly published two key papers that interested scholars finally could begin to appreciate all the complexities involved in integrating Cypriot archaeological and archaeometallurgical data, cuneiform documentary evidence, and historical interpretation (Muhly 1972, 1989). Muhly's views, at least about the identification of Alashiya with Cyprus, have found supporters (e.g., Knapp 1985; Knapp, ed., 1996, 3–13) as well as detractors (e.g., Merrillees 1972, 1987). Overall, however, a century of very uneven debate and scholarship has done nothing to persuade all Cypriot archaeologists that Alashiya was the Bronze Age name for Cyprus, nor will I attempt to do so here (but see Knapp 2008, 298–347 for full discussion of all the Alashiya evidence and its relevance for the protohistory of Cyprus).

An entirely different perspective on the identification of Alashiya emerged from comparative petrographic and chemical analyses on four cuneiform tablets sent from Alashiya to the Egyptian pharaoh and found at the site of Amarna, two tablets found at the Syrian site of Ugarit, and nine Cypro-Minoan documents from Cyprus itself (Goren et al. 2003). The notably consistent results of this extensive body of analytical work showed that five of the clay tablets sent from Alashiya (to Amarna and Ugarit) were manufactured from clays that once formed in the foothills of Cyprus's southern Troodos Mountains. Although one need not necessarily conclude from these analyses that the Alashiya tablets in question were inscribed at Alassa-*Paliotaverna* or Kalavassos–*Ayios Dhimitrios* (Goren et al. 2003, 250–252), it would seem perverse to deny that these tablets, and by extension all documents originating in Alashiya, came from Cyprus. But perversity has never deterred archaeologists from constructing their arguments, and in the rest of this paper I take the identification of Alashiya with Cyprus to have been demonstrated beyond any reasonable doubt. I leave it to the sceptics to identify any other land in the eastern Mediterranean that can be so convincingly linked, historically or materially, with Alashiya.

Copper in the Alashiya Documents

As Muhly (1972) pointed out in his pioneering summary and historical interpretation of most of the then-known documentary records mentioning Alashiya, the earliest indisputable references to Alashiya appear in 19th–17th century B.C. cuneiform texts from Mari, Alalakh, and Babylonia (Knapp, ed., 1996, 17–20, 30). Most of these references are concerned primarily with copper, bronze, or silver. At Mari, all references to Alashiya deal with the import of metals, whether copper or bronze, and distinguish between "copper," "mountain copper," and "refined, quality copper." At Alalakh, one economic text mentions silver from Alashiya, while another tablet from Babylonia records 12 *minas* of copper from Alashiya. Although we cannot otherwise prove that direct trading contacts existed between Cyprus and Syria or Babylonia, these references are contemporary with the earliest finds of Near Eastern imports into Cyprus, many of them recovered from sites near copper ore deposits in the northern and eastern Troodos Mountains. Courtois (1986, 74–75) plausibly suggested that these imports represented a Near Eastern package of goods exchanged for Cypriot copper. Moreover, as Alashiya's copper resources increasingly were sought by Near Eastern merchants, so Alashiya's merchants sought access to tin, another eastern import (Muhly 1985; 1993, 131–132).

Most written documents that refer to Alashiya stem from the Late Bronze Age and are concerned with economic or diplomatic transactions. These include the Akkadian letters found at Amarna in Egypt (Moran 1996, 21–25), Akkadian and Ugaritic royal or official letters and diplomatic texts from Ugarit (Beckman 1996a, 26–28; Walls 1996, 36–40), and at least five newly excavated Akkadian tablets sent by the king of Alashiya or two of his "viziers" to Ugarit (preliminary reports in Malbran-Labat 1999, 121; Yon 2003, 47–48; see also Singer 2006).

During the Late Bronze Age, Alashiya supplied copper upon demand to various lands. Five of the nine Amarna letters from Egypt mention copper shipments from Alashiya (tablets EA 33, EA 34, EA 35, EA 36, EA 40). Three of them give the measure, in "talents" (GUN URUDU—*bilat erê*): 10 talents (EA 33: 18), 100 talents (EA 34: 18), 3 talents of "refined" copper (EA 40: 13). The other Amarna passages omit the measure and give only a number (see Table 24.1, following Zaccagnini's analysis [1986, 414]).

The information given in one of the recently published Akkadian documents from Ras Shamra (RS 94.2475) must also be mentioned here: 33 (ingots) of copper, weighing 30 talents and 6,500 shekels (Malbran-Labat in Galliano, Calvet, and al-Maqdissi 2004, 188).

The approximate weight of one talent, or ingot, can be determined from the known measuring systems in use during the Late Bronze Age (Table 24.2; based on Zaccagnini [1986, 416, with refs.; fuller discussion in Knapp 2008, 309–310]).

Whether we take the Syrian and Hittite talent at 28.2 kg, or the average of all the talents in use in the eastern Mediterranean and Near East at this time (28.11 kg), the *average* weight of one talent seems to have been approximately 28 kg (Alberti and Parise 2005, 382, 389, pl. 83), the *approximate* (ideal standard) weight of one copper oxhide ingot. In this respect, it must be pointed out that Muhly (1979, 95) argued long ago that Late Bronze Age copper oxhide ingots had a standard weight of 29 kg and represented the equivalent of one talent. However, even if most oxhide ingots were intended to weigh one talent (i.e., about 28 kg), the actual weight of those that have been recovered throughout the Mediterranean varies considerably, ranging from 21–39 kg (Zaccagnini 1986, 416; Alberti and Parise 2005, 382).

Other Near Eastern documents that list copper coming from Alashiya include:

1) A Hittite text from Boğazköy, dated ca. 1200 B.C. (Beckman 1996b, 32), lists one talent of copper paid as "tribute" by the king of Alashiya.
2) A Hittite ritual text of the 13th century B.C. mentions "copper and bronze from Alashiya, from Mount Taggata" (Beckman 1996b, 35).
3) An Akkadian text from Ugarit mentions copper from Alashiya being sent to Emar, a major Syrian trade center situated on the Euphrates River (Beckman 1996a, 28).
4) An Akkadian letter from Ugarit of the late 13th century B.C. lists 33 ingots of copper sent by the king of Alashiya to the ruler of Ugarit as a "greeting gift" (RS 94.2177+ and RS 94.2447, see Yon 2003, 47–48).

From Egyptian hieroglyphic texts, two place names—*'a-si-ja* (Asiya) and *'á-la-sá* (Alasa)—are usually regarded as the equivalent of Alashiya (Vercoutter 1956, 179–180; Muhly 1972, 208–209). Both Asiya and Alasa sent large shipments of copper to Egypt. The "tribute" (*inw* actually means "supply"; Liverani 1990, 255–266) reportedly sent by the prince of Asiya to Tuthmosis III during the 15th century B.C. included 108.5 ingots of copper and 2,400 *dbn* of (s)melted copper (Year 34); an unspecified amount (Year 38); and 40 ingots of copper (Year 39) (Ockinga 1996, 42, texts 67–69). The 2,400 *dbn* (about 200 kg) of copper listed for Year 34 is most likely specified to denote copper in non-ingot form (Zaccagnini 1986, 414). Inscribed over the heads of 31 figures depicted in a processional scene, a 13th century B.C. hieroglyphic text (reign of Ramesses II) mentions that "silver and bronze in countless quantities, millions, hundreds of thousands" were brought to Egypt from the mountain of Asiya (inscribed over figure 21) and from the mountain of Alasa (inscribed over figure 22) (Ockinga in Knapp, ed., 1996, 45, 47). Dating from around 1200 B.C. (19th Dynasty, perhaps the reign of Seti II), the Papyrus Anastasi IV records "many ingots of raw copper" brought as gifts for pharaoh on the necks of the "children of Alasa" (Ockinga 1996, 49). Muhly (1972, 209–210) observed that the image one gets from this description seems to be reflected in New Kingdom tomb paintings that show gifts for pharaoh (metals and ingots, vessels and animals) born on the necks of emissaries.

El-Amarna Tablet	Amount of Copper
EA 33: 16	200 (ingots) of copper
EA 35: 10	500 (ingots) of copper
EA 36: 6	120 +? (ingots) of copper
EA 40: 7	9 (ingots) of copper
EA 40: 13	5 (ingots) of copper
EA 36: 6–7	70 and 30 copper (ingots) weighing (one) talent

Table 24.1. Copper from Alashiya mentioned in the Amarna Letters.

Measuring System	Weight of One Talent
Babylonian	30.3 kg
Syrian	28.2 kg
Hittite	28.2 kg
"Ashdod" (Levant)	23.5 to 23.7 kg
Mycenaean	29.5 or 31.2 kg

Table 24.2. The talent in various Bronze Age measuring systems.

Discussion

All of these documents indicate that Alashiya: (1) played a central role in eastern Mediterranean diplomacy and trade during the Late Bronze Age; (2) provided copper on demand to Egypt, Hittite Anatolia, Babylonia, Ugarit, and other northern Syrian city-states; and (3) enjoyed close economic and political relations with Egypt and the Levant. Given the constant demand for Alashiyan copper by the rulers or merchants of all these areas, the land of Alashiya clearly held an important position among the major economic powers of the time. The documentary evidence from Ugarit, in particular, reveals the extent to which Alashiya was involved in Levantine affairs during the 14th–13th centuries B.C.

Although the documentary and archaeological evidence associated with copper oxhide ingots is, admittedly, complex, we can at least offer a tentative impression of the scale of the Bronze Age trade in copper and the role of Alashiya/Cyprus in that trade. If, as argued above, we take the average weight of a copper oxhide ingot to be 28 kg, the Amarna letters alone indicate that we are dealing with shipments of Cypriot copper ranging from 140 kg (5 ingots) to 14,000 kg (500 ingots). The latter no longer appears to be an extreme or inflated amount, given the 10,160 kg of copper (nearly ten tons) recovered from the Uluburun shipwreck (Pulak 2000, 140).

But how does archaeological evidence from Cyprus dating to the 14th–13th centuries B.C. correspond to this interpretation of the documentary records? First of all, we find indisputable evidence for the production and/or processing of copper ores at several different Cypriot sites—e.g., Enkomi, Kition, Kalavassos-*Ayios Dhimitrios*, Alassa-*Paliotaverna*, Maroni-*Vournes*, and Hala Sultan Tekke. Secondly, the wealth of these sites includes notable amounts of imported goods from western Asia and Egypt (Knapp 1998, 2006). Thirdly, the importance of Cyprus's abundant copper resources for the production of bronze in many lands of western Asia and the Mediterranean, as well as the resulting prosperity of Cyprus during the Late Bronze Age, have been argued by many different archaeologists, not least James Muhly (1986, 1989; also Keswani 1993; Knapp 1994, 282–290; Peltenburg 1996; Pickles and Peltenburg 1998).

Documentary records, archaeological data, and archaeometallurgical evidence all confirm that Cyprus was involved in intense and continuous economic, political, and social relations with the Levant (especially Ugarit), Egypt, Hittite Anatolia, and the Aegean. All of the documentary evidence related to Alashiya cited above, and much more besides, not only demonstrate the extent of Cyprus's economic contacts with neighboring lands but also reveals a concern with trade emissaries, the exchange of precious goods as well as raw materials, and the economic practices that both enabled and constrained eastern Mediterranean spheres of interaction during the Late Bronze Age. Several coastal centers on Cyprus served as commercial entrepots for Aegean and Levantine trade at this time—e.g., Enkomi, Kition, and Hala Sultan Tekke in the east; *Ayios Dhimitrios*, Maroni, and Alassa in the south; *Toumba tou Skourou* in the west. Even non-indigenous products such as gold, silver, ivory, lapis lazuli, and ebony were transhipped through these ports. These coastal centers served multiple purposes—commercial, industrial, ceremonial, and administrative—and can be distinguished by their size and spatial layout from smaller inland towns, and the even smaller production sites (mining, agricultural, pottery) of the interior (Knapp 2008, 134–144).

Conclusion

A growing body of documentary evidence related to Alashiya, taken into account alongside the archaeological record of Late Bronze Age Cyprus, indicates that—despite currently prevailing opinions—economic enterprises and political power during and probably throughout the centuries 1400–1200 B.C. were invested centrally in the king of Alashiya (argued at length, with full references, in Knapp 2008, 324–341). Whether this king was the paramount ruler or a *primus inter pares* (Goren et al. 2003, 248–252), he exercised extensive control over multiple facets of production, trade, and international diplomacy, both within and beyond Cyprus. It is still not possible to determine where the king's power center was located, or if indeed it was located in a single place. As Muhly (1989, 299) long ago pointed out, however, Enkomi is still the strongest candidate, even if Kalavassos–*Ayios Dhimitrios* and Alassa-*Paliotaverna* also must have played some role in the island's geopolitical configuration. It is likely that Cyprus's ruling elite(s) resided in and controlled all three centers, at different times, and for differing reasons.

I hope to have demonstrated in this brief study the central role that James Muhly's work and ideas have played in any and all interpretations—in particular my own—of Late Bronze Age Cyprus, copper, and Alashiya.

References

Alberti, M.E., and N. Parise. 2005. "Towards a Unification of Mass-Units between the Aegean and the Levant," in *Emporia: Aegeans in the Central and Eastern Mediterranean* (*Aegaeum* 25), R. Laffineur and E. Greco, eds., Liège, pp. 381–391.

Beckman, G. 1996a. "Akkadian Documents from Ugarit," in Knapp, ed., 1996, pp. 26–28.

———. 1996b. "Hittite Documents from Hattusha," in Knapp, ed., 1996, pp. 31–35.

Courtois, J.-C. 1986. "A propos des apports orientaux dans la civilisation du bronze récent a Chypre," in *Acts of the International Archaeological Symposium: Cyprus between the Orient and Occident*, V. Karageorghis, ed., Nicosia, pp. 69–87.

Goren, Y., S. Bunimovitz, I. Finkelstein, and N. Na'aman. 2003. "The Location of Alashiya: New Evidence from Petrographic Investigation of Alashiyan Tablets from el-Amarna and Ugarit," *AJA* 107, pp. 233–255.

Galliano, G., Y. Calvet, and M. al-Maqdissi. 2004. *Le royaume d'Ougarit. Aux origins de l'alphabet*, Paris and Lyon.

Keswani, P.S. 1993. "Models of Local Exchange in Late Bronze Age Cyprus," *BASOR* 292, pp. 73–83.

Knapp, A.B. 1985. "Alašiya, Caphtor/Keftiu, and Eastern Mediterranean Trade: Recent Studies in Cypriote Archaeology and History," *JFA* 12, pp. 231–250.

———. 1994. "Emergence, Development and Decline on Bronze Age Cyprus," in *Development and Decline in the Mediterranean Bronze Age* (*Sheffield Archaeological Monograph* 8), C. Mathers and S. Stoddart, eds., Sheffield, pp. 271–304.

———. 1998. "Mediterranean Bronze Age Trade: Distance, Power and Place," in *The Aegean and the Orient in the Second Millennium* (*Aegaeum* 18), E.H. Cline and D. Harris-Cline, eds., Liège, pp. 260–280.

———. 2006. "Orientalisation and Prehistoric Cyprus: The Social Life of Oriental Goods," in *Debating Orientalisation: Multidisciplinary Approaches to Change in the Ancient Mediterranean* (*Monographs in Mediterranean Archaeology* 10), C. Riva and N Vella, eds., London, pp. 48–65.

———. 2008. *Prehistoric and Protohistoric Cyprus: Identity, Insularity and Connectivity*, Oxford.

Knapp, A.B., ed. 1996. *Near Eastern and Aegean Texts from the Third to the First Millennia* B.C. (*Sources for the History of Cyprus* 2), Altamont, NY.

Liverani, M. 1990. *Prestige and Interest: International Relations in the Near East ca. 1600–1100* B.C., Padua.

Malbran-Labat, F. 1999. "Nouvelles données épigraphiques sur Chypre et Ougarit," *RDAC* 1999, pp. 121–123.

Merrillees, R.S. 1972. "Alasia," in *Acts of the First International Cyprological Congress (Nicosia, 14–19 April, 1969)*, V. Karageorghis, ed., Nicosia, pp. 111–119.

———. 1987. *Alashia Revisited* (*Cahiers de la révue Biblique* 22), Paris.

Moran, W.L. 1996. "Akkadian/Amarna," in Knapp, ed., 1996, pp. 21–25.

Muhly, J.D. 1972. "The Land of Alašiya: References to Alašiya in the Texts of the Second Millennium B.C., and the History of Cyprus in the Late Bronze Age," in *Acts of the First International Cyprological Congress (Nicosia, 14–19 April, 1969)*, V. Karageorghis, ed., Nicosia, pp. 201–219.

———. 1979. "Cypriote Copper: Some Geological and Metallurgical Problems," in *Acts of the International Archaeological Symposium: The Relations between Cyprus and Crete, ca. 2000–500 B.C.*, V. Karageorghis, ed., Nicosia, pp. 87–100.

———. 1985. "Sources of Tin and the Beginnings of Bronze Metallurgy," *AJA* 89, pp. 275–291.

———. 1986. "The Role of Cyprus in the Economy of the Eastern Mediterranean during the Second Millennium B.C.," in *Acts of the International Archaeological Symposium: Cyprus between the Orient and the Occident*, V. Karageorghis, ed., Nicosia, pp. 45–60.

———. 1989. "The Organisation of the Copper Industry in Late Bronze Age Cyprus," in *Early Society in Cyprus*, E.J. Peltenburg, ed., Edinburgh, pp. 298–314.

———. 1993. "Metalle. B: Archäologisch," *Reallexikon der Assyriologie und vorderasiatische Archaologie* 8 (1/2), pp. 119–136.

Ockinga, B.G. 1996. "Hieroglyphic Texts from Egypt," in Knapp, ed., 1996, pp. 42–50.

Peltenburg, E.J. 1996. "From Isolation to State Formation in Cyprus, c. 3500–1500 B.C.," in *The Development of the Cypriot Economy*, V. Karageorghis and D. Michaelides, eds., Nicosia, pp. 17–44.

Pickles, S., and E.J. Peltenburg. 1998. "Metallurgy, Society and the Bronze/Iron Transition in the East Mediterranean and the Near East," *RDAC* 1998, pp. 67–100.

Pulak, C. 2000. "The Copper and Tin Ingots from the Late Bronze Age Shipwreck at Uluburn," in *Anatolian Metal* 1 (*Der Anschnitt Beiheft* 13), Ü. Yalçın, ed., Bochum, pp. 137–157.

Singer, I. 2006. "Ships Bound for Lukka: A New Interpretation of the Companion Letters RS 94.2530 and RS 94.2523," *AF* 33, pp. 242–262.

Vercoutter, J. 1956. *L'Egypte et le monde égéen préhellènique* (*Bibliotheque des Études* 22), Cairo.

Walls, N. 1996. "Ugaritic Documents from Ugarit," in Knapp, ed., 1996, pp. 36–40.

Yon, M. 2003. "The Foreign Relations of Ugarit," in *Πλόες . . . Sea Routes . . .: Interconnections in the Mediterranean, 16th–6th Centuries B.C. Proceedings of the International Symposium held at Rethymnon, Crete, September 29th–October 2nd, 2002*, N.C. Stampolidis and V. Karageorghis, eds., Athens, pp. 41–51.

Zaccagnini, C. 1986. "Aspects of Copper Trade in the Eastern Mediterranean during the Late Bronze Age," in *Traffici Micenei nel Mediterraneo: Problemi Storici e Documentazione Archeologica* (*Magna Graecia* 3), M. Marazzi, S. Tusa, and L. Vagnetti, eds., Taranto, pp. 413–424.

CHAPTER

25

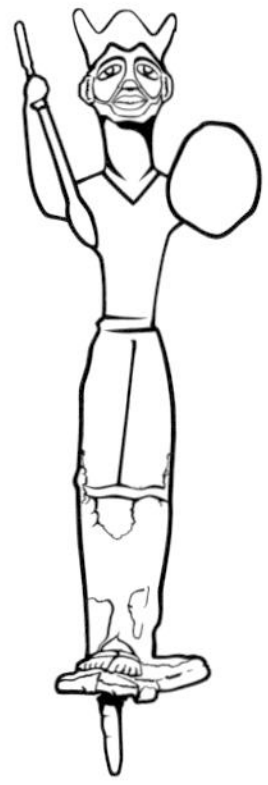

Alashiya: A Scientific Quest for Its Location

Robert S. Merrillees

Muhlenbergiam cuprumque cano.

This is not the first time I have paid tribute to Jim Muhly's contribution to ancient Near Eastern history (Merrillees 1998), nor, I suspect, will it be the last. Longevity—and active engagement—will ensure that Jim continues his valued research in this field, and we can look forward to many more years of productive work on the archaeology, especially metallurgy, of the eastern Mediterranean in the Bronze Age. Forty years ago we both had articles on Alashiya included in the published proceedings of the First International Cypriological Congress held in Nicosia in April 1969 (Merrillees 1972; Muhly 1972). Our respective views on the location of this ancient place name were diametrically opposed and have remained unchanged and unchanging since that time. The latest exchanges between us are to be found in *Archaeology Odyssey* (Merrillees 2005, 35–40; Merrillees, Muhly, and Wachsmann 2006, 25–27). The case for equating Alashiya with Cyprus continues to rest exclusively on circumstantial evidence, especially the shibboleth that because Alashiya sent copper abroad, it must have been Cyprus, as if the island were the only producer of this metal in the Bronze Age (cf. Muhly 1996, esp. 49). Empirical argument over the years has shown itself incapable of advancing, let alone resolving the question of Alashiya's whereabouts, and in the absence of a systematic philological re-examination of all the relevant textual references, for which the potpourri of translations in Knapp, ed., 1996 is no substitute, other means for making progress with the search have been increasingly sought. To this end, the recent publication of *Inscribed in Clay: Provenance Study of the Amarna Letters and Other Ancient Near Eastern Texts* (Goren, Finkelstein, and Na'aman 2004), the final report on a major research project undertaken by Yuval Goren, Israel Finkelstein, and Nadav Na'aman, has broken new ground, literally and metaphorically, in the ongoing debate. It is with a critical review of this volume, where it concerns Alashiya, that I respectfully salute Jim Muhly and urge him to reconsider his untenable position.

Locating Alashiya

The location of Alashiya, long taken for granted to be Cyprus or part of it by a majority of scholars, continues to exercise the archaeological community and defy rational consideration (cf. Merrillees 1987). The problem of supplying a definitive answer to the issue of where it lay remains the absence of incontrovertible proof that a place with this name has been uncovered in Cyprus or elsewhere in the eastern Mediterranean region. So far, Alashiya has been mentioned in contemporaneous documents of the second millennium B.C. only at sites or in areas where the place is thought not to have occurred! Today, the major empirical challenge facing ancient historians of the period is the difficulty of reconciling the written implication that Alashiya was a superpower in the Levant during the Late Bronze Age with the archaeological evidence from Cyprus, which suggests just the opposite. Over a century's worth of controlled excavations on the island have so far failed to provide substantive grounds for arguing that Cyprus had a king (or queen), a centralized administration, and commensurate wealth that would justify ranking it on a par with Egypt under the pharaohs in the New Kingdom. No palaces, archives, or other witnesses to royal or imperial power have yet turned up anywhere in Bronze Age Cyprus, and the determination of archaeologists to elevate the island's government in the second half of the second millennium B.C. to the status of the ruler of Alashiya has induced a kind of academic schizophrenia in the literature, nicely summed up by Louise Steel in her ambitious synthesis, *Cyprus before History: From the Earliest Settlers to the End of the Bronze Age* (2004). There she writes, *inter alia*, that "Alashiya appears to have been an urban polity ruled by a king who claimed to be the equal of the Egyptian pharaoh [p. 185]. . . . However, there is no evidence for a single unified state on Cyprus, which instead appears to have been organized into smaller regional polities [p. 186]. Petrographic analysis of the Amarna tablets indicates that the clay used for the Alashiya tablets is Cypriot. . . . Goren et al. 2003, 242–5 [p. 244 n. 224]. . . . At present it is impossible to demonstrate conclusively that Alashiya and Cyprus (or part thereof) were one and the same [p. 184]."

It is therefore no wonder that, in the absence of irrefutable evidence and scholarly consensus, recourse has increasingly been had to scientific methods for dealing with the conundrum of where Alashiya actually lay. The appeal of this approach lies in the underlying assumption that scientific findings are objective, exact, and unassailable, when in fact they are nothing of the kind. Not only is there an inevitable tendency on the part of the scientists concerned to invest their results with an aura of finality that is seized upon by the academic specialists to vindicate their favored opinions, but few archaeologists are equipped to master the techniques used and interpret the process with commensurate expertise. Furthermore, scientists rarely set out the parameters they have followed for their sampling, the range and consistency of the analytical procedures employed, and the degree of probability in reaching their conclusions. The volume under review is no exception. As a consequence, archaeologists have come to show a distinct inclination to accept scientific results if they conform to or support existing empirical positions, and dismiss those that appear to confront longstanding and conventional wisdom. This mindset is epitomized by the controversy over the dating of the volcanic eruption on Thera that put an end to the Late Minoan IA civilization on the island (cf. Manning 1999). The dispute over the location of Alashiya has fallen prey to the same syndrome. Unfortunately, *Inscribed in Clay: Provenance Study of the Amarna Letters and Other Ancient Near Eastern Texts* will do nothing to lessen scepticism about the scientific and empirical methods used for this purpose and allay misgivings about the results.

The work by Goren, Finkelstein, and Na'aman (henceforth, the co-authors) is far from being the first word we have heard on the subject. It has been preceded by the academic equivalent of a media blitz, worthy of a Harry Potter novel, which has, as Steel's quotation above makes clear, all but translated the outcome of the analyses into historical fact. Apart from seminars, lectures, and other unrecorded or unpublished presentations, the co-authors posted a Web site on the Internet in 2000 entitled: "Mineralogical and Chemical Study of the Amarna Tablets" with so-called "Preliminary Results," including the following: "We succeeded to come to a firm conclusion as to the location of the kingdom of Alashiya. The mineralogical composition of the tablets strongly relates it to western Cyprus. This suggests that Alashiya of the 14th century B.C. was located in western Cyprus and not in the eastern part of the island, in northwestern Syria or in Cilicia as was previously suggested." Though the Web site stated that the final results of this study would be published in a volume to appear in 1999, the next word came from Yuval Goren in a paper on "Provenance Study of the Alashiyan Tablets and Related Texts" given at the international conference on "Egypt and Cyprus in Antiquity" in Nicosia in April 2003. This coincided with the appearance of a comprehensive article by Goren and his collaborators (Goren et al. 2003) on "The Location of Alashiya: New Evidence from Petrographic Investigation of Alashiyan Tablets from El-Amarna and Ugarit." Again, the co-authors claimed, without any qualification, that "petrographic and chemical examinations indicate that the Amarna letters from Alashiya originate from the area on the margin of the Troodos Mountains in Cyprus . . . Within Cyprus, either Kalavasos–*Ayios Dhimitrios* or Alassa–*Paliotaverna/Pano Manadilaris* [sic] are identified as the source of official Alashiyan letters" (Goren et al. 2003, 233). By this stage the archaeological community had allowed itself to be so totally persuaded by the initial scientific results that Wikipedia (as recently as December 2007) could baldly assert without reference to the final report on the el-Amarna project, that "the identification of Cyprus with Alashiya was confirmed by the 2003 publication by Goren et al. of an article in the *American Journal of Archaeology* . . ."!

In the light of all this advance publicity, which served to foreclose discussion of the co-authors' results before it had even begun, is there anything new or worthwhile to be learned from the volume under review? The first observation to be made is that this handsomely produced work prejudicially and inconsistently anticipates its conclusions by overwriting the whole of the map of Cyprus inside the front cover with "Alashiya" in capital letters. The rest of the presentation comes as something of an anti-climax. The co-authors' project had as its avowed aim "through mineralogical and chemical analyses of samples from over 300 tablets housed in museums in Berlin, London, Oxford, and Paris. . . . pin-pointing their geographic origin and clarifying the geographic history of the Ancient Near East" (Goren, Finkelstein, and Na'aman 2004, xvii). The section on Alashiya is only one part of a much more extensive scientific examination of the tablets from the el-Amarna archive, but the most controversial as it purports to reach a definitive conclusion about the place's identity (Goren, Finkelstein, and Na'aman 2004, 48–75). It is, in fact, only an expanded version of the study that appeared in the *American Journal of Archaeology* the preceding year (Goren et al. 2003). After a short introduction the catalog proceeds with a description—physical and mineralogical—of four of the Alashiya tablets from el-Amarna (nos. EA 37, 33, 34, 38); four Cypro-Minoan tablets from Enkomi–*Ayios Iakovos*, a Cypro-Minoan cylinder from Enkomi–*Ayios Iakovos*, four Cypro-Minoan cylinders from Kalavassos–*Ayios Dhimitrios*; a letter allegedly from a king of Alashiya, whose source is disputed, found in Ras Shamra (RS L.1; Louvre AO 21087) (Bordreuil and Pardee 1989, 380, no. RS [Varia 16]; Cunchillos 1990, 41; Yon 1999, 118 n. 25; Malbran-Labat 1999, 122; 2004a, 375 n. 74, 2004b; Goren et al. 2003, 234, 238–240; Goren, Finkelstein, and Na'aman 2004, 55–56); and a letter from the king of Carchemish to the king of Ugarit, also from Ras Shamra (RS 8.333; Louvre AO 19955; see Bordreuil and Pardee 1989, 46; Cunchillos 1990, 29; Goren, Finkelstein, and Na'aman 2004, 57). This is followed by interpretation of the petrographic data, subdivided into "the general provenance of the Alashiya letters" and "the provenance of the Alashiya letters within Cyprus;" chemical analyses of the Alashiya letters and selected Cypriote clays; and conclusion: the Alashiya tablets and Late Bronze Age Cyprus, subdivided into "the government of Cyprus in the Late Bronze Age: an overview of current opinions," "the political-administrative center of Alashiya during the 14th–13th centuries B.C.E. according to the analytical results," and "the government of Cyprus in the Late Bronze Age: alternative models."

While the co-authors correctly emphasize the need for a multidisciplinary approach to identifying Alashiya's location, it should be noted that none is an expert on Cypriote Bronze Age archaeology or, for that matter, on the geology of Anatolia, Syria, or even Cyprus. Nearly all their sources for these inputs are published and/or secondary, and the comparative data are very uneven in scope, quality, and date. For advice on the Cypriote archaeological evidence they appear to have obtained comments only from Professor Karageorghis, a leading authority on Cypriote antiquity but a declared supporter of Alashiya's equation with Cyprus (e.g., Karageorghis 1995, 75–76), while relying heavily on the theory-based studies by Keswani and Knapp (Goren, Finkelstein, and Na'aman 2004, 70–71). For information on the island's geology they thank G. Konstantinou, Director of the Cyprus Geological Survey [sic], and "S. Vaughan who read the draft concerning the Alashiya letters for his [sic] input on Cypriote clays" (Goren, Finkelstein, and Na'aman 2004, xviii). Comments on the clay types in Syria and Lebanon were supplied by J.-P. Thalmann, who, according to the bibliography, has published nothing on the geology of this region (Goren, Finkelstein, and Na'aman 2004, 380), and E. Cooper, co-author of one short article on the clays of the pottery from Tell Atchana (Goren, Finkelstein, and Na'aman 2004, 360—L. Cooper and M. Fortin). Evidently no one was consulted on the geology of modern-day Turkey. This line-up does not inspire confidence in the authority of the co-authors' observations on the archaeology of Cyprus or the geology of Anatolia, Syria, and Cyprus.

More disturbing is their approach to the ceramic and geological data needed for comparative purposes. On the one hand, they write in the volume under review that "cuneiform tablets are in fact ceramic objects" and that to identify the sources of the clays used in the Alashiya tablets from el-Amarna and in the Cypro-Minoan tablets and cylinders, comparisons should be made with pottery and geological specimens from Cyprus, Cilicia, and northwestern Syria (Goren, Finkelstein, and Na'aman 2004, 2, 49). On the other hand, however, they state that "clay types used to produce tablets, however, are not necessarily the same as those from which vessels are manufactured. This may sometimes inhibit the use of routine chemical procedures in which a database containing the elemental composition of standard pottery from known sites is compared with the samples examined" (Goren et al. 2003, 235). Untroubled by these reservations, and indeed the latitude this offers for a selective application of analytical techniques, the co-authors opt for petrography as the primary method of investigation on the grounds that "the results can be interpreted on the basis of detailed and usually available geological data" and comparative material from investigations of pottery assemblages (Goren et al. 2003, 235; Goren, Finkelstein, and Na'aman 2004, 10). Nevertheless, it emerges that their database varies widely from region to region, comes from secondary sources rather than first hand knowledge, and is decidedly patchy.

For Cyprus, the co-authors have used petrographic information drawn from a limited typological and geographical pool in Cyprus as published by Sarah Vaughan, and none of the clays analyzed came from anywhere else in the island except the southern Troodos and south coast (Goren, Finkelstein, and Na'aman 2004, 11, 61, 65–69). They did not even consult Knapp's and Cherry's comprehensive synthesis of provenance studies on Cypriote Bronze Age ceramics (Knapp and Cherry 1994) or indeed others published before 2004 and now conveniently listed by Åström and Nys (2007, 55, table 1), and, most curiously, given their final position on the origins of the Alashiya tablets, they examined no pottery from Alassa–*Palaiotaverna/Pano Mandilaris* or Kalavassos–*Ayios Dhimitrios*! For Syrian, Lebanese, and Mespotamian assemblages, they used the collections—of pottery—in the Institute of Archaeology, University College London, and the Department of Scientific Research, now the Department of Conservation and Scientific Research, in the British Museum, together with "a reference collection of pottery from Ras Shamra and Byblos," as well as ceramic material from the Amuq Valley (Goren, Finkelstein, and Na'aman 2004, xviii, 11, 59, 88, 134). They seem to have examined no specimens of clay from Syria or Turkey, or indeed any pottery remains from Anatolia, outside Bogazköy, including Cilicia (Goren, Finkelstein, and Na'aman 2004, 31, 58–59). The omission of any petrographic or chemical analyses of pottery from southeastern Turkey is all the more surprising since Carl Knappett has, for the last decade, been conducting a scientific study into the origins of the Red Lustrous Wheel-made Ware in the eastern Mediterranean and made the preliminary results of his work available on-line since 2000 (Knappett 2000).

It is difficult to know how any firm conclusions can confidently be drawn from such an incomplete, random, and derivative assemblage of comparative evidence. No attempt was made by the co-authors to set out uniform standards for the reference database in terms of the types of materials (tablets, pottery, and/or clay samples) appropriate for inclusion, their geographical spread in each region, or the quantities required to establish a degree of statistical reliability about the results. The co-authors are nevertheless aware of the methodological problems posed by the nature of the benchmark material as they comment tellingly that "unlike Syrian and Mespotamian ceramics, Egyptian pottery has long been analyzed petrographically, hence our comparisons can rely on a solid body of data" (Goren, Finkelstein, and Na'aman 2004, 29). In the light of this observation no credibility can be given *a priori* to their categorical rejection of the Cilician coast and exclusion of Syria as possible sources of the Alashiya tablets from el-Amarna (Goren, Finkelstein, and Na'aman 2004, 58–60). In fact, the co-authors note, following their petrographic analysis of the Alashiya tablets from el-Amarna, "an ironic coincidence that ophiolitic occurrences can be found in each and every area suggested for the location of Alashiya" (Goren et al. 2003, 240; Goren, Finkelstein, and Na'aman 2004, 58). Given the unbalanced character of the comparative data from Cyprus, their decision to settle on the south coast of the island for the placement of Alashiya—Enkomi–*Ayios Iakovos* is, *pace* the French, unceremoniously ruled out of contention (Goren, Finkelstein, and Na'aman 2004, 14, 71–72)—seems more a case of inductive than deductive reasoning. And despite the co-authors' declared preference to equate Alashiya with Kalavassos–*Ayios Dhimitrios* and Alassa–*Palaiotaverna/Pano Mandilaris* (Goren, Finkelstein, and Na'aman 2004, 72–73), "Alashiya" is written across the central western part of the island on the map in figure 3.1 (Goren, Finkelstein, and Na'aman 2004, 23).

For comments on the technical side of the co-authors' presentation, I approached two experts, neither of whom has expressed any opinion in writing on the location of Alashiya. Nor do they necessarily endorse all or any of the views I have expressed in this paper. Dr. Allan Gilbert, Professor of Anthropology, Department of Sociology and Anthropology, Fordham University, New York, NY, has spent many years investigating and studying the geology of Cilicia, and has contributed the following considered observations, for which I am duly grateful. Dr. Costas Xenophontos, former Senior Geologist with the Cyprus Geological Survey Department, who is an authority on the geology not only of Cyprus but of Syria, has also contributed his opinion, and I am endebted for his observations.

A Geoarchaeologist's Technical Critique

by Allan Gilbert

Goren et al. (Goren et al. 2003; Goren, Finkelstein, and Na'aman 2004) have made an extraordinary contribution with their sampling and analysis of a substantial number of ancient tablets. Notwithstanding the valuable evidence they have produced, some of their conclusions regarding the identification of Alashiya are premature. There has simply been too little field testing in addition to their extensive examination of artifacts to rule out in any convincing way the non-Cypriote locations for the long-lost kingdom.

Much of the introductory methods section of the book (Goren, Finkelstein, and Na'aman 2004) concerns theoretical expectations regarding the procurement of clays and tempers for pottery and tablet production. The co-authors make three initial assumptions:

(1) The specifications for tablet-making were probably more exacting than, and therefore likely different from, those needed to make pottery (p. 5);

(2) Based upon analogy to several treatises on recent village pottery manufacture, clays and tempers used by ancient potters and scribes would have been collected within some catchment area surrounding the manufacturing centers (p. 6);

(3) Calculated clay-temper factors will provide a satisfactory characterization of the diversity of natural raw material combinations within such catchments (p. 7).

Each of these assumptions represents a useful working hypothesis, but none can be presumed accurate without on-the-ground, in addition to in-the-lab, testing. Scribes may have exploited their own collecting localities and prepared their own pastes in ways different from those of potters, but confirmation of this in any particular place depends upon investigation of not only the artifacts but also the potential raw materials in order to assess their variability and track their transformation as constituents in tablet and pottery manufacture. The co-authors' idea of clay-temper factors—ostensibly "ecofactual" clays and tempers that reflect the direct use of local sediments—does not acknowledge the possibility that technical processes, such as levigation, clay mixing, and grit selection, could skew the ratios of natural inclusions and thereby distort the picture of neighboring geology. Even the most detailed petrographic analyses of tablet fabrics could never guarantee that the visible paste constituents had not been filtered in some way by the application of special scribal recipes. Comparing sherds and tablets is only the first step in determining (1) the degree to which potters and scribes shared production procedures and (2) whether their raw material catchments were similar or different. The difficult process of determining this must include extensive survey, sampling, and experimentation.

Throughout their project, the co-authors opted to perform petrographic analyses of tablet crumbs and peels first, then follow up with selective chemical analyses to confirm the similarities suggested by tempers (pp. 9–10). In their estimation, the mineral constituents of a single tablet or ceramic object can be readily linked to geological provinces that contain the parent rocks. Thus, petrography offers an immediate and independent indicator of an artifact's source. In contrast, successful chemical analysis depends upon a substantial data base containing compositional profiles of known provenience because elements and their abundances by themselves do not possess the same obvious connections to physiographical regions that mineral constituents provide.

Petrography, however, does not discriminate with great exactness. Though they contain direct evidence of the geological environment, inclusions comprise a wide range of variability—and thus little precision—due to the vagaries of sedimentological deposition and the modifications created by paste preparation. Chemical profiles require more high-priced instrumental input and lots of comparative material, but their power to distinguish among discrete samples yields much greater precision and a more quantifiable measure of compositional similarity, as long as a sufficient number of elements is assayed. As the co-authors point out, there are advantages and disadvantages to each approach.

Following their own methodology, the co-authors turned first to petrography of tablet inclusions to eliminate Cilicia, the Amanus, and Hatay from further consideration as potential locations for Alashiya. They based their argument on the fact that, of the eight Alashiyan tablets examined, only EA37 from Amarna (and the closely associated RS L.1 from Ras Shamra) contain geographically specific temper, which happened to be indicative of a region where oceanic crustal rocks (igneous ophiolites) juxtapose with sedimentary limestones and sandstones. This geological combination can be found in each of the above places as well as Cyprus, but the abundant clasts of basaltic origin found among the inclusions of EA37 prompted them to rule out all but Cyprus, because the parent rocks for basalt—pillow lavas—are substantially represented only around the Troodos in Cyprus (pp. 58–60). They then expanded their investigation into chemical analysis, examining samples from what they considered to be promising clay deposits in southwestern Cyprus only. Using these data, they conducted very limited statistical clustering in which only the tablets and the Cypriote raw materials were compared. With no other sampling having been done, their dendrograph of 12 clays and three tablets (Goren et al. 2003, fig. 6; Goren, Finkelstein, and Na'aman 2004, fig. 3.4) gave the impression of close resemblances among samples and, to the unsophisticated eye, a satisfactory match.

What's wrong with this investigation? First, eliminating three of the four possible candidates for the location of Alashiya based on only one tablet is risky, but doing it without a fair sampling of clays and temper sources in those regions is riskier still. Pillow lavas do exist within the bedrock associations of Cilicia, the Amanus, and Hatay, though they are not prominently represented there. They are always the uppermost stratum of ophiolite sequences, which are slabs of oceanic seafloor crust that have been thrust up, or accreted, onto the edges of continents. Many ophiolites become fractured and shattered as they are pushed inexorably landward, and in the rapidly uplifting region around the Taurus and Amanus mountains, their uppermost layers likely experienced the earliest and heaviest weathering, resulting in their detrital residues being washed into lowland sedimentary basins, which might by now be buried under later alluvial accumulations. Within tablet RS L. 1, the co-authors identified inclusions of dolerite (known as diabase in the U.S.), the slightly coarser grade of rock that is chemically identical to basalt. Ophiolitic dolerites are usually found within dike complexes just below the pillow lavas, and they are far more commonly encountered in the ophiolites of the northeastern Mediterranean than are basalts. If raw materials used to make EA37 and RS L.1 were obtained from localities near extant basalts and underlying dolerites, or from an exposed clay layer containing erosional debris from pre-existing ones, then the absence of major pillow lava formations in the region today would be insufficient to preclude any of these locations in favor of Cyprus alone. In 1980, Thierry Juteau (cited inaccurately by the co-authors as J. Thierry) wrote about the perplexing dearth of silica-rich (i.e., tholeiitic) basalts in the Turkish ophiolites: "The puzzling fact is that the upper parts of the tholeiitic complexes have systematically disappeared, as if all the massifs without exception, but at variable degrees, had been skinned of their upper parts. . . . *Tholeiitic pillow-lavas do not appear anywhere*, and the total disparition of this tholeiitic basaltic cover remains totally unexplained, taking in account the remarkable absence of tholeiitic basaltic fragments or pebbles in the numerous detrital formations of the Taurus range" (Juteau 1980, p. 222).

Juteau's report is now nearly 30 years old. One could not then, and still cannot now, guarantee that no sedimentary sequences of the northeastern Mediterranean mainland would, upon closer investigation, yield microscopic clasts from basaltic erosion products like those the co-authors identified in tablet EA37. Such particulates might have been too small, unobtrusive, or limited in distribution to have caught the attention of the earth science community three decades ago.

Second, compositional analysis using the very small samples obtained from crumbs and peels comes with a disadvantage. Sample error tends to rise when extremely minute amounts of material are assayed, and thus much of the precision inherent in the chemical methodology is squandered. In addition, clustering 15 samples produces linkages among members of a very limited set of compositional profiles, but, as is the nature of clustering algorithms in general, the addition of new sample profiles would change the distributions as well as the scale of similarity (i.e., the amalgamation distances). Samples that are moderately alike but far from identical may begin to look progressively dissimilar as new profiles are introduced and the multidimensional spacing widens proportionately to accommodate the ever larger universe of variability. When the co-authors used only Cypriote clays in their compositional analysis, they closed the door on learning anything about the other regions, which had been previously eliminated on petrographic grounds. Adding clays from these other regions and raising the number of clustered profiles into the dozens or hundreds would have expanded the elemental ranges for a much greater number of samples, and in so doing, the connection between the Alashiyan tablets and the Cypriote clays would have been more rigorously put to the test.

Appreciation is extended to structural geologist Charles Merguerian (Hofstra University) and nuclear chemist Garman Harbottle (Brookhaven National Laboratory) for their discussions on the subject and their support of the comments expressed above.

A Geologist's Technical Critique

by Costas Xenophontos

The co-authors, in their quest for Alashiya, have chosen to use petrographic analysis as the sourcing method for the tablets and other inscribed artifacts from el-Amarna, Enkomi and Ugarit (Goren, Finkelstein, and Na'aman 2004). There is no doubt that this analytical method can give very useful information if the operator is fully conversant with both the local and regional geology.

Cyprus, Cilicia and north-western Syria are considered as possible sources, but the co-authors have concentrated almost exclusively on Cyprus as if they had made up their minds as to the result at the outset. Whereas an albeit simplified geological map of Cyprus is given together with the sites mentioned in the text (p. 60, fig. 3.3), no such information is provided for the other two possible source areas. The reader is therefore unable to decide on the validity of any statements or comparisons made with such a skewed database. As one example, tablet EA 37 is described as having a carbonatic matrix considered to be clay from the Moni Mélange whose clayey component contains no carbonate. The inclusions are dominated by lava clasts, some of which show quench textures. How this information on its own hints at a pillow-lava complex is, to say the least, enigmatic (p. 49). The presence of an unknown abundance of low-grade metamorphic greywacke among the inclusions firmly puts it in the Moni Mélange terrane except that—to our knowledge—no greywacke outcrops are shown on any geological map of the area, although serpentinite outcrops are extensive but not represented in the inclusions.

Tablet RS L.1, a letter supposedly of the king of Alashiya from Ugarit, has a tan-colored matrix with a silt fraction (~ 4%) containing quartz, plagioclase, biotite, epidote, calcite and serpentine. The assemblage of inclusions is dominated by diabase (dolerite), gabbro and serpentine, and fragments of their constituent minerals. Sedimentary lithologies are represented by fossiliferous limestone micrite and corallinean algal fragments. Significantly, no lava clasts are mentioned as being present, and, being rather friable, they may have been removed in the sediment transport. The assemblage of inclusions also argues in favor of long transportation, almost as far as the coastal area. Despite this, however, the matrix of this tablet has no carbonate—that is, it has picked up coarser carbonate clasts but none of the extensive finer carbonate sediments (marls) it passed through.

Without a geological map and a site location, it is difficult to define which part of the Baër-Bassit ophiolite contributes sediment to the area of the Amuq Valley. Studies in the last seven years have proved the ophiolite to be identical in its lithologies to the Troodos ophiolite, but it has been dismembered and interleaved with older ocean floor rocks (sediments and igneous rocks) not dissimilar to those found in the Mamonia complex of Cyprus. The presence of greywacke (quartzitic sandstone) was considered by the co-authors to be sufficient evidence to regard this carbonatic clay as being derived from the Moni Mélange. Such rock types, however, are also present in the Bassit, and carbonatic clays (marls) cover extensive areas adjacent to and extending south of the ophiolite.

Much has been made of a skewed and small database, including statements that are, to say the least, contentious; for example, the Troodos ophiolite was at no time covered by the Mamonia Complex, which was removed prior to the deposition of the supra-ophiolitic Tertiary pelagic carbonates. In describing the tablets from Enkomi, the Pedieos River is assumed to be the only contributor to the igneous fraction of the inclusions. Serpentine, however, is not exposed anywhere in the drainage area of this river, and yet it is found in most of the artifacts described from Enkomi.

Petrographic analysis as used in chapter 3. VI on Alashiya in Goren, Finkelstein, and Na'aman 2004 is by itself not sufficient to favor a Cypriote origin, let alone a specific location within Cyprus for Alashiya. An expanded database that would definitely exclude any rival locations is the proper avenue for future work.

Concluding Remarks

Far from settling the problem of Alashiya's identity once and for all, Goren, Finkelstein, and Na'aman have added one more layer of dissension to the body of argument for and against its equation with Cyprus. They have nevertheless amplified another dimension to the evidence for resolving the issue by carrying out an extended, but by no means exhaustive, petrographic and chemical analysis of the Alashiya tablets from el-Amarna. That this approach has the potential to further refine our knowledge of the source of these letters is undeniable, and the co-authors' work opens a new chapter in the ongoing saga to which this review is hopefully a constructive contribution. Not only is there scope, in fact, need to add substantially to the database for comparative purposes, but there are other tablets from Alashiya that should also be submitted to appropriate scientific examination. Of the Alashiya tablets from el-Amarna, four were included in the co-authors' project (EA 33, EA 34, EA 37, and EA 38) (Goren, Finkelstein, and Na'aman 2004, 49–51). EA 35 in the British Museum (WA 29788) was unavailable at the time of sampling (Goren et al. 2003, 234 n. 8), and the other three (EA 36, EA 39, and EA 40) are all in the Cairo Museum, from which no reply was received in response to the co-authors' request to examine the letters (Goren, Finkelstein, and Na'aman 2004, 3). Even then, EA 36 does not mention the name of Alashiya (Knapp, ed., 1996, 23). Despite the co-authors' presumption that a tablet written in Akkadian from Ras Shamra (RS L.1) was sent by the king of Alashiya to the king of Ugarit (see above), no letter found in Ras Shamra with Alashiya specified as the source was examined by the co-authors.

In fact only five, possibly six, texts from Ras Shamra were actually sent by someone in Alashiya:

1) from Eshuwara, the "Grand Intendant" of Alashiya: RS 20.018 (Damascus Museum DO 5281) (Bordreuil and Pardee 1989, 230; Cunchillos 1990, 35; Yon 1999, 118 n. 24; Malbran-Labat 2004a, 375 n. 74);
2) from the "Grand Intendant" of Alashiya: RS 94.2173 (Damascus Museum) (Malbran-Labat 1999, 121 n. 2, 122, fig. 1; 2004a, 372 n. 49);
3) from Kushmeshusha, king of Alashiya: RS 94.2177 + 2491 or 2 (Damascus Museum DO 7849) (Malbran-Labat 1999, 121 n. 2, 122–123; 2004a, 374 nn. 68, 69; 2004b, 108, no. 80);
4) from the "grand MAŠKIM, Šangiwa" of Alashiya: RS 94.2447 + 2588 + 2590 (Damascus Museum) (Malbran-Labat 1999, 121 n. 2, 122; 2004a, 373 n. 63, 374 n. 71);
5) from Kushmeshusha, king of Alashiya: RS 94.2475 (Damascus Museum) (Malbran-Labat 1999, 121 n. 2, 122; 2004a, 374 n. 65; 2004c, 56); and
6) from Alashiya(?): RS 94.2561 (Malbran-Labat 1999, 121 n. 2, 123).

The unconscionable delay in fully publishing the last six texts is much to be regretted. Indeed, the failure of the academic community to address all the archaeological, lexicographical, and scientific aspects of the issue in a comprehensive and authoritative manner renders all partial treatments partial, if not partisan, and an impediment rather than a catalyst to resolving the question of Alashiya's location. In the end, circumstantial evidence will never be enough to persuade the skeptics to drop their objections and achieve scholarly consensus, but at least fresh approaches—like those by Goren, Finkelstein, and Na'aman—will progressively narrow the range of possible options and sharpen the focus on geographical areas and analytical processes most likely to produce convincing and definitive results. Mere repetition of the mantra that an answer has finally been found will only magnify the historical factoid that has been created and the revision that will be necessary to excise it.

References

Åström, P., and K. Nys. 2007. *Hala Sultan Tekke.* 12: *Tomb 24, Stone Anchors, Faunal Remains and Pottery Provenance* (*SIMA* 45 [12]), Sävedalen.

Bordreuil, P., and D. Pardee. 1989. *La trouvaille épigraphique de l'Ougarit.* 1: *Concordance* (*Ras Shamra-Ougarit* 5 [1]), Paris.

Cunchillos, J.-L. 1990. *La trouvaille épigraphique de l'Ougarit.* 2: *Bibliographie* (*Ras Shamra-Ougarit* 5 [2]), Paris.

Goren, Y., S. Bunimovitz, I. Finkelstein, and N. Na'aman. 2003. "The Location of Alashiya: New Evidence from Petrographic Investigation of Alashiyan Tablets from El-Amarna and Ugarit," *AJA* 107, pp. 233–255.

Goren, Y., I. Finkelstein, and N. Na'aman. 2004. *Inscribed in Clay: Provenance Study of the Amarna Letters and other Ancient Near Eastern Texts*, Tel Aviv.

Juteau, T. 1980. "Ophiolites of Turkey," *Ofioliti* 2, pp. 199–235.

Karageorghis, V. 1995. "Relations between Cyprus and Egypt—Second Intermediate Period and XVIIIth Dynasty," *Egypt and Levant* 5, pp. 73–77.

Knapp, A.B., ed. 1996. *Sources for the History of Cyprus.* 2: *Near Eastern and Aegean Texts from the Third to the First Millennia B.C.*, Altamont, NY.

Knapp, A.B., and J.F. Cherry. 1994. *Provenience Studies and Bronze Age Cyprus: Production, Exchange and Politico-Economic Change* (*Monographs in World Archaeology* 21), Madison.

Knappett, C. 2000. "The Provenance of Red Lustrous Wheel-made Ware: Cyprus, Syria, or Anatolia?" in *Internet Archaeology* 9, http://intarch.ac.uk/journal/issue9/knappett_index.html.

Malbran-Labat, F. 1999. "Nouvelles données épigraphiques sur Chypre et Ougarit," *RDAC* 1999, pp. 121–123.

———. 2004a. "Alašiya et Ougarit," *Res antiquae* 1, pp. 365–377.

———. 2004b. "Un scribe ougaritain à Chypre," in *Le Royaume d'Ougarit. Aux origines de l'alphabet*, G.G. Galliano and Y. Calvet, eds., Paris, p. 108.

———. 2004c. "Les archives de la 'Maison d'Ourtenou' sont-elles des archives privées?" *Dossiers d'Archéologie* 10, pp. 54–57.

Manning, S.W. 1999. *A Test of Time: The Volcano of Thera and the Chronology and History of the Aegean and East Mediterranean in the Mid Second Millennium B.C.*, Oxford.

Merrillees, R.S. 1972. "Alasia," in *Acts of the First International Cypriological Congress (Nicosia, 14–19 April 1969)*, vol. 1, V. Karageorghis and A. Christodoulou, eds., Nicosia, pp. 111–119.

———. 1987. *Alashia Revisited* (*Cahiers de la revue Biblique* 22), Paris.

———. 1998. "Vote of Thanks to James D. Muhly," in *The Aegean and the Orient in the Second Millennium. Proceedings of the 50th Anniversary Symposium, Cincinnati, 18–20 April 1997* (*Aegaeum* 18), E.H. Cline and D. Harris-Cline, eds., Liège and Austin, pp. 215–216.

———. 2005. "Don't Be Fooled! Despite What Many Scholars Say, Ancient 'Alashiya' Was Not Cyprus," *Archaeology Odyssey* 8 (5), pp. 35–40, 50–51.

Merrillees, R., J.D. Muhly, and S. Wachsmann. 2006. "Alashiya Redux," *Archaeology Odyssey* 9 (1), pp. 25–27.

Muhly, J.D. 1972. "The Land of Alashiya: References to Alashiya in the Texts of the Second Millennium B.C. and the History of Cyprus in the Late Bronze Age," in *Acts of the First International Cypriological Congress (Nicosia, 14–19 April 1969)* vol. 1, V. Karageorghis and A. Christodoulou, eds., Nicosia, pp. 201–219.

———. 1996. "The Significance of Metals in the Late Bronze Age Economy of Cyprus," in *The Development of the Cypriot Economy from the Prehistoric Period to the Present Day*, V. Karageorghis and D. Michaelides, eds., Nicosia, pp. 45–60.

Steel, L. 2004. *Cyprus before History: From the Earliest Settlers to the End of the Bronze Age*, London.

Yon, M. 1999. "Chypre et Ougarit à la fin du Bronze Récent," *RDAC* 1999, pp. 113–119.

CHAPTER

26

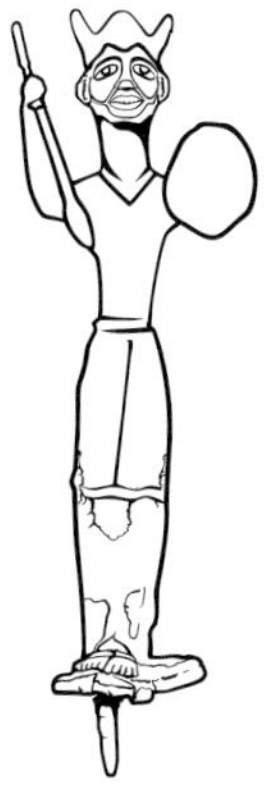

Hittite Metals at the Frontier: A Three-Spiked Battle Ax from Alalakh

K. Aslıhan Yener

Introduction

It is both a pleasure and a pleasant surprise to have been invited to contribute an article to a festschrift for James Muhly: a pleasure because it is gratifying to have the opportunity to pay tribute to a prolific and inspiring scholar; and a pleasant surprise because he has been a worthy opponent on the issue of tin sources in Turkey, about which we have disagreed for so long. The question of tin sources in Turkey has been dormant for 15 years, but new discoveries by the Turkish Geological Survey (MTA) have strengthened the notion that considerable tin deposits of value to Bronze Age exploitation existed in proximity to the appearance of early tin bronzes in the third millennium B.C. One of the startling recent announcements has been the Hisarcik source of tin close to Kayseri and the Kültepe/Kanesh site. Moreover, the MTA has initiated a new project to flesh out the sources of tin reported in the bend of the Kizilirmak (Halys) river basin. Yet again, the translation of *anaku* as tin is being challenged by Heather Snow's dissertation (2005).

In the course of a distinguished career, Professor Muhly touched on many aspects of Bronze Age politics, trade, and exploitation of vital metal resources in the ancient Near East, and it is one of these that I would like to consider in this paper, Hittite gods and their bronze weapons at the edges of the Empire.

In the *Anatolian Studies* publication "Iron in Anatolia and the Nature of the Hittite Iron Industry," James Muhly and his colleagues presented a wonderful description of bronze in the Late Bronze Age (LBA): "Iron was also a hateful metal because it lacked the warm, lustrous beauty of bronze. Moreover, whereas bronze took on a lovely greenish color as a surface corrosion or patina, iron soon was reduced to a pile of worthless reddish powder." They go on to say, "The museums of the world are full of countless numbers of well preserved ancient bronze artifacts of unsurpassed beauty, but there are very few specimens of ancient iron that have survived in anything resembling their original state" (Muhly et al. 1985, 68).

Indeed, these splendid bronze artifacts are often the products of Hittite palace patronage and fit into the general category of the special, numinous accoutrements of the gods prevalent in Anatolia. As with the later medieval "magic" and "supernatural" swords, these weapons were profoundly associated with the gods and were generally depicted in ritual ceremonies. For example, during the Middle Bronze Age (MBA), texts reveal that the lower town of Kanesh had a sacred precinct where verdicts were given "before the dagger of Aššur in the *hamrum*" (Larsen 1976, 59 n. 42).

During the Hittite kingdom, special weapons were often embellished with gold or silver and decorated with relief designs, animals, and gods in three dimensions, and they often had inscriptions incised on them. It is with the expansion and spread of the Hittite Empire that the intellectual perception of weapons is transformed. By the empire period, the sword is deified and worshipped as a god in its own right as typified by the Yazılıkaya relief in Chamber B (Bittel 1976, fig. 254). I would like to focus attention upon a find of a three-spiked shaft-hole ax discovered at Tell Atchana, ancient Alalakh in the Amuq Valley of Hatay, Turkey. It is the continuity of such special bronze weaponry over a period of several millennia and the transformative layers of meaning they may have attained that I will review below. In the course of this enquiry I will also touch upon their ceremonial attributes and how they functioned as expressions of royal and religious iconography during the second millennium B.C.

Description of the Ax

The copper-based shaft-hole ax with three spikes (Fig. 26.1) was found just below disturbed topsoil in Phase 1 of Square 64.83 during the 2007 excavation season.

The new series of surveys and excavations at Tell Atchana (ancient Alalakh) began in 2000 and is under the directorship of the author (Yener 2005, 2010). Since 2006 the excavation is under the sponsorship of the Ministry of Culture and Tourism of Turkey, and the Mustafa Kemal University in Antakya, and now Koç University in Istabul. The site was previously excavated between 1936–1939 and 1946–1949, under the direction of Sir Leonard Woolley (1955). Although metallographic analysis has not been conducted on the ax, it appears to be fabricated with a two-part closed mold and has a green patina on the surface caused by the oxidation of the copper-based metal. Two flanges flank the shaft hole, above and below, which served to anchor the ax to the handle, thus stabilizing it. Three prominent spikes, split longitudinally and vertically and then joined, possibly reflect the method of casting. Two of the spikes splay out leaving the central spike to face forward in an axial direction. All three would have delivered a fatal blow upon contact with the head, making it a very effective weapon: small but deadly. What sets this spiked weapon apart from other spiked axes is the rare, hammer-like knob making the spikes the business end of the weapon rather than a flat ax. Three raised tubular bands separate the spikes from the knob end. The tip of the hammer-like protuberance is grooved, perhaps for decorative purposes. A similar knob hammer appears on an ax from Gaziantep, the neighboring region to the north (Müller-Karpe 1994, pl. 98:4), unfortunately from an unprovenienced context.

The concept of an emblematic weapon takes on another level of importance, both symbolic and economic, as evidenced by the extensive corpus of sumptuous bronzes at Alalakh. Along with a whole array of more or less utilitarian bronze weapons, Woolley (1955, pl. 70) published a prestigious bronze sword with gold-plated hilt and a spearhead decorated with lion protomes. That the ax comes from Alalakh, the capital of the Amuq Valley, then known as the Kingdom of Mukish, is

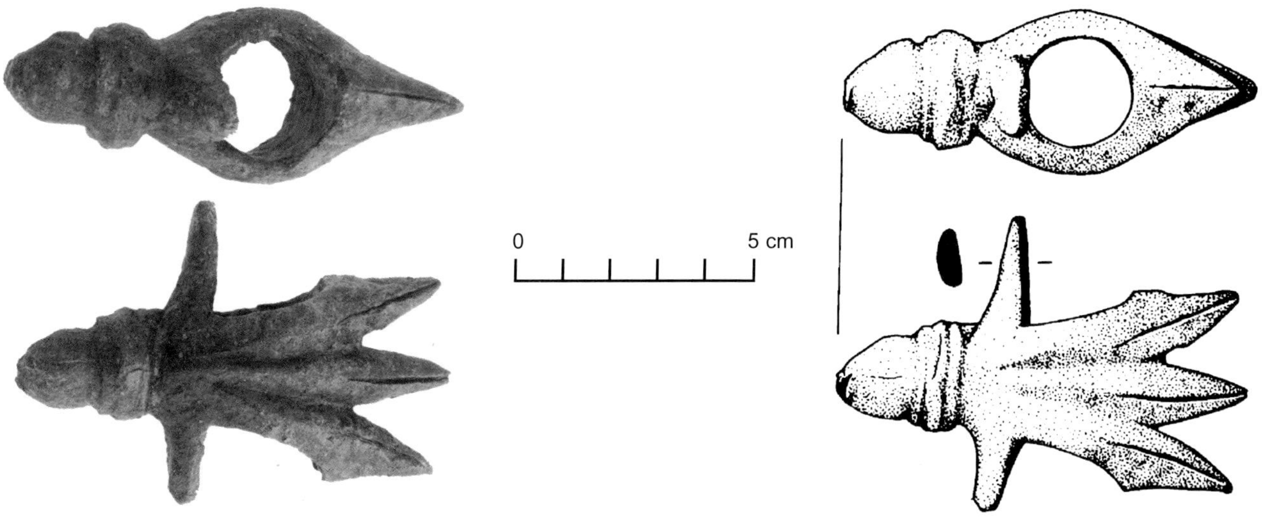

Figure. 26.1. Photo and drawing of the copper-based shaft-hole ax with three spikes (AT1889).

relevant given that this region was incorporated into the Hittite Empire during the Late Bronze Age. Exhibiting strong stylistic ties to a spiked shaft-hole ax carried by the King's Gate "War God" relief, it is consistent with other finds of sacred weapon production and differs stylistically from the rest of the axes found at Alalakh.

History of Prestige and Divine Bronze Weaponry in Anatolia

Shaft-hole weapons are a class of object that have had a long ranging ubiquity in Anatolia and its neighboring regions going back to the Chalcolithic period (for Mesopotamia, Syria, Greece, and Iran, see Deshayes 1960, pls. 56–67; Maxwell-Hyslop 1949, pls. 35, 36). An early example of a shaft-hole weapon comes from early Chalcolithic Can Hasan (Yalçın 1998). This massive lump of copper in the shape of a mace was not cast but perforated manually for hafting the handle. The next advance in ax technology came with the proliferation of open molds for casting a variety of tools and weapons, one of which was the flat ax. This simple shape could further be annealed and folded over on the edge to create the hafting shaft (for a late example, see Woolley 1955, pl. 72:AT/8/37). Utilitarian flat axes and ax-adzes continued to be used in the second millennium, often with the addition of a lug on both sides of the weapon (Maxwell-Hyslop 1953, 84–86; Müller-Karpe 1993, pl. 64:1).

But the fabrication of a shaft-hole ax takes on another level of technological and conceptual sophistication with the development of the two-piece mold. Simple and complex shaft-hole axes were cast using solid investitures (for easy reference, see Müller-Karpe 1994, pl. 61:6–11) to create a hollow for hafting the handle. These axes have enjoyed unabated continuity throughout the Early Bronze Age and proliferated widely during the Iron Age in certain areas of the Near East (Stronach 1957, 117–120; Curtis 1983, 79–80; 1989). Achieving stronger, sharper, and more durable levels of use with alloying (variously with arsenic, tin, antimony, and other element combinations), these weapons were truly remarkable in their craftsmanship. This is best exemplified by the third millennium B.C. silver and gold ceremonial shaft-hole ax from Tomb E at Alaca Höyük (Koşay 1951, pl. 166:1). At the turn of the second millennium B.C., shaft-hole axes were further embellished and reinforced with ribbing

(Bilgi 2005, figs. 73–75), which was directly carved into the mold and transferred onto the ax with casting (Woolley 1955, pl. 72:AT/48/20). Two such molds for ribbed axes are illustrated from Acemhöyük and Kültepe (Müller-Karpe 1994, pl. 45:1, 2). Raised bands, flanges, and tubular rims further accomplished linear decorations, while at the same time achieving strength with the extra support provided by the extra metal (Woolley 1955, pl. 72:AT11).

The addition of a set of sharp, pointed spikes and saw-toothed edges on axes were a type of shaft-hole weapon called the halberd. These distinctive weapons with a hammer-like butt characterized the shaft-hole ax repertoire of Horoztepe and other Early Bronze Age north-central Anatolian kingdoms. Contemporary spiked picks and axes, sometimes embellished with animals, were found in Iran and Early Dynastic Ur (Maxwell-Hyslop 1949, 99, 106). These third millennium B.C. spiked weapons were perhaps earlier antecedents of the spiked ax from Alalakh. Although no spiked axes were previously found at Alalakh, other spiked artifacts seem to be quite common. Five-spiked implements with ribbed stem and seven-rayed star head were found in graves in Levels VII–V and in the Level VII Palace (Woolley 1955, 276, pl. 70:AT/38/274), making the spike a ubiquitous feature of the metals in the MBA/LBA. A roughly contemporary spiked tool (part chisel, part spike) was found at Boğazköy Hattusha Temple 20, Level 3–4 (Müller-Karpe 1994, pl. 65:10).

However, it was not until the Hittite period in Anatolia that weapons appeared to be suffused with sacred powers and depicted with deities in fantastic, and, for us today, difficult to understand imagery. Of most relevance to the Alalakh spiked weapon is the depiction of a shaft-hole ax with four spikes on the Hittite relief on the inside entrance stone slab at the King's Gate (Bittel 1976, pl. 268). Here, the deified image is depicted in warrior garb (thus, perhaps a War God) clutching the spiked ax in his right hand and wearing a horned helmet with ear flaps. In this case, both ends of the weapon, the spikes as well as the splayed ax, were functional. In Anatolia the supreme symbols of significance were male and female deities who were displayed with their attributes, usually a weapon or tool. Thus, the association of lightning bolts and axes with a god became part of the easily recognized identity and attribute of the Storm God, Teshup, in standard iconography. In the case of the Hattusha relief, the depiction is the War God carrying the supreme weapon of destruction, the spiked ax in the hand of a god. The War God is variously identified as Nergal or Zababa in Mesopotamia. This diety was absorbed into the Hittite pantheon together with the local Hattic warrior god, called Wurunhatte and Sharruma.

The relationship of the spiked weapon with the divine realm extends beyond the confines of actual bronze examples or depiction on a large-scale relief. The three-spiked shaft-hole ax is portrayed on two ritual artifacts. The first is an Old Hittite silver fist vessel with relief decoration (Güterbock and Kendall 1995), the second is a cylinder seal impression on a pithos jar (Uzunoğlu 1989) dated to the Late Bronze Age. According to the silver vessel, the War God or Storm God (holding the reigns of a bull) raises the spiked weapon in a belligerent smiting stance above his head. A figure dressed like a king is pouring an offering libation and faces him over an altar. This menacing smiting stance is repeated on the second example, a sealing on a pithos found at Imikuşağı in eastern Turkey. Stylistically more similar to seals from sites such as Alalakh, Ugarit, and Byblos, a three-spiked shaft-hole ax is wielded by the Storm God (holding a lightning fork), who wears an Egyptianizing skirt. Again, as the previous example above, the long-robed figure of the king pours a libation and stands to the left. Another horned god wielding a long sword with a crescent-shaped hilt grabs the enemy by the hair. Both weapons of these gods were found at Alalakh (Woolley 1955, pl. 70:AT/36/4)

Throughout the ancient Near East, weapons metaphorical of the great warrior are emblems that become codified in religion even today. Best known is the legendary and highly distinct scimitar sword, Zulfiqar, belonging to the Prophet Mohammed's grandson, Ali. Often portrayed with a split V-shaped blade and engraved with the inscription "La fatà illa Ali, la saif illa Zulfiqar"—"There is no hero but Ali, and no sword except Zulfiqar"—the sword became a battle cry for Shiite Muslims. I would argue that the positively-charged images of Hittite deities and divine weapons in metal would have been displayed as emblematic of a victorious Hittite warrior in religious contexts. The Sumerian word *tukul* (a general designation of weapon) also connotated a metaphorical sense of power, deities, and troops. Accordingly, the texts reveal that weapons used in victorious battle were decorated with precious metals and placed on the altar of the protecting deity (Popko 2002, 8). Consequently, the relevant anchor points for the spiked ax should be its association with the War God and the fact that it was imbued with sacred power.

Perhaps profoundly associated with the divine were the composite hybrid creatures, part human, part animal embellishments, which adorned ceremonial weapons. Powerful totemic representations of hybrid animals and inscriptions imbued these special weapons with the attributes of specific deities thereby attaining an iconic identification. A good example of this is the elaborately decorated, three-spiked Hittite ax from Şarkışla. Griffons, sphinxes, eagles, and gods share the space on this weapon, giving evidence for particular ways of viewing and organizing the netherworld. Eagles function as flanges on an unprovenienced ceremonial shaft-hole ax (Porada 1989, 442), which is dated to the 14th/13th century B.C. Vertically tiered, the weaponized gods are intermingled with composite creatures and animals. The vertical arrangement recalls the tiered nature of the Kanesh Anatolian-style seals where the throned gods are held aloft by up to 3 tiers of supernatural creatures and lesser deities. These stunning series of metal weapons were created during the period of the Hittite ascendancy (ca. 1600–1200 B.C.) and demonstrated shared cognitive codes and ritual experience. This is a world where the weapons of war are afforded a special place and depicted on the same plane with the supernatural realm of sacred animals and gods.

Throughout the first half of the second millennium B.C., inscriptions were incised onto a variety of weapons such as spearheads, swords, pikes, and axes. It is in this numinous, symbol-ridden world that the origins of sacred, inscribed weapons could be found at Kültepe-Kanesh, Kuşaklı-Shapinuwa, Mama, and Yazılıkaya-Hattusha. The best known example from the Middle Bronze Age is the inscription *E.GAL A-ni-ta ru-ba-im* [Palace of Anitta, the King] on the bronze slotted Anitta spearhead that declares ownership of the palace at Kanesh (Özgüç 1956; 1986, pl. 107:1a–c). A crescent symbol is set next to the inscription. The style of the spearhead demonstrates a long continuity from Early Bronze Age (EBA) III examples found in burials in north-central Anatolia at sites such as Alaca Höyük and Ikiztepe in the Black Sea region.

Inscribed in Old Assyrian cuneiform and given as a votive offering to a temple of the blood-stained warrior god Nergal, the god of the underworld, is a sword from Diyarbakır with divine burial associations (Güterbock 1965). Two antithetical lions decorate the hilt, which recalls the lions decorating the hilt of a copper-based spearhead from Alalakh (Woolley 1955, pl. 70:AT/39/305) (Fig. 26.2). Old Assyrian inscriptions were also inscribed on two copper-based pikes acquired by the Gaziantep museum (Donbaz 1998) and said to come from Hasancık village near Maraş. Anum-Hirbi, king of Mama, who was well known from the Kanesh texts, declares ownership of these two weapons. Adding to this list of MBA royal and sacred weapons are two examples from Soli on the Cilician coast. The first is a sign on a bronze blade and an inscribed "foot" symbol on a flat ax (Bittel 1940, 186, fig. 4, and S3462, pl. V). Several flat axes from Anatolia were inscribed with the symbol known as the "signe royal," also interpreted as the motif identified as Ishtar. Examples of this royal branding on axes come from Konya-Karahöyük (see Erkanal 1977, pl. 1:f and others, no. 7, pl. 2:24). Often stamped on large storage jars at Kanesh (Özgüç 1986, pl. 102:3–5), other examples include a stone stamp (Özgüç 1986, pl. 92:6) and a sealing motif at Acemhöyük. Several examples of the "signe royal" pendants in ivory and gold-plated ivory were found at Alalakh dating to the Level IV Palace (Woolley 1955, pl. 77:AT/38/78).

Weapons were also used for political purposes, given the fact that they were seen as the attribute of a god. Popko (2002, 8) writes that "during the reign of Zimri-Lim of Mari (18th century B.C.), weapons were moved to the temple of the god Dagan in the city of Terqa on the Euphrates. In return, the priests of Addu (Storm God of Halab/Aleppo) demanded that their god "be given primacy." The policy proved effective, because it led to the spread of the cult of the Storm God of Halab far beyond Syria.

The worship of weapons as emblems of various deities and inscribed with the names of the king continued in the Hittite empire period. Thus, a fairly compelling lineal development can be observed in the powerful language on another inscribed sword. Written in Akkadian cuneiform and attributed to 1430 B.C., the sword was found near the Lion's Gate in Hattusha and is dedicated to the Storm God. A victorious King Tuthaliya II declares victory over

Assuwa-land (Ünal 1993). Another blade said to be stylistically similar to Mycenaean spearheads was found inscribed with heiroglyphic signs "*walwaziti* GREAT SCRIBE" (Bilgi and Dinçol 1989), well known as an important official during the time of Hittite Great King Hattusili III and his Queen Puduhepa. The blade has been stylistically attributed to the 14th–13th century B.C. The prevailing pattern of numinous weapons can be found in eastern Anatolia as well. At Ortaköy, Middle Kingdom ancient Shapinuwa, a clay model used in bivalve casting demonstrates the production of a three-spiked shaft-hole ax, but in this case the ax or knob side of the weapon is absent (A. Süel 2005, pl. 6). Along with this, the site yielded an inscribed lugged ax. Found among a hoard of military arms in Building D, the inscription reads GREAT KING in hieroglyphic Luwian (M. Süel 2005, 698). These symbolic renderings and augmented attention to weapons and their embellishment foreshadow the worship of the divine Sword God as depicted in the Yazılıkaya relief. The relief sword god has a double-edged blade with a distinct midrib and is shown stuck into the ground of the sacred chamber. Two symmetrical lions decorate the hilt, yet again tying the lion decorated sword iconography (Fig. 26.2) with Alalakh (Woolley 1955, pl. 70:AT/39/305).

Further south in the eastern Mediterranean and southern Levant, very few examples of inscribed weapons are found (see Philip 1989), although beautifully decorated prestige weapons do indeed typify the corpus at Ugarit and Byblos. Maxwell-Hyslop (2002, 211) published a curved and inscribed sickle sword found in eastern Turkey dating to the 13th century B.C. The Middle Assyrian king Adad-Nirari claims ownership of the palace in cuneiform and lists his geneology. As Liverani (1989) pointed out, the weapons such as the ones described above are not standard infantry issue, but may reflect what he sees as a change in the nature of prestige weapons. With the coming of the Hittites, warfare became institutionally controlled, thus superseding the status of the individual warrior. Moreover, it is interesting to note the coincidence of these numinous weapons with regions conquered by the Hittites, a most formidable, aggrandizing warrior society.

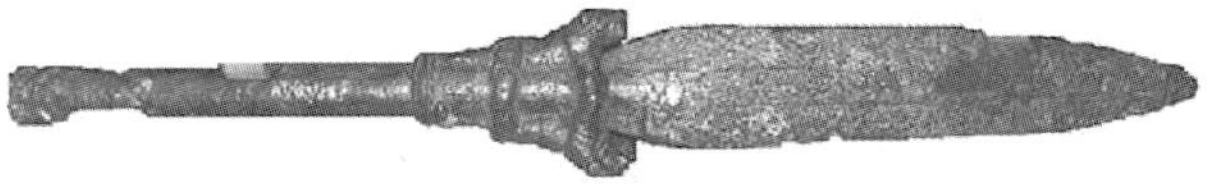

Figure 26.2. Bronze blade with antithetical lions, Alalakh (Woolley 1955, pl. 70:AT/39/305).

Conclusions

I have argued that the occurrence of special numinous and sacred weapons of Anatolia expands considerably with the notion of aggrandizing states, formalizing religion, and shared iconography during the Middle and Late Bronze Ages. The social significance of these special weapons may have been radically altered over the Bronze Age and are very much embedded in sociopolitical phenomena. While their origins go back to more imperceptible royal iconography during the Early Bronze Age, as in the case of the silver ceremonial shaft-hole ax from Alaca Höyük and the gold-plated iron sword from the same site, inscribed weapons and the worship of divine swords in the subsequent millennia demonstrate the artistic retention of particular traits and tendencies in the portrayal of deities brandishing their weapons. Branding by creating distinctive weapons in association with particular gods, such as the spiked shaft-hole battle ax wielded by the War God, which would be instantly recognizable, has been a tactic empires have often used. Divine weapons never lose their primary association with deities in ritual depictions, however, as they became more widely distributed in the Late Bronze Age, I suggest that the weapons themselves were transformed so that these powerful artifacts were loaded further with profound associations of city, ruler, and formal religion. What emerges from this general theme is that bronze weapons such as the Alalakh spiked battle ax were not only preferred as a highly effective weapon, but they also could have been valued as symbols of state, the king, and appropriate religion.

References

Bilgi, Ö. 2005. "Orta Karadeniz Bölgesi M.Ö.2. Binyılı Metal Silahları," in *Acts of the Vth International Congress of Hittitology*, A. Süel, ed., Çorum, pp. 119–149.

Bilgi, Ö., and A.M. Dinçol. 1989. "A Unique Spearhead from Sadberk Hanım Museum," in *Anatolia and the Ancient Near East: Studies in Honor of Tahsin*, T. Özgüç, K. Emre, B. Hrouda, M. Mellink, and N. Özgüç, eds., Ankara, pp. 29–31.

Bittel, K. 1940. "Der Depotfund von Soloi-Pompeiopolis," *ZA* 46, pp. 183–205.

———. 1976. *Les Hittites*, Paris.

Curtis, J.E. 1983. "Some Axe-Heads from Chagar Bazar and Nimrud," *Iraq* 45, pp. 73–81.

———. 1989. *Bronzeworking Centres of Western Asia c. 1000–539 B.C.*, London.

Deshayes, J. 1960. *Les outils de bronze, de l'Indus au Danube (IV au II Millenaire)*, Paris.

Donbaz, V. 1998. "Inscribed Spear Heads and Some Tablets at the Gaziantep Archaeological Museum," *Altorientalische Forschungen* 25 (1), pp. 173–185.

Erkanal, H. 1977. *Die Äxte und Beile des 2, Jahrtausends in Zentralanatolien* (*Prähistorische Bronzefunde Abteilung* 9, Band 8), Munich.

Güterbock, H.G. 1965. "A Votive Sword with Old Assyrian Inscription," *Studies in Honor of Benno Landsberger on His Seventy-Fifth Birthday* (*Institute of Assyriological Studies* 16), Chicago, pp. 175–181.

Güterbock, H.G., and T. Kendall. 1995. "A Hittite Silver Vessel in the Form of a Fist," in *The Ages of Homer: A Tribute to Emily Townsend Vermeule*, J.B. Carter and S.P. Morris, eds., Austin, pp. 45–60.

Koşay, H.Z. 1951. *Alaca Hüyük Kazısı (Les fouilles d'Alaca Hüyük. Rapport préliminaire sur les travaux en* 1937–39) (*Türk Tarih Kurumu Series* V, no. 5), Ankara.

Larsen, M.T. 1976. *The Old Assyrian City-State and Its Colonies* (*Copenhagen Studies in Assyriology* 4), Copenhagen.

Liverani, M. 1989. "Economy of Ugaritic Royal Farms," in *Production and Consumption in the Ancient Near East*, D.C. Zaccagnini, ed., Budapest, pp. 127–68.

Maxwell-Hyslop, R. 1949. "Western Asiatic Shaft-hole Axes," *Iraq* 9, pp. 90–129.

———. 1953. "Bronze Lugged Axe or Adze Blades from Asia," *Iraq* 15, pp. 69–87.

———. 2002. "Curved Sickle-Swords and Scimitars," in *Of Pots and Pans. Papers on the Archaeology of Mesopotamia and Syria Presented to David Oates on His 75th Birthday*, L. al-Gailani Werr, J. Curtis, H. Martin, A. McMahon, J. Oates, and J. Reade, eds., London, pp. 210–214.

Muhly, J.D., R. Maddin, T. Stech, and E. Özgen. 1985. "Iron in Anatolia and the Nature of the Hittite Iron Industry," *AnatSt* 35, pp. 65–84.

Müller-Karpe, A. 1993. "Hethitische Dechsel," *Istanbuller-Mitteilungen* 43, pp. 227–234.

———. 1994. *Altanatolisches Metallhandwerk* (*Offa-Bücher* 75), Neumünster.

Özgüç, T. 1956. "The Daggar of Anitta," *Belleten* 77, pp. 33–36.

———. 1986. *Kültepe-Kaniş* II: *Eski Yakındoğ'nun Ticaret Merkezinde Yeni Araştırmalar* [*New Researches at the Trading Center of the Ancient Near East*] (*Türk Tarih Kurumu Yayınları*, 5°, 41), Ankara.

Philip, G. 1989. *Metal Weapons of the Early and Middle Bronze Ages in Syria-Palestine* (*BAR* 526), Oxford.

Popko, M. 2002. "Mace-Heads and Similar Objects in the Ancient Near East According to Written Sources," *Baltic-Pontic Studies* 11, pp. 7–11.

Porada, E. 1989. "An Anatolian Axhead in the Honolulu Academy of Arts," in *Anatolia and the Ancient Near East. Studies in Honor of Tahsin Özgüç*, K. Emre, B. Hrouda, M. Mellink, and N. Özgüç, eds., Ankara, pp. 441–445.

Snow, H. 2005. *The Role of Metals in the Early Bronze Age Economics of North Syria and Anatolia*, Ph.D. diss., University of Toronto.

Stronach, D.B. 1957. "The Development and Diffusion of Metal Types in Early Bronze Anatolia," *AnatSt* 7, pp. 90–125.

Süel, A. 2005. "Ortaköy/Şapinuwa Ören Yerinde 2003 Yılı Çalışmaları [2003 Season at the site of Ortaköy/Şapinuwa]," *Kazı Sonuçları Toplantısı* 26, pp. 243–250.

Süel, M. 2005. "Ortaköy-Şapinuva 'D' Yapısı Hitit Dini Mimarisinde Değişik bir Yorum [A New Interpretation of Hittite Sacred Architecture at

Building 'D' Ortaköy-Şapinuva]," in *Acts of the Vth International Congress of Hittitology*, A. Süel, ed., Çorum, pp. 687–700.

Ünal, A. 1993. "Boğazköy Kılıcının Üzerindeki Akadca Adak Yazısı Hakkında Yeni Gözlemler [New Observations on the Akkadian Votive Inscription on the Boğazköy Sword]," in *Aspects of Art and Iconography: Anatolia and Its Neighbors. Studies in Honor of Nimet Özgüç*, M.J. Mellink, E. Porada, and T. Özgüç, eds., Ankara, pp. 727–730.

Uzunoğlu, E. 1989. "Fragment eines Bemalten Gefässes mit Libationsdarstellung aus Imamoglu Höyük bei Malatya," in *Anatolia and the Ancient Near East. Studies in Honor of Tahsin Özgüç*, K. Emre, B. Hrouda, M. Mellink, and N. Özgüç, eds., Ankara, pp. 497–503.

Woolley, C.L. 1955. *Alalakh: An Account of the Excavations at Tell Atchana in the Hatay, 1937–1949* (*Report of the Research Committee* 18), London.

Yalçin, Ü. 1998. "Der Keulenkopf von Canhasan (TR) naturwissenschaftliche Untersuchung und neue Interpretation," in *Metallurgica Antiqua* (*Der Anschnitt Beiheft* 8), T. Rehren, A. Hauptmann, and J. Muhly, eds., Bochum, pp. 279–289.

Yener, K.A. 2005. *The Amuq Valley Regional Projects. Volume One. Surveys in the Plain of Antioch and Orontes Delta from the Years 1995–2002*, Chicago.

———. 2010. *Tell Atchana, Ancient Alalakh* I: *The 2003–2004 Excavation Seasons*, Istanbul.

CHAPTER

27

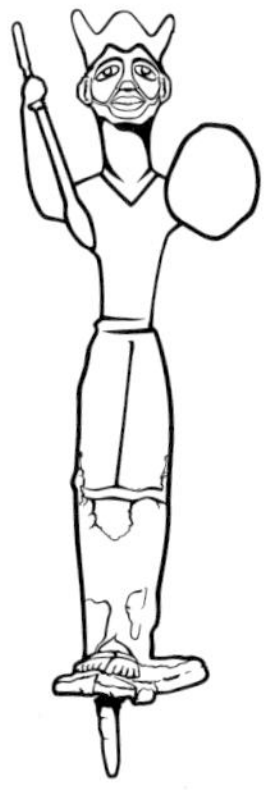

Sources of Tin and the Tin Trade in Southwest Asia: Recent Research and Its Relevance to Current Understanding

Vincent C. Pigott

Tin-Bronze as a Pan-Asian Technological Phenomenon

Current archaeological evidence suggests that during the early 3rd millennium B.C. in Southwest Asia, the earliest known use of tin-bronze occurred in Anatolia and Mesopotamia (e.g., Weeks 1999, 2003; Kaniuth 2007).* This alloy of copper and tin was used initially mainly for decorative items. Over the next two millennia, tin-bronze became the alloy of choice for ornaments, vessels, tools, and weapons across the Asian continent—from Beirut to Beijing to Bangkok. This was not to change until iron began to become widely available ca. 1000 B.C. and in the centuries immediately following, at which point tin-bronze reverted back to being a mainly decorative metal. While archaeologists today, benefiting hugely from the opening of Central Asia and China to international research teams, are only now observing the

---

* It is a remarkable thing in scholarship today when a volume published in 1973 continues to stand as a fundamental reference work in any field of study. Such is the case with James D. Muhly's *Copper and Tin: The Distribution of Mineral Resources and the Nature of the Metals Trade in the Bronze Age*. This seminal publication appeared in the heyday of the emergence of archaeometallurgy, i.e., not long after Theodore A. Wertime's metallurgical expeditions across Southwest Asia in which the likes of Cyril Stanley Smith, R.F. Tylecote, Radomir Pleiner, Beno Rothenberg, and Robert Brill began the scientific quest for the beginnings of metallurgy. Although *Copper and Tin* was important at the time for scholars of the Ancient Near East and the Eastern Mediterranean, in particular, it proved to be equally significant as an example *par excellence* for the successful integration of archaeological, metallurgical, ancient historical, and textual data.

spread of tin-bronze as a pan-Asian technological phenomenon with ramifications in the economic, political, and social lives of the ancient peoples who adopted it during this period (e.g., Mair 1998; Sherratt 2006; Anthony 2007; Kohl 2007; Kuzmina 2007, 2008), Muhly's *Copper and Tin* explored this same phenomenon and laid the foundations for its discussion over much of the Eurasian landmass. Exactly how and why the initial production and use of tin-bronze came about remains unclear, and, curiously, this all important technological innovation, on current evidence, appears to have occurred in Southwest Asia, a region with a long heralded dearth of tin deposits.

Rather than focusing on origins, however, the discussion to follow offers a generalized overview of current thinking on which tin sources—now multiple—may have supplied tin-bronze production and trade in Southwest Asia. With new answers, of course, come new questions. One question that can be posited at the outset here is: are we any closer today to understanding the locations of the source(s) of the tin that was traded across Southwest Asia and into the Eastern Mediterranean world during the Bronze Age?

Uluburun Shipwreck: A Ton of Tin

The discussion begins in the later centuries of the Bronze Age in Southwest Asia. There is little question that research since *Copper and Tin*, over the last two decades, has yielded a wealth of new information on ancient tin. Among the most remarkable discoveries is the wreck of the ca. 15-meter-long cargo ship (ca. 20-ton capacity) that sank off Uluburun on the southern coast of Turkey ca. 1300 B.C. The Uluburun shipwreck site was excavated by the Institute of Nautical Archaeology (INA) in 1984 and 1986 by George Bass (Bass 1986; Bass et al. 1989) and Cemal Pulak (1988, 1997, 1998; see also Yalçin, Pulak, and Slotta, eds., 2005). Among its rich and diverse cargo, the wreck yielded an astonishing 10 tons of copper ingots and approximately one ton of tin ingots, the largest Bronze Age find of its kind (Pulak 2000, 137; see also Maddin 1989; Yalçin 2005, 19). As an indication of the magnitude of this find, prior to this discovery, finds of archaeological tin in the Old World were measured in mere tens of kilograms.

At some point in its voyage along the Eastern Mediterranean coast (Yalçin 2005, fig. 4; Aruz, Benzel, and Evans, eds., 2008, fig. 97) the ship was loaded with its valuable cargo of ingots. Fairly strong arguments, based mainly on lead (Pb) isotope analysis (which, it should be said, is a method that continues to struggle with both analytical and interpretive problems), support the island of Cyprus as the source for most of the copper ingots (Pulak 2000, 147; see also Kassianidou 2003). Identifying the source(s) of the tin from which the tin ingots were produced is much more problematic, given especially the relative rarity of tin ore deposits in Southwest Asia. Scholarly speculation is rife as to the sources for the Uluburun tin

---

By the late 1970s, Muhly, an ancient historian at the University of Pennsylvania, had teamed up with a Penn colleague, the eminent metallurgist Robert Maddin, of the Laboratory for Research on the Structure of Matter and, not long after that, with Near Eastern archaeologist Tamara Stech-Wheeler. This fortuitous collaboration produced many significant contributions to emerging archaeometallurgical research (e.g., Maddin, Stech-Wheeler, and Muhly 1980; Muhly et al. 1985; Maddin, Muhly, and Stech 1999). In addition, it had a role in fostering the formation of the ancient metals research group at the Museum Applied Science Center for Archaeology (MASCA) at the University of Pennsylvania Museum, which, for more than two decades, served as the research base for the author and Stuart J. Fleming (Scientific Director), as well as over time for the likes of Reed Knox, Stephen M. Epstein, Elizabeth Hamilton, Samuel K. Nash, Harry Rogers, and William W. Vernon among others.

Despite its longevity as a seminal volume in ancient metallurgy, *Copper and Tin* and its *Supplement* (Muhly 1976) are less frequently cited today as there has been an intervening three decades of research building upon the foundation it laid. At the same time, Muhly's discussion of potential sources of tin and the tin trade—most especially his arguments surrounding tin "from the East" and the role of *Meluhha* and the Indus Valley Harappans in moving tin westward—all remain remarkably timely and convincing. Discussion of the tin trade comprises only half of his volume, with the other half devoted to copper resources and trade in the Mediterranean and adjacent areas.

Copper and Tin constitutes a scholarly *tour-de-force* that merits repeated consultations by scholars interested in Bronze Age metals and metallurgy in the broadest possible context. In pursuing the question of what we do know today about ancient tin, tin sources, and the movement of tin in the Bronze Age of Southwest Asia, I returned again to this remarkable volume written at the start of Muhly's long and distinguished career. In innumerable instances, his insightful observations on ancient tin presage what scholarship was later to reveal.

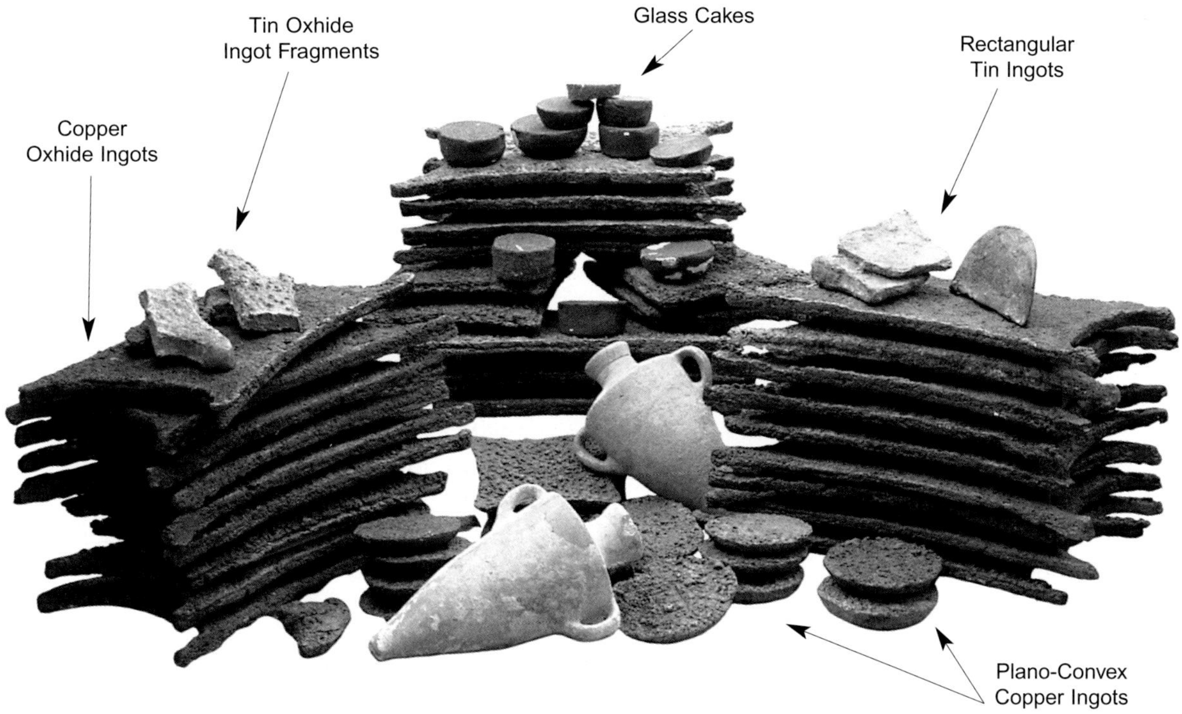

Figure 27.1. Cargo excavated from the 14th century B.C. shipwreck at Uluburn, Turkey (after Yalçin 2005, fig. 2).

ingots, and it reiterates the time-honored question asked by the Cambridge University metallurgist J.A. Charles (1975) in the *Antiquity* journal article, "Where is the Tin?" Thus, in seeking to source the Uluburun tin ingots, we have the opportunity to assess how far we have come since Charles posed his question.

Some 35 years ago, archaeologists could not speak with any certainty on "known sources" of tin essential to the vast quantities of tin-bronze made in Southwest Asia and neighboring regions. Muhly (1973a, 248–262) reviewed the literature concerning sources of ancient tin for the Mediterranean and neighboring regions from Cornwall to India, a review well worth reading in the light of more recent discoveries. However, at that time there was only the barest inkling of geologically documented tin deposits anywhere within reasonable proximity of Southwest Asian communities of the 3rd and 2nd millennia B.C. Guiding much of the modern scholarly thinking about tin sources has been Muhly's (1976, 97) cautious statement, "[t]he Near Eastern texts dealing with the tin trade, both in the Old Assyrian and in the Old Babylonian periods, still seem to make best sense in terms of a source of tin coming from the east. . . ." Modern archaeological research has now shed some light on actual tin sources in the East, some found only in the past 5–10 years. These discoveries continue to refine our understanding of potential sources, and at the same time make it clear that there were *multiple* sources of tin supplying the bronze workers of the ancient Near East and neighboring regions.

Returning to the Uluburun tin ingots, they occur in various shapes: oxhide (three were intact, with the remainder cut into quarters and halves), rectangle, plano-convex "bun", wedge, and bar (Fig. 27.1). In this regard, the Pb-isotope ratios for the 99 tin ingots analyzed thus far indicate that they fall into at least two distinct clusters *suggesting* a minimum of two separate sources for the tin (see Pulak 2000, 153–154, fig. 23; also Stos-Gale et al. 1998, 119, 123). Pulak (2000, 153, 155) cautions that these analyses are preliminary and that too much should not be made of the results until further study is undertaken, in part because, with analyses of additional tin ingots, the clusters may in fact "somewhat merge." Pertinent here is the fact that these two isotopically defined clusters also have distinctly different lead compositions, reinforcing the notion that two clusters are in fact observable. One group of 68 ingots has a high Pb concentration of greater than 200 parts per million (ppm; nearly 1% in some specimens), while the other group of 31 ingots immediately adjacent has a low amount of Pb, below 200 ppm and as low as 40 ppm. Interestingly, the so-called oxhide ingot shape is, for all intents and purposes, not encountered east of the Levantine Mediterranean world (see map of oxhide ingot finds in the Mediterranean, fig. 10, in Lo Schiavo 2005, 409; also Pulak 2000, 137), so the question of where such ingots were cast to their distinctive shapes remains unresolved at present. The discussion now turns to the import of the two potential tin ingot clusters.

Kestel Tin Mine and Göltepe in Bolkardağ, Taurus Mountains, Turkey

In interpreting the Pb-isotope results for the ingot cluster with high amounts of Pb, there appears to be an overlap with the isotopic field (Taurus 1A) that encompasses the Bolkardağ mining district of the Taurus Mountains in south-central Turkey (Pulak 2000, 155; see Yener et al. 1991, 546–548; Sayre et al. 1992, 74–75). It is in this region that Aslıhan Yener and her team from the Oriental Institute of the University of Chicago, working with a number of Turkish colleagues, conducted the excavations in the later 1980s at what she termed a "first-tier industrial production complex in the highlands" (Yener 2000, 71). This complex is comprised of the Early Bronze Age tin mine at Kestel (ca. 3rd millennium B.C.) and its associated tin-processing settlement at Göltepe (Yener 2000, 14–15). Research at the Kestel mine indicated that among known ancient mines it was not particularly small, with 1.5 km of tunnels known thus far, measuring a minimum of 4,600 cubic meters. Such an area would have yielded about 115 tons of tin based on a low-grade ore (1%). Estimates of what the ore grade may have been when Kestel was an active mine run as high as 2% tin, with the high grade ores being moved to Göltepe for processing (Yener 2000, 73–74, 112; see also Earl 1994; Earl and Özbal 1996).

The site of Göltepe, in promixity to the Kestel mine, has a 5-hectare area of the densest habitation with scattered extramural settlement extending over an additional 10 ha. When linked with Kestel, Yener (2000, 101) estimates a total of 60 ha for the entire "man-mine system," not a small industrial complex by any standard in the Bronze Age. Göltepe's surface yielded an impressive 50,000 ground stone tools used primarily to crush ore, while from pit-house structures at the site an additional 5,000 were excavated (Yener 2000, 74, 112).

Interestingly, while the news of this discovery of a tin source within the confines of the ancient Near East proper was greeted with great interest, the identification of Kestel as a major tin source having supplied much of the tin for the Near East was met with considerable disagreement, with James Muhly at the vanguard of the critics (e.g., Muhly et al. 1991). A somewhat heated debate ensued among scholars (see also, e.g., Hall and Steadman 1991). Since then, and with the completion of considerable geological and metallurgical laboratory analyses as well as the publication of Yener's *The Domestication of Metals* (2000), much of the controversy is now laid to rest as Yener has taken the main points of criticisms and answered them one by one (Yener 2000, 72–76). Analyses now indicate that at Göltepe a higher grade tin-rich hematite ore was being crushed and ground to a powder, which, in turn, was being smelted to tin metal in crucibles (see Vandiver, Yener, and May 1992; Vandiver et al. 1993; Yener and Vandiver 1993a, 1993b; Adriaens 1996; Earl and Özbal 1996; Adriaens et al. 1997; Adriaens et al. 1999). Offering strong support for the identification of an industrial tin production complex at Kestel-Göltepe, Yener (2000, 127) mentions that a ton of tin-rich vitrified ceramic material was excavated at Göltepe. There is little question that this complex was in the 3rd millennium B.C. meeting a certain, though inestimable, portion of the needs of those metalworkers producing bronze across the ancient Near East and Eastern Mediterranean worlds.

Yener does acknowledge that the Kestel tin mine is but one of a number of potential sources for tin in these regions (Yener 2000, 72; see also Yener and Vandiver 1993a, 1993b). It should also be said that the remarkable mineralogical richness of the Bolkardağ mining district and elsewhere in the Taurus Mountains means that the potential is high for additional discoveries of tin sources to be made. This in turn will offer further support for the notion that it was most certainly a small deposit, one of multiple sources that were supplying much needed tin across the greater region at least in the earliest stages of the use of tin-bronze on any scale. One crucial final point needs to be made here, namely that the Kestel-Göltepe complex, by all indications, was not operating in the 2nd millennium B.C.

(Yener 2000, 75), which would rule it out as the possible source for the Uluburun tin ingots. But what of the possibility of another Taurus source(s) being exploited in the 2nd or 1st millennia B.C.? Only future fieldwork combined with continuing Pb-isotopic patterning of ores and artifacts from the region is going to tell the tale.

At least where tin mining is concerned, the exploitation of innumerable small scale "ma and pa" type of mining "claims" typical of the 3rd millennium B.C., it would seem, had seen its day (see Yener 2000, 75). It is during the 2nd millennium B.C. that tin-bronze comes into its own as the hegemonic alloy, as a major trade in tin is documented in the cuneiform texts, and the tin being traded is discussed in ever increasing and substantial amounts. Tin ore sources capable of considerable production had by this time been located and were being accessed by or under the direct control of state centered economies participating in an "international" and "long distance" metals trade (from the East, but how far east?)—a trade so substantial that it could over time afford to consign any number of shiploads of copper and tin ingots to the bottom of the sea.

It is worth noting at this juncture that in the historical record of the ancient Near East, tin as a commodity never became common, yet it was in constant demand. As late as the 8th century B.C., tin was still being listed as a precious material along with gold, silver, iron, elephant hides, ivory, garments, purple-dyed wool, and precious woods (Muhly 1973b, 406). Indications are that people went to remarkable lengths to acquire it. Thus, it appears that tin sources were limited, and tin ores and/or tin metal was a scarce and much sought commodity—essential to craftsman, artisan, and armorer alike. Such demand dictated an active and lucrative trade across the greater region—as said, one conducted at the level of empires. The ton of tin on the Uluburun ship was but a single shipload; over the centuries, how many ships must have completed their journeys successfully, delivering their cargoes to waiting ports? Other shipwreck and undersea finds of tin ingots have been made over the years, but none as massive as the Uluburun site. To reiterate, while it can be argued that based on current, albeit preliminary, Pb-isotope results, the high-lead tin ingots from Uluburun may have had a Near Eastern origin (perhaps in south-central Anatolia), the low-lead cluster of tin ingots *does not* overlap with the Anatolian or any known isotopic field. Some portion of that shipload of tin ingots therefore apparently came from somewhere other than Anatolia, perhaps by trade from points far afield—east of Mesopotamia according to the texts—the question remains, however, just how far east?

Tin Trade in the Ancient Near East. . . and beyond?

Current scholarly understanding of the tin trade in the ancient Near East has been much discussed and need not be repeated here, although it is important to note that it is based primarily on the later 3rd or early 2nd millennium B.C. textual record initially published by Oppenheim (1954) and Leemans (1960; see also Weeks 2003, 178–180; Dercksen 2005; Kaniuth 2007). Muhly's 1973 *Copper and Tin*, his 1973 "Tin Trade Routes of the Bronze Age" in *American Scientist*, and his 1985 article "Sources of Tin and the Beginnings of Bronze Metallurgy" in the *American Journal of Archaeology* offer among the best synthetic overviews of this earlier evidence for the tin trade. The two routes for the trade in tin are the northern or "overland" route and the southern or "coastal" route, both bringing tin to the west from sources to the east of Mesopotamia (Fig. 27.2). Most certainly these routes had been established for long-distance trade long before the advent of tin-bronze production.

Southern or Coastal Tin Trade Route

From at least the 3rd millennium B.C. onward, in particular with the emergence of the Harappan civilization of the Indus Valley (ancient Meluhha), a well-documented sea-going trade moved commodities including luxury materials such as carnelian from this region to the ancient gulf entities of Dilmun (region around Bahrain) and Magan (Oman) (Fig. 27.3; e.g., Crawford 1974; Possehl 1997; D.T. Potts 1993; T.F. Potts 1993). Here again Muhly (1973a) offers a detailed overview and an excellent starting point for understanding the nature of this

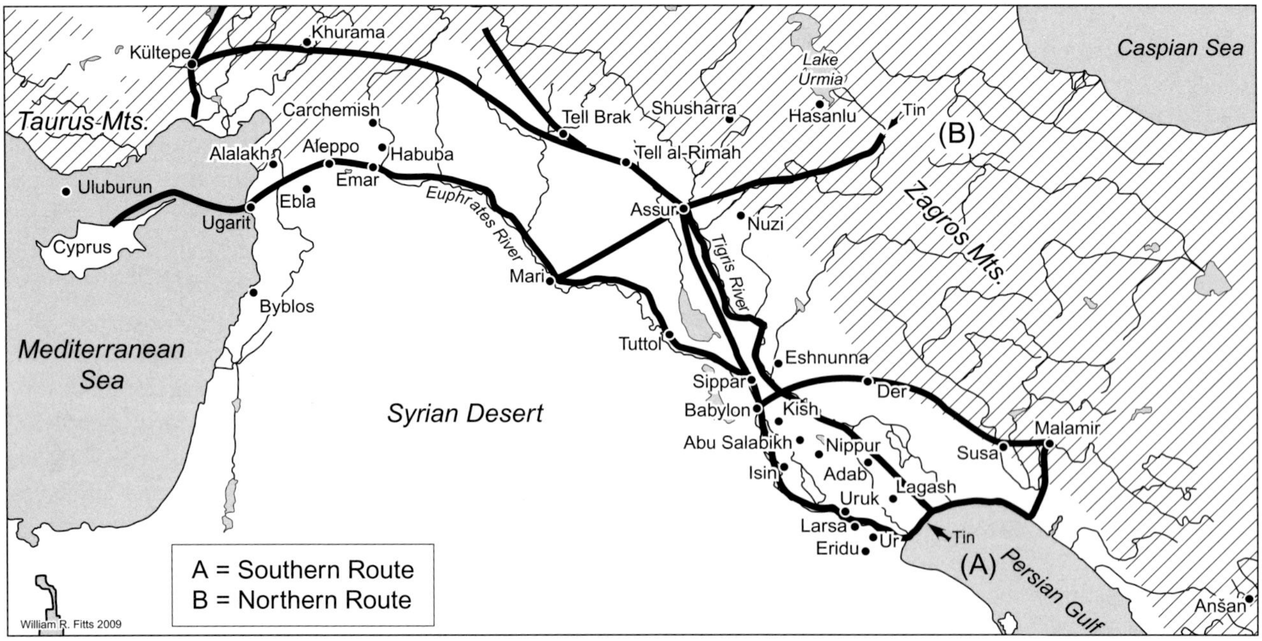

Figure 27.2. Map showing the southern or "coastal" (A) and the northern or "overland" (B) routes used to bring tin to the west from sources to the east of Mesopotamia (map by William R. Fitts, MASCA, Penn Museum).

sea-going trade as it integrates all data sets available at this time into a comprehensive summary.

The mute testimony of the contents of the Royal Tombs of Ur bespeaks the want for and the movement of luxury raw materials from well beyond the Mesopotamian alluvium. As research has progressed, however, the picture of the long distance trade movements of these luxury materials is perhaps less certain than it once was. Be that as it may, a sea-going trade between Meluhha, the gulf, and ultimately Mesopotamia remains supported by the data. The issue is just what was being traded.

Over the centuries "Sumerian and Babylonian merchants went to Dilmun to buy copper and tin, while traders from Dilmun came to Mesopotamia, for example, to Sargonic Umma, Lagash, and Agade as well as to Susa when it was under Sargonic rule" (Foster 1997, 59). Extreme southern Iraq with its Mesopotamian delta, marshes, and waterways is the presumed entry point and conduit northward for the copper (considerable amounts coming from Oman) and tin—available *in the gulf* according to textual evidence. At present, however, there is *no textual evidence* indicating that tin or lapis was coming from Meluhha (Possehl 1996), a point that clouds the discussion considerably. A sea-going trade in tin thus can be supported only circumstantially. What we do know of the land-based tin trade is that from its entry point, the tin traveled from extreme southern Mesopotamia to central Anatolia either by boat and/or by donkey caravan. We must ask ourselves, therefore, if tin was being transshipped to Mesopotamia via the gulf, where were these seafaring merchants obtaining it?

While it is the case that the land of Meluhha is not specifically mentioned as a source of tin or lapis, these commodities may have arrived in the gulf on Meluhhan ships or "sold" to gulf middlemen via the ports of Dilmun, who in turn moved them on to Mesopotamia. One can at least speculate that it was the Harappans, themselves, who served as middlemen in this trade. Why? The Harappans were well positioned in the Indus Valley to exploit highly desirable raw materials to be found in abundance in the neighboring highlands of Afghanistan, including the three that comprise the famed "Sumerian trio" of gold, lapis, and tin (in tin-bronze) known from major urban sites such as Ur after ca. 3000 B.C. (e.g., Muhly 1977; for the occurrence of tin-bronze, see Hauptmann and Pernicka, eds., 2004). This co-occurrence, of course, does not mean that the three materials came from the exact same source area; but the fact that all

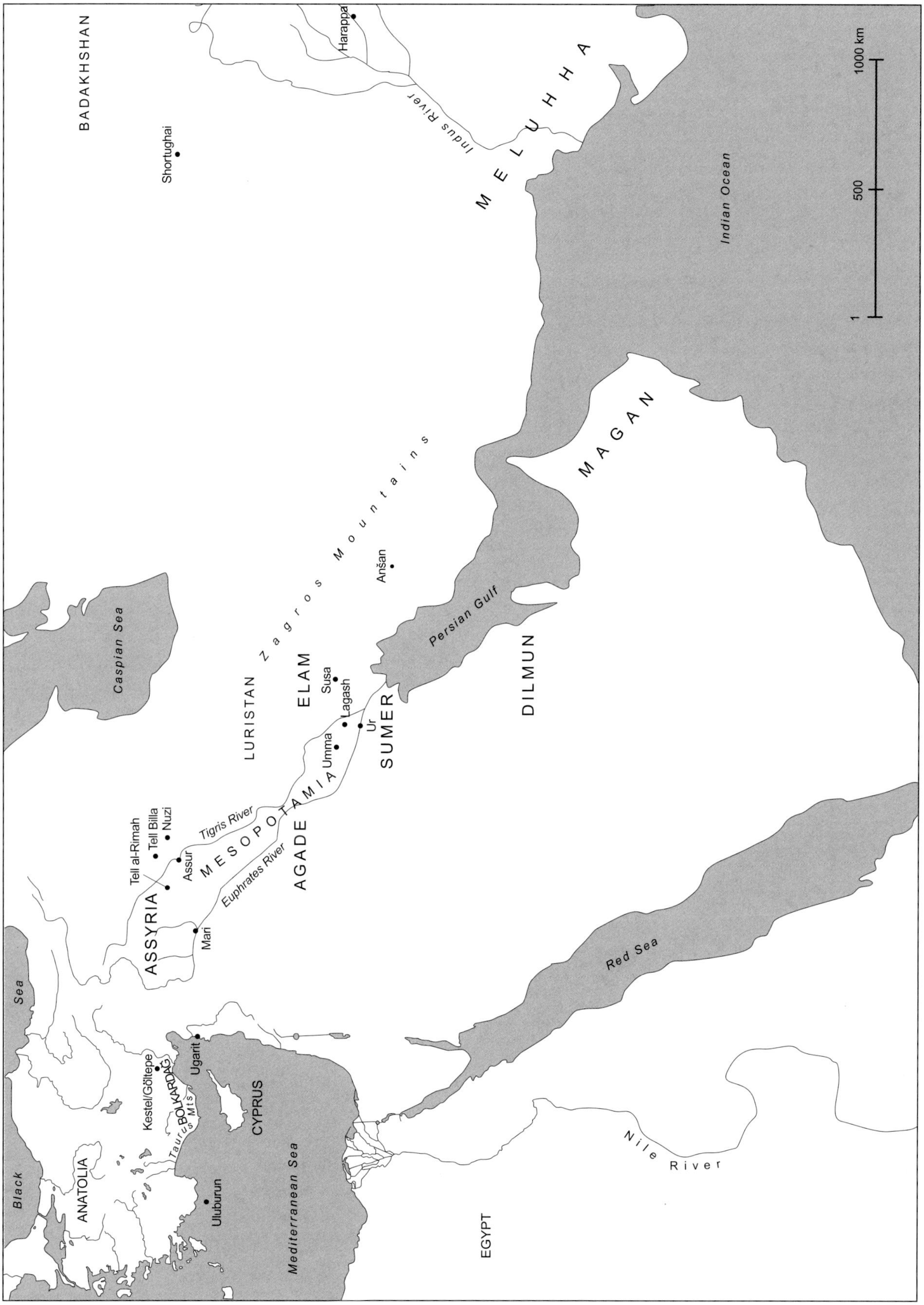

Figure 27.3. Map of western and central Asia showing the sites mentioned in the text.

three might enter Mesopotamia via the same trade route—overland or by sea—would have the same effect. The Harappans had an outpost in northeastern Afghanistan at Shortughai (Francfort 1989) close to the lapis source in Badakhshan, they had spread across Baluchistan (between Indus and Iran) to control the overland route from Afghanistan to the coast, and, as stated, they had a considerable presence in the gulf linking them directly or indirectly to Mesopotamian consumers (Fig. 27.4; Sarianidi 1971).

Soviet geologists working in Afghanistan documented its enormous mineral reserves including making it verifiably clear that this country has substantial placer and hard-rock tin and gold reserves (e.g., Shareq et al. 1977; see summary in Stech and Pigott 1986; also Rossovsky, Mogarovsky, and Chmirev 1987). An archaeometallurgical field survey was conducted by the French in the late 1970s in Iran and western Afghanistan (e.g., Berthoud 1979; Cleuziou and Berthoud 1982; Berthoud et al. 1982). They documented by panning the presence of the rich ore of tin cassiterite (SnO_2) in a stream bed in the Sarkar Valley in western Afghanistan, a region dotted with Bronze Age sites (Cleuziou

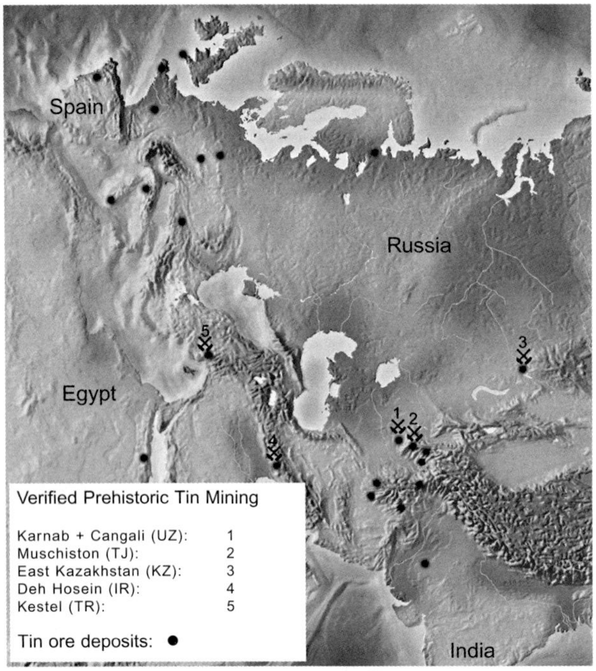

Figure 27.4. Prehistoric tin mines in western and central Asia (after J. Cierny, T. Stöllner, and G. Weisgerber 2005, fig. 1). Courtesy of the Deutsches Bergbau-Museum, Bochum.

and Berthoud 1982). This is one indication that Afghanistan's innumerable placer deposits would have been readily accessible, but regrettably, placer mining would leave little or no trace of having been conducted. Even lapis has been documented as occurring in stream bed nodules in Badakhshan (H.-P. Francfort, pers. comm.). Gold panning continues to this day (Knauth 1972, 26–27). Though tin adds the qualities of toughness, durability, and ease of casting to bronze, there is little evidence that the Sumerians chose it for such material properties. Instead, it has been argued elsewhere that the Sumerians sought tin-bronze as a high status material based, perhaps in part, on its rarity and exclusivity, much like the gold and lapis so widely used in similar elite circumstances (e.g., Stech and Pigott 1986; see also Weeks 2003, 188–191, Hauptmann and Pernicka, eds., 2004; Thornton et al. 2005).

Afghanistan, with its abundant mineral resources, constitutes a potential major source on the grounds alone of available and substantial reserves (e.g., Stech and Pigott 1986; Weeks 2003; Kaniuth 2007). Regrettably, due to current political circumstances, the possibility of "ground truthing" ancient mining activity is a dim but not impossible prospect (see Dalton 2007).

The reasonable proximity of the Indus Valley to the ore-rich river valleys that slice across Afghanistan from northeast to southwest offers at least the opportunity for an informed hypothesis. Though at this stage it remains purely speculative, we should not underestimate the role of the gulf polities and their Harappan trading partners as middlemen and/or principals in the movement of tin, lapis, copper, and tin-bronze into and through the gulf. German Mining Museum archaeologist Gerd Weisgerber, who, with his colleagues, documented the major Bronze Age copper industry of Oman, suggested that some of the Omani copper may have gone east to the Indus Valley in return for tin and other commodities (Weisgerber 1981). The tin had to get to the gulf by some means, and the Harappans offer at least one likely candidate.

Around 1800 B.C., the Harappan civilization's westward influence began to decline, yet tin kept coming from somewhere, and there is another player which merits mention in the complex trading scenario of the 2nd millennium B.C.—namely the Elamites of southwestern Iran. This relatively small empire appears to have played an important role in this trade as texts of the 18th century B.C. from Mari, a major trading entrepôt in Syria situated on the southern tin route, associate the acquisition of lapis with the procurement of tin indirectly through Elamites or more directly in Elam itself. Muhly (1973a) makes repeated references to Elam and how it may have had a role in maintaining access to at least some of the tin that was moving into the ancient Near East. This access to tin may also explain why the Elamite capitals of Susa (in Khuzestan/Mesopotamian lowlands) and Anšan (Tal-e Malyan, in Fars/highland Iran) are among the earliest users of tin-bronze in ancient Iran—they had access to tin that others did not (e.g., Tallon 1987; Pigott, Rogers, and Nash 2003).

While there is good archaeological and textual evidence for the transshipment of lapis across the plateau, texts also suggest that a trade in tin along the northern edge of the plateau was in operation (Muhly 1973a, 1973b). There is surprisingly little evidence, however, for tin-bronze anywhere on the plateau before the very end of the 3rd millennium (e.g., Stech and Pigott 1986; Thornton et al. 2005). For example, analyses indicate that the site of Tepe Hissar, located directly on the "Great Khorasan Road" in northeastern Iran, is almost entirely devoid of tin-bronzes (Muhly 1973a, 339–342; Pigott, Howard, and Epstein 1982; see also Thornton 2009). At Shahr-i Sokhta, a major third-millennium city on Iran's eastern border located on a trade route from the tin fields of central Afghanistan to the gulf, it has recently become apparent that there are not even traces of tin in the large number of metal artifacts that have been studied (e.g., Artioli et al. 2005; see also the description of Claudio Giardino's recent study in Thornton and Rehren 2007, 316). For this reason, the overland route for 3rd-millennium tin, as well as the use of Afghan tin in the Early Bronze Age, has been called into question recently at a workshop on the early metallurgy of Iran, held in 2007 at the University of Nottingham. Weeks summarized the main points made during the discussion session on tin-bronze as follows (Weeks 2008, 343):

> "Tin-bronze: the issue of tin sources was discussed. It was considered by Chris Thornton that a number of factors—most notably the lack of evidence for tin-bronzes on the Iranian plateau before the late 3rd millennium B.C. and the re-analyses of material from Mundigak (Benoit Mille, pers. comm.), which showed no evidence for the previously reported early (4th millennium) tin-bronze use—meant that Afghanistan should no longer be cited as a possible source of the earliest tin in SW Asia. Afghanistan should not be re-habilitated as a tin source unless controlled archaeological fieldwork and archaeometallurgical analyses conclusively demonstrate early tin or tin-bronze production in the region."

Thus, until such time as archaeologists can return to fieldwork in Afghanistan, the jury must remain out as to its role (or lack thereof) as a major tin source for greater Southwest Asia and its peripheries. In order to conclude this speculative discussion, one final point is raised. The last 20 years have brought to light much new archaeological evidence for copper and tin in the gulf notably via a major program in Pb-isotope analysis of regional copper-based metal conducted by Lloyd Weeks of the University of Nottingham, UK. Based on his extensive research, he is of the opinion that sources of tin for the gulf and greater Southwest Asia "almost certainly" are to be found in the Afghan and Central Asian deposits (Weeks 2003, 200). In addition, he maintains that "if tin sources are very scarce, one or a very limited number of sources could have supplied a very large area, and . . . isotopic matches [*as noted by him and other researchers*] could be a reflection of shared provenance" (Weeks 2003, 163; emphasis added). Thus, in my estimation, Week's meticulous research lends much needed support to the suggestion that tin, perhaps along with lapis, coming through the gulf en route to Mesopotamia, could have come from Afghanistan and was assembled along with Gujarati carnelian onboard Harappan reed boats that were caulked with bitumen—as excavated on the coast of Oman (Cleuziou and Tosi 2000).

Northern or Overland Tin Trade Route

Textual sources indicate that tin may have arrived in Mesopotamia via an overland crossing from a northern route (Muhly 1973a, 1973b). Many of our assumptions about an Afghan source of Bronze Age tin are based on the well-studied trade in lapis moving from the major source in the northeastern corner of Afghanistan to elite Sumerian consumers in distant Mesopotamia during the 3rd millennium B.C. and later (e.g., Herrmann 1968; Sarianidi 1971; Tosi 1974; Majidzadeh 1982; T.F. Potts 1993). Muhly discusses the lapis trade at length and indicates that it did not necessarily travel, *at all times*, with tin (Muhly 1973a, 303). Those interested in the lapis trade are encouraged to reread Muhly's detailed assessment (for a succinct overview of Bronze Age lapis lazuli and its trade, see also Casanova 2008).

Muhly discusses the evidence for a "northern or northeastern" source and route for tin moving into Mesopotamia in the 3rd and 2nd millennia B.C. (Muhly 1973a, 289–292). Tin, he argues, is moving "south" from Iran (i.e., from the north of southern Mesopotamia/Sumer) from two general directions, namely via northwestern Iran or via more southerly central western Iran. Tin had to move into Mesopotamia via the passes of the Zagros that penetrate this broad swath of mountains ranging from south of Lake Urmia to Luristan (Muhly 1973a, 302–303). Routes could vary according to which pass was chosen to transit, and they could include, for example, the so-called "Great Road" linking points east of Iran with Mesopotamia cutting across the northern part of the Iranian plateau through Hamadan and Kermanshah (Young 1967, 12; Muhly 1973a, 304; see also Young 1985). As mentioned, cogent arguments over the years suggest that Afghan lapis, among other commodities, moved to Mesopotamia via such routes. Tin could have traveled the same route, and while strong circumstantial arguments can be made (e.g., Stech and Pigott 1986), direct evidence does not exist at present. Additionally, there is no current evidence concerning tin sources in northwestern Iran, a possibility that Muhly raises based on textual interpretations, or, as he also indicates, some texts detailing tribute suggest the possibility of tin coming from the Nairi Lands to the northwest of Assyria (Muhly 1973a, 290). It is worth reiterating that in ancient Iran, tin-bronze does not appear in any quantity until the early centuries of the second millennium B.C., and even then it is linked primarily to the Elamite capitals of Susa and Anšan in southwestern Iran and somewhat later to the Luristan bronze industry in central western Iran.

In terms of a northern/overland movement of tin, recent research over the last decade has painted a new picture with regards to potential tin sources. Thus, at this juncture we depart western Iran, where I will return in the conclusion, and shift our attention farther east to Central Asia and to assessing the import of the recent documentation of tin sources with ancient mining evidence vis á vis their involvement with the ancient Near East.

Sources of Tin in Central Asia

Eurasia is crossed by a major tin belt (Fig. 27.5) that coincides with the border zone between the steppe and the cultivated oases of Central Asia, thereby serving as a natural conduit for movement of peoples and goods both to the East and to the West (see map in Bourgeois and Gheyle 2006, 24–25). Central Asian tin deposits were cited in Soviet literature (e.g., Litvinskij 1950), and there were other scholarly hints over time (e.g., de Jesus 1978), but it took the collapse of the U.S.S.R. and the opening of Central Asia for fieldwork by a German team to locate and document the prehistoric exploitation of these deposits along the Zeravshan River Valley (e.g., Alimov et al. 1998; Boroffka et al. 2002; Parzinger and Boroffa 2003). Through extensive fieldwork, the team composed of experts from the German Mining Museum, the German Archaeological Institute, and the Freiberg Technical University discovered prehistoric tin mines in the Central Asian countries of Uzbekhistan at Karnab, discussed as but one example below, Tadjikistan at Muschiston, and elsewhere in the region.

Tin Mine at Karnab, Uzbekhistan

The tin mining locus known as Karnab lies 2.5 km northeast of the modern village of Karnab in the Samarkand region (Boroffka et al. 2002; Cierny and Weisgerber 2003). It was studied by B.A. Litvinskij in 1946. The mine area itself extends across an area ca. 1.5 x 0.5 km with about 20 groups of ancient workings identified. These include slot trench open mines, mines with lateral galleries, underground caverns, and one sizeable open-cast mine resulting from collapsed ancient underground workings (Boroffka et al. 2002, 145). Modern prospecting trenches have disturbed the site significantly, but at the same time they have helped to reveal the ancient

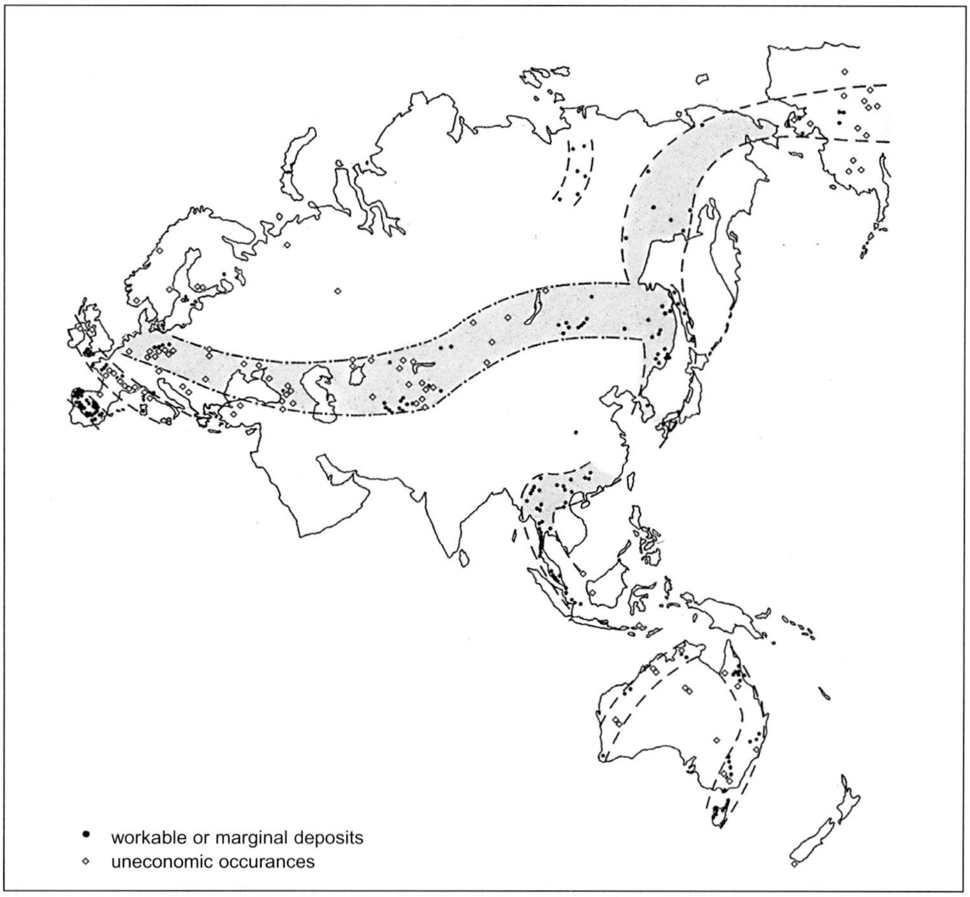

Figure 27.5. Tin belts in Eurasia and Australia (after de Jesus 1978).

mines themselves. The slot trench mines have a distinct vertical aspect due to the placement of the tin-bearing quartz veins (ca. 0.40–1.20 m thick) in a granite matrix. Such veins can extend over several hundred meters. Cassiterite (SnO_2) mineralizations were only ca. 5–10 cm thick, and tin content could range as high as ca. 3% tin with an estimated average tin content of ca. 0.5% (Boroffka et al. 2002, 145; see Cierny and Weisgerber 2003, 25, fig. 3.) While in antiquity these veins, it seems, would have been readily apparent at the surface, the cassiterite would not have been easily seen. Even though geological circumstances of contact metamorphism did reduce the hardness of the granite, the hardness induced by the high quartz content of the ore veins themselves would have made mining extremely arduous. Moreover, the excavators are of the opinion that no adult (Central European), using heavy tools (local limestone mining mauls), could have done the mining. It is suggested that it was done by children and/or women and that fire setting was used to crack the hostrock and ore veins utilizing the ". . . arising tensions of the different minerals in the rock" (Cierny and Weisgerber 2003, 26).

According to the ^{14}C dates taken from the samples from mines themselves, the exploitation of these Central Asian sources dates to the very late 3rd or early 2nd millennium B.C. The pottery found in the excavations of mining pits and galleries is of the Andronovo Culture, a Eurasian Steppe culture known to have interacted with the early-2nd-millennium-B.C. oasis farmers of the Bactrian-Margiana Archaeological Complex (BMAC) in the greater Turkmenistan region. It seems that Central Asian tin, therefore, was being heavily exploited in the early 2nd millennium B.C. by the Andronovo peoples, whose archaeological assemblage ranged from Mongolia and western China to the northern Caspian region and down into southern Central Asia. The people appear to have been nomadic tribes from the northeastern Eurasian Steppe who, in the late 3rd millennium B.C., expanded quickly across the entire landmass (e.g., see arguments in Kuzmina 2007, 2008; also Koryakova 2008). One reason for their sudden movement may have been to exploit resources such as the Central Asian tin mines strung along the massive Eurasian tin belt (see Fig. 27.5; Kohl 2007).

The critical question at this juncture concerns whether there is any evidence to suggest that these Central Asian mines were supplying tin to the Eastern Mediterranean world. Regular donkey caravans along the northern overland route could have moved tin and other commodities to consumers in the ancient Near East and Eastern Mediterranean. Cogent arguments set forth by Kaniuth (2007) suggest that while the mines were supplying local needs ca. 2000 B.C., Central Asian tin was *probably not* reaching consumers well to the west.

To sum up here, on the one hand there is little question that the substantial evidence for tin and ancient tin mining from Central Asia is going to shed considerable light on the development and spread of copper/bronze metallurgy across Eurasia (see concluding discussion). On the other hand, this evidence is not going to help resolve the issue of where the ancient Near Eastern and Eastern Mediterranean worlds obtained the enormous amounts of tin that were utilized in their bronze production.

Arguments have been presented above for Afghanistan as a potential source area for eastern tin (cf. summary of Thornton in Weeks 2008, 338) that could have moved overland along the southern and northern tin routes including by sea via Meluhhan seafaring merchants. At the same time tin-bronze is noticeably rare on the Iranian plateau in the 3rd millennium B.C., with Susa in lowland Khuzistan as a regional exception. The Elamite capitals of Susa and Anšan, which have some of the earliest tin-bronzes in Iran, are mentioned with considerable frequency in the texts related to tin trade, and lapis is implicated in the Elamite context as well (Muhly 1973a). Circumstantially this evidence supports the sea trade model. Interestingly, neither the Indus Valley nor the gulf has yielded much evidence for tin-bronze before the transition centuries between the 3rd and 2nd millennia B.C. The 2nd millennium ascendance of tin-bronze as the alloy of choice across the ancient Near East allows one to at least raise the question: could there be another tin source closer than Afghanistan, but one to which Elam would have had access or for which Elam could have acted the role of the middleman in moving tin (and lapis) to western consumers?

New Evidence for " . . . a source of tin coming from the east . . . "

In Muhly's *Supplement to Copper and Tin* (1976, 97), he suggests, "[t]he Near Eastern texts dealing with the tin trade, both in the Old Assyrian and in the Old Babylonian periods, still seem to make best sense in terms of a source of tin coming from the east, *most likely from Iran*." While for years scholars have argued that there were no tin deposits on the Iranian plateau that could have been exploited in prehistory, remarkable new evidence from relatively recent Iranian-German fieldwork (Momenzadeh, Nezafati, and Pernicka 2002; Nezafati 2006; Nezafati, Pernicka, and Momenzadeh 2006, 2008; Weeks 2008) has documented what, by all indications, appears to be a major tin source in west-central Iran. At the Deh Hosein deposit, located roughly 45 km southwest of the modern city of Arak, there are copper, arsenic, tin, and lead sources at a point situated on the northeastern border of Luristan. Not only is there abundant evidence of ancient opencast mining, but lead-isotope analysis by the Max Planck and Tübingen research teams (Begemann et al. 2008; see also Pigott 2008, 2009) indicates that "[t]wenty-five Late Bronze Age bronze artefacts from Luristan (analysed by EDXRF) show high concentrations of As, Pb, Zn, and Fe, consistent with the ores from Deh Hosein. Additionally, the lead isotope ratios of 18 ore samples from Deh Hosein exhibit a narrow range from 18.415–18.547 ($^{206}Pb/^{204}Pb$), 0.8438–0.8494 ($^{207}Pb/^{206}Pb$) and 2.0901–2.0959 ($^{208}Pb/^{206}Pb$). These results are in very good agreement with the Pb ratios of the bronze artifacts from Luristan and other third millennium B.C.E. bronzes from the Persian Gulf, the Aegean and Mesopotamia" (Weeks 2008, 335–336). Thus, this current research on Deh Hosein links this deposit and Luristan copper-based artifacts of the Early–Middle Bronze Age as well as the Late Bronze–Iron Age, respectively.

Surface survey of the ancient open-pit workings near Deh Hosein yielded large mauls used in mining and ore crushing as well as potsherds of the 1st millennium B.C. In addition, a single ^{14}C sample from an intermediate level in one mining pit was dated to the early to mid 2nd millennium B.C. Thus, earlier mining activity is quite likely. These workings (70+) are spread along mineralized horizons over an area 4.5 x 6 km^2. One location has large funnel-shaped mining pits arranged in two rows up to 500 m long. The pits range up to ca. 70 x 50 m in breadth and 15 m deep, yet they are so far unexcavated. Large mauls were found in association with the mining pits. The deposit is rich in minerals such as native copper, arsenopyrite, chalcopyrite, galena, and most importantly, cassiterite (SnO_2). This rich oxidic ore of tin was analyzed by Nezafati and Pernicka: "[a]nalysis of 17 ore samples from Deh Hosein by neutron activation showed up to 6.72% Sn, 10% Cu, 23.9% As, 3.7% Pb, 0.75% Zn, and 13.3 ppm Au" (Weeks 2008, 336). Tin ore was found mixed with copper ores in veined deposits and as nuggets in local streambeds. Smelting the mixed copper/arsenic/tin ores would result in a natural bronze, low in arsenic, but of uncontrolled tin content. Alternatively, the alluvial tin could be collected by hand or panned (like gold) and then mixed with local low-arsenic copper ore in a smelting furnace to yield tin-bronze. Such arsenical-tin bronzes were found among 3rd millennium artifacts from Luristan that were analyzed both by the MASCA team (Fleming et al. 2005; Fleming et al. 2006; see also Pigott 2008, 2009) and by the Max Planck team (Begemann et al. 2008).

On the basis of the above evidence and their laboratory research, the Max Planck team has argued that: (a) the idea of copper coming to Luristan from Mesopotamia or tin from Afghanistan are both ruled out on the basis of lead isotope and elemental analyses; and (b) though with some qualifications, "a single ore deposit has supplied, over an extended period of time, a good fraction of the copper utilized in Luristan" (Begemann et al. 2008, 25). Nezafati (2006, 105–106) summarized his views based on the numerous categories of evidence, including Pb-isotope analysis of the Max Planck team as well as his own, and he too maintains that Deh Hosein was a major supplier for ancient Iran and Mesopotamia and perhaps beyond from the 3rd to 1st millennium B.C.

Concluding Discussion

Presented above is a rather abbreviated overview of the decidedly complex body of evidence for geological sources of tin in and around Southwest Asia that may have supplied Bronze Age and, perhaps even, Iron Age metalworkers across the greater region. There is no need to review this information further. Moreover, various avenues of research have not been discussed, including that of the origins of tin-bronze. In this regard two related issues merit brief mention as they represent important new directions for research. First, among the curiosities of tin-bronze's presence in Southwest Asia is its virtual absence in ancient Iran up until the early 2nd millenium B.C., when it appears at the two Elamite capitals at lowland Susa and highland Tal-e Malyan (Anšan). Additional evidence from Tepe Yahya in southern Iran demonstrates the presence of tin-bronze (and brass) in the southern area excavations at the site by ca. 1700 B.C. (Thornton et al. 2005, 396). The material culture of this area of the site has strong connections to the Bactrian-Margiana Archaeological Complex (BMAC) of southern Central Asia (Hiebert and Lamberg-Karlovsky 1992). In short, Thornton and colleagues (Thornton et al. 2005, 397) argue "that copper-tin alloys were adopted at this site in the second millennium B.C. in relation to the influx of populations and/or associated material culture from the BMAC." Thus, they see "origins of Iranian tin bronzes and the other exotic alloys (i.e., brass) may lie among the Steppe Cultures of Central Asia." This is a particularly intriguing hypothesis, and one which merits specific attention as it is apt to shed light on a host of significant cultural issues not the least of which concerns the beginning of the "bronze age" on the Iranian plateau (Thornton et al. 2005, 397) in Southwest Asia. This development in Iran also reflects similar socio-technical transitions under way at the eastern boundaries of Central Asia and the steppes, along ancient China's western and northern borders, where tin-bronze was making its presence felt at exactly this same point in time

(Thornton et al. 2005, 397; see also, e.g., Mair 1998; Mei 2000; Sherratt 2006; Anthony 2007; Kohl 2007; Kuzmina 2007, 2008; Pigott and Ciarla 2007; White and Hamilton 2009).

Second, with regard to the sporadic occurrences of the zinc-copper alloy brass, in Southwest Asia from as early as the 3rd millennium B.C., Thornton (2007, 123) has noted that, when found, it tends to co-occur with tin-bronze and other tin-based metals. Not only does he link the fact that the two metals can share a golden color (depending on their alloy compositions), but he also suggests that the two alloys were not seen as different from one another. Early on they may well have been confused with each other and, at times, inadvertently mixed. Only metalworkers, Thornton surmizes (2007, 131), would have had an understanding of how the alloys were prepared and of their distinct material properties, e.g., hardness and durability. He suggests that prior to the first millenium B.C., the two metals—tin-bronze and brass—may well have been grouped as part of the same ethnocategory, i.e., that of golden-colored, copper-based alloys. At a minimum, this hypothesis has important ramifications for our need to understand tin-bronze and related metals as "cultural alloys" (Hamilton 1991) "whose adoption and use was conditioned by the social contexts in which it was produced, exchanged, and utilized" (Weeks 2003, 200). Thus, there is little question that we need to see the coming of tin-bronze not only as a regional phenomenon, but one that at the same time must now be seen in pan-Asiatic terms—a worldview of which J.D. Muhly would most certainly approve.

In closing, the discussion concludes where it began, i.e., with reference to Muhly's *Copper and Tin* and the insights it contains. In this instance, it is with particular regard to the recent revelation of the presence of what is, by all indications, a major tin source in western Iran at Deh Hosein and in proximity to Luristan.

There is little question, as Muhly (1973a, 290) indicates:

> "It was to Iran, to the Zagros mountains and the highlands beyond, that the inhabitants of Mesopotamia traditionally looked for the natural resources that they themselves lacked. The economic dependence of Mesopotamia upon Iran and the subsequent desire of the former to dominate the latter politically, is a theme which runs throughout all periods of Mesopotamian history."

In speaking of what he terms as among the evidence for the northern or northeastern source (i.e., north of the southern Mesopotamian heartland) for Sumerian tin, Muhly (1973a, 289) says:

> "The texts of the Šū-sîn speak of campaigns in the Zagros against the Šū-people and the land of Zabšali. Šū-sîn defeated the land of Zabšali and the lands of the Šū-people and forced them to pay what is referred to as the 'tribute of Anšan.' From the campaign against Zabšali a rich collection of booty, including tin, was brought back to be dedicated before the God Enlil: . . . 'Gold and silver . . . copper and tin, the artifacts made out of (their) bronze he loaded on pack donkeys. (All of these) he devoted eternally to the house of Enlil and the house Ninlil, and the house of the great gods'."

Additional texts of Šū-sîn mention quantities of precious metal, gold in particular, brought from this same area as tribute. Maintaining connections to the land of Zabšali over time must have been of such importance to Mesopotamian rulers that, for example, near the end of the Third Dynasty of Ur, Ibbi-Sin married his daughter to the ensi of Zabšali (Muhly 1973a, 320). Mountain peoples in the Zagros in both the 3rd (e.g., Gutians and Lullubaeans) and 2nd millennia B.C. (e.g., Turukkaeans) interdicted supplies of tin coming from northern and or western Iran or at least through the Zagros (Muhly 1973a, 320). Their ability to do so suggests that a tin source may have been close. If the tin was coming overland through the Zagros from a considerable distance farther east (e.g., Afghanistan), then it would seem that those involved with the trade could simply redirect the trade around the areas of unrest.

Most importantly, Muhly states that the land of Zabšali is thought to be located in the mountains of Luristan (perhaps near modern Khorramabad). Unless the inhabitants of Zabšali were but middlemen in the tin trade and having access to traded tin that others did not (i.e., tin moving across the plateau from Afghanistan to Elam?), then the indications are that they controlled a local tin source. If this was the case then, given the reference to Anšan, the Elamite capital in Fars on the Marv Dasht plain (modern Tal-e Malyan), one wonders if it refers broadly to a region under Anšan's rule and hence Elamite hegemony. Did the Elamites have a hand in controlling access to nearby tin deposits, as it is hard to imagine a more sought after and strategic material? Muhly (1973a, 293) states with reference to the early 2nd millennium B.C., textually documented Old Babylonian Period tin trade in which the trading entrepot of Mari is heavily implicated:

> "Together with the reference in the 'itinerary' of tin sent to Mari from Šeplarpak (to be identified with Siwe-palar-huppak, king of Anšan) it suggests an eastern source for tin, coming from Susa to Ešnunna via Der and then from Ešnunna to Mari via Sippar. *Presumably the tin came to Susa from somewhere farther east or to the north . . .*" [emphasis my own].

As a second relevant example, but from somewhat later in time, Muhly (1973a, 291) says that "[t]he quantities of tin mentioned in the Neo-Assyrian annals, often in the amounts of 100 talents, are far in excess of those mentioned in the Middle Assyrian texts." Texts excavated from Middle Assyrian sites of Nuzi, Tell Billa, and Tell al-Rimah and Assur discuss loans of tin and make it clear that tin was available in considerable quantities in the second half of the 2nd millennium B.C. (Muhly 1973a, 290). What is important here is that tin was readily available and in play commercially in Assyria in this period. Tin had to be coming from somewhere and the latest research demonstrates that immediately to its east lies at least one important tin source that could have been supplying consumers to the west.

Muhly (1973a, 331) argues that tin could have traveled south to Susa, the point being that Susa is repeatedly mentioned as having a controlling access to tin. Arguments suggest that tin came to Elam and Susa via the gulf trade. At the same time, we now have a more local and substantial source for tin and precious metals like gold in Luristan immediately to the north of Elam. The economic geology of the Deh Hosein ore bodies indicates that both tin and gold were available in appreciable quantities. Moreover, Deh Hosein is but one locus in the enormous NW–SE trending regional folds of central western Zagros that share common geological features, and thus the potential for additional ore bodies is considerable.

Muhly's discussion provides strong support for tin coming out of a western Iranian (possible Luristan) source and moving west. In fact, there is no direct evidence to link such tin to that found on the Uluburun ship at ca. 1300 B.C. What is needed is a program of Pb-isotope analysis to compare the Deh Hosein regional isotopic profile to that group of Uluburun tin ingots which are not linked isotopically to the Bolkardağ region of the Taurus.

To close on a somewhat uncertain note, Muhly (1973a) indicates that ca. 1600 B.C. and after in the ancient Near East and Eastern Mediterranean worlds a shift in tin sources (away from the East) appears to have ensued. Tin begins to come into the Eastern Mediterranean from sources in Europe, potentially Cornwall, and as part of the extensive trade in amber. Be that as it may, it remains difficult to envision that such tin was to be found on the Uluburun ship given its course along the Levantine littoral stopping at entrepots in Syria like Ugarit, which for centuries had supplied eastern tin to points farther west. Only further research is going to answer the new questions about tin arriving in the ancient Near East that have now arisen. For now, from where the Uluburun ingot tin came remains unclear. At the same time, since Muhly's 1973 landmark publication of *Copper and Tin* much new evidence has come to light regarding where the tin is to be found and how it is necessary to be viewed in context—albeit historical, archaeological, cultural, philological, or technological. Such recent revelations in many ways were not only presaged, but guided by, his early insights, and there remains much to be gleaned from the contents of this all important volume.

References

Adriaens, A. 1996. "Elemental Composition and Microstructure of Early Bronze Age and Medieval Tin Slags," *Mikrochimica Acta* 1124, pp. 89–98.

Adriaens, A., P. Veny, F. Adams, R. Sporken, P. Louette, B. Earl, H. Özbal, and K.A. Yener. 1999. "Analytical Investigation of Archaeological Powders from Göltepe (Turkey)," *Archaeometry* 41, pp. 81–90.

Adriaens, A., K.A. Yener, F. Adams, and R. Levi-Setti. 1997. "SIMS Analyses of Ancient Ceramic Crucibles and Slags from Turkey," in *Secondary Ion Mass Spectrometry. Proceedings of the Tenth International Conference on Secondary Ion Mass Spectrometry (SIMS X), University of Münster, Münster, Germany*, A. Benninghoven, B. Hagenhoff, and H.W. Werner, eds., Chichester, pp. 877–880.

Alimov, K., N. Boroffka, M. Bubnova, J. Burjakov, J. Cierny, J. Jakubov, J. Lutz, H. Parzinger, E. Pernicka, V. Radililovsky, V. Ruzanov, T. Sirinov, and G. Weisgerber. 1998. "Prahistorischer Zinnbergbau in Mittelasien: Vorbericht der ersten Kampagne 1997," *Eurasia Antiqua* 4, pp. 137–199.

Anthony, D.W. 2007. *The Horse, the Wheel, and Language: How Bronze-Age Riders from the Eurasian Steppe Shaped the Modern World*, Princeton.

Artioli, D., C. Giardino, G. Guida, A. Lazzari, and M. Vidale. 2005. "On the Exploitation of Copper Ores at Shahr-i Sokhta (Sistan, Iran) in the 3rd Millennium B.C.," in *South Asian Archaeology, 2003. Proceedings of the Seventeenth International Conference of the European Association of South Asian Archaeologists, 7–11 July 2003, Bonn (Forschungen zur Archäologie Aussereuropäischer Kulturen* 1), U. Franke-Vogt and H.-J. Weisshaar, eds., Bonn, pp. 179–184.

Aruz, J., K. Benzel, and J.M. Evans, eds. 2008. *Beyond Babylon: Art, Trade, and Diplomacy in the Second Millennium B.C.*, New York.

Bass, G.F. 1986. "A Bronze Age Shipwreck at Ulu Burun (Kaş): 1984 Campaign," *AJA* 90, pp. 269–296.

Bass, G.F., C. Pulak, D. Collon, and J. Weinstein. 1989. "The Bronze Age Shipwreck at Ulu Burun: 1986 Campaign," *AJA* 93, pp. 1–29.

Begemann, F., E. Haerinck, B. Overlaet, S. Schmitt-Strecker, and F. Tallon. 2008. "An Archaeo-Metallurgical Study of the Early and Middle Bronze Age in Luristan, Iran," *IrAnt* 43, pp. 1–66.

Berthoud, T. 1979. *Etude par l'analyse de traces et la modelisation de la filiation entre minerai de cuivre et objets archéologiques du Moyen-Orient (IVème et IIIème millénaires avant notre ère)*, Ph.D. diss., Université Pierre et Marie Curie, Paris.

Berthoud, T., S. Cleuziou, L.P. Hurtel, M. Menu, and C. Volfovsky. 1982. "Cuivres et alliages en Iran, Afghanistan, Oman au cours des IVe et IIIe millénaires," *Paléorient* 8, pp. 39–55.

Boroffka, N., J. Cierny, J. Lutz, H. Parzinger, E. Pernicka, and G. Weisgerber. 2002. "Bronze Age Tin from Central Asia: Preliminary Notes," in *Ancient Interactions: East and West in Eurasia*, K.V. Boyle, C. Renfrew, and M. Levine, eds., Cambridge, pp. 135–159.

Bougeouis, J., and W. Gheyle. 2006. *The Frozen Tombs of the Altai Mountains*, Ghent.

Casanova, M. 2008. "Lapis Lazuli," in Aruz, Benzel, and Evans, eds., 2008, pp. 68

Charles, J.A. 1975. "Where is the Tin?" *Antiquity* 49, pp. 19–24.

Cierny, J., and G. Weisgerber. 2003. "Bronze Age Tin Mines in Central Asia," in *Le problème de l'étain à l'origine de la métallurgie. Acts of the XIVth UISPP Congress, University of Liège, Belgium, 2–8 September 2001* (*BAR-IS* 1199), A. Giumlia-Mair, and F. Lo Schiavo, eds., Oxford, pp. 23–31.

Cierny, J., T. Stöllner, and G. Weisgerber. 2005. "Zinn in und aus Mittelasien," in Yalçin, Pulak, and Slotta, eds., 2005, pp. 431–448.

Cleuziou, S., and T. Berthoud. 1982. "Early Tin in the Ancient Near East," *Expedition* 25 (1), pp. 14–19.

Cleuziou, S., and M. Tosi. 2000. "Ra's al-Jinz and the Prehistoric Coastal Cultures of the Ja'alan," *JOS* 11, pp. 19–73.

Crawford, H.E.W. 1974. "The Problem of Tin in Mesopotamian Bronzes," *WorldArch* 6, pp. 242–247.

Dalton, R. 2007. "Geology: Mine Games," *Nature* 449 (Oct. 22), pp. 968–971.

de Jesus, P. 1978. "Considerations on the Occurrence and Exploitation of Tin Sources in the Ancient Near East," in *The Search for Ancient Tin*, A.D. Franklin, J.S. Olin, and T.A. Wertime, eds., Washington D.C., pp. 33–38.

Dercksen, J.G. 2005. "Metals According to Documents from Kültepe-Kanish Dating to the Old Assyrian Colony Period," in *Anatolian Metal* III (*Der Anschnitt Beiheft* 18), Ü. Yalçin, ed., Bochum, pp. 17–34.

Earl, B. 1994. "Tin from the Bronze Age Smelting Viewpoint," *Journal of the Historical Metallurgy Society* 28, pp. 117–120.

Earl, B., and H. Özbal. 1996. "Early Bronze Age Tin Processing at Kestel-Göltepe," in *Archaeometry '94. Proceedings of the 29th International Symposium on Archaeometry, Ankara, 9–14 May 1994*, S. Demirci, A.M. Özer, and G.D. Summers, eds., Ankara, pp. 447–449.

Fleming, S.J., V.C. Pigott, C.P. Swann, and S.K. Nash. 2005. "Bronze in Luristan: Preliminary Analytical Evidence from Copper/Bronze Artifacts Excavated by the Belgian Mission in Iran," *IrAnt* 40, pp. 35–64.

Fleming, S.J., V.C. Pigott, C.P. Swann, S.K. Nash, E. Haerinck, and B. Overlaet. 2006. "The Archaeometallurgy of War Kabud, Western Iran," *IrAnt* 41, pp. 31–57.

Foster, B. 1997. "A Sumerian Merchant's Account of the Dilmun Trade," *Acta Sumerologica Japonica* 19, pp. 53–62.

Francfort, H.-P. 1989. *Fouilles de Shortughaï. Recherches sur l'Asie centrale protohistorique* (*Mémoires de la Mission archéologique française en Asie centrale* 2), Paris.

Hall, M., and S. Steadman. 1991. "Anatolia and Tin: Another Look," *JMA* 4 (1), pp. 77–90.

Hamilton, E. 1991. "Metallurgical Analyses and the Bronze Age of Bohemia: Or, Are Cultural Alloys Real?" *Archeomaterials* 5, pp. 75–89.

Hauptmann, H., and E. Pernicka, eds. 2004. *Die Metallindustrie Mesopotamiens von den Anfängen bis zum 2. Jahrtausend v. Chr.* (*Orient-Archäologie* 3), Rahden.

Herrmann, G. 1968. "Lapis Lazuli: The Early Phases of Its Trade," *Iraq* 30, pp. 21–57.

Hiebert, F.T., and C.C. Lamberg-Karlowsky. 1992. "Central Asia and the Indo-Iranian Borderlands," *Iran* 30, pp. 1–15.

Kaniuth, K. 2007. "The Metallurgy of the Late Bronze Age Sapalli Culture (Southern Uzbekhistan) and Its Implications for the 'Tin Question,'" *IrAnt* 42, pp. 24–40.

Kassianidou, V. 2003. "The Trade of Tin and the Island of Copper," in *The Problem of Early Tin* (*BAR-IS* 1199), A. Giumlia-Mair and F. Lo Schiavo, eds., Oxford, pp. 109–119.

Knauth, P. 1972. *The Metalsmiths*, New York.

Kohl, P.L. 2007. *The Making of Bronze Age Eurasia*, Cambridge.

Korayakova, L. 2008. Review of *The Origins of the Indo-Iranians* by E.E. Kuzmina, *Antiquity* 82 (316), pp. 509–510.

Kuzmina, E.E. 2007. *The Origin of the Indo-Iranians*, J.P. Mallory, ed., Leiden.

———. 2008. *The Prehistory of the Silk Road*, V.H. Mair, ed., Philadelphia.

Leemans, W.F. 1960. *Foreign Trade in the Old Babylonian Period*, Leiden.

Litvinskij, B.A. 1950. "K istorii dobytschi olova v Uzbekistane" [Toward a History of Tin Extraction in Uzbekistan], in *Trudy Sredneaziatskogo Gosudarstvennogo Universiteta* (*Archeologija srenej Azii* n.s. 9), pp. 51–67.

Lo Schiavo, F. 2005. "Metallhandel im zentralen Mittelmeer," in Yalçin, Pulak, and Slotta, eds., 2005, pp. 399–414.

Maddin, R. 1989. "The Copper and Tin Ingots from the Kas Shipwreck," in *Archäometallurgie der Alten Welt. Beiträge zum Internationalen Symposium "Old World Archaeometallurgy," Heidelberg 1987* (*Der Anschnitt Beiheft* 7), A. Hauptmann, E. Pernicka, and G.A. Wagner, eds., pp. 99–105.

Maddin, R., J.D. Muhly, and T. Stech. 1999. "Early Metalworking at Çayönü," in *The Beginnings of Metallurgy* (*Der Anschnitt Beiheft* 9), A. Hauptmann, E. Pernicka, T. Rehren, and U. Yalçin, eds., Bochum, pp. 37–44.

Maddin, R., T. Stech-Wheeler, and J.D. Muhly. 1980. "Distinguishing Artifacts Made of Native Copper," *JAS* 7, pp. 211–225.

Mair, V.H., ed. 1998. *The Bronze Age and Early Iron Age Peoples of Eastern Central Asia* 1–2 (*JIES* 26), Washington, D.C.

Majidzadeh, Y. 1982. "Lapis Lazuli and the Great Khorasan Road," *Paléorient* 8, pp. 59–69.

Mei, J. 2000. *Copper and Bronze Metallurgy in Late Prehistoric Xinjiang: Its Cultural Context and Relationship with Neighboring Regions* (*BAR-IS* 865), Oxford.

Momenzadeh, M., N. Nezafati, and E. Pernicka. 2002. "First Indication of Tin at the Ancient Mining Site near Deh Hosein," in *Proceedings of the 33rd International Symposium on Archaeometry, 22–26 April 2002, Amsterdam* (*Geoarchaeological and Bioarchaeological Studies* 3), H. Kars and E. Burke, eds., Amsterdam, pp. 116–117.

Muhly, J.D. 1973a. *Copper and Tin: The Distribution of Mineral Resources and the Nature of the Metals Trade in the Bronze Age* (*Transactions of the Connecticut Academy of Arts* 43), Hamden, CT.

———. 1973b. "Tin Trade Routes of the Bronze Age," *American Scientist* 61 (4), pp. 404–413.

———. 1976. *Supplement to Copper and Tin: The Distribution of Mineral Resources and the Nature of the Metals Trade in the Bronze Age* (*Transactions of the Connecticut Academy of Arts* 46), Hamden, CT.

———. 1977. "The Copper Ox-Hide Ingots and the Bronze Age Metals Trade," *Iraq* 39, pp. 73–82.

———. 1985. "Sources of Tin and the Beginnings of the Bronze Metallurgy," *AJA* 89, pp. 275–291.

Muhly, J.D., F. Begemann, Ö. Öztunali, E. Pernicka, S. Schmitt-Strecker, and G.A. Wagner. 1991. "The Bronze Age Metallurgy of Anatolia and the Question of Local Tin Sources," in *Archaeometry '90. Proceedings of the 27th Symposium on Archaeometry Held in Heidelberg, April 2–6, 1990*, E. Pernicka, and G.A. Wagner, eds., Basel, pp. 209–220.

Muhly, J.D., R. Maddin, T. Stech, and E. Özgen. 1985. "Iron in Anatolia and the Nature of the Hittite Iron Industry," *AnatSt* 35, pp. 67–84.

Nezafati, N. 2006. *Au-Sn-W-Cu-Mineralization in the Astaneh-Sarband Area, West Central Iran: Including a Comparison of the Ores with Ancient Bronze Artifacts from Western Asia*, Ph.D. diss., University of Tübingen.

Nezafati, N., E. Pernicka, and M. Momenzadeh. 2006. "Ancient Tin: Old Question and a New Answer," *Antiquity* 80 (308), pp. 308–311.

———. 2008. "Iranian Ore Deposits and Their Role in the Development of the Ancient Cultures," in *Anatolian Metal* IV (*Der Anschnitt Beiheft* 21), Ü. Yalçin, ed., Bochum, pp. 77–90.

Oppenhiem, A.L. 1954. "The Seafaring Merchants of Ur," *JAOS* 74, pp. 6–17.

Parzinger, H., and N. Boroffka. 2003. *Das Zinn der Bronzezeit in Mittelasien.* I: *Die siedlungsarchäologischen Forschungen im Umfeld der Zinnlagerstätten* (*Archäologie in Iran und Turan* 5), Mainz.

Pigott, V.C. 2008. "Le bronze et le fer au Luristan Nouvelle confirmation d'une tradition technologique ancienne," in *Bronzes du Luristan. Énigmes de l'Iran ancien IIIe–Ier millénaire av. J.-C.*, N. Engel, ed., Paris, pp. 55–65.

———. 2009. "'Luristan Bronzes' and the Development of Metallurgy in the West-Central Zagros, Iran," in *Metals and Societies. Studies in Honour of Barbara S. Ottaway* (*Universitätsforschungen zur prähistorischen Archäologie* 169), T.K. Kienlin and B.W. Roberts, eds., Bonn, pp. 369–382.

Pigott, V.C., and R. Ciarla. 2007. "On the Origins of Metallurgy in Prehistoric Southeast Asia: The View from Thailand," in *Metals and Mines: Studies in Archaeometallurgy*, S. La Niece, D.R. Hook, P.T. Craddock, eds., London, pp. 76–88.

Pigott, V.C., S.M. Howard, and S.M. Epstein. 1982. "Pyrotechnology and Culture Change at Bronze Age Tepe Hissar (Iran)," in *Early Pyrotechnology: The Evolution of the First Fire-Using Industries*, T.A. Wertime and S.F. Wertime, eds., Washington D.C., pp. 215–236.

Pigott, V.C., H.C. Rogers, and S.K. Nash. 2003. "Archaeometallurgical Investigations at Malyan: The Evidence for Tin-Bronze in the Kaftari Phase," in *Yeki bud, yeki nabud: Essays on the Archaeology of Iran in Honor of William M. Sumner* (*UCLAMon* 48), N.F. Miller and K. Abdi, eds., Los Angeles, pp. 161–175.

Possehl, G.L. 1996. "Meluhha," in *The Indian Ocean in Antiquity*, J. Reade, ed., London, pp. 133–208.

———. 1997. "Seafaring Merchants of Meluhha," in *South Asian Archaeology 1995. Proceedings of the 13th Conference of the European Association of South Asian Archaeologists, Cambridge, 5–9 July*, 1995, F.R. Allchin and B. Allchin, eds., Cambridge, pp. 87–100.

Potts, D.T. 1993. "Rethinking Some Aspects of Trade in the Arabian Gulf," *WorldArch* 24, pp. 423–439.

Potts, T.F. 1993. "Patterns of Trade in Third Millennium B.C. Mesopotamia and Iran," *WorldArch* 24 (3), pp. 379–402.

Pulak, C. 1988. "The Bronze Age Shipwreck at Ulu Burun, Turkey: 1985 Campaign," *AJA* 92, pp. 1–37.

———. 1997. "The Uluburun Shipwreck," in *Res Maritimae: Cyprus and the Eastern Mediterranean from Prehistory to Late Antiquity. Proceedings of the Second International Symposium "Cities on the Sea," Nicosia, Cyprus 18–22 October 1994* (*American Schools of Oriental Research Archaeological Reports* 4), S. Swiny, R.L. Hohlfelder, and H.W. Swiny, eds., Atlanta, pp. 233–262.

———. 1998. "The Uluburun Shipwreck: An Overview," *IJNA* 27, pp. 188–224.

———. 2000. "The Copper and Tin Ingots from the Late Bronze Age Shipwreck at Uluburun," in *Anatolian Metal* I (*Der Anschnitt Beiheft* 13), Ü. Yalçin, ed., Bochum, pp. 137–157.

Rossovsky, L.N., V.V. Mogarovsky, and V.M. Chmirev. 1987. "The Metallogeny of Tin and Rare Metals in Eastern Part of the Mediterranean Folded Belt," in *Mineral Deposits of the Tethyan Eurasian Metallogenic Belt between the Alps and the Pamirs: Selected Examples* (*IGCP Project* 169), S. Jankovic, ed., Belgrade, pp. 170–177.

Sayre, E.V., K.A. Yener, E.C. Joel, and I.L. Barnes. 1992. "Statistical Evaluation of the Presently Accumulated Lead Isotope Data from Anatolia and Surrounding Regions," *Archaeometry* 34, pp. 73–105.

Sarianidi, V. 1971. "The Lapis Lazuli Route in the Ancient East," *Archaeology* 24, pp. 12–15.

Shareq, A., V.M. Chmyriov, K.F. Stazhilo-Alexseev, V.I. Dronov, D.J. Gannon, G.K. Lubemov, A.K. Kafarshiy, and E.P. Malyarov. 1977. *Mineral Resources of Afghanistan*, 2nd ed., Kabul.

Sherratt, A. 2006. "The Trans-Eurasian Exchange: The Prehistory of Chinese Relations with the West," in *Contact and Exchange in the Ancient World*, V.H. Mair, ed., Honolulu, pp. 30–61.

Stech, T., and V.C. Pigott. 1986. "The Metals Trade in Southwest Asia in the Third Millennium B.C," *Iraq* 48, pp. 39–64.

Stos-Gale, Z.A., N.H. Gale, G. Bass, C. Pulak, E. Galili, and J. Sharvit. 1998. "The Copper and Tin Ingots of the Late Bronze Age Mediterranean: New

Scientific Evidence," in *Proceedings of the Fourth International Conference on the Beginning of the Use of Metals and Alloys (BUMA-IV), January 16–17, 1996, Shimane, Japan*, Tokyo, pp. 115–126.

Tallon, F. 1987. *Métallurgie Susienne* I. *De la fondation de Suse au XVIIIe avant J.-C. (Notes et documents des musées de France* 15), Paris.

Thornton, C.P. 2007. "Of Brass and Bronze in Prehistoric Southwest Asia," in *Metals and Mines: Studies in Archaeometallurgy*, S. La Niece, D.R. Hook, P.T. Craddock, eds., London, pp. 123–135.

———. 2009. *The Chalcolithic and Early Bronze Age Metallurgy of Tepe Hissar, Northeast Iran: A Challenge to the 'Levantine Paradigm,'* Ph.D. diss., University of Pennsylvania.

Thornton, C.P., C.C. Lamberg-Karlovsky, M. Liezers, and S.M.M. Young. 2005. "Stech and Pigott Revisited: New Evidence Concerning the Origins of Tin Bronze in the Light of Chemical and Metallographic Analyses of the Metal Artefacts from Tepe Yahya, Iran," in *Proceedings of the 33rd International Symposium on Archaeometry, 22–26 April 2002, Amsterdam (Geoarchaeological and Bioarchaeological Studies* 3), H. Kars and E. Burke, eds., Amsterdam, pp. 395–398.

Thornton, C.P., and T. Rehren. 2007. "Report on the First Iranian Prehistoric Slag Workshop," *Iran* 45, pp. 315–318.

Tosi, M. 1974. "The Lapis Lazuli Trade Across the Iranian Plateau in the 3rd Millennium B.C.," in *Gururājamañjarikā: Studi in onore di Guiseppe Tucci*, Naples, pp. 3–22.

Vandiver, P.B., K.A. Yener, and L. May. 1992. "Third Millennium B.C. Tin Processing Debris from Göltepe (Anatolia)," in *Materials Issues in Art and Archaeology III. Symposium Held April 27–May 1, 1992, San Francisco, California, U.S.A.*, P.B. Vandiver, J. Druzik, and G.S. Wheeler, eds., Pittsburgh, pp. 545–569.

Vandiver, P.B., R. Kaylor, J. Feathers, M. Gottfried, K.A. Yener, W.F. Hornyak, and A. Franklin. 1993. "Thermoluminescence Dating of a Crucible Fragment from an Early Tin Processing Site in Turkey," *Archaeometry* 35, pp. 295–298.

Weeks, L.R. 1999. "Lead Isotope Analyses from Tell Abraq, United Arab Emirates: New Data Regarding the 'Tin Problem' in Western Asia," *Antiquity* 73, pp. 49–64.

———. 2003. *Early Metallurgy of the Persian Gulf: Technology, Trade, and the Bronze Age World*, Boston.

———. 2008. "Archaeological Reports: The 2007 Early Iranian Metallurgy Workshop at the University of Nottingham," *Iran* 46, pp. 335–345.

Weisgerber, G. 1981. "Makan and Meluhha—Third Millennium B.C. Copper Production in Oman and the Evidence of Contact with the Indus Valley," in *South Asian Archaeology 1981. Proceedings of the Sixth International Conference of the Association of South Asian Archaeologists in Western Europe, held in Cambridge University, 5–10 July 1981*, B. Allchin, ed., Cambridge, pp. 196–201.

White, J.C., and E.G. Hamilton. 2009. "The Transmission of Early Bronze Technology to Thailand: New Perspectives," *Journal of World Prehistory* 22.4, pp. 357–397.

Yalçin, Ü. 2005. "Ein Schiff macht Geschichte," in Yalçin, Pulak, and Slotta, eds., 2005, pp. 19–24.

Yalçin, Ü., C. Pulak, and R. Slotta, eds. 2005. *Das Schiff von Uluburun: Welthandel vor 3000 Jahren*, Bochum.

Yener, K.A. 2000. *The Domestication of Metals*, Leiden.

Yener, K.A., and P.B. Vandiver. 1993a. "Tin Processing at Göltepe, an Early Bronze Age Site in Anatolia," *AJA* 97, pp. 207–237.

———. 1993b. "Reply to J.D. Muhly, 'Early Bronze Age Tin and the Taurus,'" *AJA* 97, pp. 255–264.

Yener, K.A., E.V. Sayre, E. Joel, H. Özbal, I.L. Barnes, and R.H. Brill. 1991. "Stable Lead Isotope Studies of Central Taurus Ore Sources and Related Artifacts from the Eastern Mediterranean Chalcolithic and Bronze Age Sites," *JAS* 18, pp. 541–577.

Young, T.C., Jr. 1967. "The Iranian Migration into the Zagros," *Iran* 5, pp. 11–34.

———. 1985. "Early Iron Age Iran Revisited: Preliminary Suggestions for the Re-analysis of Old Constructs," in *De l'Indus aux Balkans. Recueil à la mémoire de Jean Deshayes*, J.-L. Huot, M. Yon, and Y. Calvet, eds., Paris, pp. 361–377.

CHAPTER

28

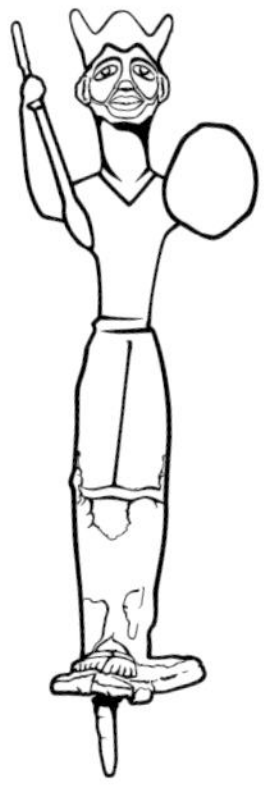

Three Copper Oxhide Ingots in the Şanlıurfa Archaeology Museum, Turkey

Cemal Pulak

Introduction

In September 2000, Tufan Turanlı, of the Institute of Nautical Archaeology (INA) in Bodrum, Turkey, informed me of three unpublished copper ingots in the typical Late Bronze Age oxhide shape that he had seen by chance in the storeroom of the Şanlıurfa Archaeology Museum. I was first told about these ingots in 1994 by the then director of the Şanlıurfa Museum, Adnan Mısır, who had invited me to study and publish them, but his untimely death in 1998 prevented me from doing so. At my request and with the kind permission of the museum's director, Eyyüp Bucak, Tufan Turanlı and Berta Lledo-Turanlı visited the museum again in early October to measure, weigh, photograph, and draw the three ingots. The ingots had been found in 1991 during dredging in the Göksu River in southeastern Turkey. Intrigued by the landlocked location of the find, Mr. Turanlı and I traveled to Şanlıurfa in August 2001 to investigate the circumstances surrounding the discovery of the ingots, to sample them for chemical analysis, and study them for publication. Before the study and chemical analysis could be completed, however, an excellent article on the ingots was published in the catalog on the Late Bronze Age Uluburun shipwreck exhibition in Bochum, Germany, in 2005–2006 (Sertok and Güllüce 2005). Although some of the previously published information is repeated here for continuity purposes, an effort has been made to present mostly new and supplementary information about these ingots, including the results of lead-isotope analysis. It is with great pleasure, therefore, that I contribute this paper in honor of James D. Muhly, a scholar who has contributed so much to our understanding of these large and curious but otherwise seemingly unpretentious slabs of raw copper traded all across the Late Bronze Age Mediterranean.

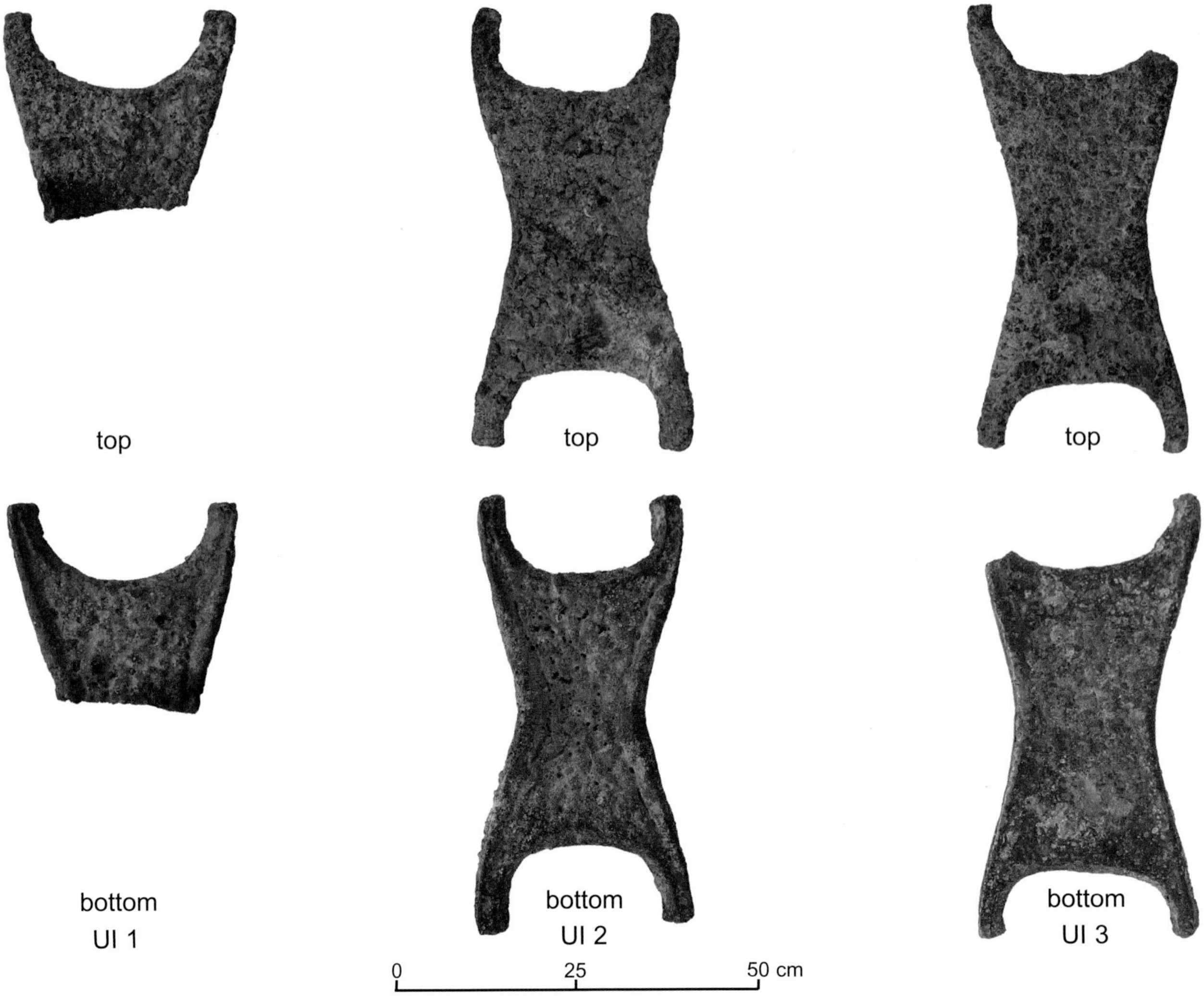

Figure 28.1. Three ingots, showing both faces of each, found at Göksu River in the Şanlıurfa Archaeology Museum.

Discovery of the Ingots and the Location of the Site

During summer 1991, while dredging for sand and gravel on the eastern bank of the Göksu River, dredge workers recovered two large metal objects, later identified as copper oxhide ingots, among the load of gravel they had just dumped on the river bank. Located in the province of Adıyaman, the Göksu River is a small, western tributary of the Euphrates River, which flows southward to converge with the Euphrates just within the borders of the adjacent province of Şanlıurfa. Additional dredging in the river uncovered more oxhide ingots, which were then divided among the workers present. The exact number of ingots recovered is unclear. At the time, Hamza Güllüce, an assistant archaeologist at Şanlıurfa Archaeology Museum, who tried to locate the dispersed ingots for purchase by the museum, reported a total of seven ingots, consisting of six complete ingots and one half of an ingot (Sertok and Güllüce 2005, 393). This total is likely to be correct, since Güllüce interviewed the workers within about a year of the discovery, when all was still fresh in their memory. Yet, when we interviewed some of the same workers 10 years later, the total number of ingots found seemed less certain, with the lowest estimate at 4.5 ingots and the highest at 7.5 ingots. Ismail Polat, the bucket-dredge operator who first discovered the ingots recalls receiving two or three ingots as part of his share; Mehmet Başıbüyük, a front-end-loader driver loading gravel onto the trucks, received two complete ingots; and Reşit

Demirbilek, a machinery maintenance worker, received a half ingot. The gravel loaded onto the trucks was dumped into mechanized sifters nearby where the sand, gravel, and cobbles were separated. Two other intact ingots are said to have been recovered from the sifter by its unidentified operator, but no one we interviewed was certain about the number of these last ingots or even if any had been found by this person at all, although I subsequently heard rumors that some oxhide ingots appeared on the antiquities market at Aleppo, Syria soon afterward.

The three ingots were brought to the Şanlıurfa Archaeology Museum between 1991 and 1993. They were subsequently purchased by the museum but have not yet been inventoried. In their publication, Sertok and Güllüce (2005, 394) numbered the ingots in the order they were accessioned by the museum, a sequence that is also retained here to prevent any possible confusion. The Urfa Ingots (UI), their accession dates, accession numbers, and previous owners are as follows (Fig. 28.1): UI 1 (23 June 1992, no. 56/23, Reşit Demirbilek), UI 2 (11 December 1992, no. 59/9, Mehmet Başıbüyük), and UI 3 (22 June 1993, no. 60/26, Yasin Arıcı). Ingot UI 3 originally also belonged to Başıbüyük, but changed hands twice before it was sold to the museum. Ismail Polat sold his two (or three) ingots to a local junkman in Urfa, who then sold them to an out-of-town recycler. These ingots were traced by Güllüce to a metal recycling facility in the city of Gaziantep, where they had been dumped into the melting pot only hours before. If Ismail Polat owned yet a third ingot, its fate is unknown, although Polat stated that he had sold all of his ingots to the same junkman.

Ismail Polat took us to the site where he had discovered the ingots a decade earlier. After driving 72 km northwest on the Şanlıurfa-Adıyaman highway, we turned left onto an unpaved road just after the town of Kuyulu (formerly Turus but also referred to as Turunç) and headed to the eastern bank of the Göksu River, about 0.5 km beyond the village of Kikan and situated on the right-hand side of the road. The village of Malpınar, also on the eastern bank, is approximately 1.3 km downriver from the site. The general landscape around the river had been completely altered by the intense sand and gravel reclamation activity. It was reduced to ponds about 2 m deep, with many huge mounds of discarded cobbles and river stones from the mechanized sieving operations. The present contractor working the river bed had also diverted the river in order to build a bridge, and the river no longer flowed in its original course where the ingots were

Figure 28.2. The general area where the ingots were discovered in 1991 during dredging of the Göksu riverbed. The discarded river stones are heaped into towering mounds on the river bank. The river was diverted recently to build a bridge (at center left) to access the eastern bank for additional sand and gravel reclamation, leaving the original river bed dry (at center right). The ingots were found in the area of artificial ponds at center right of the photograph.

found (Fig. 28.2). Polat pointed to an area about 50 x 20 m wide along what was left of the old river bed as the approximate location where he had found the ingots; the coordinates of its center were registered with a portable GPS instrument as 37°31.2181' N and 38°08.9888' E. The ingots had been dredged somewhere between the bottom of the river at 6–8 m deep and a shallower depth along the river bank, in an area of the river bed that mostly dried out during summers and became inundated in spring with melted snow and seasonal rains. All the ingots were found more or less together and on the same day; no other finds were made after that, even though the river had been dredged 40–50 m on either side of the original location in the hope of finding more ingots.

The Oxhide Ingots

The Urfa ingots are of the typical oxhide shape known from other Late Bronze Age sites. They are flat, oblong slabs of "blister" or raw copper, roughly 3–4 cm thick at the center, which is also the thinnest part of the ingots. Each ingot has a slightly inward-curving, thick protrusion at each of the four corners. They average 63.5 cm in length and 33.3 cm in width at the widest point, which is between the tips of the projecting corners. The ingots have a fairly pronounced narrow area in the middle to form a waist. One face of the ingot is more rough and bubbly than the other face. The smoother face, usually containing low mounds and small gas holes, corresponds to the surface in contact with the mold in which the ingot was cast. The edges of the mold face are outlined by a raised ridge. The opposite face of the ingot would have been exposed to air and rendered rough and bubbly by the effervescing gasses escaping from the molten metal during casting.

Urfa Ingot Catalog

The descriptions of the ingots appear elsewhere (Sertok and Güllüce 2005, 394–395) and will be mentioned here only briefly. The abbreviated catalog below gives their linear measurements and weights, which were taken with a digital scale for accuracy (resolution 2 g); all three ingots are lighter than previously indicated, where UI 1 is published as 12 kg, UI 2 as 27.7 kg, and UI 3 as 24.7 kg (Sertok and Güllüce 2005, 394).

UI 1. One half ingot. Pres. L. 0.307 m, pres. min. L. 0.175 m, w. 0.336 m, min. w. 0.205 m, min. th. 0.034 m (0.029 m at center of broken or cut edge), max. th. 0.052 m (at one corner protrusion). Weight 11.148 kg. No marks. The ingot was deliberately broken into halves in antiquity, as indicated by the hammer blow impressions on its rough face. The cross-section of the broken edge reveals a thin, elongated, and irregular cavity or line at the center of the ingot extending from one side to the other, indicating that the ingot was cast in at least two separate pours of molten copper. Although an impressed slash-like (/) mark is reported near the center of the broken edge of the ingot's rough face (Sertok and Güllüce 2005, 394), no primary mark was observed on the ingot.

UI 2. Complete ingot. L. 0.636 m, min. L. 0.394 m, w. 0.336 m, min. w. 0.200 m, min. th. 0.030 m, max. th. 0.052 m (at one corner protrusion). Weight 24.662 kg. Impressed mark L near one edge of rough face (cf. Bass 1967, 57, 59, 72, Gelidonya ingot no. In 28, mark 36A, fig. 90). The mark is ca. 0.075 x 0.078 m, groove width ca. 0.017–0.019 m, max. depth of groove 0.017 m.

UI 3. Ingot. L. 0.644 m, min. L. 0.441 m, w. 0.328 m, min. w. 0.192 m, min. th. 0.032 m, max. th. 0.054 m (at one corner protrusion). Complete except for one corner protrusion. Weight 23.436 kg. Impressed mark "+" near one edge of rough face (cf. Bass 1967, 53, 54, 56, 72, Gelidonya ingot nos. In 3, 17, mark 24A, fig. 90). The mark is ca. 0.072 x 0.055 m, groove width ca. 0.010 m. The corner protrusion was broken off in antiquity, possibly accidentally or when the ingots were lost in the river, as there does not appear to be any impressions from hammer blows to suggest a deliberate breakage.

Discussion

In overall shape and size, the Urfa ingots are nearly identical to some of the ingots from the late 13th-century B.C. Cape Gelidonya shipwreck (Bass 1967, 52–78), but they are less like the larger and wider ingots recovered from the late 14th-century B.C. Uluburun shipwreck (Pulak 1998, 193–198; 2000, 137–143; 2001, 18–22; 2008; Yalçın, Pulak, and Slotta 2005, 560–568). All three Urfa ingots are of the same type and nearly identical both in shape and dimensions to Bass's (1967, 53, fig. 55) Type 2b ingots from Cape Gelidonya. Bass follows

Sample	Description	$^{208}Pb/^{206}Pb$	$^{207}Pb/^{206}Pb$	$^{206}Pb/^{204}Pb$
UI 1	Oxhide	2.07145	0.84007	18.565
UI 2	Oxhide	2.07184	0.84036	18.517
UI 3	Oxhide	2.07121	0.83997	18.542
APLKOR3	Chalcopyrite	2.07068	0.84077	18.522
AOL 43	Pyrite	2.07172	0.84055	18.528
AOLKOR4	Chalcopyrite	2.07172	0.84064	18.514
ALPKOR	Chalcopyrite	2.07069	0.84096	18.514
ALPKOR2	Chalcopyrite	2.07192	0.84045	18.528
APL 27	Pyrite	2.07083	0.84044	18.530

Table 28.1. Lead-isotope data for the three Urfa oxhide ingots listed together with data from four chalcopyrite and two pyrite samples from the Apliki mine in northwestern Cyprus. Ore data courtesy of Zofia Stos-Gale.

Buchholz's (1959, 4–7) ingot typology but expands on his Type 2 by adding three subtypes that he denotes as 2a–c. Bass's subtypes 2a and 2b are identical in terms of ingot profiles but differ in the details of the ingots' mold surface, with the latter exhibiting a more pronounced rim on the smoother mold surface. The two complete Urfa ingots were cast in separate molds. Based on the surviving portion of the half-ingot UI 1, it seems somewhat similar in its contour to that of ingot UI 3, suggesting the use of two very similar molds, if not the same mold.

On their upper or rough faces, the two complete Urfa ingots bear "primary marks" stamped or impressed into the newly poured, soft metal. Twenty-four of the at least 34 ingots found on the Cape Gelidonya shipwreck bear primary marks on their rough faces as well, but a few of them also include "secondary marks," cut or incised on the ingot's surface with a chisel after each had completely cooled and hardened. The latter marks, without exception, occurred on the smoother face. The two primary marks on the Urfa ingots are represented among the repertoire of the Gelidonya ingot primary marks. Both marks are also found in Cypro-Minoan syllabary, and may indicate the Cypriot origin of the ingots. The Uluburun shipwreck ingots, in contrast, carry only secondary marks, all of which were incised on the smoother face of the ingots.

The lead-isotope ratios for the Urfa ingots, along with those of six ore samples from Apliki for comparison purposes, are listed in Table 28.1 and plotted against ingots from the Cape Gelidonya and Uluburun shipwrecks, as well as other Cypriot ores, and slags from Enkomi and Kalavassos-*Ayios Dhimitrios*, in Figure 28.3. Data on other ores and oxhide ingots originating from the same Cypriot sources are listed in Stos-Gale et al. 1997 and Gale et al. 1997.

All three Urfa ingots have practically identical lead-isotope compositions and are fully consistent with those of copper ores from the Apliki mine in the Solea axis of northwestern Cyprus. The lead-isotope compositions of the Cape Gelidonya ingots also match those of the Solea axis ores, and most of them form a fairly tight cluster around the isotopic data of the Apliki mine ores, while a few overlap with those of the Mavrovouni and Skouriotissa mine ores (Gale and Stos-Gale 2005, 119–120; Stos 2009, 166–170). The lead-isotope compositions of the Uluburun ingots, on the other hand, form a relatively tight cluster that only partly overlaps the isotopic field of the Solea-axis ores, with some of them being close to that of the Apliki ores, and a few others are close to those of the Mavrovouni and Skouriotissa deposits (Pulak 2000, 147–150; Gale and Stos-Gale 2005, 119–124). The lead-isotope fingerprint of copper ore samples from the Phoenix mine, which is also located near the Apliki and Skouriotissa copper mines, is particulary close and overlaps well with the Uluburun lead-isotope data (Pulak 2000, 150; Stos-Gale et al. 1998, 118). It appears, therefore, that the isotopic compositions of nearly all of the Uluburun oxhide ingots fall near, but only partly overlap, the isotopic compositions of ores from the same deposits as those from which almost all of the later Late Bronze Age oxhides ingots were made. That the isotopic signatures of the Urfa ingots and most of the Cape Gelidonya ingots are consistent with those of the Apliki ores comes as no great surprise, not only because of the nearly

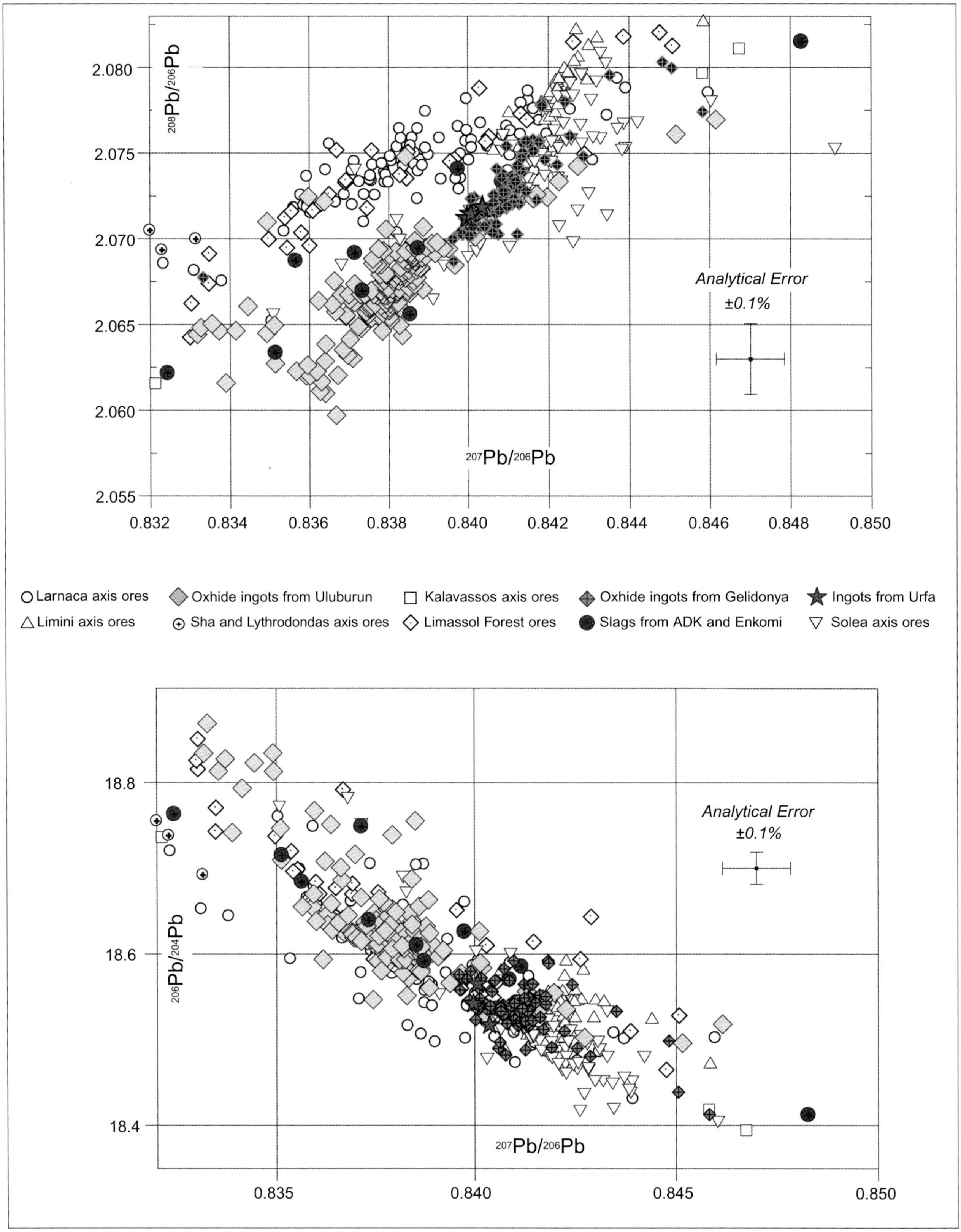

Figure 28.3. Lead-isotope ratio plots of oxhide ingots from the Uluburun and Cape Gelidonya shipwrecks, Urfa ingots, copper-smelting slags from Kalavassos–*Ayios Dhimitrios* (ADK) and Enkomi, and Cypriot copper ores, including those of the Solea axis, which includes the Apliki mine. Graph courtesy of Zofia Stos-Gale.

identical shapes of the ingots from both sites, but also because nearly all 13th-century (or later) copper oxhide ingots that have been found appear to have been made from Cypriot ores, especially those of the Solea axis, and primarily from Apliki.

Oxhide ingots are widely distributed across the Mediterranean, and they are found in some numbers in regions west of Cyprus and as far west as places such as Sicily, Sardinia, and Corsica (Lo Schiavo 2005, 400–408), southern France (Domergue and Rico 2002, 141–144; Lo Schiavo 2005, 408), and even southern Germany (Primas 2005, 389). Besides those found on Cyprus itself and on the Cape Gelidonya and Uluburun shipwrecks, only the three ingots from the Göksu River and three other copper oxhide ingots are known from locations east of Cyprus: two from the Near East and one from Egypt. One of these is a quarter ingot, cut in antiquity and probably of 14th-century date, found at the Hittite capital of Boğazköy (Hattuša) in central Turkey (Müller-Karpe 2005, 491, fig. 10; for the overlap of its lead-isotope composition with those of Apliki ores, see Gale 1999, 117, fig. 10). An intact oxhide ingot was found in an early 12th-century context—but the ingot itself may have been older when it was deposited—in the Kassite city of Dur-Kurigalzu, about 30 km west of Baghdad, in Iraq. Although the ingot was misplaced or lost in the museum, its presence was first mentioned by John Brinkman in 1970 (Muhly 1973, 370, n. 141), who later published information about the ingot based on the original excavation field register and the accession register of the Baghdad Museum (Brinkman 1987, 35–36). The field register gives the ingot's dimensions as 56 x 32 cm, and the museum register states it as 45 x 33 x 6 cm, with the discrepancy in the length of the ingot undoubtedly due to measuring it from the tips of the corner protrusions in the former, and between the corner protrusions along the ingot's central axis (its minimum length) in the latter. A sketch of the ingot copied from the museum register by Brinkman (1987, 36; pers. comm., December 2, 2004), depicts it as an oxhide type similar to Buchholz's Type 2 ingot. Although the ingot's weight is not mentioned, its dimensions strongly suggest that it is similar in size to the Urfa and Cape Gelidonya ingots. Because of their physical similarity to and common lead-isotope characteristics with the Cape Gelidonya ingots, the Urfa ingots probably also date to the late 13th or early 12th century. A late 14th- or early 13th-century quarter ingot from Qantir (ancient Piramesse) in the eastern Nile Delta of Egypt (Pusch 1996, 123) has a lead-isotope composition that also overlaps with Apliki ores (Gale 1999, 117, fig. 10). In addition to these ingots, a large corner fragment of an oxhide ingot found in a bronze hoard at Şarköy (Harmankaya 1995, 232–233, 254, fig. 17) in Thracian Turkey, and tentatively dated to the 12–11th centuries, also plots with the Apliki ores (Gale 1999, 117, fig. 10). As with the Urfa ingots, the lead-isotope compositions of these ingot fragments are all consistent with the lead-isotope data of Apliki ores. The Dur-Kurigalzu ingot, unfortunately, is unavailable for study, but judging from the dimensions of the ingot and its early 12th-century date, it is likely that it also originated from Apliki.

The Urfa Ingots and the Metals Trade

Large-scale trade in the base metals of copper and tin represents one of the most important aspects of Bronze Age trade in the Mediterranean. This metals trade flourished and peaked in the Late Bronze Age with the appearance of copper ingots in the typical oxhide shape and continued as one of the major aspects of long-distance trade down to the very end of the Bronze Age. Although the oxhide ingot, consisting of raw blister copper and not involving any alloying or remelting (Hauptmann, Maddin, and Prange 2002, 1–15), has been generally associated with Cyprus, because the ingots could not be provenienced conclusively in the past, their source has remained suspect (Muhly 1982, 255). The importance of oxhide ingots, however, was recognized long ago by James Muhly (1977, 81) as a means of providing our best archaeological evidence for determining the development of smelting technology and trade patterns based on the ore sources from which these ingots were made and the consumption centers to which they were shipped between the 16th and 12th centuries B.C. Consequently, for more than a quarter of a century, oxhide ingots have been subjected to a battery of chemical analyses; those based on determining the ingot's lead-isotope fingerprints, preferably combined with certain trace-element analysis, have produced our most reliable data for attempting to determine the sources of these ingots. With the exception of some of the earliest oxhide ingots from the 16th century B.C., these analyses suggest that nearly all the copper oxhide ingots originate from Cyprus, and most come from the Solea-axis ores in northwestern Cyprus, which include the Apliki mine among others (for sites mentioned in this section, see Fig. 28.4). Such attributions have not gone unchallenged over the years, and while there still are many problems to be sorted out, especially in the interpretation of the lead-isotope results, a clear picture of Bronze Age copper trade in the Mediterranean is beginning to emerge (Gale 1989; 1991; Gale and Stos-Gale 2005; Muhly 2005, 507–513).

Lead-isotope analysis of all copper ingots found in eastern Mediterranean shipwreck and terrestrial sites, as well as those found in the Near East, and a single fragmentary ingot from Qantir, Egypt, are consistent with Cypriot copper ores, mostly from the Solea-axis region, which includes the Apliki mine. These ingots would have been transported from Cyprus to the mainland by ships, such as those that sank at Uluburun at the end of the 14th century and at Cape Gelidonya at the very end of the 13th century. The port of entry to the mainland for the Urfa ingots is not known, but if they were imported into Hittite territory directly, one possibility is the major port of Ura, likely located near modern-day Silifke in southern Turkey. Alternatively, the Urfa ingots may have been imported through the port-city of Ugarit on the Syrian coast, a prominent maritime city-state controlled by the Hittite city of Carchemish to its northeast. Once unloaded at a mainland port, the ingots would then have been transported overland by donkey caravans, often over harsh steppes, through rugged mountain passes, and across rivers. That the Urfa ingots were found in the Göksu River near Şanlıurfa, about 195 km from the Mediterranean Sea, reveals that they, like the Uluburun and Cape Gelidonya ingots, were lost in transit. The Urfa ingots are therefore unique inasmuch as they represent the only assemblage found on land lost during transit. What cannot be determined, however, is the direction of travel at the time of their loss. If the ingots were being transported on a river craft, shipping would have been in an upstream (south-to-north) or, less likely, downstream (north-to-south) direction; if they were lost while fording the river, however, then their probable direction of travel would have been west to east or vice versa.

Our visit to the site where the ingots were discovered was in August when the water in the Göksu River water is near its lowest level, but still maintains a reasonable flow. Even so, a brief survey of the general region did not suggest an easily navigable river conducive for regular transportation by water, unlike the year-round favorable conditions prevalent on the Euphrates River, into which the Göksu River flows about 11 kilometers (17 km downstream) to the south.

Since the discovery of the ingots in 1991, intensive sand and gravel-dredging operations have continued at several locations along the river, one of which we visited about one kilometer farther upstream from the site to inquire about chance finds of ancient artifacts. The workers informed us that the dredging operations had revealed only the occasional iron object. We were shown a representative pile of finds from the river that included several wrought-iron plow shares, tools, harness fittings, and even a bundle of bars or spits. With the exception of a single reported bronze dagger of unknown date, no copper or bronze objects had been found in the 10 years following the discovery of the oxhide ingots. This paucity of evidence appears to support our view that the Göksu River was not likely to have been used on a regular basis as a conduit for transporting heavy loads by river craft in antiquity. The oxhide ingots were most likely lost while they were being carried across the river, especially since this part of the river is wide and correspondingly shallow, offering a suitable ford or crossing point. Although it is reported that the ingots were discovered only a few hundred meters to the west of a Roman bridge (Sertok and Güllüce 2005, 395), the only Roman bridge recorded in the region is located approximately 8.5 km to the south of the discovery site—or 3.5 km (4.5 km upstream) from the confluence of the Göksu and Euphrates Rivers (Comfort, Abadie-Reynal, and Ergeç 2000, 101, 108, 110, figs. 2, 9; Comfort and Ergeç 2001, 24, 40, fig. 3). The bridge would have provided the necessary access over the Göksu River for a south–north Roman route following the west bank of the Euphrates. Unless an

Figure 28.4. Map of sites mentioned in the text. X indicates location of the Urfa ingots.

unrecorded second bridge exists near the ingot discovery site, the reference to a Roman bridge there must be an oversight.

The ingots were possibly lost when several caravan donkeys slipped while fording the river and were washed downstream into deeper water with their heavy load. The 19th-century B.C. Old Assyrian texts from *kārum* of Kültepe, ancient Kaniš near Kayseri in central Anatolia, suggest that the total load carried by a caravan donkey was at least 90 kg (Veenhof 1972, 45). Of this amount approximately 73 kg was in the form of tin ingots, with the remainder corresponding to the weight of the harness, packing materials, and other items. The tin ingots were of small size and therefore wrapped in cloth to form two packs, one placed on either side of the harness. Since the only complete oxhide ingot from Urfa weighs just under 25 kg, it would appear that each donkey could have carried up to three ingots, although this would require that the third ingot be lashed to the top of the harness in order to maintain a balanced load, rather than to one side. Depending on the actual number of ingots discovered at Göksu River and the lading of each donkey, two to four donkeys carrying either two or three ingots each would be required for carrying all the ingots. Alternately, the donkeys may have forded the river without their loads, and the ingots possibly were taken across on a float of inflated skins lashed together (*kelek*), but accidentally lost in the river.

The region of Şanlıurfa (ancient Edessa) was an important crossroads for caravan routes throughout history. Routes proceeding in the east–west direction eventually had to cross the Euphrates River. Since the Middle Ages, and for hundreds of kilometers up and downriver, the major crossing point of the Euphrates was at the town of Birecik, only 10 km east of ancient Zeugma (Comfort, Abadie-Reynal, and Ergeç 2000, 107) and about 50 km (104 km downstream) from the confluence of the Göksu River with the Euphrates. Even much later, in the 16th and 17th centuries A.D., the main west–east Ottoman caravan route joining Aleppo with Diyarbakır in southeastern Turkey, and Iran beyond, ran through Şanlıurfa after crossing the Euphrates at Birecik. Birecik also served as a way station for goods coming from all directions for destinations east and west. From Birecik, goods could even be shipped on small wooden river boats and *keleks* southward to Baghdad and onward to Basra on the Persian Gulf (Üner 2002, 76–77; for a brief account of trade and communications along and on the Euphrates for all periods, see Comfort and Ergeç 2001, 20–27). Birecik's importance as a crossing, however, did not come about until that at Zeugma, which had become established as a major crossing in the 4th century B.C., had declined probably in the early Middle Ages (Comfort, Abadie-Reynal, and Ergeç 2000, 107). During the Iron Age, and perhaps also in the Bronze Age, these routes were mostly in use (Birmingham 1961, 193–194), but a major crossing of the Euphrates seems to have been at Samsat (ancient Samosata), about 76 km (130 km upstream) from Birecik. From Latakia on the Mediterranean coast, which is near the Bronze Age port-city of Ugarit, a caravan road reached Aleppo that was located about 145 km to its east. Thus, goods unloaded on the Mediterranean coast could be taken by caravan through the Latakia–Aleppo–Carchemish/Birecik/Samsat–Urfa route to areas farther east, with the crossing of the Euphrates taking place at Carchemish, Birecik, or Samsat. In addition to these crossings there were many other minor crossings of the Euphrates (Comfort, Abadie-Reynal, and Ergeç 2000, 113–121). Conversely, upon reaching Birecik or Samsat, goods coming from the west could have been shipped southward by caravan or downstream on river craft along the Euphrates to destinations in Mesopotamia. In this context, it is noteworthy that the Dur-Kurigalzu ingot, the easternmost copper oxhide ingot ever discovered, was found near the east bank of the Euphrates River, near Baghdad.

If the Urfa oxhide ingots were coming directly from a Mediterranean port, and especially if they had originated from Ugarit, then they may have been transported overland on pack animals from Ugarit, through Aleppo, to Carchemish, which is located on the western bank of the Euphrates River just within the Turkish border with Syria. Instead of crossing the Euphrates at Carchemish or Birecik, the caravan may have proceeded north along the western bank of the Euphrates toward Samsat when disaster struck while crossing the Göksu River west of Samsat. Had the caravan reached Samsat, it may have continued northward into the Anatolian heartland, perhaps ultimately on to the Hittite capital of Hattuša, or, it could have crossed the Euphrates at Samsat and proceeded to an unknown eastern destination. If, on the other hand, the ingots were not imports originating directly from the Mediterranean coast, but instead represented trade items or a form of payment such as taxes or levies from a center that had received them at an earlier time, then the direction from which the ingots were coming or going at the time of their loss cannot be deduced.

The three oxhide ingots from the Göksu River are the most important indicators of the wide-reaching Late Bronze Age trade networks through which Cypriot copper and other goods reached the Anatolian hinterland and other Near Eastern destinations. Furthermore, they are a stark reminder that what is logical or expedient in trade did not always prevail. In this instance raw copper from Cyprus was imported from a great distance involving maritime and overland routes when it could have been obtained more readily from the largest copper deposit in Anatolia at Ergani-Maden, located only 170 km to the northeast of the area where the Urfa ingots were discovered, and less than a mere one-third of the distance from Cyprus. Whether this preference for Cypriot copper, characterized by its readily recognizable typical oxhide shape, was due primarily for the expediency and low cost of obtaining Cypriot copper through well-established maritime and overland trade routes and efficiently run procurement networks, or for its esteemed regard, perhaps one representing a certain prestige for its owner and, therefore, having a symbolic significance beyond the intrinsic trade value of the copper the ingot contained, remains to be determined.

Acknowledgments

I would like to thank the late Director of Şanlıurfa Museum, Mr. Adnan Mısır, for informing me about the discovery of the Urfa ingots and inviting me to publish them, the former Director of Şanlıurfa Museum, Mr. Eyyüp Bucak, for allowing us to study, sample, access museum records, and publish the ingots, Mr. Hamza Güllüce, for sharing with us his detective story about the sad fate of some of the ingots, and Mr. Nedim Dervişoğlu for re-checking the accession dates and numbers of the ingots. I am most grateful to Dr. Zofia Stos-Gale, the former Director of the Isotrace Laboratory, University of Oxford, for interpretating and commenting on the lead-isotope ratios of the Urfa ingots, for providing the additional data included in Table 28.1, and for preparing Figure 28.3, as well as for her all help on my queries regarding lead-isotopes. Thanks are also due to the Institute of Nautical Archaeology at Texas A&M University for contributing toward travel and accommodation costs to Şanlıurfa, as well as for the use of supplies and equipment from their center in Bodrum.

References

Bass, G.F. 1967. *Cape Gelidonya: A Bronze Age Shipwreck* (*TAPS* 57 [8]), Philadelphia.

Birmingham, J.M. 1961. "The Overland Route across Anatolia in the Eighth and Seventh Centuries B.C.," *AnatSt* 11, pp. 185–195.

Brinkman, J.A. 1987. "Twenty Minas of Copper," *Language, Literature, and History: Philological and Historical Studies Presented to Erica Reiner* (*American Oriental Series* 67), F. Rochberg-Halton, ed., New Haven, pp. 33–36.

Buchholz, H.G. 1959. "Keftiubarren und Erzhandel im zweiten vorchristlichen Jahrtausend," *PZ* 37, pp. 1–40.

Comfort, A., C. Abadie-Reynal, and R. Ergeç. 2000. "Crossing the Euphrates in Antiquity: Zeugma Seen from Space," *AnatSt* 50, pp. 99–126.

Comfort, A., and R. Ergeç. 2001. "Following the Euphrates in Antiquity: North–South Routes around Zeugma," *AnatSt* 51, pp. 19–49.

Domergue, C., and C. Rico. 2002. "À propos de deux lingots de cuivre antiques trouvés en mer sur la côte languedocienne," in *Vivre, produire et échanger: reflets méditerranéens. Mélanges offerts à Bernard Liou*, Montagnac, pp. 141–152.

Gale, N.H. 1989. "Archaeometallurgical Studies of Late Bronze Age Ox-hide Copper Ingots from the Mediterranean Region," in *Archäometallurgie der Alten Welt* (*Der Anschnitt* 7), A. Hauptmann, E. Pernicka, and G.A. Wagner, eds., Bochum, pp. 247–268.

———. 1991. "Copper Oxhide Ingots: Their Origin and Their Place in the Bronze Age Metals Trade in the Mediterranean," in *Bronze Age Trade in the Mediterranean* (*SIMA* 90), N.H. Gale, ed., Jonsered, pp. 197–239.

———. 1999. "Lead Isotope Characterization of the Ore Deposits of Cyprus and Sardinia and Its Application to the Discovery of the Sources of Copper for Late Bronze Age Oxhide Ingots," in *Metals in Antiquity* (*BAR-IS* 792), S.M.M. Young, A.M. Pollard, P. Budd, and R.A. Ixer, eds., Oxford, pp. 110–121.

Gale, N., and Z.A. Stos-Gale. 2005. "Zur Herkunft der Kupferbarren aus dem Schiffswrack von Uluburun und der spätbronzezeitliche Metallhandel im Mittelmeerraum," in Yalçın, Pulak, and Slotta 2005, pp. 117–131.

Gale, N.H., Z.A. Stos-Gale, G. Maliotis, and N. Annetts. 1997. "Lead Isotope Data from the Isotrace Laboratory, Oxford: *Archaeometry* Data Base 4, Ores from Cyprus," *Archaeometry* 39, pp. 237–246.

Harmankaya, N.S. 1995. "Kozman Deresi Mevkii (Şarköy, Tekirdağ) Maden Buluntuları," in *Readings in Prehistory: Studies Presented to Halet Çambel*, Istanbul, pp. 217–254.

Hauptmann, A., R. Maddin, and M. Prange. 2002. "On the Structure and Composition of Copper and Tin Ingots Excavated from the Shipwreck of Uluburun," *BASOR* 328, pp. 1–30.

Lo Schiavo, F. 2005. "Metallhandel im zentralen Mittelmeer," in Yalçın, Pulak, and Slotta 2005, pp. 399–414.

Muhly, J.D. 1973. *Copper and Tin: Distribution of Mineral Resources and the Nature of the Metals Trade in the Bronze Age* (*Transactions of the Connecticut Academy of Arts and Sciences* 43), New Haven.

———. 1977. "The Copper Ox-hide Ingots and the Bronze Age Metals Trade," *Iraq* 39, pp. 73–82.

———. 1982. "The Nature of Trade in the LBA Eastern Mediterranean: The Organization of the Metals' Trade and the Role of Cyprus," in *Early Metallurgy in Cyprus, 4000–500 B.C.*, J.D. Muhly, R. Maddin, and V. Karageorghis, eds., Nicosia, pp. 251–266.

———. 2005. "Kupfer und Bronze in der spätbronzezeitlichen Ägäis," in Yalçın, Pulak, and Slotta 2005, pp. 503–513.

Müller-Karpe, A. 2005. "Metallbarren bei den Hethitern," in Yalçın, Pulak, and Slotta 2005, pp. 485–492.

Primas, M. 2005. "Ochsenhautbarren in Europa," in Yalçın, Pulak, and Slotta 2005, pp. 385–391.

Pulak, C. 1998. "The Uluburun Shipwrecck: An Overview," *IJNA* 27, pp. 188–224.

———. 2000. "The Copper and Tin Ingots from the Late Bronze Age Shipwreck at Uluburun," in *Anatolian Metal* I (*Der Anschnitt* 13), Ü. Yalçın, ed., Bochum, pp. 137–157.

———. 2001. "The Cargo of the Uluburun Ship and Evidence for Trade with the Aegean and Beyond," in *Italy and Cyprus in Antiquity 1500–450 BC*, L. Bonfante and V. Karageorghis, eds., Nicosia, pp. 13–60.

———. 2008. "Cargo: Selection of Ingots," in *Beyond Babylon: Art, Trade, and Diplomacy in the Second Millennium B.C.*, J. Aruz, K. Benzel, and J.M. Evans, eds., New York, pp. 307–310.

Pusch, E.B. 1996. "High Temperature Industries in the Late Bronze Age Capital of Piramesse (Qantir): II. A Quasi-industrial Bronze Factory Installation, Tools, and Artifacts," in *Proceedings of the First International Conference on Egyptian Mining and Metallurgy and Conservation of Metallic Artifacts*, F.A. Esmael, ed., Cairo, pp. 121–132.

Sertok, K. and H. Güllüce. 2005. "Ochsenhautbarren aus dem mittleren Euphratgebiet," in Yalçın, Pulak, and Slotta 2005, pp. 393–398.

Stos, Z.A. 2009. "Across the Wine Dark Seas . . . Sailor Tinkers and Royal Cargoes in the Late Bronze Age Eastern Mediterranean," in *From Mine to Microscope: Advances in the Study of Ancient Technology*, A.J. Shortland, I.C. Freestone, and T. Rehren, eds., Oxford, pp. 163–180.

Stos-Gale, Z.A., G. Maliotis, N.H. Gale, and N. Annetts. 1997. "Lead Isotope Characteristics of the Cyprus Copper Ore Deposits Applied to Provenance Studies of Copper Oxhide Ingots," *Archaeometry* 39, pp. 83–123.

Stos-Gale, Z.A., N.H. Gale, G. Bass, C. Pulak, E. Galili, and J. Sharvit. 1998. "The Copper and Tin Ingots of the Late Bronze Age Mediterranean: New Scientific Evidence," in *Proceedings of the Fourth International Conference on the Beginnings of the Use of Metals and Alloys (BUMA-IV), May 25–27, 1998, Kunibiki Messe, Matsue, Shimane, Japan*, Aoba, pp. 115–126.

Üner, M.E. 2002. "Urfa-Haleb Kervan Yolu Üzerinde Bir Derbend: Çarmelik Derbendi," in *Anadolu'da Tarihi Yollar ve Şehirler Semineri; 21 Mayıs 2001 Bildiriler*, Istanbul, pp. 73–78.

Yalçın, Ü., C. Pulak, and R. Slotta. 2005. *Das Schiff von Uluburun: Welthandel vor 3000 Jahren* (*Deutsches Bergbau-Museum Bochum* 138), Bochum.

Veenhof, K.R. 1972. *Aspects of Old Assyrian Trade and Its Terminology*, Leiden.